Femtochemistry VII
Fundamental Ultrafast Processes in Chemistry, Physics, and Biology

Cover graphic produced by L. Poth.
M. Rouhi, Chem. Eng. News 76 (1998) 17.

Femtochemistry VII
Fundamental Ultrafast Processes in Chemistry, Physics, and Biology

VIIth International Conference
on Femtochemistry
Fairmont Washington, Washington, DC, USA
July 17-22, 2005

Edited by

A. Welford Castleman, Jr.
and
Michele L. Kimble

Departments of Chemistry and Physics
Pennsylvania State University
University Park, PA 16802, USA

ELSEVIER

Amsterdam – Boston – Heidelberg – London – New York – Oxford
Paris – San Diego – San Francisco – Singapore – Sydney – Tokyo

Elsevier
Radarweg 29, PO Box 211, 1000 AE Amsterdam, The Netherlands
The Boulevard, Langford Lane, Kidlington, Oxford OX5 1GB, UK

First edition 2006

Library of Congress Cataloging-in-Publication Data
A catalog record for this book is available from the Library of Congress

British Library Cataloguing in Publication Data
A catalogue record for this book is available from the British Library

ISBN-13: 978-0-444-52821-6
ISBN-10: 0-444-52821-0

For information on all Elsevier publications
visit our website at books.elsevier.com

Printed and bound in The Netherlands

06 07 08 09 10 10 9 8 7 6 5 4 3 2 1

PREFACE

About seventeen years ago, Femtochemistry emerged as a new and important field for the study, in real time, of the dynamics of chemical and biological reactions. Since then remarkable progress has been made and femtosecond spectroscopy devoted to Chemistry and Biology has become a well established field of research, giving rise to Femtochemistry and Femtobiology. In 1999, the scientific community was extremely delighted by the remarkable achievement in this field of research with the award of the Nobel Prize in Chemistry to Ahmed H. Zewail who pioneered the subject of Femtochemistry.

The significance of the field is recognized through its impact around the world, giving rise to numerous conferences and publications. The extensive activity and interest in the subject spawned one particular series of conferences devoted entirely to Femtochemistry which is now in its twelfth year. The first Femtochemistry Conference was organized by Jörn Manz and Ludger Wöste which took place in Berlin in March 1993. The success of the Berlin Conference and the rapid growth of Femtochemistry and Femtobiology gave rise to this series of Femtochemistry Conferences which has become a major one in this field of science. In the summer of the years 1995, 1997, and 1999, the second, the third and fourth Femtochemistry Conferences were organized, respectively, by Majed Chergui in Lausanne, Villy Sundstrom in Lund, and Franz de Schryver in Leuven. The contributions made in these conferences have crystallized in a series of books and Special Issues of the *Journal of Physical Chemistry.*

The first Femtochemistry Conference of the 21st century took place in Toledo, Spain during the period of September 2 – 6, 2001. The conference had some 250 participants from more than 30 countries, with a total of 45 lectures and 130 poster presentations in wide-ranging topics: gas phase, clusters, liquids, coherent control, new techniques, biological systems, polymers, nanostructures, surfaces and interfaces. The second Femtochemistry Conference of the 21st century took place in Paris, France during the period July 6 – 10, 2003. The conference had some 272 participants from more than 26 countries, with a total of 40 lectures and 157 poster presentation devoted to wide-ranging topics: hydrogen bond dynamics, coherent control of chemical reactions, reaction dynamics in isolated molecules, gas phase and clusters, protein dynamics, dynamics in conjugated polymers, molecular structural dynamics, primary chemical events in photobiology, surface reaction dynamics, DNA photodynamics, and reaction dynamics in liquids and solutions. We are delighted to have completed the third Femtochemistry Conference of the 21st century and are very exited to be holding this in the USA. Femtochemistry VII was held in Washington, DC, July 17 – 22, 2005 at the Fairmont Washington. This is the first time the Femtochemistry Conference series was held in the USA. The stimulating conference was opened by Professor Ahmed Zewail

(1999 Nobel Prize Winner), and as was evident by the attendees at the conference, we had a very active program with the presentation of numerous talks and a large number of posters. These covered subjects such as imaging, structural dynamics, and spectroscopies, fundamentals of reaction dynamics, solvation phenomena, liquids and interfaces, aggregates/particles/surfaces, protein dynamics and photobiology, quantum control, and intense laser – matter interactions. We are grateful to all speakers, chairpersons, presenters, and authors who provided stimulating contributions.

We would also like to acknowledge several individuals who helped in the organization of Femtochemistry VII. We first recognize Majed Chergui, Marcus Dantus, Abderrazzak Douhal, J. T. Hynes, Jörn Manz, Monique Martin, Franz de Schryver, Villy Sundström, Ludger Wöste, and Ahmed Zewail, members of the International Organizing Committee. Their suggestions helped in ensuring the quality of the program. We also appeciate the help of the local committee members, Nicholas J. Bianco, Daniel E. Blumling, Heide G. Castleman, Darren P. Hydutsky, Charles E. Jones, Jr., Kenneth L. Knappenberger, Jr., Scott G. Sayers, Mohamed A. Sobhy, and Nicole R. Stroud; their assistance helped make certain that everything ran smoothly. The work of the web designer and master, Clifton C. Castleman, is also greatly appreciated. We also express our sincere gratitude to Connie M. Smith, Administrative Assistant and Conference Coordinator. Her tireless efforts helped make Femtochemistry VII a great success.

Finally, we gratefully acknowledge financial support form the following Agencies, Companies, and Institutions: U. S. Air Force Office of Scientific Research, Coherent, Inc., U. S. Department of Energy, ELS Lasers, Elsevier B. V., IMRA America, Inc., Intense Light, Kapteyn-Murnana Laboratories, Inc., U. S. National Science Foundation Engineering Section, Ocean Optics, Inc., U. S. Office of Naval Research, Philip Morris USA, Pennsylvania State University – Eberly College of Science, Pennsylvania State University – Department of Chemistry, and Newport/Spectra-Physics.

A. W. Castleman, Jr.
Departments of Chemistry and Physics, Pennsylvania State University, Pennsylvania, USA

Michele L. Kimble
Department of Chemistry, Pennsylvania State University, Pennsylvania, USA

April 2006

CONTENTS

1. Imaging, Structural Dynamics, Spectroscopies

4D Structural Dynamics 3
Ahmed H. Zewail

Ultrafast Wave Packet, Coherent Control and Dissociation Dynamics 14
Kenneth L. Knappenberger, Jr., Stefan Gilb, Vasilios G. Stavros, Eliza-Beth W. Lerch, Elva A. Torres, Daniel Strasser, Lana Lau, and Stephen R. Leone

Ultrafast Laser-pump X-ray Probe Measurements of Solvated Transition Metal Complexes 23
Taewoo Lee, Frank Benesch, Christian Reich, and Christoph Rose-Petruck

Electron Dynamics of Metal-Carbon Clusters 33
K. L. Knappenberger, Jr., C. E. Jones, Jr., M. A. Sobhy, and A. W. Castleman, Jr.

Time Resolved CARS Measurements of I_2 in Solid Kr: Vibrational State Dependent Dephasing Times Between 2.6 – 32 K 39
Tiina Kiviniemi, Jukka Aumanen, Pasi Myllyperkiö, and Mika Pettersson

The Ultrafast Photofragmentation Pathway of N,N-Dimethylisopropylamine 44
M. P. Minitti, J. L. Gosselin, T. I. Sølling, and P. M. Weber

Ultrafast Photofragmentation Dynamics of Gas Phase CH_2BrCl 49
Rebeca de Nalda, Jesús G. Izquierdo, Daniel Irimia, Maurice H. M. Janssen, and Luis Bañares

Determination of Charge State of Tungsten During Ultrafast Hard X-Ray Generation 53
Fang Shan, Vernon A. Couch, William H. Fink, and Ting Guo

Multi-Dimensional fs Spectroscopy of an Excitonic Dimer, Described by Redfield Relaxation Theory 57
Pär Kjellberg and Tõnu Pullerits

High Power Synchronized Femtosecond Ti:Sapphire Laser with Sub-Femtosecond Timing Jitter 62
Jinrong Tian, Zhiyi Wei, Peng Wang, Hainian Han, Zhaohua Wang, Jie Zhang, Tao Yang, and Jianwei Pan

Ultrafast Vibrational Evolution of Excited-State Coumarin 153 Using fs/ps CARS as a Transient Raman Probe 66
Benjamin D. Prince, Abhijit Chakraborty, Beth M. Prince, Tse-Ming Hsin, and Hans U. Stauffer

Development of fs/ps CARS as a Sub-100 fs Transient Raman Probe 70
Benjamin D. Prince, Abhijit Chakraborty, Beth M. Prince, and Hans U. Stauffer

Ultrafast Fluorescence Parametric Amplification for Femtosecond Molecular Dynamics Research 75
Shufeng Wang, Fengming Li, Hong Yang, and Qihuang Gong

A Density-Functional Study of the Structural, Electronic, and Vibrational Properties of Ti_8C_{12} Metallocarbohedrynes with Relevance to Ultrafast Time-Resolved Spectroscopy 80
M. A. Sobhy, J. O. Sofo, and A. W. Castleman, Jr.

2. Fundamentals of Reaction Dynamics

Sub 5 fs Spectroscopy of Polydiacetylene 87
Takayoshi Kobayashi, Mitsuhiro Ikuta, and Yoshiharu Yuasa

Dark Intermediate States in Photochemical Cis-Trans Isomerization and Pericyclic Reactions 96
Werner Fuß, Sergei A. Trushin, Wolfram E. Schmid, Constantine Kosmidis, and Subhasis Panja

Multi-Component Fitting of Cluster Dynamics 105
K. L. Knappenberger, Jr. and A. W. Castleman, Jr.

Photodissociation of SO_2 at 200 nm: Evidence of an Anisotropy Effect 109
D. P. Hydutsky, N. J. Bianco, and A. W. Castleman, Jr.

Energy Transfer Dynamics in Zn-Porphyrin-Appended Dendrimers 113
Jane Larsen, Ben Brüggemann, Tomáš Polívka, Joseph Sly, Maxwell J. Crossley, Villy Sundström, and Eva Åkesson

Vectorial Energy Transfer: Direct Measurement of an Intermediate Transfer Step 118
Evrim Atas, Zhonghua Peng, and Valeria D. Kleiman

Ultrafast Dynamics of Strongly Confined Water 123
Adriaan M. Dokter, Sander Woutersen, and Huib J. Bakker

Femtosecond-Domain Dynamical Studies of Energy-Transfer Cassettes 128
T. G. Kim, J. C. Castro, A. Loudet, G.-S. Jiao, K. Burgess, R. M. Hochstrasser, and M. R. Topp

Fluorescence of Polythiophenes: The Effect of Intra- and Interchain Processes on the Excited State Dynamics 132
Ignác Bugár, Michal Žitnan, Dušan Velič, and Gabriel Čík

Probing and Controlling Ultrafast Dynamics of a Molecular Switch 136
Janne Savolainen, Riccardo Fanciulli, Niels Dijkhuizen and Jennifer L. Herek

3. Solvation Phenomena

Solvation and Photochemical Funnels: Environmental Effects on Conical Intersection Structure and Dynamics 143
Irene Burghardt, Riccardo Spezia, and James T. Hynes

Simulation of Resonance Raman Spectra of the Solvated Electron in Water and Methanol 154
Stefanie Neumann, Wolfgang Eisfeld, Andrzej L. Sobolewski, and Wolfgang Domcke

Dynamics Following Charge-Transfer to Solvent in $I^-(water)_n$ (n = 3- 28) Clusters Observed with Photoelectron Imaging 163
Aster Kammrath, Jan R. R. Verlet, Graham B. Griffin, Arthur E. Bragg, Daniel M. Neumark

Ultrafast Photoelectron Imaging of the Electronic Relaxation of a Molecule Deposited at the Surface of an Argon Cluster 174
Lionel Poisson, Marc-André Gaveau, Eric Gloaguen, Jean-Michel Mestdagh, Benoît Soep, and David H. Parker

Solution Femtochemistry of Two Selected H-bonded Systems 183
A. Douhal, J. A. Organero, M. Sanz, L. Tormo, and L. Santos

Solvent Dependent Conformational Relaxation of cis-1,3,5-Hexatriene 189
D. Ahmasi Harris and Roseanne J. Sension

4. Liquids and Interfaces

Vibrational Echo and Pump-Probe Spectroscopic Studies of the Dynamics of Water Molecules Confined to Nanoscopic Dimensions 195
Ivan R. Piletic, Howe Siang Tan, David E. Moilanen, Ben Spry, and Michael D. Fayer

Excited-State Dynamics of the PYP Chromophore in Solution. Environment and Structure Effects 204
Agathe Espagne, Pascale Changenet-Barret, Jean-Bernard Baudin, Pascal Plaza, and Monique M. Martin

Second Harmonic and Sum Frequency of Molecules at Aqueous Interfaces 215
Eric A. McArthur, Yi Rao, Matthew Comstock, and Kenneth B. Eisenthal

Solvation Dynamics in Ionic Liquids, Results from ps and fs Emission Spectroscopy 225
Sergei Arzhantsev, Hui Jin, Gary A. Baker, Naoki Ito, and Mark Maroncelli

Femtosecond Relaxation Dynamics of Solvated Electrons from Liquid Metal-Ammonia Solutions: Temperature Jump versus Local Density Jump 235
Jörg Lindner and Peter Vöhringer

Multiplex Electronic Sum Frequency Generation Spectroscopy of Dye Molecules at the Air/Water Interface 245
Shoichi Yamaguchi, Kentaro Sekiguchi, and Tahei Tahara

Photodetachment of Aqueous Iodide: Variation of the Excitation Energy and Solution Temperature 249
H. Iglev, A. Trifonov, I. Buchvarov, T. Fiebig, and A. Laubereau

A Combined Femtosecond Fluorescence and TD-DFT Study of Uracil Derivatives in Aqueous Solution 254
Thomas Gustavsson, Elodie Lazzarotto, Dimitra Markovitsi, Ákos Bányász, Roberto Improta, and G. Scalmani

5. Aggregates/Particles/Surfaces

Ultrafast Electronic and Lattice Processes of Plasmonic Nanoparticles of Different Shape 260
Wei Qian, Wenyu Huang, Qusai Darugar, and Mostafa A. El-Sayed

Relaxation of Photoexcited States in Mass-Selected Ag_n^- and $Ag_nO_2^-$ Clusters 274
Marco Niemietz, Markus Engelke, Young Dok Kim, and Gerd Ganteför

Time-Resolved Single Molecule Spectroscopy of Conjugated Polymer Nanoparticles 284
Young Jong Lee, Rodrigo E. Palacios, and Paul F. Barbara

Influence of Different Anchor-Bridge Groups on Electron Transfer from the Molecular Chromophore Perylene to the Wide Gap Semiconductor TiO_2 295
R. Ernstorfer, L. Gundlach, S. Felber, R. Eichberger, and F. Willig

Electron Transfer from Molecular Chromophores to Semiconductors Probed with Two-Photon-Photoemission 299
Lars Gundlach, Ralph Ernstorfer, Silke Felber, Rainer Eichberger, and Frank Willig

Interfacial Electron Transfer or Concentration Quenching? 304
Yu-Xiang Weng, Xin-Hai Chen, and Lu-Chao Du

6. Protein Dynamics and Photobiology

Femtosecond Spectroscopy for the Study of Initial Reactions in Protein Folding 311
Wolfgang J. Schreier, Tobias E. Schrader, Florian O. Koller, Arne Sieg, Christian Renner, Luis Moroder, Josef Wachtveitl, and Wolfgang Zinth

The Photochemistry of Novel Flavin-Binding Photoreceptors 321
Magdalena Gauden, Wouter Laan, Jason M. Key, Ivo H.M. van Stokkum, Rienk van Grondelle, Klaas J. Hellingwerf, Peter Hegemann, and John T. M. Kennis

Electronic 2D Spectroscopy of Light Harvesting 331
Tobias Brixner, Jens Stenger, Harsha M. Vaswani, Minhaeng Cho, Robert E. Blankenship, and Graham R. Fleming

Intermediates in the Ultrafast Repair of DNA by DNA Photolyase 337
Zhanjia Hou, Goutham Kodali, Madhavan Narayanan, Kongsheng Yang, and Robert J. Stanley

Ultrafast Protein Dynamics 346
Dongping Zhong

A Mechanism of Photo-Labile (De)-Protection in Chemical and Biological Systems Studied by Femtosecond Vibrational Spectroscopy 357
Stefan Laimgruber, Wolfgang Schreier, and Peter Gilch

Photochemistry and Dynamics of Base Pairs 367
Elena Samoylova, V.R. Smith, H.-H. Ritze, W. Radloff, I.V. Hertel, Thomas Schultz

2D Electronic Spectroscopy of the B800-B820 LH3 Light-Harvesting Complex 372
Donatas Zigmantas, Elizabeth L. Read, Tomáš Mančal, Tobias Brixner, Alastair T. Gardiner, and Graham R. Fleming

Spectrally Resolved Femtosecond Nonlinear Spectroscopy for the Study of Vibrational Dynamics of Carotenoid Light Harvesting Molecules 377
Lap Van Dao, My Thi Tra Do, Peter Hannaford, Junji Akahane, and Yasushi Koyama

Ultrafast Studies of the Electronic Structure and Dynamics of B_{12} Cofactors 382
Andrew B. Stickrath, D. Ahmasi Harris, and Roseanne J. Sension

A Simple Artificial Light Harvesting Dyad as a Mimic of Nonphotochemical Quenching in Green Plants 387
Rudi Berera, Ivo H. M. van Stokkum, Christian Herrero, Mikas Vengris, Gerdenis Kodis, Rodrigo E. Palacios, Herbert van Amerongen, Rienk van Grondelle, Devens Gust, Thomas A. Moore, Ana L. Moore, and John T. M. Kennis

Ultrafast Dynamics of Fluorescent DNA Intercalators 391
Alexandre Fürstenberg, Marc D. Julliard, Todor G. Deligeorgiev, Nikolai I. Gadjev, Aleksey A. Vassilev, and Eric Vauthey

Two-Photon Excited-State and Conformation Dynamics of NADH Binding with Dehydrogenases 396
Yuexin Liu, Florly S. Ariola, Hye-Ryong Kim, Qianru Yu, Ronn Walvick, and Ahmed Heikal

Transient IR and 2D-IR Spectroscopy of Thiopeptide Isomerization 402
Valentina Cervetto, Rolf Pfister, Jens Bredenbeck, Peter Hamm, and Jan Helbing

Direct Observation of DNA Repair by Photolyase 407
Chaitanya Saxena, Ya-Ting Kao, Lijuan Wang, Aziz Sancar, and Dongping Zhong

Ultrafast Hydration Dynamics in Protein Conformational Transitions 411
Weihong Qiu, Luyuan Zhang, Lijuan Wang, and Dongping Zhong

Intramolecular Electron Transfer Beyond Solvent Control: Probing Sub-100 fs Photoinduced Electron Transfer in Triarylmethane Lactones 415
Jerzy Karpiuk, Stefan Lochbrunner, and Eberhard Riedle

Fluorescence Dynamics of Coumarin on Montmorillonite Structure 420
M. Zitnan, J. Bdzoch, I. Bugar, V. Szocs, T. Palszegi, M. Janek, D. Chorvat, and D. Velic

Fluorescence Anisotropy of Photoactive Yellow Protein (PYP) and Related Systems: Is the Low-Lying Excited State of the PYP Quasi-Degenerate? 424
Seiji Taniguchi, Haik Chosrowjan, and Noboru Mataga

Light-Driven Enzymatic Reaction in Thermophilic Protochlorophyllide Oxidoreductase 428
O. Sytina, M. L.Groot, R. van Grondelle, D. J. Heyes, and C. N. Hunter

7. Quantum Control

Converting Concepts and Dreams of Coherent Control into Applications 434
Marcos Dantus, Vadim V. Lozovoy, Igor Pastirk, and Johanna M. Dela Cruz

Design of Circularly Polarized Laser Pulses for Periodic Electron Circulation in Mg-Porphyrin 441
Ingo Barth and Jörn Manz

Learning Feedback Control Analysis 455
J. L. White, B. J. Pearson, E. C. Carroll, A. C. Florean, P. H. Bucksbaum, and R. J. Sension

Analysis and Control of Ultrafast Dynamics in Clusters 466
Vlasta Bonačić-Koutecký, Roland Mitrić, and Jiri Pittner

Controlling the Dynamical Pathways in the Ionization Processes of NaK Dimers 479
Albrecht Lindinger, Stefan M. Weber, Andrea Merli, Franziska Sauer, Mateusz Plewicki, and Ludger Wöste

Coherent Control of Isotope Selective Ionization and Fragmentation of Potassium 490
Franziska Sauer, Andrea Merli, Mateusz Plewicki, Stefan M. Weber, Ludger Wöste, and Albrecht Lindinger

Understanding Learning Control in a Molecular Family 495
David Cardoza, Mark Baertschy, and Thomas Weinacht

Controlling the Dissociative Ionization of DCl by Femtosecond Laser Pulses 500
Hans Georg Breunig, Alexandra Lauer, and Karl-Michael Weitzel

Spectrum Control in Femtosecond Time-Resolved Spectroscopy 505
J. Konradi, A.V. Scaria, A.K. Singh, and A. Materny

Pulse Shaping Control of Spatially Aligned Rotational Wavepackets of N_2 and O_2 510
Rebeca de Nalda, Christian Horn, Marc Krug, Florian Ausfelder, Matthias Wollenhaupt, Luis Banares, and Thomas Baumert

"Radiationless" Photon Echo 514
Boris D. Fainberg

Exciton Control in the FMO Complex: Influence of Laser Pulse Polarization 518
Ben Brüggemann, Tõnu Pullerits, and Volkhard May

Spin-Orbit Coupling Under Control 522
Jesús González-Vázquez, Ignacio R. Sola, and Jesus Santamaria

Quantitative Phase Characterization and Compensation by Multiphoton Intrapulse Interference Phase Scan (MIIPS) 526
Jess M. Gunn, Bingwei Xu, Johanna M. Dela Cruz, Vadim V. Lozovoy, and Marcos Dantus

8. Intense Laser – Matter Interactions

Ultrafast Cluster Electron and Coulomb Explosion Dynamics Driven by Ultraintense Laser Fields 532
Isidore Last, Andreas Heidenreich, and Joshua Jortner

Nonadiabatic Multielectron Dynamics in (Moderaterately) Strong Laser Fields: C_{60} a Model Case for Large Finite System 543
Tim Laarmann, Ihar Shchatsinin, Mark L. Boyle, Gero Stibenz, Günter Steinmeyer, C. P. Schulz, and Ingolf V. Hertel

Chirped Photoelectron Attochemistry 553
Andre D. Bandrauk, Szczepan Chelkowski, and Gennady L. Yudin

Intense Femtosecond Laser Driven Explosions of Heteronuclear Clusters 563
T. Ditmire, M. Hohenberger, D. R. Symes, K. W. Madison, F. Buersgens, R. Hartke, J. Osterhoff, A. Henig, and A. Edens

Spectral Phase and Detuning Effects in High-Power Chirped Pulse Excitation of a Dye Solution 573
E. C. Carroll, A. C. Florean, K. G. Spears, P. H. Bucksbaum, and R. J. Sension

Squeezing Rb_2 Wave Packets with Mixed Adiabatic and Dynamic Strategies 578
Ignacio R. Sola, Bo Y. Chang, Jesús González-Vázquez, and Jesus Santamaria

Ultrafast Conformational Changes in Carbonmonoxy-Myoglobin After Photolysis Observed by Time-Resolved Circular Dichroism 583
Thibault Dartigalongue and François Hache

Author Index 588

1. Imaging, Structural Dynamics, Spectroscopies

Femtochemistry VII
Fundamental Ultrafast Processes in Chemistry, Physics, and Biology
A.W. Castleman, Jr. and M.L. Kimble

4D Structural Dynamics

Ahmed H. Zewail

Laboratory for Molecular Sciences, and Physical Biology Center for Ultrafast Science & Technology
California Institute of Technology
Pasadena, CA 91125 U.S.A.

Abstract

We highlight the progress made in the development of 4D structural dynamics in physical, chemical, and biological systems. The joint atomic-scale resolutions in space and time, and sensitivity reached, make it possible to determine complex transient structures and molecular assemblies in different phases. These applications include studies of isolated chemical reactions (molecular beams), interfaces, surfaces and nanocrystals, self-assembly, and 2D crystalline fatty-acid bilayers. With the recent development of 4D microscopy, we are now able, using timed, single-electron packets, to image nano-to-micro scale structures of materials and biological cells. The scope of phenomena is elucidated with prototypical examples.

1. Introduction

When molecular systems, or assemblies of them, are those with hundreds or thousands of atoms and the chemical or biological changes involve many possible conformations, one must not only resolve the temporal behavior but also determine the three-dimensional (3D) molecular structures during the change. Such combined atomic-scale resolutions *in space and time* constitute the basis for a new field of study in what we referred to elsewhere as four-dimensional (4D) ultrafast electron diffraction, crystallography, and microscopy.

After the discovery of the electron in 1897, the development of electron diffraction/microsocopy has made possible (static) structural determination in the gas-phase, of crystal surfaces, and of biological systems. In fact the first membrane protein crystal structure of bacteriorhodopsin (crystal thickness of ca. 10 nm) was determined using electron diffraction/microscopy. From the 3D structures determined by electron or x-ray diffraction, one pictures the static spatial arrangements of atoms, but the mechanism of the function cannot be directly unraveled without knowledge of structural dynamics.

Macromolecular structures have the additional complexity of changes involving numerous possible conformations, with some that are "active" and others "inactive" in the biological function. Such non-equilibrium structures on a complex energy landscape, if determined, can provide an understanding of the origin of reduced-coordinate space for the motion and the real enthalpic and entropic contribution to the free energy. Prime examples of these energy landscapes are those describing protein folding and molecular recognition. The function and dynamics are strongly correlated.

In this laboratory, we have studied the dynamics of the elementary processes involved in molecular recognition, protein hydration, and electron transfer. The time scales were established, but the transient structures at the critical stage of the function remained unknown. As an example, the recognition of myoglobin to dioxygen, and similarly of a model picket fence to dioxygen, were studied on the femtosecond and up to the millisecond time scale. We were able to obtain the time scales for oxygen liberation, rebinding, and escape from the macromolecular structure. We also measured the diffusion-controlled rates (k_{on}) and the dissociation rates (k_{off}) of the complexes. From the results we inferred the global nature of the landscape. But we still know little about the active intermediate structure(s) critical to the different pathways in the protein. These structures cannot be optically resolved.

For structural dynamics, the method of choice, in our laboratory at Caltech, has been *ultrafast electron diffraction and microsopy,* for several reasons. First, electrons can be focused to obtain images in microscopy. Second, the experiments are 'tabletop' scale and can be implemented with ultrafast (femtosecond and picosecond) laser sources. Third, the cross-section for electron scattering is orders of magnitude larger than that of x-ray scattering, and electrons are less damaging to specimens per useful elastic scattering event. And finally, electrons can reveal transient structures of gases, surfaces, and (thin) crystals, and with properly timed sequences of electron pulses—frame referencing—we are able to "isolate" in time the evolving transient structure(s).

It is important to realize two concepts involved in structural dynamics—the time-energy uncertainty and the relevant time scales. First, there should be no concern about the uncertainty principle in limiting the information to be gained from femtosecond-to-picosecond time resolution, as all structures are prepared coherently. Second, for any dynamical process, the change with time is continuous, and although some global events may occur at longer times—so-called "relevant time scales"—these events are triggered by changes at early times. The primary events are an essential part of any complete description of the landscape and the dynamics. Thus the notion that "relevant" biological events occur only far beyond the ultrafast time domain gives an incomplete picture and may prove, as in early notions of chemical reactions, to be inaccurate.

2. From Static Structures to Stuctural Dynamics

For studies with electrons, Young's interference (1801), J. J. Thomson's corpuscle (1897), de Broglie's postulate of the electron's wave nature (1924; Doctoral Thesis), and Debye's work on diffraction and structure of gases (1915, 1929) provided the foundation. Gas phase electron diffraction (GED) was observed first by Mark and Wierl in 1930, only three years after the discovery of electron diffraction by Davisson and Germer for a crystal of nickel and by G. P. Thomson (the son of J. J.) and Reid for thin films of aluminum, celluloid, and other materials. Methods and applications have progressed over the following decades, culminating in studies of (time-averaged) structures that were—since Pauling's days—and still are important for the understanding of the nature of the chemical bond . Thousands of static structures were obtained using GED. For solids, progress has been made in many areas of studies, including those of low-energy electron diffraction (LEED) and reflection high-energy electron diffraction (RHEED). For biological studies, as mentioned above, microscopy and diffraction are powerful tools in structural determination.

Determining transient molecular structures on the ultrafast timescale demands not only the marriage of ultrafast probing techniques with those of conventional diffraction but also the development of new concepts for reaching simultaneously the temporal and spatial resolutions of atomic scale. Following the development of femtochemistry in the mid 1980s, we embarked upon a new challenge, that of achieving time-resolved ultrafast electron diffraction (UED) for structural determination. In 1991, we proposed that replacing the 'spectroscopic probes' of femtochemistry with ultrashort electron pulses would open up a new field for studying the nature of transient structures and their coherent atomic motions. A year later, we reported diffraction patterns with picosecond electron pulses (~10 ps) but without recording the temporal evolution of the reaction in the gas phase. Since those first images, the technical and theoretical machinery had to be developed, evolving from the first-generation apparatus (UED-1) to the second one (UED-2) and culminating in the third-generation apparatus (UED-3). We then constructed UED-4 for the sole purpose of ultrafast electron crystallography (UEC). In 2004, we embarked on the development of ultrafast electron microscopy (UEM), with capabilities for single-electron, femtosecond (≤ 100 fs) diffraction and microscopy.

For UED and UEC, the leap forward came from the integration of new 2D digital processing with CCD cameras, the generation of ultrashort electron packets using femtosecond lasers and high extraction fields, and the in situ pulse sequencing and clocking—all of which gave us unprecedented levels of sensitivity and spatiotemporal resolution to perform real experiments. Perhaps the most critical advance was the development of the frame-reference method. When properly timed frame referencing is made, before and during the change, the evolving transient structures can be determined. Armed with these developments, we have studied a variety of complex molecular structures, resolved their temporal evolution, and studied phenomena both in the gas and condensed phases and on surfaces, as discussed below. For UEM, the paradigm shift, after many failed attempts to reach space-charge free focusing and avoid temporal broadening, was the development of timed, single-electron packets for imaging of materials structures and biological cells.

The earliest efforts at introducing time resolution into electron diffraction were not developed for the ultrashort time domain. As in flash photolysis, studies of radicals with electron diffraction were made using electronic circuitry ("shutters") on the milli-to-micro second timescale. The patterns, if properly deconvoluted from the contribution of the precursor, give the structure of the radical, not its molecular history or the reaction dynamics. As such, studies of this type provide valuable information on final structures (stable or metastable), but both the approach and time resolutions are inadequate for obtaining direct information of the dynamics on the ultrashort time scale, and similarly for condensed phase studies. Elsewhere we have reviewed these and other studies (see section, Relevant References), and here we mention our efforts to reach the 4D Ultrafast (ps & fs) structural dynamics of complex systems, chemical and biological.

The *ultrafast* limit of structural determination, particularly molecular, with diffraction/microscopy was reached through the above-mentioned five generations of table-top instruments. Our current resolutions are: spatial, picometer to nanometer (UED and UEC to UEM) and temporal, 1 picosecond (UED), 300 femtosecond (UEC, transmission), and less than or equal to 100 femtosecond (UEM, single electrons). Only with frame-referencing at different times could we reach as low as 1% for the sensitivity of change and temporally isolate the transient structures. This development of the frame-reference methodology, combined with the high sensitivity achieved by the design of our CCD detection system, provides the impetus for investigating diverse molecular phenomena.

However, several other conceptual challenges had to be surmounted. First, there had not previously been a way to determine in situ the zero-of-time (clocking), and we developed electron lensing methodology for this purpose. Second, for an ultrafast electron pulse, electron–electron repulsion takes place and broadens the pulses over time. With a Bradley-Sibbett streak camera design, introduced for diffraction by the groups of Mourou and Elsayed-Ali, we determined pulse characteristics while recording patterns. Third, the orders-of-magnitude lower density, especially in the gas-phase, results in much weaker scattering intensities, and our detection system had to be designed for the required sensitivity. Last, the limited extent of reaction or change requires unprecedented sensitivity in the number of transients detected. These challenges were major hurdles that had to be circumvented in order to reach the current state of the art.

For microscopy, we built on the machinery of UED and UEC but with a conceptual fundamental difference, namely the realization of the significance of timed, single-electron pulses for imaging in UEM. In the design of 4D UEM, we integrate a femtosecond optical system with a modified transmission electron microscope. Space-charge broadening and defocusing of electron pulses and stability of the electron flux during imaging are major issues that we had to resolve. To circumvent space-charge induced broadening and the concomitant decrease in the ability of the microscope optics to focus these electrons, we have redesigned the transmission electron microscope by directly illuminating the photocathode with extremely weak femtosecond pulses, but as a high-frequency train of pulses separated by nanosecond or longer intervals. This was a crucial advance for reaching the femtosecond resolution in 4D UEM. In Berlin, Bostanjoglo and his coworkers had utilized a single giant pulse with $\sim 10^8$ electrons, and the pulses were photoelectrically generated as in UED and UEC. Such a pulse (~20ns), which was used to study laser-induced melting in metals, is detrimental to achieving ultrashort pulse imaging. Moreover, because the time window is nanoseconds, the uncertainty in spatial resolution due to noise statistics is on the order of micrometers, as pointed out by Bostanjoglo.

3. Applications

3.1. Ultrafast Electron Diffraction (UED)

The temporally and spatially resolved transient structures elucidated by UED have now been studied in different reactions, for excited states and for conformations at non-equilibrium geometries. A "textbook" case is that of the non-concerted elimination reaction of dihaloethanes. It demonstrates the UED methodology of using different electron pulse sequences to isolate the reactant, intermediates-in-transition, and product structures. The specific reaction studied involves the elimination of two iodine atoms from the reactant (ethanes) to give the product (ethylenes). The structures of all intermediates were unknown, and the challenges lay in determining the structural dynamics of the entire reaction. As detailed elsewhere this was achieved by referencing the diffraction to different time frames. The temporal evolution of the two steps of the reaction was also recorded. Moreover, the molecular structure of the C_2F_4I intermediate was determined from the frame referencing at a time after the fs breakage of the first bond. Both the bridged and classical C_2F_4I structures were considered in the analysis of the diffraction data. The theoretical curves for the classical structures reproduce the experimental data very well, whereas the fit provided by the theoretical bridged structure is vastly inferior. Thus, we concluded that the structure of the C_2F_4I radical intermediate is, in fact, classical in nature—the iodine atom does not bridge the two carbons. The structures and dynamics reported for this reaction are vital in describing the retention of stereochemistry in this class of reactions, and this is the first example of resolving such complex structures during the transition to final products, with the changes of bond lengths and angles observed in real time.

Another example in the applications of UED is that of excited-state structures, bifurcating to undergo radiationless transitions and chemical reactions. For example, in a recent publication from this laboratory, four prototypical hetero-aromatic (pyridine, 2-methylpyridine and 2,6-dimethylpyridine) and aromatic-carbonyl (benzaldehyde) organic molecules have been studied. For these molecules and others studied, we determined the initial ground-state structure and followed upon excitation the changes in the diffraction pattern with time. The depletion of old bonds and emergence of new bonds elucidate the structural origin of dark transitions in the hetero-aromatics and the quinoid-like excited-state structure of triplet benzaldehyde. Of significance is the understanding of the influence of parent structure on the dynamical evolution of relaxation pathways, and their relative time scales, and the possible bifurcation into physical and chemical channels on the energy landscape.

For the studies of the pyridine series, we focused on the so-called 'channel three' phenomenon. On the basis of the determined transient structures, we concluded that upon excitation, pyridine and picoline undergo C-N bond scission to open the aromatic ring and form a diradical structure. Lutidine does not undergo ring opening but instead gives vibrationally-hot/electronic-state species. The refined structures of ring-opened pyridine and picoline show alternating single (near 1.45 Å) and double-bond (near 1.35 Å) character for the skeletal distances, indicating disruption of the aromaticity of the parent ring structure. The disruption of the ring, or lack thereof, was observed in recent calculations of charge density maps by Shorokhov *et al.* in this laboratory and is fundamentally due to the nature of $n\pi^*/\pi\pi^*$ excitation and energetics of the state involved.

For benzaldehyde, we determined a quinoid structure, which is optically dark for the ($\pi\pi^*$) triplet excited state, formed as a result of intersystem crossing. This excited-state structure exhibits well-defined single and double bonds, indicating disruption of aromaticity in the ring. On the other hand, benzene formed in the dissociation pathway is in its ground electronic state, as indicated by the refined C–C bond distance. The simultaneous emergence of the photophysical and photochemical products in the diffraction data indicates a bifurcation on the excited singlet surface with apparent rise, through an intermediate. This observed bifurcation resolves some long-standing issues. The quantum chemical calculations of charge density maps by Shorokhov *et al.* in this laboratory support the conclusion regarding the electronic distribution of the structures involved in the bifurcation.

In order to extend these studies to systems of more complex energy landscapes and to structures with no heavy atoms, we studied organometallic structures, which react through multiple pathways and yield products of different spin multiplicity, and a series of hydrocarbons that undergo thermal/photochemical reactions. More recently, we have reported our study of the nature of the hydrogen bond in acetylacetone and the structural dynamics of its elimination reactions. We also concerned ourselves with the theoretical treatment of diffraction on the ultrashort time scale with particular focus on the role of coherence and the effect of orientational order on diffraction of reacting populations.

3.2. Ultrafast Electron Crystallography (UEC)

In our UEC apparatus, which includes three interconnected UHV chambers, the crystal is mounted on a computer-controlled goniometer for high-precision (0.005°) angular rotation, and the substrate can be cooled to low temperatures and characterized with LEED and Auger spectroscopy. Molecules were studied as adsorbates on the surface either as physisorbed or chemically functionalized entities, and thin crystals were studied in the reflection or transmission mode.

First we studied the substrate crystals. Determination of surface structural dynamics, using frame referencing, was achieved for crystalline solids (GaAs), following the temperature rise of the crystal. From the change of Bragg diffraction (shift, width, and intensity), we showed the change in non-equilibrium structure as the temperature rises and the restructuring at longer times. The 'transient temperature' reaches its maximum value (1565 K) in 7 ps. These results were compared with those of nonthermal fs optical probing reported by Mazur's group, and the agreement for the temperature response from the fluence dependence of the dielectric function is impressive. Structural dynamics were classified into three regimes, and we are still exploring with this system because of its richness and some unexpected results. Similarly, we studied surface and bulk crystals of silicon, with and without adsorbates. Because we can vary the fluence of the initiating pulse, we also studied the structural changes involved in phase transitions when the temperature of the lattice is sufficiently high to cause large amplitude disorder. The structural change is a phase transition to the liquid-like state.

Molecular assemblies were our next target, and water at interfaces is perhaps the most challenging chemical structure. The directional molecular features of hydrogen bonding and the different structures possible, from amorphous to crystalline, make the interfacial collective assembly of water, on the mesoscopic scale, much less understood. Structurally, the nature of water on a substrate is determined by forces of orientation at the interface and by the net charge density, which establishes the hydrophilic or hydrophobic character of the substrate. However, the transformation from ordered to disordered structure and their coexistence critically depends on the time scales for the movements of atoms locally and at long range. Therefore, it is essential to elucidate the nature of these structures and the time scales for their equilibration.

We reported using UEC the determination of the structural dynamics of interfacial water following substrate infrared temperature jump. Interfacial water was formed on a hydrophilic surface (silicon, chlorine-terminated) or hydrophobic surface (silicon, hydrogen terminated) under controlled ultrahigh vacuum (UHV) conditions. We identified the interfacial and ordered (crystalline) structure from the Bragg diffraction and the layered and disordered (polycrystalline) structure from the Debye-Scherrer rings. The temporal evolution of interfacial water and layered ice after the temperature jump was studied with monolayer sensitivity. On the hydrophilic surface substrate, the structure is cubic (I_c), not hexagonal (I_h), and structural dynamics are distinctive. On the hydrophobic surface, the structure is still cubic but very different in order. The structural dynamics are also different. The interface is dominated by polycrystalline I_c, but coexisting in this phase are crystallite structures, not adjacent to the surface of the substrate. The change in the structure of I_c has different temporal behaviors, reflecting the distinct difference in the transfer of energy to polycrystalline and crystallite I_c. We note that for the hydrophilic substrate studied, water adjacent to the surface is mostly crystalline, while that away from the surface is polycrystalline. The issues of interest are the coexistence of these structures, their different dynamics, and the time scales for energy transfer and disruption of the hydrogen bond network.

On the way to biological complexity of membrane-type structures, we decided to first investigate a bilayer of fatty acids deposited on a hydrophobic surface substrate, invoking the well known Langmuir-Blodgett (LB) technique. It allows for a controlled layer-by-layer deposition of ordered molecular films and has been used to create model biological membranes for studies under controlled conditions. The molecular chains extend up to ~50Å.

We studied UEC of one bilayer (two chains of $C_{19}H_{39}COOH$) of arachidic (eicosanoic) fatty acid and determined the structure, thus establishing the orientation of the aliphatic chains and obtaining the sub-unit cell molecular $-CH_2-CH_2-$ dimensions. The observation of the "coherent and anisotropic" dynamical expansion in the bilayer structure, and the restructuring toward equilibration at longer times, were also studied. The time scales involved following the temperature jump were obtained from the diffraction frames taken every one picosecond. From these studies a structural dynamics picture emerges for the bilayer on the substrate and for its atomic motions; the sub-unit cell parameters of the bilayer were determined to be $a_0 = 4.7$Å, $b_0 = 8.0$Å, and $c_0 = 2.54$Å.

Because there is no absorption resonance for the fatty acids at the wavelength of the heating pulse, the pulse is absorbed only by the substrate and the bilayer is practically transparent. The heating is through the phonon temperature described above in our studies of Si and GaAs crystals (here substrate). Although the energy is directly transferred into the bilayer from the substrate, the expansion in the bilayer is along the aliphatic chains of the molecules, indicating an efficient and "wire-like" energy transfer within the fatty acid assembly. The covalent bondings in the chains, as opposed to the weak interactions across chains, facilitate such directional motion, besides the 2D stacking interaction in the crystal-like structure. This apparent non-equilibrium behavior must be handled cautiously. At the maximum extension of the fatty acids, the effective temperature is above the melting point of the assembly (in the range of 100°C), but the structure is far from equilibrium primarily in modes of the stretch and scissor (and twist) vibrations of the chains. Thus the motions along the chains should reflect the behavior of coupled oscillators – the forming of wave motion by local change in velocity and amplitude. If true, the behavior is that of a soliton-type motion, assuming a coherent and nondiffusive motion. MD simulations and theoretical modeling of anharmonically-coupled bonds are work in progress to address the time scales involved, the nature of coherent structural dynamics, and the incoherent energy dissipation in the restructuring at long times. So far we have studied multiple-layered fatty acid systems and phospholipids. There is another dimension to this work, namely the origin of self-assembly and crystallization, and UEC is our method of choice for the reasons given above.

3.3. Ultrafast Electron Microscopy (UEM)

In a recent publication, we reported the development of 4D UEM, which provides the ability to image complex structures with the spatial resolution of transmission electron microscopy but as snapshots captured with ultrafast electron packets derived from a train of femtosecond pulses. The images and diffraction patterns were obtained at 120 keV for materials (single crystals of gold, amorphous carbon, and polycrystalline aluminum) and for biological cells of rat intestines. The strobing packets contain on average one electron per pulse, and the specimen dose is a few electrons per $Å^2$, but the pulses are now fully controlled in space and time.

For calibration, we also obtained 'images' when the femtosecond pulses entering the microscope were blocked. The fact that no images were observed when the light pulses were blocked indicates that the electrons generated in the microscope were indeed those obtained optically and that thermal electrons were negligible. The microscope can operate in UEM and TEM modes, and for atomic-scale investigations, we operated UEM in the diffraction mode by adjusting the intermediate lens to select the back focal plane of the objective lens as its object.

The novel attributes of UEM encouraged us to test the viability of imaging biological cells. UEM images of rat intestinal cells were obtained in a few seconds, and both the microvilli as well as the sub-cellular vesicles of these epithelial cells were visualized in the UEM images. As in the case of ultrafast electron diffraction and crystallography, UEM is now poised for measurements of structural dynamics. Applications are numerous and are the focus of the new UST center at Caltech.

4. Conclusion

Four-dimensional ultrafast electron diffraction, crystallography, and microscopy are at a turning point for exploitation in applications in physics, chemistry, and biology. The major hurdle in the use of electron diffraction and imaging in microscopy was the space charge problem, and with single-electron coherent packets, this problem is over. Moreover, local structures of a limited number of unit cells, as opposed to an average over trillions of unit cells, can be explored and their constituents imaged in real space. From isolated molecular stuctures to structures in biological cells, 4D structural dynamics is fundamental to the understanding of the function.

5. Acknowledgments

We gratefully acknowledge the generous support by the National Science Foundation for building the new generations of UED and UEC, and Caltech for some of the initial help in building UEM. Partial support was provided by the Air Force Office of Scientific Research for studies in UED. I am grateful for the timely support granted by the Gordon and Betty Moore Foundation for the UST Center devoted to the new horizons of ultrafast microscopy in physics, chemistry, and biology.

Selected References

This paper is based on the plenary lecture at *Femtochemistry VII* and is an adapted summary from a recent annual review by the author. The work presented is the collective efforts by members of the Caltech group (see publications).

1. A. H. Zewail, Phil. Trans. R. Soc. A. 364 (2005) 315, and references therein.
2. R. Srinivasan, V. A. Lobastov, C.-Y. Ruan, and A. H. Zewail, Helv. Chim. Acta. 86 (2003) 1763, and references therein.
3. S. Zou, J. S. Baskin, and A. H. Zewail, Proc. Natl. Acad. Sci. U.S.A. 99 (2002) 9625.
4. Y. Wang, J. S. Baskin, T. Xia, and A. H. Zewail, Proc. Natl. Acad. Sci. U.S.A. 101 (2004) 18000.
5. T. Xia, C. Wan, R. Roberts, and A. H. Zewail, Proc. Natl. Acad. Sci. U.S.A., in press.
6. A. H. Zewail, Nature (London) 412 (2001) 279.
7. A. H. Zewail, Faraday Discuss. Chem. Soc. 91 (1991) 207.

8. H. Ihee, V. A. Lobastov, U. M. Gomez, B. M. Goodson, R. Srinivasan, C.-Y. Ruan, and A. H. Zewail, Science 291 (2001) 458.
9. C.-Y. Ruan, F. Vigliotti, V. A. Lobastov, S. Chen, and A. H. Zewail, Proc. Natl. Acad. Sci. U.S.A. 101 (2004) 1123.
10. V. A. Lobastov, R. Srinivasan, and A. H. Zewail, Proc. Natl. Acad. Sci. 102 (2005) 7069.
11. R. Srinivasan, J. S. Feenstra, S. T. Park, S. Xu, and A. H. Zewail, Science 307 (2005) 558.
12. D. Shorokhov, S. T. Park, and A. H. Zewail, Chem. Phys. Chem. 6 (2005) 2228.
13. F. Vigliotti, S. Chen, C.-Y. Ruan, V. A. Lobastov, and A. H. Zewail, Angew. Chem. Int. Ed. 43 (2004) 2705.
14. C.-Y. Ruan, V. A. Lobastov, F. Vigliotti, S. Chen, and A. H. Zewail, Science 304 (2004) 80.
15. S. Chen, M. T. Seidel, and A. H. Zewail, Proc. Natl. Acad. Sci. U.S.A. 102 (2005) 8854.
16. J. S. Baskin and A. H. Zewail, Chem. Phys. Chem. 6 (2005) 2261.
17. J. M. Thomas, Angw. Chem. Int. Ed. 43 (2004) 2606.
18. J. M. Thomas, Angw. Chem. Int. Ed. 44 (2005) 5563.
19. S. J. Xu, S. T. Park, J. S. Feenstra, R. Srinivasan, and A. H. Zewail, J. Phys. Chem. A 108 (2004) 6650.

Femtochemistry VII
Fundamental Ultrafast Processes in Chemistry, Physics, and Biology
A.W. Castleman, Jr. and M.L. Kimble

Ultrafast Wave Packet, Coherent Control and Dissociation Dynamics

Kenneth L. Knappenberger, Jr., Stefan Gilb, Vasilios G. Stavros, Eliza-Beth W. Lerch, Elva A. Torres, Daniel Strasser, Lana Lau and Stephen R. Leone

Departments of Chemistry and Physics and Lawrence Berkeley National Laboratory, University of California, , Berkeley, CA, USA, 94720

1. Introduction

In this invited talk several examples are presented of the use of broad-band ultrafast lasers to initiate and control excited state coherent processes in atomic and molecular species. In this contribution the discussion is focused on the utility of electron kinetic energy analysis as a technique to study electronic and vibrational wave packet motion, coherent control, and state-resolved dissociation. With well-defined initial and final states, analysis of the outgoing electron kinetic energies and angular distributions enables an understanding of the mediating processes that evolve on an ultrafast timescale. In polyatomic molecules the dynamics of excited potential surfaces become complicated, due to overlapping resonances of multiple states with broad-band excitation. To decipher the multi-component dynamics of these systems a kinetic energy slicing method is employed, which provides a spectral filter for the common final ion state. This method has been developed during the analysis of Rydberg wave packets in krypton and has recently also been applied to molecular bromine and to polyatomic species, including methyl iodide (CH_3I) and carbon disulfide (CS_2).

The remainder of this manuscript is outlined as follows: In section 2 an experimental overview is given, including a description of the laser system and various electron detection schemes. In section 3 electronic wave-packet dynamics in krypton, induced by excitation in strong laser fields, are explored. In section 4 time-dependent photoelectron angular distributions during wave packet recurrences are revealed by photoelectron imaging. In section 5 these methods developed to analyze atomic excitation are extended to diatomic molecular systems, and very brief preliminary results of polyatomic dissociation are presented. Conclusions drawn from this work are given in section 6.

2. Experimental Overview

In the experiments presented here, a sample is injected into the vacuum by a pulsed molecular beam. The excitation source consists of a regeneratively amplified Ti:sapphire laser, which can be used to pump a non-collinear optical parametric amplifier. Both the molecular beam and laser system operate at a 1 kHz repetition rate. Three different electron collection methods are used, and specific details can be found in the references indicated below. In the study of krypton wave packets presented in Section 3, electrons are collected in a 36 cm long field-free flight tube that is shielded by mu-metal [1]. The signal from micro-channel plates is amplified prior to counting with a multichannel scalar. In Section 4 electrostatic lenses are applied in the interaction region to implement the velocity map imaging design of Eppink and Parker [2]. This design enables measurement of the photoelectron angular distribution (PAD). The photo emitted electrons are incident on an imaging MCP assembly, and the resulting image is captured by a CCD camera and is subsequently inverted using the basis set expansion method (BASEX) [3]. The results obtained from molecular targets presented in Section 5 are obtained by analyzing the electron kinetic energy with a "magnetic bottle" time-of-flight photoelectron spectrometer [4, 5,]. After the electrons impinge on z-stack mounted MCPs, the signal is amplified and sent to a multichannel scalar.

3. Electronic Wave Packets formed by ac-Stark shifting

A two color (3+1′) pump-probe REMPI scheme is combined with photoelectron spectroscopy to study Rydberg wave packets created in krypton atoms. In this scheme absorption of three 270 nm photons results in a resonance with the 9d Rydberg level, which converges to the 3/2 spin-orbit pair of the ion core. By varying the pump laser power, while maintaining a low probe pulse power, the extent to which ac-Stark shifting may be used to coherently excite otherwise non-resonant Rydberg levels is demonstrated. Photoelectron spectra are collected with pump laser powers spanning 9.3-65 TW/cm^2 [1]. The photoelectron time-of-flight spectra presented in Figure 1A shows the influence of increasing the pump photon density from 9.3 to 51 TW/cm^2. The photoelectron spectra shown in Figure 1A have been truncated to include only the electron energy range corresponding to the pump-probe REMPI response when the probe pulse is delayed from the pump by more than 1 ps. The $^2P_{1/2}$ and $^2P_{3/2}$ spin-orbit components of the krypton ion state are separated by 5370 cm^{-1}.

This simplifies the peak assignment process as has been discussed elsewhere [1]. Many of the individual states of interest in this study display a preferential kinetic energy in the $^2P_{3/2}$ component, and the attention is mainly focused there. As is evident in Figure 1A, with increasing pump power an increasing number of states are accessed by the pump pulse. This is shown directly in going from the top panel where only one peak is observed for ionization to the $^2P_{3/2}$ core of the ion to the bottom panel where three peaks appear. The temporal analysis of the quantum beat recurrences shows that these intermediate resonances are from lower-lying states that are shifted by as much as 3000 cm^{-1} to lower energy. Thus coherent superpositions are formed during the strong transient Stark shifting of the energy levels during the high field of the pump pulse, and these superpositions remain intact after the pump pulse is over. It is important to note that these intermediate states are only discerned when the REMPI method is combined with photoelectron kinetic energy analysis. Experiments employing mass detection alone would not reveal the effect of the ac-Stark shift, which is clear with electron kinetic energy resolution.

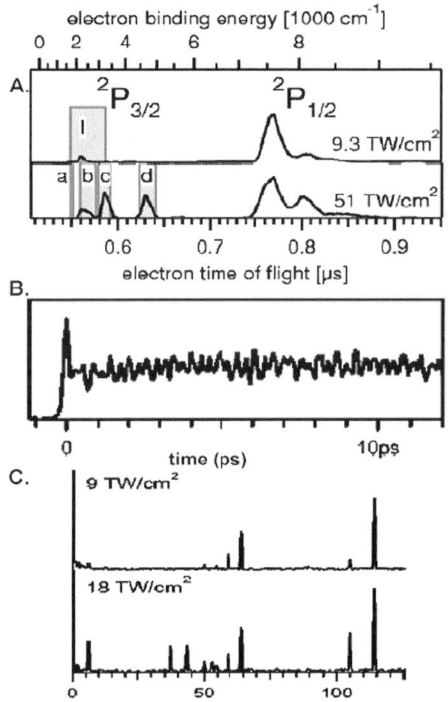

Figure 1A. Time-of-flight photoelectron spectrum of krypton generated by 3 + 1 REMPI scheme at pump pulse energies of 9.3 and 51 TW/cm^2. B. Oscillatory time-dependent $^2P_{3/2}$ electron signal at 9.3 TW/cm^2 pump energy. C. Beat frequencies obtained at two different pump powers

The temporal dependence of the intermediate state reached with pump powers of 9.3 TW/cm^2 following ionization to the $^2P_{3/2}$ threshold is shown in Figure 1B. The pump-probe delay is extended to longer than 100 ps; however the delay is only shown out to 12 ps. Oscillations in the temporal profile are clearly evident, indicating the presence of wave packet recurrences. By applying a Hanning window and taking the Fourier transform, the beat frequencies of the states involved are extracted. The beat frequencies recovered from excitation pulse energies of 9.3 and 18 TW/cm^2 are shown in Figure 1C. It is also clear that different wave packet states are created with stronger excitation fields. The beat frequencies determined from the 9.3 TW/cm^2 data are not consistent with those anticipated for the 9d Rydberg state. Rather they track well with what would be expected from the lower 8d band. The 8d origin is located approximately 600 cm^{-1} lower than the 9d state. This shows that even at the more modest photon densities, typical for three-photon excitation, the state levels can be perturbed quite significantly during the laser pulse.

As noted, the excitation with higher intensity pulses produces even more significant level shifts. This is reflected in the time-dependent dynamics arising from excitation, for example, with 51 TW/cm^2 pulses. The photoelectron spectra become more complicated as more energy regions are shifted into resonance. In this case the temporal dependence of the peaks is readily determined after energy-slicing selected electron time-of-flight regions, which are denoted, for example, in sections a-d in the bottom panel of Figure 1A. These beat frequencies clearly identify states that are shifted into resonance and populated coherently in the strong field excitation step. The beat frequencies reveal these states as the 6-8d Rydberg energy levels. By integrating a narrow slice of the kinetic energy region, a 9d level is also revealed. With fine electron kinetic energy slicing and Fourier transformation of the temporal response, coherent population of several previously unobserved ng states are resolved.

4. Time-dependent electron orbital alignment studied by photoelectron imaging

In this section the photoelectron detection capability is extended to include measurement of the photoelectron angular distribution (PAD) as a function of pump-probe delay during wave packet dynamics. The method is applied to Rydberg wave packets in Kr atoms. These measurements provide insight into the orbital configurations resulting in the outgoing electrons, after promotion to intermediate excited wave packet states. In a typical pump-probe experiment using ultrafast laser pulses, several intermediate states are coherently excited. A time-dependent interference between these states results, which can be observed if the participating states are mapped onto a common final state in the ionization step. Here a method is developed to disentangle the outgoing electron angular directions as a function of time by slicing the photoelectron image along various radial angles. This is shown schematically in Figure 2. The (3+1′) REMPI scheme is shown in Figure 2A. The 5730 cm^{-1} final state separation between the $^2P_{3/2}$ ground state and the higher $^2P_{1/2}$ component of the ion enables identification of intermediate resonances, since each state has a specific cross-section for ionization to one of these cores. The branching ratios of these final states, and their angular distribution, are distinguished experimentally as shown in Figure 2B. Electrons ionized to the $^2P_{3/2}$ component have more kinetic energy than those ionized to $^2P_{1/2}$. This results in greater displacement from the center of the image for electrons going to the ionic ground state. The angular dependence of the time-resolved photoionization signal is extracted by analyzing the image at specific electron kinetic energies and angles with respect to the plane of polarization of the pump and probe laser pulses, denoted ε as shown in Figure 2B. This is accomplished by collecting the PAD at a specific time step, slicing the image at an angle, θ_k in Figure 2B, and analyzing the electron kinetic energy of interest.

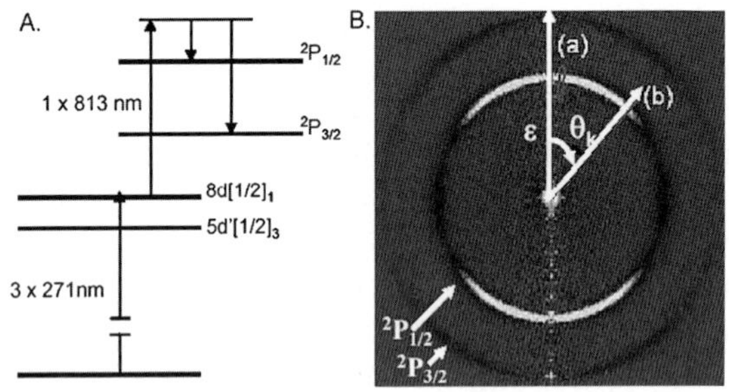

Figure 2.A. Energy level diagram depicting REMPI scheme. This diagram shows the separation of the $^2P_{1/2}$ and $^2P_{3/2}$ ion cores. B. Generalized schematic of PAD showing slicing method.

This method has been successfully applied to the study of electronic wave packets formed by promoting electrons to the 5d'[5/2]$_3$ and 8d[1/2]$_1$ Rydberg states of krypton, which are separated by 52.9 cm^{-1}. The work described above that investigated the influence of strong laser fields on the excitation of krypton showed that the coherent population of the states of interest can be controlled by ac-Stark shifting, and this is used to select specific pairs of levels for study. The wave packet is described in a general form as two entangled Bell states [6]. The dynamics involve a time-dependent population transfer between these Bell states [6]. This process is a consequence of the change in the relative orientation of the angular momenta of the core and of the excited Rydberg electron spin, which switches between parallel and antiparallel. The PADs switch between two spherical harmonic shapes for the outgoing electron, $Y_{3,0}$ and $Y_{3,\pm1}$. This particular pair of states shows directly a change in angular distribution.

Experimental observation of the time-dependent population transfer is realized only with the photoelectron imaging. By integrating the PAD at several angles (where θ_k = 5°, 41° and 89°), the temporal dependence shows an oscillatory behavior, corresponding to the electronic wave packets formed in the intermediate excited states. The beat frequency recovered at all radial slices fit to a value of 53 cm^{-1}, in excellent agreement with the energetic separation of the two target states, 5d'[5/2]$_3$ and 8d[1/2]$_1$. This finding shows that by careful adjustment of the pump laser field the intermediate levels that are coherently excited in the population step are controlled. Further, by controlling the final state distribution that is monitored (in this case $^2P_{1/2}$), the dynamic information from only the desired states is extracted. By analyzing the PAD at delay times separated by 300 fs, corresponding to half of the revival time of the measured 53 cm^{-1} quantum beat, it is found that the angular distribution changes in time along with the probability of ionization. This provides direct experimental evidence for the switching of the population between the well-defined Bell states. The combination of laser control fields and photoelectron imaging has permitted the analysis of the time-evolving states.

5. Extension of methods to molecular vibrational wave packet analysis

A two-color (2+1′) ionization scheme is employed with photoelectron kinetic energy analysis to study vibrational wave packets in molecular bromine (Br_2). In this experiment two-photon UV-excitation with 266 nm photons creates a superposition of two vibrationally excited Rydberg states. The superposition is composed of the $v' = 3$ and $v' = 4$ levels of the 4d-Rydberg state [7]. The time-dependence of the wave packet is probed by a delayed 800 nm photon that promotes the superposition to the $v^+ = 3$ or $v^+ = 4$ levels of the $n = 8$ and 9 Rydberg levels of the ½ spin-orbit core [8]. These states are autoionized by transitions to the less energetic, lower spin-orbit state, in which the ionic core relaxes to the lower $^2\Pi_{3/2}$. This ionization scheme is shown schematically in Figure 3A. Electrons ionized from either the $n = 8$ ($v^+ = 4$) or $n = 9$ ($v^+ = 3$) levels appear in the photoelectron spectrum with either 0.14 or 0.18 eV kinetic energy, respectively, the autoionization energies for these states. The peaks from the autoionization of these two levels are highlighted in Figure 3B.

The temporal evolutions of the $n = 8$ and 9 states are shown in Figure 3C. The electron signals from these channels clearly have an oscillatory time-dependence, with a revival time of 93 fs, corresponding to 360 cm^{-1}. The spectral bandwidth of the pump laser pulse is greater than 420 cm^{-1}, yet with well-defined final states serving to detect the exit channels, higher resolution of the molecular excited states is achieved. This is realized by analysis of the photoelectron kinetic energies, which provides a finer spectral filter than would be possible with conventional methods such as time-of-flight mass spectrometry. The 360 cm^{-1} beat frequency reveals the energetic separation of the two intermediate levels.

The inset of Figure 3C shows that while both channels have a similar quantum beat, they are π-phase shifted with respect to each other, appearing as a 45 fs shift in the peak amplitude between the two. For detection of a two-state superposition, there are four optical transitions, two that create the superposition, and two that reach the final common state. The signs of the transition amplitudes determine whether the observed recurrences are phase shifted by 0 or π radians. For example, if one or three of the transition amplitudes become negative relative to a reference superposition, a π shift will be observed. The observed phase shift is a consequence of the intermediate 4d vibra- tional wave function states as they are mapped onto the final n = 8 and 9 vibrational states. The system is modeled by simple overlap factors. It is assumed that the final n = 8 and 9 levels can be approximated well by the Br_2 ionic ground state. With this assumption, the vibrational wavefunction of the 4d levels are determined by fitting the experimental data to a theoretically predicted phase shift. Wave packet motion is only observed when the $\Delta v = 0$ propensity of Rydberg to Rydberg transitions breaks down. The peak appearing at 0.14 eV kinetic energy is due to $v' = 3 \rightarrow v^+ = 4$ (n = 8), with v′ = 4 to v+ = 4 unchanged. Similarly, the 0.18 eV kinetic energy peak is attributed to a $v' = 4 \rightarrow v^+ = 3$ (n=9) transition, with $v' = 3$ to $v^+ = 3$ unchanged. The observed change of phase of the recurrences is due to a change in the sign of the product of the four

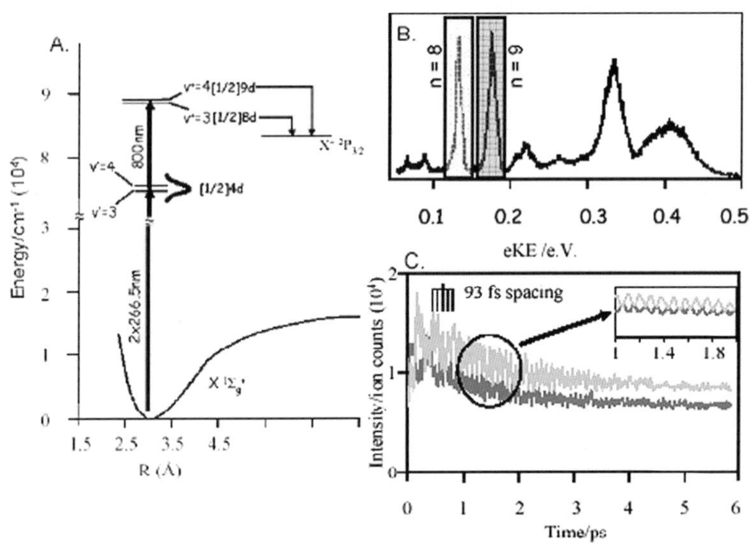

Figure 3 A. REMPI scheme employed to generate and detect vibrational wave packets generated in the 4d-Rydberg state. B. Electron kinetic energy spectrum truncated to show only low kinetic energy electrons. Only the energy slices belonging to the n = 8 and 9 states, that reflect the vibrational wave packets, are highlighted. C. Time-integrated photoelectron signal of the n = 8 and 9 features. The inset shows the π-phase shift of the dynamics.

overlap factors to form and probe the two wave packet superposition states to the $v^+ = 3$ to $v^+ = 4$ final states. Again, such dynamics are only observable with well defined final states resolved by photoelectron spectroscopy.

Recently, the dynamics of coherently excited states in more complex polyatomic systems are being probed. Currently, the dissociation processes of methyl iodide and carbon disulfide are being studied. The bending and asymmetric stretching vibrations of CS_2 are resolved, and it is found that they influence the decay rate of the excited Rydberg states differently. The interest is in elucidating how a specific molecular vibrational mode influences excited state dynamics such as those going to repulsive surfaces. Time-resolved mass spectra reveal only a multi-component response, but by analyzing the photoelectron spectra it is possible to assign the complicated dynamics with state-specificity.

6. Conclusions

We have presented results from several experiments where electronic and vibrational wave packets created by multi-photon excitation are resolved by outgoing electron energies and angular distributions. In all cases the dynamics are treated in greater detail due to the detection method of photoelectron spectroscopy. In one example a superposition of excited electronic states in krypton atoms is created. By measuring the time-resolved photoelectron spectrum we were able to extract beat frequencies due to states that are coherently populated. These results demonstrate the ability to employ strong pump laser fields to significantly perturb the electronic states, creating well-defined coherent superpositions that persist after the strong field pulse is finished. The selectivity of this method is verified in Fourier transformation of the time-dependent signal and shows that the beat frequencies are well-resolved.

The electron detection capability is extended to include measurements of the PAD of krypton. In this case we employed an excitation scheme that selectively populated two intermediate levels. By radially slicing the PAD image at various angles we were able to measure the response of the two states during the time resolved wave packet motion to show that the inner core and outer electron angular momenta are coupled. This technique shows promise to advance understanding in the area of quantum computation.

We have also investigated vibrational wave packet motion in molecular bromine (Br_2). In this experiment a superposition of vibrational states, $v' = 3$ and 4, is formed in the 4d Rydberg level. The sign of the overlap factors of the wave packet states as they are created and mapped onto the final state is observed. This determination is achieved with a spectral filter made possible by photoelectron spectroscopy. Currently this method is being applied to polyatomic systems and is likely to provide a more complete picture of the nature of multi-dimensional potential energy surfaces.

Acknowledgements

The authors gratefully acknowledge support of this research by the director, Office of Science, Office of Basic Energy Science, of the U.S. Department of Energy under contract number DE-AC02-05CH11231, the Army Research Office (43789-PH), the National Science Foundation through the Information Technology Research program and the Chemistry Division, and the Air Force Office of Scientific Research.

References

1. S. Gilb, V. Nestorov, S. R. Leone, J. C. Keske, L. Nugent-Glandorf, and E. R. Grant, Phys. Rev. A. 71 (2005) 042709.
2. A. Eppink and D. H. Parker, Rev. Sci. Instrum. 68 (1997) 3477.
3. V. Dribinski, A. Ossadtchi, V.A. Mandelshtam, and H. Reisler, Rev. Sci. Intrum. 73 (2002) 2634.
4. P. Kruit and F. H. Read, J. Phys. E. 16 (1983) 313.
5. B. J. Greenblatt, M. T. Zanni, and D.M. Neumark, Science 276 (1997) 1675.
6. S. Gilb, E. A. Torres, and S. R. Leone, manuscript in preparation.
7. R. J. Donovan, A. C. Flexen, K. P. Lawley, and T. Ridely, Chem. Phys. 236 (1998) 227.
8. V.G . Stavros, L. Lau, D. Strasser, and S. R. Leone, manuscript in preparation.

Femtochemistry VII
Fundamental Ultrafast Processes in Chemistry, Physics, and Biology
A.W. Castleman, Jr. and M.L. Kimble

Ultrafast laser-pump x-ray probe measurements of solvated transition metal complexes

Taewoo Lee,[a] Frank Benesch,[b] Christian Reich,[a] and Christoph G. Rose-Petruck[b]

[a] *Department of Chemistry, Box H, Brown University, Providence, RI 02912*
[b] *Division of Engineering, Box D, Brown University, Providence, RI 02912*

1. Introduction

Ultrafast laser pump – x-ray absorption fine structure (XAFS) probe spectra of $[Fe(CN)_6]^{4-}$ solvated in water have been measured using table-top ultrafast laser driven plasma x-ray sources. The principles of XAFS have been discussed extensively in the literature [1, 2]. Close to the x-ray absorption edge, in the region of x-ray absorption near edge structure (XANES), large electron scattering cross sections cause intense modulations and favor multiple photoelectron scattering events with several atoms. This, in principle, permits the measurements of not just metal-ligand distances but also bond angles. As shown in Figure 2 and in the literature [3, 4], the modulation depth for highly symmetric molecules can reach 100% of the absorption edge jump. At higher x-ray energies, corresponding to the region of Extended X-ray Absorption Fine Structure (EXAFS) [1], spectra are determined by single scattering events between the x-ray absorbing atom and surrounding atoms. Thus, the EXAFS spectra are a measure of the first-order atomic pair correlation function, which can be approximated by Fourier-transforming the EXAFS spectrum.

Few experiments with temporal resolutions of at least several nanoseconds have been done using synchrotron sources [5-13]. For instance, Chen *et al.* measured the transient molecular structure of a reaction intermediate after photo excitation of NiTPP-L_2 (NiTPP: nickeltetraphenylporphyrin; L: piperidine) [11]. They achieved a temporal resolution of 14-ns using the Advanced Photon Source at Argonne National Laboratory. A few years later, the MLCT state dynamics of a copper(I) diimine complex was measured with 100 ps temporal resolution and a data acquisition time of 20 hours per delay step [14]. The chemical shift of the absorption edge of photoexcited aqueous $[Ru^{II}(bpy)_3]^{2+}$ was measured with similar temporal resolution [15-17], and chemical shifts of the XAFS spectra of the charge transfer to solvent of $[Fe(CN)_6]^{4-}$ in water have been measured with 200-ps temporal resolution using synchrotron sources [18, 19].

Higher temporal resolution can, in principle, be achieved using laser-driven plasma x-ray sources. Laser-pump soft-x-ray-probe measurements with laser-driven x-ray sources have been used for measurements in material science [17] and chemistry [18]. In another experiment, laser-plasma generated x-ray pulses have been used to observe the dissociation of SF_6 in the gas phase after intense laser pulse illumination [19]. A single-picosecond temporal resolution was achieved in this gas-phase experiment. The disappearance of a shape resonance line was monitored, but a XANES spectrum was not measured.

2. Ultrafast XAFS measurements

Ligand substitution through 400-nm photolysis of $[Fe(CN)_6]^{4-}$ in aqueous solutions was carried out. It should yield $[Fe(CN)_5H_2O]^{3-}$ through the reaction $[Fe(CN)_6]^{4-} + h\nu_{400nm} \rightarrow [Fe(CN)_6]^{4-*} + H_2O \rightarrow [Fe(CN)_5H_2O]^{3-} + CN^-$. Subsequently, dimeric cluster ions, such as $[Fe_2(CN)_{10}]^{6-}$ and $[Fe_2(CN)_{11}]^{7-}$

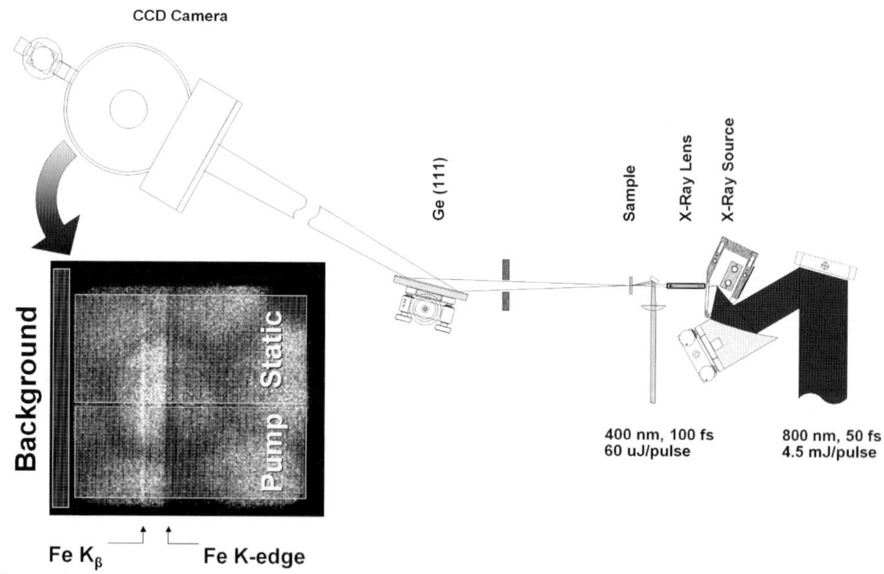

Figure 1: Experimental setup; X-rays generated in the laser focus are transmitted through the x-ray lens and imaged into the liquid sample. A flat Ge(111) crystal disperses the x-radiation. They are subsequently detected by a liquid nitrogen cooled x-ray CCD camera. The 400-nm pump laser pulse overlaps the x-ray beam in the sample solution. A typical CCD image and superimposed regions of interest are shown.

might form [20, 21]. These chemical processes are accompanied by structural changes that are observable by ultrafast x-ray absorption fine structure (UXAFS) measurements. We measured UXAFS spectra of solvated $[Fe(CN)_6]^{4-}$ at pump-probe delay times tens of picoseconds before and after photoexcitation using the experimental setup shown in Figure 1. The x-radiation was generated by interacting 4.5-mJ, 800-nm, 50-fs laser pulses with a brass-coated steel wire target at 2-kHz repetition rate. Further details of the source and the laser and x-ray optical system are discussed elsewhere [22-26].

The average x-ray flux at the sample position was 5×10^5 photons per second within between 7-keV and 8-keV. Approximately 50 ph / s were detected behind the liquid sample beam within the 1-keV spectral range around the iron K-edge. The emitted x-radiation can be expected to have a pulse length on the timescale of several hundred femtoseconds [27-32]. A fiberoptic x-ray lens focused the emitted x-radiation into a 100-μm diameter spot. Spatial overlap between this spot and the pump laser focus was ensured by simultaneously transmitting both beams through a 50-μm platinum pinhole placed at the position of the liquid sample beam. The relative timing between pump and probe pulses was electronically calibrated with about 5 ps accuracy

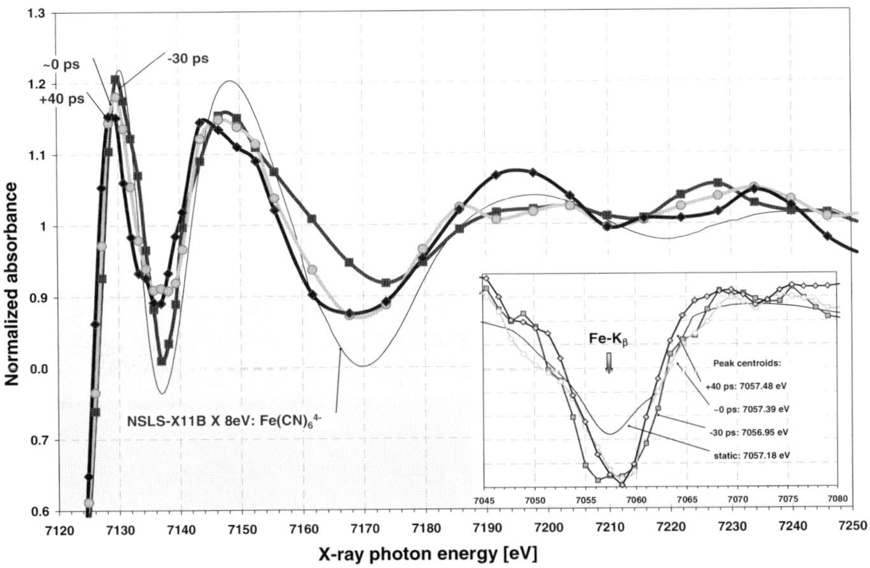

Figure 2: Ultrafast XAFS spectra measured using the laser plasma x-ray source at −30 ps (squares), approximately 0 ps (circles) and +40 ps (diamonds) after photoexcitation. This line: $Fe(CN)_6^{4-}$ reference spectrum convoluted with the spectral resolution of the laser-driven x-ray spectrometer.

by illuminating a photodiode (GaAs PIN detector ET-4000, Electro-Optics Technology, Inc.) with a pulse response of 200-ps (fwhm) sampled by a 3-GHz oscilloscope (LeCroy, Wavepro 7300) with a sampling interval of 5 ps. Subsequently, the pinhole was replaced by a liquid jet of 400-μm thickness.

A typical CCD image, reproduced in Figure 1, shows the far-field pattern of the fiber optic lens. This spectroscopically irrelevant structure was removed during the data analysis using a flat-field image measured without sample beam. Laser pump and x-ray beam were adjusted to overlap in the lower section of the interaction region. The CCD image was segmented into regions for the measurements of the background, the time-dependent XANES signal from the laser pumped volume, and the static XANES signal. A plasma emission line, the iron K_β-line, serves as a spectral reference. The insert in Figure 1 verifies that spectral drifts during the experiments were insignificant. Furthermore, since the time-dependent signal and all reference data are on the same image, possible slow drifts of the apparatus are not influencing the final data analysis.

$K_4[Fe(CN)_6]\cdot 3H_2O$ was purchased from Strem Chemicals, Inc. and dissolved in distilled water. The concentration was about 500 mM. The sample solution was excited by 100-μJ, 400-nm, approximately 100-fs laser pulses. The

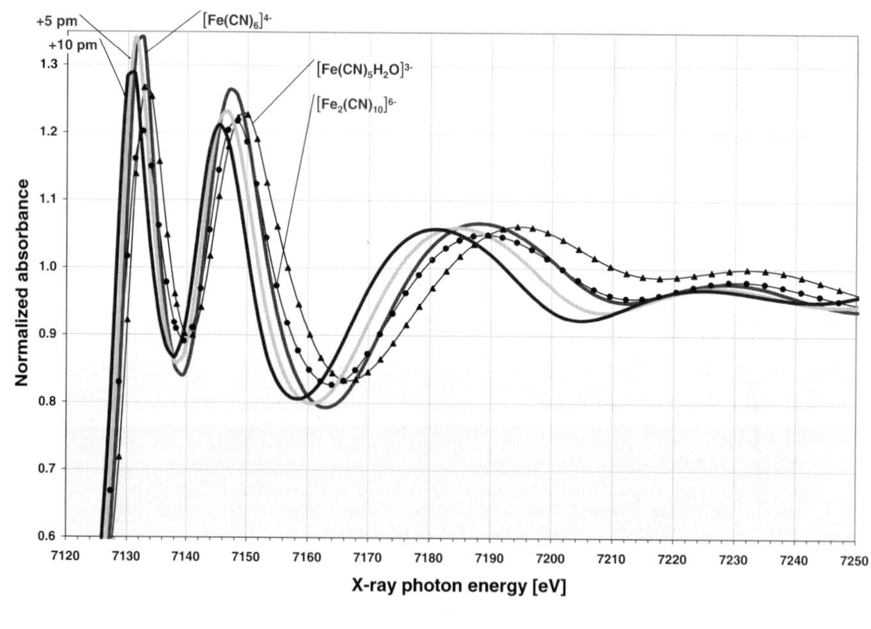

Figure 3: Theoretical XANES spectra of $Fe(CN)_6^{4-}$ for +5 pm and +10 pm elongated Fe-C distances but fixed C-N bond length. Additionally, theoretical XANES spectra two possible product complexes are shown.

focus diameter was about 50 μm. The resulting light intensity, i.e., 2×10^{13} W/cm^2, caused photo-excitation of the solute as well as ionization of the solution. Thus, the induced dynamics do not only comprise ligand substitution but most likely also include higher fragmentation of the chemical system. The samples were kept in a light protected container prior to the measurements in order to prevent premature photochemical conversion. Figure 2 shows a sequence of experimental XANES spectra at pump-probe delays of -30 ps, approximately 0 ps, and +40 ps. The data acquisition time for each XANES spectrum was 24 hours. For comparison, a XANES spectrum measured at the National Synchrotron Light Source was convoluted with a Gaussian line shape with a width of 8-eV (fwhm) [22]. It is shown in Figure 2 as a thin line. Within the precision of the laser-plasma driven measurement, the XANES spectrum measured before photoexcitation, at -30 ps, is in agreement with the synchrotron-measured spectrum. This implies that the spectral accuracy of our apparatus is comparable to that of a synchrotron source while its precision is lower due to its lower spectral resolution and lower signal-to-noise.

Photoexcitation changes the XANES spectrum. The iron K-edge position measured after photoexcitation is lower than that for unexcited $[Fe(CN)_6]^{4-}$, which can be explained through an elongation of the Fe-C bond lengths. Photoexcitation is likely to instantaneously increase the average metal-ligand bond length *en route* to ligand substitution. It has been theoretically recognized before that an increase of the iron oxidation state increases the iron K-edge energy while a lengthening of the iron-ligand distances lowers the K-edge position by about 40 meV / pm [33]. This effect can be observed in theoretical XANES spectra reproduced in Figure 3. An elongation on the order of 5 to 10 pm causes a shift comparable to the one observed in our measurements. The currently used pump-probe delay steps of tens of picoseconds are too large to observe atomic motions of ligand substitution reactions which are expected to proceed on the single-picosecond time scale. Each measured spectrum requires a data acquisition time of about 24 hours. As a consequence, only a few spectra were measured, and the delay time steps were chosen such that they transverse zero pump-probe time. The thermally induced metal-ligand bond elongation of reactants and products would results in a K-edge shift to lower energy. Additionally, Figure 3 shows XANES spectra for two of the possible reaction products, $[Fe(CN)_5H_2O]^{3-}$, $[Fe_2(CN)_{10}]^{6-}$. At high concentration, about 0.5M, of $[Fe(CN)_6]^{4-}$, the average center-to-center distance is 130 to 140 pm compared to 100 pm for the sum of the crystal radii which is a relatively favorable inter-ionic distance to form a dimeric product. For these compounds, the depth of the XANES modulations decreases. Thus, the XANES spectra of photolysed $[Fe(CN)_6]^{4-}$ continuously change in a way that is consistent with increasing iron-ligand distances.

3. Theoretical characterization of heavy-metal x-ray sources

The measurements discussed above were carried out using a brass-coated steel wire. The emitted spectrum consisted of continuum and line radiation. For x-ray absorption measurements, the continuum portion of the spectrum is used, and the positions of the emission lines are essentially irrelevant. Since long data acquisition times were required for the measurements of even a single XAFS spectrum, x-ray flux improvements are highly desirable. This continuum radiation flux can be substantially increased by using a target with a higher atomic number. In this section, some parameters of the x-ray source used in above measurements are discussed. These data are compared to theoretical data of a newly constructed x-ray source.

The new source is based on a liquid mercury jet that is irradiated by the laser pulses at 45 degrees incidence angle and p-polarization. Liquid mercury as the target material has the advantages that it can be circulated by a pump, yielding a very stable target. The plasma debris is liquid and can be automatically wiped off the chamber window. Furthermore, the plasma-physical properties of mercury make it very well suited for sub-100-fs hard x-ray pulse generation, as discussed in below.

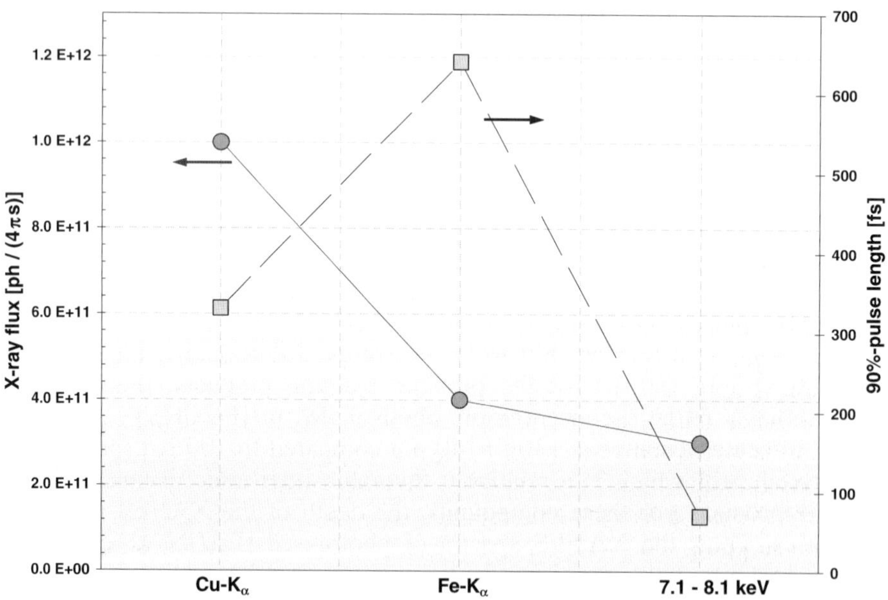

Figure 4: Simulated x-ray flux and pulse length emitted from a copper-coated steel wire target at 10^{17} W/cm^2, 5 mJ / pulse, 4 kHz repetition rate.

3.1. Simulation of sub-100-fs x-ray pulse emission from metal targets

The x-ray pulses that will be emitted from the target have been theoretically investigated by a combination of particle-in-cell (PIC) and Monte-Carlo (MC) simulations of the laser target interaction and the resulting electron dynamics. A two-step approach was applied to determine the x-ray emission. First, 1D, oblique incidence PIC simulations [34] of the laser-plasma interaction were performed to obtain the energy distribution of the hot electrons generated during the laser plasma interaction. All calculations were performed for p-polarized, 100-fs, 800-nm laser pulses with an incidence angle of 45 degrees. An exponential plasma density profile with a scale length $L/\lambda = 0.2$ for a laser wavelength $\lambda = 800$ nm was used. This scale length corresponds approximately to an interaction where a laser pre-pulse or pedestal generates a small amount of preformed plasma. Second, a Monte-Carlo electron-photon transport code [35] was used to compute the electron trajectories in the mercury target. The generated relativistically corrected continuum intensity for an energy interval around the iron K edge (7.1 to 7.2 keV) was calculated using the K-shell ionization cross sections from Casnati *et al.*[36]. Fluorescence yields, relative line intensities, and absorption lengths for emitted K_α radiation given by Zschornack [37] were used.

Temporal information on electrons and photons was calculated taking into account the electron entry time into the solid as obtained from the PIC calculations, the photon generation time, and the time of flight of the photons to the detector calculated with the MC code. The x-radiation was observed normal to the target front side. For calibration purposes, we calculated the x-ray emission properties for the wire-target used for the measurements briefly discussed in reference [38]. This target consists of a 200-μm steel wire, coated with 10-μm brass. The emitted Fe-K_α, Cu-K_α, and continuum flux around the Fe-K edge is shown in Figure 4 for a laser pulse intensity of 10^{17} W/cm^2. The calculated flux is larger than that used for the measurements presented above because these experiments have been performed at 2 kHz and at laser pulse intensities around 10^{16} W/cm^2. The calculated x-ray pulse length that contains 90% of the pulse energy depends greatly on the x-ray wavelength and the depth in the target at which the radiation is generated. For instance, the Cu-K_α radiation is generated in a layer between the target surface and 10 μm depth, and the pulse length is several hundred femtoseconds. In contrast, the Fe-K_α radiation is generated in a much thicker volume, and the diffusion time of the electrons into the target and of the x-radiation out of the target substantially broadens the pulse length. This effect demonstrates that the target with a larger electron stopping cross section and a larger x-ray absorption cross section could potentially reduce the x-ray pulse length substantially. This was evaluated for mercury, the target material for which the x-ray source was designed. Figure 5 shows a simulated x-ray pulse emitted from a mercury target illuminated with an 800-nm, 100-fs laser pulse with an intensity of 10^{17} W/cm^2 and 5 mJ pulse

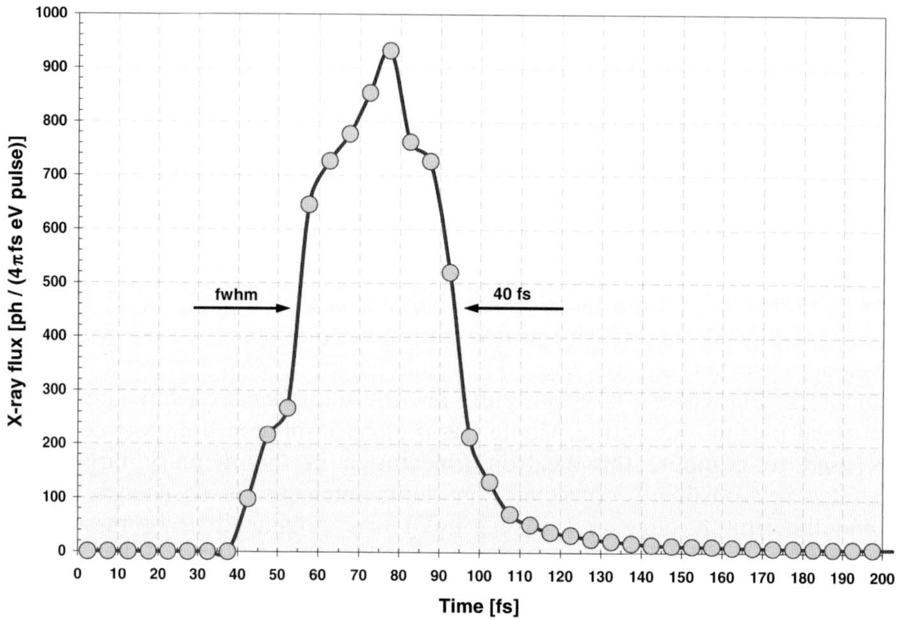

Figure 5: Simulated x-ray pulse in the spectral range from 7.1 to 7.2 keV emitted from a mercury target illuminated with an 800-nm, 100-fs laser pulse with an intensity of 10^{17} W/cm^2 and 5 mJ pulse energy.

energy. The full width at half maximum (fwhm) of the emitted x-ray pulse in the spectral range between 7.1 and 7.2 keV is 40 fs. This short pulse duration is due to the short electron penetration depth into the target material.

The average x-ray flux of the continuum radiation in the spectral range from 7.1 to 8.1 keV is shown in Figure 6. While the shortest pulse is obtained close to 10^{18} W/cm^2, higher intensities seem to increase the flux substantially but do not significantly lengthen the x-ray pulse width. This can be most likely explained by the fact that the continuum radiation generation efficiency monotonically increases with increasing electron energy. However, the deeper electron penetration into the target causes only slight x-ray pulse lengthening because the high x-ray absorption cross section of mercury largely suppresses contributions of "deep" x-ray photons to the overall pulse. Such a trend cannot be expected for fluorescence radiation. However, our x-ray applications are mostly in the area of chemical dynamics in solutions, which is best investigated using the continuum radiation for ultrafast x-ray absorption fine structure spectroscopy.

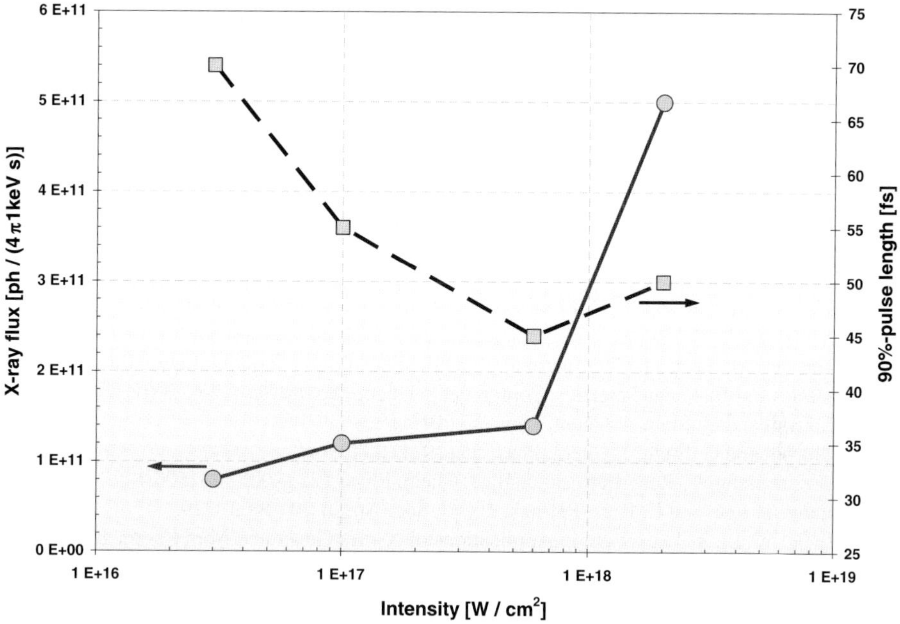

Figure 6: Simulated x-ray flux and pulse length emitted from a mercury target at illuminated with 800-nm, 100-fs laser pulses with 5-mJ pulse energy and a repetition rate of 4 kHz.

4. Acknowledgments

The authors gratefully acknowledge partial funding and support for this work by the National Science Foundation, Grant No. CHE-9984890, the U.S. Department of Energy, Award No. DE-FG02-03ER15413, the Research Corporation, Award No. RI0455, and equipment support by Fastpulse Technologies, Inc.

5. References

1. B. K. Agarwal, X-ray Spectroscopy, Springer-Verlag, Berlin, 1991.
2. D. C. Koningsberger and R. Prins, in Chemical Analysis, J. D. Winefordner (Ed.) 92, John Wiley & Sons, New York, 1988.
3. P. J. Durham, in X-Ray Absorption, D. C. Koningsberger and R. Prins (Eds.) 92, John Wiley & Sons, New York, 1988, p. 53
4. A. Bianconi, in X-Ray Absorption D. C. Koningsberger and R. Prins (Eds.) 92, John Wiley & Sons, New York, 1988, p. 573
5. C. Bressler and M. Chergui, Chem. Rev. 104 (2004) 1781.
6. H. W. Huang, W. H. Liu, T. Y. Teng, et al., Rev. Sci. Instr. 54 (1983) 1488.
7. T. Y. Teng, H. W. Huang, and G. A. Olah, Biochemistry 26 (1987) 8066.
8. A. Bianconi, S. D. Longa, I. Ascone, et al., Synchrotron Radiat. Biosci. 294 (1994).

9. L. X. Chen, G. Jennings, T. Liu, et al., J. Am. Chem. Soc. 124 (2002) 10861.
10. L. X. Chen, Faraday Discussions 122 (2002) 315.
11. L. X. Chen, W. J. H. Jager, G. Jennings, et al., Science 292 (2001) 262.
12. L. X. Chen, P. L. Lee, D. Gosztola, et al., J. Synchrotron Radiation 6 (1999) 403.
13. L. X. Chen, P. L. Lee, D. Gosztola, et al., J. Phys. Chem. B 103 (1999) 3270.
14. L. X. Chen, G. B. Shaw, I. Novozhilova, et al., J. Am. Chem. Soc. 125 (2003) 7022.
15. M. Saes, C. Bressler, R. Abela, et al., Phys. Rev. Lett. 90 (2003) 047403/1.
16. C. Bressler, M. Saes, M. Chergui, et al., J. Chem. Phys. 116 (2002) 2955.
17. C. Bressler, M. Chergui, P. Pattison, et al., Proc. SPIE-The International Society for Optical Engineering 3451 (1998) 108 (1998).
18. C. Bressler, M. Chergui, F. V. Mourik, et al., Paul Scherrer Institut, Switzerland, 2001, p. 94.
19. C. Bressler, M. Saes, M. Chergui, et al., Femtochemistry and Femtobiology: Ultrafast Dynamics in Molecular Science, [Conference], Toledo, Spain, Sept. 2-6, 2001, 449.
20. M. Shirom and G. Stein, J. Chem. Phys. 55 (1971) 3379 .
21. M. Shirom and G. Stein, J. Chem. Phys. 55 (1971) 3372.
22. T. Lee, Y. Jiang, C. Rose-Petruck, et al., J. Chem. Phys. 122 (2005) 084506/1.
23. Y. Jiang, T. Lee, and C. Rose-Petruck, J. Opt. Soc. Am. B 20 (2003) 229.
24. Y. Jiang, W. Li, T. Lee, et al., in Applications of X-rays Generated from Lasers and Other Bright Sources, G. A. Kyrala and J.-C. J. Gauthier (Eds.) 42504, SPIE Publishing, Bellingham, WA, 2001, p. 42.
25. T. Lee, Y. Jiang, F. Benesch, et al., in Laser-Generated and Other Laboratory X-Ray and EUV Sources, Optics, and Applications, G. A. Kyrala, A. M. J.-C. Gauthier, C. A. MacDonald and A. M. Khounsary (Eds.) 5196, SPIE Publishing, Bellingham, WA, San Jose, CA, 2003, p. 352.
26. F. Benesch, T. Lee, Y. Jiang, et al., Optics Lett. 29 (2004) 1028.
27. M. M. Murnane, H. C. Kapteyn, and R. W. Falcone, Phys. Fluids B 3 (1991) 2409.
28. J. C. Kieffer, M. Chaker, J. P. Matte, et al., Phys. Fluids B 5 (1993) 2676.
29. C. Rischel, A. Rousse, I. Uschmann, et al., Nature 390 (1997) 490.
30. C. Rose-Petruck, R. Jimenez, T. Guo, et al., Nature 398 (1999) 310.
31. C. Reich, P. Gibbon, I. Uschmann, et al., Phys. Rev. Lett. 84 (2000) 4846.
32. I. Uschmann, J.-P. Geindre, E. Foerster, et al., Proc. SPIE-The International Society for Optical Engineering 4143 (2001) 38.
33. A. Bianconi, M. Dell'Ariccia, P. J. Durham, et al., Phys. Rev. B 26 (1982) 6502.
34. P. Gibbon, A. Andreev, E. Lefebvre, et al., Physics of Plasmas 6 (1999) 947.
35. C. Reich, in Physikalisch-Astronomische Fakultaet,Friedrich-Schiller Universitaet, Jena, 2003, p. 104.
36. E. Casnati, A. Tartari, and C. Baraldi, J. Phys. B: Atom. and Molec. Phys. 15 (1982) 155.
37. G. Zschornack, Atomic Data for X-ray Spectral Analysis, 1988.
38. P. Gibbon and A. R. Bell, Phys. Rev. Lett. 68 (1992) 1535.

Femtochemistry VII
Fundamental Ultrafast Processes in Chemistry, Physics, and Biology
A.W. Castleman, Jr. and M.L. Kimble

Electron dynamics of metal-carbon clusters

K. L. Knappenberger, Jr., C. E. Jones, Jr., M. A. Sobhy, and A. W. Castleman, Jr.

Departments of Chemistry and Physics, The Pennsylvania State University, University Park, Pa, USA, 16802

1. Introduction

The interaction between carbon and transition metals has implications to many areas of chemistry. As such, the nature of metal-carbon bonding has been of considerable interest. The study of cluster assemblies has provided significant advances in the understanding of metal-carbon bonding. In fact, research in our laboratory has revealed the existence of a very stable class of molecular clusters known as metallocarbohedrenes, or "Met-Cars" for short [1]. The prototypical titanium Met-Car cluster was discovered during experiments intended to investigate the catalytic properties of titanium in the dehydrogenation of methane. The cluster distribution revealed an especially stable "magic" peak in the mass spectrum corresponding to a stoichiometry of Ti_8C_{12}. Since the first reporting of the Met-Car, it has been shown experimentally to form from several other metals, including V, Cr, Fe, Zr, Nb, Mo, and Hf [2,3]. Binary metal Met-Cars also show enhanced stability in the gas phase with stoichiometry $Ti_{8-x}M_xC_{12}$, where M includes: Y, Zr, Nb, Mo, Hf, Ta, W, and Si [4]. The discovery of Met-Cars has stimulated a wealth of research in both theoretical and experimental chemical physics [5]. However, many questions still remain regarding the geometric and electronic properties of Met-Cars. Additionally, further information regarding the formation mechanism of these clusters is needed. In this report, we review advances from our group, as well as an overview of ongoing projects toward resolving these questions.

2. General properties of Met-Cars

In our initial report on the existence of Met-Cars we suggested that Met-Cars were likely to be of icosohedral symmetry (T_h). Subsequently, ion mobility experiments provided support that a cage structure was likely for Met-Cars. However, our most recent theoretical results indicate that the most stable structure is likely of the D_2d point group. This geometric structure consists of an inner and outer tetrahedron with the inner being highly ionic in nature and the outer covalent.

Studies from our group to measure the photoionization efficiency of metal-carbon clusters have revealed that the ionization potential of Met-Cars is quite low [6]. For example, the ionization potential of carbon is known to be 11.260 eV [7] and that of zirconium is 6.84 eV [7]; yet the I.P. of the zirconium Met-Car is 3.99 eV. We have found that these very stable clusters also undergo appreciable delayed ionization, the analog to bulk phase thermionic emission. Collectively, the low ionization energies and the observed delayed ionization of Met-Car clusters suggest they have extensive free electron character.

Since their discovery, Met-Cars have been proposed as candidates for a range of applications, spanning the fields of semi-conductor science, catalysis, and hydrogen fuel storage. The potential implications of these unique clusters has led us to investigate their properties by many methods. Currently we are taking a bottom up approach to elucidate the growth mechanisms and better understand the many interesting phenomena these clusters display. Primarily we have investigated them by studying their electron dynamics.

3. Cluster dynamics laboratory

We have developed a laboratory devoted to the study of electron dynamics in metal-carbon cluster assemblies with an emphasis on Met-Cars. This laboratory is centered around a 50 fs regeneratively amplified Ti:Sapphire laser system. The output of the amplifier may be split and directed to either an optical parametric amplifier or harmonic generation. As such, we have the ability to generate 50 fs pulses ranging from 240 nm to 3.1 μm through frequency mixing. We reported the first ultrafast relaxation dynamics of electronically excited Met-Cars and other vanadium-carbon species [8,9]. In these experiments, we excited a band of states by the absorption of one 620 nm photon. The transients were recorded by probing with one 400 nm photon, enabling a REMPI mass spectrometry detection scheme.

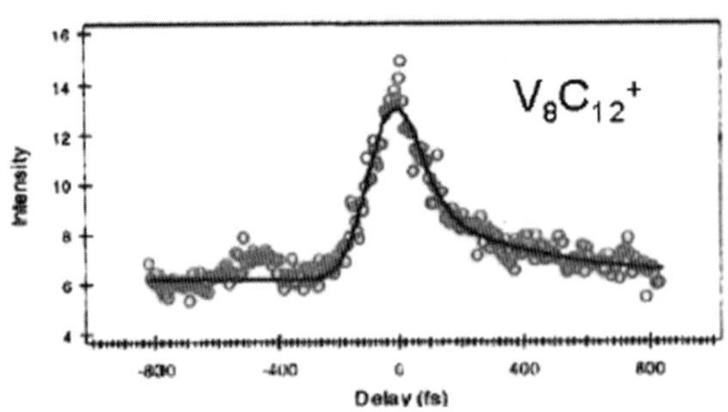

Figure 1. Pump-probe transient of V_8C_{12} Met-Car. (pump = 620 nm, probe = 400 nm)

While experiments from our group with nanosecond lasers clearly showed that Met-Cars undergo significant delayed ionization, only prompt ions are detected with femtosecond multiphoton ionization. We show the fitted femtosecond pump-probe response of the vanadium Met-Car excited with 2.0 eV in Figure 1. When compared to the instrument response function generated taken from a non-resonant zero marker, the Met-Car transient clearly displays electronic relaxation dynamics. Since the dissociation threshold of Met-Cars is well above the one photon pump energy, we can rule out the possibility of fragmentation. All metal-carbon clusters containing four or more vanadium atoms display similar pump-probe transients, with lifetimes of a few hundred femtoseconds. This finding suggests that a common chromophore is present in cluster sizes ranging from the half-cage (V_4C_6) to the V_8C_{12} Met-Car species. We have also found that the FWHM of the pump-probe responses narrows as more metal atoms are incorporated into the cluster structure. We believe the increase in the rate of electronic relaxation is indicative of the free electron character of these clusters.

4. Anion photoelectron spectroscopy of V_mC_n

Recently we have extended the cluster dynamics lab to include electron kinetic energy analysis. This added capability will provide more insight into the electronic and geometric structure of clusters and shared chromophore geometries between cluster sizes. In our first study, we have investigated the influence of the structural motif on the electronic properties of vanadium carbon clusters. Several classes of vanadium carbon clusters have been successfully generated in our laboratory by employing a range of laser vaporization source conditions. Clusters formed with lean carbon concentrations favor the generation of larger species that include many metals such as Met-Cars. Conversely, clusters formed in a carbon rich environment typically only incorporate up to two vanadium atoms. We determined the electron affinities by measuring the kinetic energies of photodeted electrons for clusters generated from all source conditions.

We have found that when high carbon V_mC_n clusters are formed in a carbon rich environment, the clusters exhibit a weak even-odd alternation of their electron affinity. As a result, it was concluded that the structure of these clusters are likely that of a monocyclic ring. The vertical electron affinities of these clusters is relatively low with respect to carbon. For most of the metal carbon clusters, the electron affinity is less than 2 eV, whereas the electron affinities of carbon clusters are generally greater than 3 eV [10]. The electron affinity of vanadium on the other hand is less than 0.6 eV [11]. The low electron affinity of the metal carbon clusters is likely due to the excess electron residing in a vanadium 3d orbital. The electron affinity of high carbon vanadium-carbon clusters does not change significantly with increasing cluster size. This is suggestive that the binding environment does not change significantly as additional carbon atoms are incorporated into the network. This finding is consistent with the idea that the excess electron occupies a 3d orbital of a vanadium atom in a planar monocyclic ring.

The electron kinetic energy spectra measured for clusters formed under lean carbon source conditions are drastically different than those formed under high carbon conditions. We have found that for these clusters the lowest electron affinities determined are assigned to the VC_2, V_2C_3, and V_4C_6 clusters. The vertical electron affinities of all clusters formed under lean source conditions are summarized in Figure 2. The VC_2 has long been perceived to be a building block of the larger three-dimensional clusters that ultimately result in the formation of the Met-Car. The V_2C_3 and V_4C_6 correspond to the one-quarter and one-half cage of the Met-Car, respectively. In a separate study, a similar effect was observed for many of the 8,12 Met-Car clusters. This finding, connecting the smaller divisions to the Met-Car, point to their importance in the formation of the larger cluster. Also, the determination of lower vertical electron affinities for the clusters belonging to the Met-Car class may point to their metal like properties.

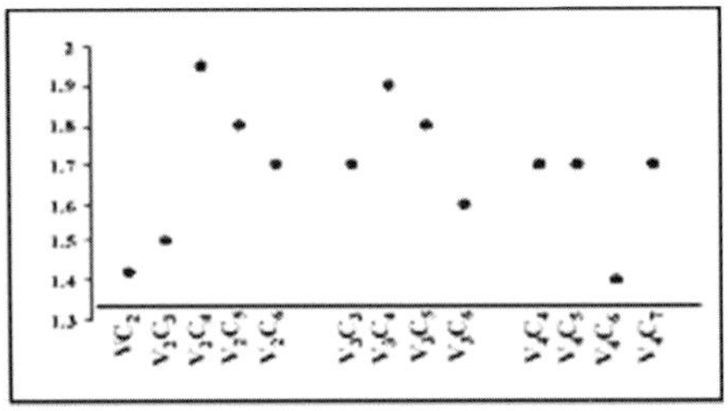

Figure 2. Summary of electron affinities determined for clusters generated in low carbon environments.

We have also found that the electron kinetic energy spectrum of the V_4C_6 displays a broader distribution of kinetic energies than other cluster sizes. Often, this is taken as a sign of thermalization processes, such as delayed emission, preceeding electron detachment. It is known that neutral Met-Cars undergo delayed ionization. According to the Richardson-Duschman equation, the rate of bulk thermionic emission increases exponentially with decreasing work function (the bulk analog of electron affinity). The finding that the half-cage has an especially low electron affinity is then consistent with the bulk model of thermionic emission.

5. Anion photoelectron spectroscopy of NbC

Our group has demonstrated control of cluster formation in niobium-carbon complexes by systematically altering source parameters. We have recently determined the vertical electron affinity of niobium-carbon clusters generated in a carbon rich environment, and report them in Figure 3. Our studies show that these clusters exist in multiple isomeric forms. We have been able to distinguish between the electron kinetic energy peaks that arise from

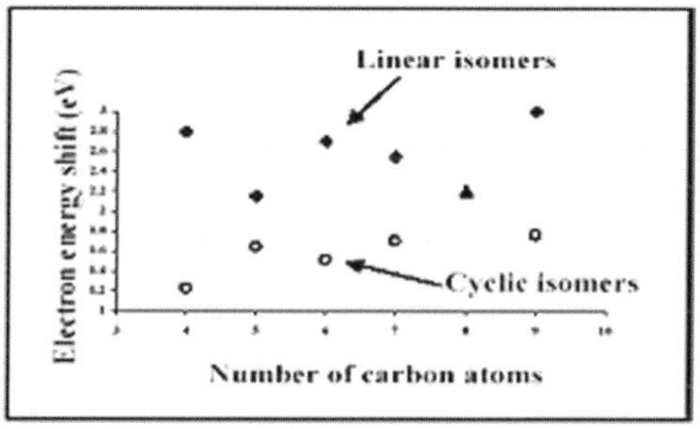

Figure 3. Summary of vertical electron affinities determined for Nb_2C_m clusters.

detachment of electrons from both linear chains and planar rings. In fact, the photoelectron spectra of the niobium-carbon clusters closely mirror the spectra obtained from pure carbon clusters. Further, we have found that smaller clusters, containing six or less atoms, prefer the cyclic geometry, and a transition to linear structure is observed when the cluster contains seven carbon atoms. Transition back to a monocylic ring is observed at nine carbon atoms. We have verified the experimentally determined structures with DFT calculations at the B3-LYP level. The calculations have confirmed that the planar and linear structures are of lower energy than the three-dimensional species we explored. Further we have demonstrated that the Nb-C interaction is covalent, unlike the ionic interaction between vanadium and carbon.

The trends in electron affinity of the niobium-carbon species track well with pure carbon clusters. However, the electron affinity of the metal containing clusters is slightly lower than that of the pure carbon species, and this effect increased progressively from clusters containing one and two niobium atoms. As was the case for vanadium-carbon clusters, this is likely due to the HOMO electron occupying a d orbital of the metal atom. In fact, the electronic configuration of niobium is $5s^14d^4$; so, in forming the anionic cluster, it is likely that the additional electron would form the stable half filled d-shell.

6. Conclusions

In conclusion, we have studied the ionization and electron detachment dynamics of several transition metal-carbon clusters with time resolved multiphoton ionization mass spectrometry and anion photoelectron spectroscopy. While nanosecond experiments on Met-Car clusters exhibit significant delayed ionization, we find that prompt ionization is preferential with femtosecond ionization. Femtosecond dynamics experiments reveal a "common chromophore" effect for V_mC_n clusters when m is larger than four.

Our photodetachment experiments on anionic vanadium carbides have provided insight to the growth mechanism of Met-Cars and other large metal carbon networks. We have been able to make a connection between VC_2, V_2C_3, and V_4C_6 based on the electron affinities that we determined. These clusters display the lowest electron affinities of all clusters considered in this study. A similar effect was observed for the Met-Car with respect to similar sized metal-carbon clusters. We have also investigated the influence of laser vaporization conditions on the geometric and electronic properties of the clusters generated. We find that carbon rich environments do not mediate the formation of large vanadium-carbon networks; rather, planar monocylic rings are formed. Lean carbon concentration results in the formation of clusters with 3-D structures.

On the other hand, we have found that carbon does indeed mediate the formation of large niobium-carbon clusters. Our studies clearly reveal the existence of multiple co-existing isomers, namely linear chains and planar monocyclic rings. We have found a transition in the preferred geometry from ring to linear and back to ring in going from the 2,6 to 2,7 and again for the 2,9, respectively. The structures determined spectroscopically for both vanadium and niobium containing clusters was verified by density functional calculations.

Acknowledgements

We gratefully acknowledge financial support from Air Force Office of Scientific Research, grant number FA 9550-04-1-0066.

References

1. B. C. Guo, K. P. Kerns, and A. W. Castleman, Jr., Science 255 (1992) 1411.
2. B. C. Guo, S. Wei, J. Purnell, S. Buzza, and A. W. Castleman, Jr., Science 256 (1992) 515.
3. J. S. Pilgrim and M. A. Duncan, J. Am. Chem. Soc. 115 (1993) 6958.
4. S. F. Cartier, B. D. May, and A. W. Castleman, Jr., J. Am. Chem. Soc. 116 (1994) 5295.
5. M.-M. Rhomer, M. Benard, and J. M. Poblet, Chem. Rev. 100 (2000) 495.
6. H. Sakurai and A. W. Castleman, Jr., J. Phys. Chem. A 102 (1998) 10486.
7. D. R. Lide, Handbook of Chem. and Phys. (1992) 10.
8. S. E. Kooi, B. D. Leskiw, and A. W. Castleman, Jr. Nano Letters 1 (2001) 113.
9. B. D. Leskiw, K. L. Knappenberger, Jr., and A. W. Castleman, Jr., J. Chem. Phys. 117 (2002) 8321.
10. D. W Arnold, S. E. Bradforth, T. N. Kitsopoulos, and D. M. Neumark, J. Chem. Phys. 95 (1991) 8753.
11. H. Hotop and W. C. Lineberger, J. Phys. Chem. Ref. Data. 14 (1985) 731.

Femtochemistry VII
Fundamental Ultrafast Processes in Chemistry, Physics, and Biology
A.W. Castleman, Jr. and M.L. Kimble

Time resolved CARS measurements of I_2 in solid Kr: vibrational state dependent dephasing times between 2.6 – 32 K.

Tiina Kiviniemi, Jukka Aumanen, Pasi Myllyperkiö, Mika Pettersson

Department of Chemistry, Nanoscience Center, P.O. Box 35, FIN-40014 University of Jyväskylä, Finland

1. Introduction

Dephasing is a central concept in condensed phase spectroscopy. It determines how long a system will maintain its coherence. The dephasing time of a system is determined by dynamic intermolecular interactions, and therefore, measurements of dephasing time can provide information on interactions and couplings between a molecule and its environment.

The theory of electronic and vibrational dephasing has been considered within different approaches [1-5]. The often used results include the T^7 dependence for coupling with acoustic phonons and exponential T dependence for coupling with pseudolocal phonons. Higher vibrational states lead to faster dephasing, and for a harmonic oscillator, quadratic dependence of the dephasing rate on the vibrational quantum number v has been derived [2,6].

Impurity-doped rare gas solids are useful model systems for studying fundamental processes in the condensed phase due to their simple structure and optical transparency in a wide spectral region [7]. Molecular iodine isolated in solid rare gases can be considered as a prototype of an anharmonic oscillator in a bath. Recent experiments on time-resolved coherent anti-Stokes Raman scattering (TR-CARS) of these systems have yielded accurate information, for example, on vibrational frequencies and dephasing times of various vibrational states and their dependence on temperature [8-12].

In this work, we have applied the femtosecond CARS method to investigate the vibrational dephasing of I_2 in solid krypton. Dephasing of vibrational states between v = 2 and v = 16 have been studied in the temperature range T = 2.6 – 32 K.

2. Experimental Details

The full description of the experimental details can be found in Ref. 12. In short, three laser beams are used in the CARS experiments. The three beams are arranged in a BOXCARS geometry, and the emitted CARS beam is detected with a photomultiplier tube. The CARS beam is filtered spatially with an iris and spectrally with an optical filter and a monochromator. The first and second pulses (pump and Stokes) are temporally overlapped, and the delay between the second and the third pulse (probe) is varied with a motorized delay stage. The wavelength of the pump and probe is varied between 565 and 580 nm, and the wavelength of the Stokes pulse is varied between 600 and 685 nm. The temporal width of the pump and probe pulses is ~80 fs, and the width of the Stokes pulse is ~40 fs.

The samples were prepared by continuous deposition of a gas mixture (I_2/Kr = 1/2600) on a thin (100 μm) sapphire substrate kept at 40 K in a liquid helium flow cryostat (Janis). The thickness of the sample is varied between 100-300 μm. After the deposition, the samples were cooled slowly (0.5 K/min) to the measurement temperature.

3. Results and Discussion

The measured CARS signal from a 10 K sample using $\lambda_{pump,\ probe}$ = 580 nm, λ_{Stokes} = 600 nm is shown in Figure 1a. The signal persists for more than 200 picoseconds, showing that the average dephasing time is in this range. The inset shows an expanded view of the signal showing the oscillations due to quantum beating. In Figure 1b, the Fourier transform of the time-domain signal is presented. It shows well resolved bands in the fundamental vibrational region of iodine. In addition, a band is observed in the first overtone region, indicating directly that the vibrational superposition contains at least three vibrational eigenstates.

Several different measurements were carried out in the same manner varying the Stokes shift corresponding to preparation of different vibrational superpositions. Each superposition was measured at 32, 20, 10, 4.1, and 2.6 K. The dephasing times were determined from time-domain signals by directly fitting the signal by a sum of exponentially damped cosines. In addition, the dephasing times were determined from the frequency-domain power spectra by fitting the bands with Lorentzian lineshape. Some of the results of this analysis are shown in Figure 2 for different vibrational states at 32 K and as a function of temperature for a superposition consisting of v = 14,15.

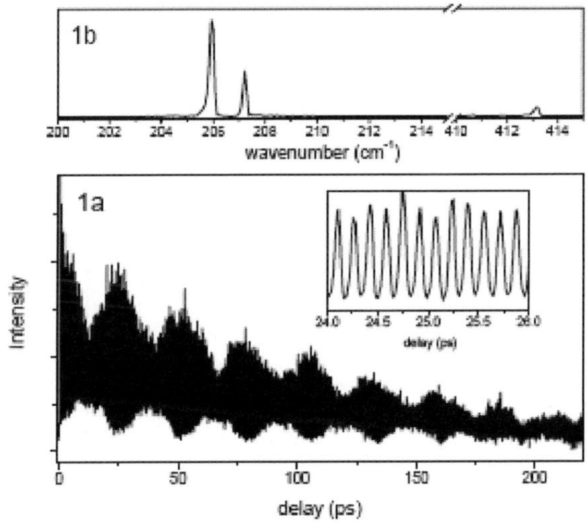

Figure 1. (1a): CARS signal from I_2 in solid Kr at 10 K. Pump and probe wavelength 580 nm, Stokes wavelength 600 nm. In (1b) a Fourier transform of the signal is shown.

An expression for the dephasing rate in the case of weak coupling is [13]:

$$g(v,T) = \frac{W^2}{p}\int_0^{\infty} dwn(w)\left[n(w)+1\right]G_P(w)^2 \qquad (1)$$

where n(ω) is the phonon occupation number and $\Gamma_P(\omega)$ is the phonon spectral density. W is the coupling parameter. For a harmonic oscillator, the v dependence is quadratic [2,6]. This is in agreement with the observed v dependence of Figure 2.

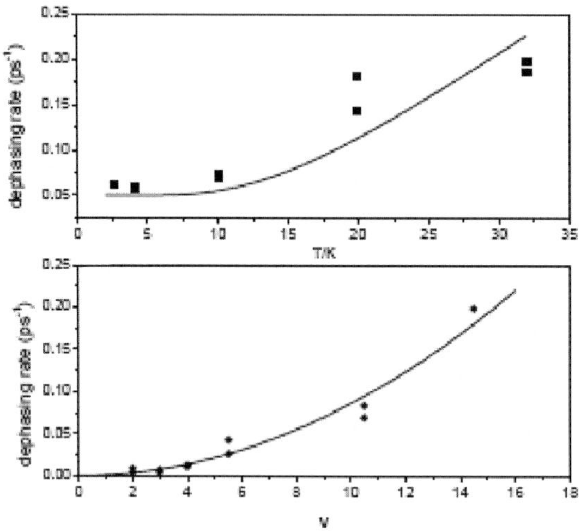

Figure 2. Lower part: dephasing rate as a function of vibrational quantum number v at a temperature of 32 K along with a quadratic fit. Upper part: dephasing rate of a v = 14,15 superposition as a function of temperature. The solid line is according to Eq. 1 (see text).

The temperature dependence was calculated using Eq. 1 with a Gaussian-type of spectral density with a frequency of 34 cm^{-1} and a FWHM of 5 cm^{-1}, describing a pseudolocal phonon [8]. The analysis suggests that dephasing is caused by coupling to pseudolocal phonons rather than to acoustic phonons that would show a T^7 dependence.

4. Conclusions

Time resolved CARS technique has been applied to an iodine molecule isolated in solid Kr in order to investigate the vibrational dephasing on the ground electronic state. The low vibrational states show dephasing times on the order of a hundred picoseconds. The v-dependence of the dephasing rate is roughly quadratic up to v = 16. The temperature-dependence of the dephasing rate indicates dephasing by coupling to pseudolocal phonons.

Acknowledgements

We thank V. A. Apkarian and M. Karavitis for numerous discussions. J. Korppi-Tommola and H. Kunttu are thanked for their support to this project. This project was supported by the Academy of Finland.

References

1. J. L. Skinner, Ann. Rev. Phys. Chem. 39 (1988) 463.
2. D. J. Diestler and A. H. Zewail, J. Chem. Phys. 71 (1979) 3103.

3. D. E. McCumber and M. D. Sturge, J. Appl. Phys., 34 (1963) 1682.
4. P. DeBree and D. A. Wiersma, J. Chem. Phys. 70 (1979) 790.
5. R. M. Shelby, C. B. Harris, and P. A. Cornelius, J. Chem. Phys. 70 (1979) 34.
6. A.H Zewail and D. J. Diestler, Chem. Phys. Lett. 65 (1979) 37.
7. V. A. Apkarian and N. Schwentner, Chem. Rev. 99 (1999) 1481.
8. M. Karavitis, T. Kumada, I. Goldschleger, and V. A. Apkarian, Phys. Chem. Chem. Phys. 7 (2005) 791.
9. M. Karavitis, R. Zadoyan, and V. A. Apkarian, J. Chem. Phys. 114 (2001) 4131.
10. M. Karavitis and V. A. Apkarian, J. Chem. Phys. 120 (2004) 292.
11. M. Karavitis, D. Segale, Z. Bihary, M. Pettersson, and V. A. Apkarian, Low Temp. Phys. 29 (2003) 814.
12. T. Kiviniemi, J. Aumanen, P. Myllyperkiö, V. A. Apkarian, and M. Pettersson, J. Chem. Phys. 123 (2005) 064509.
13. J. L. Skinner and D. Hsu, Chem. Phys. 128 (1988) 35.

Femtochemistry VII
Fundamental Ultrafast Processes in Chemistry, Physics, and Biology
A.W. Castleman, Jr. and M.L. Kimble

The Ultrafast Photofragmentation Pathway of *N,N*-Dimethylisopropylamine

M. P. Minitti,[a] J. L. Gosselin,[a] T. I. Sølling[b] and P. M. Weber[a]

[a] *Department of Chemistry, Brown University, Providence R.I. 02912, USA*
[b] *Department of Chemistry, Univ. of Copenhagen, DK-2100 Copenhagen, Denmark*

1. Introduction

The spectroscopy, energy flow between excited states, and photochemical reaction dynamics of aliphatic amines illustrate the exquisitely complex interplay of competing phenomena on multiple time scales. Several investigations have attempted to unravel the spectroscopy and dynamics of aliphatic amines upon excitation to Rydberg states [1,2]. In particular the recent time-resolved mass-spectrometric study by Sølling et al. [3] illustrates the importance of the reactive α-cleavage pathway for fragmentation and its interplay with nonreactive internal conversion channels of the highly excited molecules. The impetus for our work stems from the desire to learn about the photochemistry at lower intramolecular energies, as well as the need to clarify the precise surface on which the α-cleavage takes place. We investigated the femtosecond time-resolved dynamics of *N,N*-dimethylisopropylamine (DMIPA) using pump-probe mass spectrometry and photoelectron spectroscopy.

The binding energy of Rydberg electrons is inherently sensitive to the molecular structure, a discovery that we exploit to characterize the shape of molecules in our recently developed Rydberg Fingerprint Spectroscopy (RFS) [4,5]. The spectra of binding energies of Rydberg states can be analyzed and assigned in terms of their principal quantum numbers and constants called quantum defects. The latter have been found to be sensitive to the size, shape and charge distributions of molecules, making RFS a useful tool to address questions that could previously not be answered. For example, we investigated the location of the ionization center in the bifunctional molecule 2-phenylethyl-*N,N*-dimethylamine [6]. We were able to not only reveal the location of the ionization center but found that the laser wavelength provides a large amount of control over this location. In the present work, time resolved RFS serves as a probe to determine if and when DMIPA fragments, allowing us to identify the time scale and the electronic surface on which the fragmentation occurs.

2. Experimental

The femtosecond laser system and molecular beam apparatus outfitted with time-of-flight mass- and photoelectron spectrometers have been described in detail elsewhere [5,6]. Briefly, the output from a femtosecond Ti:Sapphire oscillator was regeneratively amplified (834 nm, 50 kHz, 100 fs) and subsequently upconverted to the second and fourth harmonic. The upconverted outputs were then separated, individually time-delayed, spatially combined, and focused into the apparatus. The pump and probe beams, serving as the excitation and ionization sources, respectively, intersected a molecular beam carrying the DMIPA molecules seeded in helium. Spectra were measured with a time-of-flight analyzer that can deliver an energy resolution of better than 10 meV. For the present spectra, the resolution is limited by the laser bandwidth, ~20-30 meV. The adiabatic ionization potential for DMIPA is 7.5 eV. Thus, by employing a resonant 208.5 nm pump and time-delayed 417 nm probe, we were able to conveniently access and ionize the low-*n* Rydberg states of DMIPA.

3. Results and Discussion

Representative mass and photoelectron spectra of DMIPA, each taken around time zero (±67 fs) and at 4 ps delay time, are shown in Fig. 1. From the mass spectrum, we notice that at time zero, the ionization produces predominantly the parent ion, while at later time, the fragment ion results from the α cleavage.

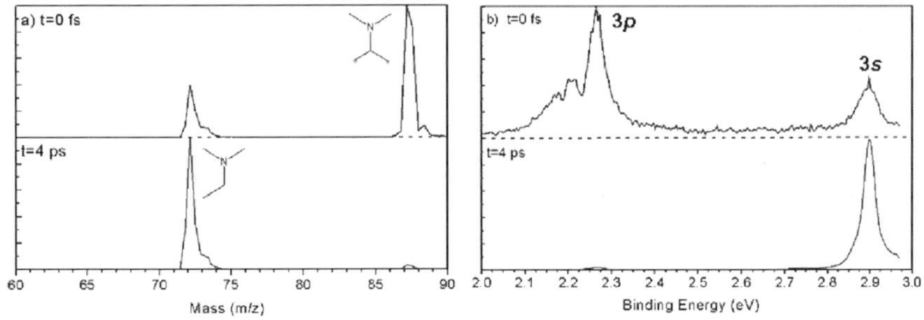

Fig. 1. Mass a) and photoelectron spectra b) of DMIPA taken at zero and 4 ps time delays.

The mass spectra can be explained using either of two scenarios. In the first, the molecule is excited to the Rydberg state and dissociates to the fragment ion while in the Rydberg level. The probe photon then ionizes the fragment, and its ion is observed. This is the model with which the experiments of reference 3 were interpreted. In the second scenario, the molecule is ionized out of the Rydberg state before it falls apart. The fragmentation then occurs subsequent to the photoionization, giving rise to the observed fragment ions. To determine which scenario is at play in DMIPA, we look at the photoelectron spectra.

The photoelectron spectrum at zero time delay is dominated by the initially populated 3*p* level (Fig. 1). The spectra show the splittings between the levels with different magnetic quantum numbers. However, at later times the 3*s* level becomes the dominant peak. The molecule decays from its initial 3*p* level to the lower 3*s* state. Recent work has revealed the energies of Rydberg levels are sensitive to the molecular structure [4,5]. The fact that the energy of the 3*s* Rydberg level at long delay times is unchanged from that at time zero is the first indication that the molecule does not fragment on the Rydberg surface.

The time dependence of the relevant mass and photoelectron peaks is shown in Fig. 2. It is apparent that the appearance of the parent is synchronous with the 3*p* peak, while the fragment has a temporal profile that matches the 3*s* level.

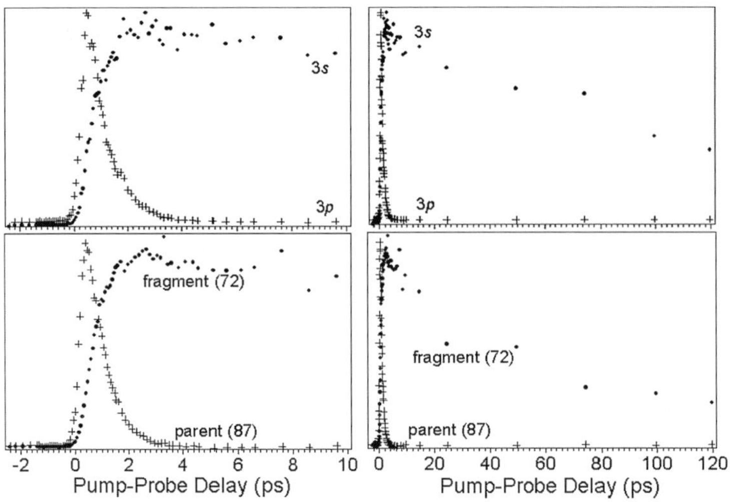

Fig. 2. Time dependent photoelectron (top) and mass spectra (bottom) of DMIPA obtained by resonant two-photon ionization with excitation to the 3*p* Rydberg level.

The combination of the mass and photoelectron spectra provides proof that the fragmentation dynamics of DMIPA follow that of the second scenario outlined above; the molecule is initially excited to the 3*p* Rydberg level, and ionization from that level produces ions with a small amount, about 0.8 eV, of internal vibrational energy. As this energy content is insufficient to induce fragmentation of the ion, we observe the parent mass in the mass spectrometer.

Subsequent to the excitation of the 3*p* Rydberg level, rapid internal conversion to 3*s* deposits an additional 0.6 eV into vibrational degrees of freedom. The total internal energy of the molecule while in the 3*s* state, 1.4 eV, is now sufficient to cause α-cleavage. However, the fragmentation does not proceed on a time scale fast enough to be observed in our experiment. Instead, the molecule leaves the 3*s* Rydberg state without generating a fragment molecule in a Rydberg level. Moreover, no new photoelectron peaks can be associated with the appearance of the neutral fragment. The most likely process for depletion of the 3*s* Rydberg level is a fluorescence to the ground electronic state on a time scale of tens of picoseconds.

Photoionization out of the 3*s* Rydberg level by the probe photon produces initially the parent ions. However, these parent ions are born with a sizeable amount of energy deposited in vibrational degrees of freedom, about 0.8 eV from the preparation of the 3*p* state and another 0.6 eV obtained during the internal conversion to 3*s*. This energy remains with the molecule during the ionization out of 3*s* because the potential energy surfaces of the Rydberg and the ion states are nearly identical. Since this energy is sufficient to cleave the α bond, the ion fragments on its way to the mass spectrometer detector. As a result, the mass spectrum reveals exclusively the fragment ion.

The time scale of the dynamics observed in our work is slower than that observed in reference 3. For DMIPA, those experiments had determined the decay times of the parent and fragment molecules to be 65 fs and 1.7 ps, respectively. The difference is possibly due to the smaller excitation energy used in our work.

4. Conclusion

In summary, we determined the pathway of the fragmentation dynamics of *N,N*-dimethylisopropylamine (DMIPA) via time-resolved mass and photoelectron spectroscopy. The dominant fragmentation mechanism is the cleavage of the α C-C bond. Upon excitation to 3*p*, a rapid internal conversion to 3*s* ensues, which deposits the energy needed for the fragmentation into vibrational degrees of freedom. At this energy, the bond cleavage does not, however, compete efficiently with the fluorescence decay, so that the bond does not break in the 3*s* state. The fragment ions are observed as a result of the bond cleavage of the molecular ions on their way to the mass spectrometer detector.

Acknowledgements

This research was supported by the Chemical Sciences, Geosciences and Biosciences Division, Office of Basic Energy Sciences, Department of Energy, grant number DE-FG02-03ER15452.

References

1. S. Hammerum, J. B. Christensen, H. Egsgaard, E. Larsen, P. J. Derrick, and K. F. Donchi, Int. J. Mass. Spectrom. Ion Phys. 47 (1983) 351.
2. S. Hammerum, K. Norrman, T. I. Sølling, P. E. Andersen, L. B. Jensen, and T. J. Vulpius, J. Am. Chem. Soc. 127 (2005) 6466.
3. T. I. Sølling, C. Kötting, and A. H. Zewail, J. Phys. Chem. A. 107 (2003) 10872.
4. N. Kuthirummal and P. M. Weber, Chem. Phys. Lett. 378, (2003) 647.
5. J. L. Gosselin and P. M. Weber, J. Phys. Chem. A. 109 (2005) 4899.
6. W. Cheng, N. Kuthirummal, J. L. Gosselin, T. I. Sølling, R. Weinkauf, and P. M. Weber, J. Phys. Chem. A. 109 (2005).

Femtochemistry VII
Fundamental Ultrafast Processes in Chemistry, Physics, and Biology
A.W. Castleman, Jr. and M.L. Kimble

Ultrafast photofragmentation dynamics of gas phase CH_2BrCl

Rebeca de Nalda[a], Jesús G. Izquierdo[a], Daniel Irimia[b], Maurice H. M. Janssen[b] and Luis Bañares[a]

[a]*Departamento de Química Física I , Facultad de Ciencias Químicas, Universidad Complutense de Madrid, Avda. Complutense s/n, 28040 Madrid, Spain.*
[b]*Laser Center and Chemistry Department, Vrije Universiteit, de Boelelaan 1083, 1081 HV, Amsterdam, The Netherlands*

1. Abstract

The ultrafast multichannel photofragmentation dynamics of gas phase CH_2BrCl have been studied by means of femtosecond laser pulses in combination with velocity map ion imaging. Two laser pulses of 150 fs at 266 nm and 400 nm have been employed in a pump-probe scheme to measure ion images of the different fragments produced under irradiation. The analysis of the images provides valuable information on the molecular multiphoton fragmentation mechanisms operating at these wavelengths.

2. Introduction

UV photofragmentation dynamics of polyatomic molecules are complex, in general, due to the large number of electronically excited states involved and to the existence of several bond-breaking pathways. The photodissociation of many polyatomic molecules, and in particular halocarbons, has been studied by means of nanosecond lasers in combination with multiphoton ionization and time-of-flight spectrometric techniques and velocity map imaging (see, for instance [1,2] for the title molecule). However, the use of nanosecond lasers has not been able to resolve the real time dynamics since, in most cases, the fragmentation occurs in a subpicosecond time scale. The use of femtosecond lasers in combination with the pump-probe technique has been applied in many laboratories to elucidate the femtosecond photodissociation dynamics of polyatomics. However, the existence of multiphoton processes and the possibility of fragmentation in both the neutral and ionic ladders makes it difficult to interpret the measured time-of-flight and time-resolved spectra. A good understanding of the photofragmentation mechanisms involved in such molecules has become particularly important, since it has been proposed theoretically [3] and demonstrated experimentally [4,5] that the fragmentation dynamics can be controlled, favouring the yield of a given fragment against others, by shaping the laser pulses.

Femtosecond velocity map ion imaging has proved to be an ideal technique to obtain time-resolved recoil energy and angular distributions of the fragments and to distinguish multiple photoexcitation and fragmentation pathways leading to the same ionic fragment. Although the complete understanding of the fragmentation mechanisms requires the performance of photoion-photoelectron coincidence experiments [6], it is sufficient, as a first approximation, to measure ion images in order to obtain information on the photofragmentation dynamics. In this work, we have applied pump-probe femtosecond laser pulses in combination with velocity map ion imaging to elucidate the fragmentation pathways involved in the photofragmentation of CH_2BrCl employing 266 nm and 400 nm laser pulses.

3. Experiment

Linearly polarized pump and probe laser pulses are produced by a chirped-pulse-regenerative amplified Ti:Sapphire laser system running at 1 kHz, resulting in 800 nm pulses with typical energy of 0.8 mJ and time duration of 150 fs. The amplified pulses are doubled and tripled in BBO crystals producing pulses around 400 nm and 266 nm, with energies in the range 2-25 μJ/pulse. The CH_2BrCl sample is seeded in Ar (1%) and expanded through a pulsed piezo valve running at 1 kHz, which is inside the source chamber of a molecular beam machine [7]. After passing two skimmers, the formed molecular beam enters the ionization chamber. The pump and probe laser beams are recombined and focused into the ionization chamber, where they intersect the pulsed molecular beam at a position in the middle between the repeller and extractor plates of a set of ion lenses. The ion optics are designed to map the velocity of the produced ionic fragments on a mass-gated MCP detector coupled to a phosphor screen, and the phosphorescence is imaged in a CCD camera. The polarization of both laser pulses was maintained parallel to the plane of the detector. Images were measured at different time delays between the pump and probe beams, and the background coming from each laser alone was subtracted.

4. Results and Discussion

Time-of-flight mass spectra have been measured when CH_2BrCl molecules interact with both pump (266 nm) and probe (400 nm) pulses at different time delays, $\Delta\tau$, and different laser intensities. A typical TOF mass spectrum taken at $\Delta\tau=0$ shows five peaks corresponding to the parent ion and to the CH_2Br^+, CH_2Cl^+, Br^+ and CH_2^+ fragment ions. The most intense peak in the mass spectrum corresponds to the CH_2Cl^+ fragment, followed by the parent ion and CH_2Br^+ fragment. No clear evidence of Cl^+ and $BrCl^+$ fragment ions was observed. Transients were measured for the parent and fragment ions by integrating each ion signal as a function of the pump-probe time delay. In all cases, the transients are characterized by a strong peak near $\Delta\tau=0$ which rapidly decays on a femtosecond time scale. In the case of Br^+, the transient is shifted towards positive time delays with respect to the others and shows an enhancement for later delay times corresponding to the possible excitation of a long-lived electronic state after absorption of one or more photons of 266 nm (pump pulse). No clear enhancements are observed in the transients measured for the rest of ions. Similar results were obtained previously for other halocarbons, such as CF_3I [7] and CH_2BrI [8] at these same wavelengths. Abel-invertible ion images were measured for all the fragments using parallel polarization of the pump and probe laser pulses with respect to the plane of the detector. Figure 1 shows three Abel-inverted images corresponding to Br^+ at three different time delays between the pump and probe pulses. Similar images have been measured for the other fragment ions. As can be appreciated, the Br^+ images are substantially anisotropic and do not depend strongly on the time delay. Time-resolved product translational energy distributions and energy-selected angular distributions have been obtained from the analysis of the Abel-inverted images for each fragment ion.

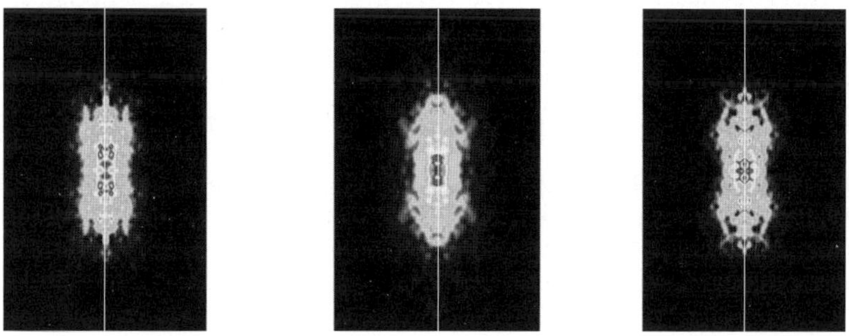

Figure 1. Br^+ fragment Abel-inverted images from CH_2BrCl at three different time delays between the 266 nm and 400 nm pump and probe femtosecond laser pulses. (Left) $\Delta\tau$=-200 fs; (Center) $\Delta\tau$=0 fs; (Right) $\Delta\tau$=+300 fs.

For CH_2BrCl, the shape of the energy distributions is rather independent on the time delay and fragment ion studied, at variance with previous studies on CF_3I [7]. The anisotropy parameters obtained from the analysis of the angular distributions indicate that the fragmentation is dominated by parallel transitions involving the absorption of two (pump or probe) photons. These results indicate that the photoframentation of CH_2BrCl at 266/400 nm occurs in a femtosecond time scale yielding mainly Br fragments from neutral potential energy surfaces. Dissociative ionization is minor in this case. More experiments at other wavelengths and photoion-photoelectron coincident experiments are timely to confirm these preliminary conclusions on the photofragmentation mechanism of CH_2BrCl.

5. Acknowledgements

RDN and JGI acknowledge financial support through the Ramón y Cajal and FPI programs, respectively, from the Spanish Ministry of Education and Science. RDN, JGI and LB thank the EU for financial support for access to the experimental facilities of the Laser Centre Vrije Universiteit through the EU program HPRI-CT-1999-00064. This research is supported by the council for Chemical Sciences of the Dutch Organisation for Scientific Research (NWO-CW).

References

1. W. S. McGivern, R. Li, P. Zou, and S. W. North, J. Chem. Phys. 111 (1999) 5771.
2. S.-H. Lee, Y.-J. Jung, and K.-H. Jung, Chem. Phys. 260 (2000) 143.
3. R. S. Judson and H. Rabitz, Phys. Rev. Lett. 68 (1992) 1500.
4. A. Assion, T. Baumert, M. Bergt, T. Brixner, B. Kiefer, V. Seyfried, M. Strehle, and G. Gerber, Science 282 (1998) 919.
5. N. H. Damrauer, C. Dietl, G. Krampert, S.-H. Lee, K.-H. Jung, and G. Gerber, Eur. Phys. J. D 20 (2002) 71.
6. A. M. Rijs, M. H. M. Janssen, E. T. H. Chrysostom, and C. C. Hayden, Phys. Rev. Lett. 92 (2004) 123002.
7. W. G. Roeterdink and M. H. M. Janssen, Phys. Chem. Chem. Phys. 4 (2002) 601.
8. A. M. Rijs, Ph.D. Thesis, Vrije Universiteit Amsterdam, 2003.

Femtochemistry VII
Fundamental Ultrafast Processes in Chemistry, Physics, and Biology
A.W. Castleman, Jr. and M.L. Kimble

Determination of charge state of tungsten during ultrafast hard x-ray generation

Fang Shan, Vernon A. Couch, William H. Fink, Ting Guo

Department of Chemistry, University of California, Davis, CA 95616 USA

1. Introduction

Ultrafast x-rays can be generated by acceleration or deceleration of electrons such as that in synchrotrons, recombination of cations after initial ionization, or fluorescence from electron bombarded neutral targets similar to how x-rays are generated in conventional x-ray tubes. The second scenario, if happens in solid density materials, can result in the production of ultrafast (sub 500 fs) kiloelectronvolt x-rays, and it has been used to produce these x-ray pulses for more than a decade [1,2]. The fluorescence-based methods commonly rely on the use of laser accelerated electron pulses generated outside the target with prepulses or pedestals. In this scenario, ultrafast hard x-ray pulses of several hundred femtoseconds have been produced by many research groups [3,4].

Although it is generally accepted that those externally produced/accelerated electrons interact with a solid target, no data is available to describe the exact species from which ultrafast x-rays are produced. For example, it is unknown whether the emitting species are neutrally charged, moderately ionized, or highly ionized. One thing is certain that the emitting species are not extremely highly ionized species as those used in producing softer ultrafast x-rays, which sometimes can reach several tens of oxidation state. Before this work, no direct measurements were available to confirm the charge state of the ultrafast x-ray emitting substance.

One way to determine the charge state is to measure the oxidation state. This can be done by measuring the chemical shift of the species, and normally simultaneous high resolution (~0.5 eV) and high accuracy (~0.5 eV) data are needed to determine oxidation state from +1 to +5. However, this is extremely difficult. In order to overcome this problem, we have devised a new method, ultrafast selected energy x-ray absorption spectroscopy, or USEXAS. As explained below, it takes advantage of using the emission spectrum of ultrafast x-ray pulses from one element to measure the absorption edge of another element. This allows one to obtain precise chemical shifts of the x-ray emitting species if a reference material is used. Vice versa, one can measure the absorption edge if emission is calibrated. This self-referencing method can potentially provide the highest combined energy resolution and spectral accuracy for x-ray absorption measurements.

In this paper, we will describe USEXAS and present the results, which will show that the emitting species of ultrafast hard x-ray pulses are neutral tungsten atoms. The current paper is written based on the results from three previous published manuscripts [5-7].

2. Experimental

The x-ray target was tungsten and the reference material was $ZnSO_4$ (aq. 0.25 M). The sample was mounted on an iris outside the x-ray generation chamber described below. The line emission was L lines from tungsten, and the K absorption edge of Zn was measured. Since the absorption edge of these compounds should be the same under any kind of measurements, they can be used to calibrate the peak positions of these L lines. The x-ray target apparatus was described in detail elsewhere [7]. In brief, a high purity tungsten rod (3/8-inch diameter, 9-inch long, with 0.0005-inch RMS surface finish) was suspended on one end by attaching to a combination of two step motors that rotated and moved the rod down at the same time. The assembly was housed in a vacuum x-ray chamber.

The laser system, also described elsewhere in detail, produced terawatt pulses at 800-nm central wavelength and at 50-Hz repetition rate [8]. The focused laser beam on a fresh tungsten surface through a 2-inch gold coated parabola mirror (4-inch focal length) resulted in the production of ultrafast x-rays in the energy range of a few keV to a few hundred keV. The keV x-rays were measured with a Si crystal spectrometer and an x-ray CCD camera. Under optimal conditions, line emission from the tungsten target with little bremsstrahlung radiation was obtained. The data analysis routine was described elsewhere, and normally a few thousand images were processed to produce the final absorption spectra [6].

Using the error propagation method and known signal-to-noise ratios, we determined the spectral precision of x-ray absorption measurements with the current x-ray apparatus to be approximately 0.4 eV.

3. Theory

In order to estimate the oxidation state based on the absorption edge measurements, we calculated the spectral positions of various L lines from tungsten as a function of oxidation state. The ground state energy and associated eigenvalues were calculated by solving the Dirac-Fock equation using a program called Dirac. An averaging method was used to calculate open-shell ions. In these calculations and the estimation of transition energies, the initial hole created by electron ionization before the event of x-ray fluorescence was not included. We found that this assumption is correct, and the hole will not affect the transition energies because there is no time for energy levels to resettle after the creation of the initial hole.

4. Results

The absorption edge measurements of $ZnSO_4$ (aq) are shown in Figure 1. Since the edge is very close to line $L_{\beta 1}$, we can use the position of this K edge to calibrate the spectral position of $L_{\beta 1}$. The linearity of the spectrometer (Si crystal and the CCD camera) is excellent over a large energy range (1000 eV), and it was calibrated with other x-ray lines in the 5-10 keV range.

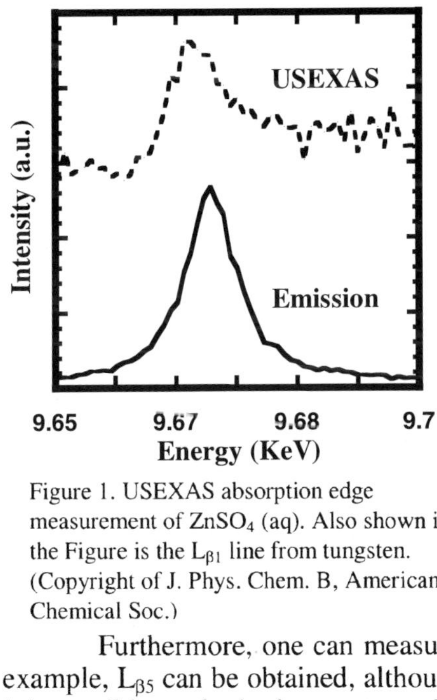

Figure 1. USEXAS absorption edge measurement of $ZnSO_4$ (aq). Also shown in the Figure is the $L_{\beta 1}$ line from tungsten. (Copyright of J. Phys. Chem. B, American Chemical Soc.)

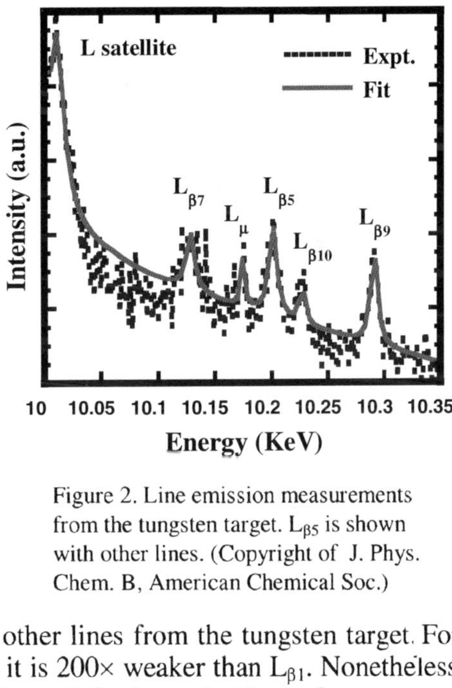

Figure 2. Line emission measurements from the tungsten target. $L_{\beta 5}$ is shown with other lines. (Copyright of J. Phys. Chem. B, American Chemical Soc.)

Furthermore, one can measure other lines from the tungsten target. For example, $L_{\beta 5}$ can be obtained, although it is 200× weaker than $L_{\beta 1}$. Nonetheless, it is possible to obtain the spectrum. This result is shown in Figure 2.

5. Comparison of Theoretical and Experimental Data

As the calculations show, the changes of the spectral peak positions of various L lines from tungsten are relatively small (<0.4 eV for all lines except $L_{\beta5}$), too small for using these L lines and the current setup to determine the oxidation state of the emitting tungsten species. Fortunately, $L_{\beta5}$ experiences a large shift upon changing from charge neutral to oxidation state +1, mainly because the transition involves the O_V level, which is very close to 6*s* valence orbitals. The change is estimated at 1.9 eV for oxidation state +1. Such a big change to the transition energy should be detectable with the current setup.

Based on Figure 2, it is clear that $L_{\beta5}$ is very close to the theoretically calculated values for neutral tungsten. All other lines are also close to their theoretical values from neutral tungsten. As a result, although it is impossible to tell whether the tungsten species from other L lines, $L_{\beta5}$ can be used to conclusively determine the oxidation state of tungsten to be zero because it has not changed within the detection limit (0.4 eV).

6. Conclusion

It is thus obvious that the charge state of tungsten in our ultrafast hard x-rays apparatus is zero. The high resolution and high accuracy of USEXAS measurements allow the determination of spectral peak positions of x-ray line emission from such a source.

Acknowledgement: This work is supported by the National Science Foundation (CAREER Award 0135132).

References

1. M. M. Murnane, H. C. Kapteyn, M. D. Rosen, and R. W. Falcone, Science 251 (1991) 531.
2. J. C. Kieffer, Z. Jiang, A. Ikhlef, C. Y. Cote, and O. Peyrusse, J. Opt. Soc. Am. B-Opt. Phys.13 (1996) 132.
3. A. Rousse, P. Audebert, J. P. Geindre, F. Fallies, J. C. Gauthier, A. Mysyrowicz, G. Grillon, and A. Antonetti, Phys. Rev. E 50 (1994) 2200.
4. T. Guo, C. Spielmann, B. C. Walker, and C. P. J. Barty, Rev. Sci. Instrum. 72 (2001) 41.
5. F. Shan, J. Carter, V. Ng, and T. Guo, Phys. Rev. E 71 (2005).
6. F. Shan, V. Couch, and T. Guo, J. Phys. Chem. A 109 (2005) 4216.
7. F. Shan and T. Guo, J. Chem. Phys. 122 (2005).
8. G. J. Cheng, F. Shan, A. Freyer, and T. Guo, Appl. Optics 41 (2002) 5148.

Femtochemistry VII
Fundamental Ultrafast Processes in Chemistry, Physics, and Biology
A.W. Castleman, Jr. and M.L. Kimble

Multi-dimensional fs spectroscopy of an excitonic dimer, described by Redfield relaxation theory.

Pär Kjellberg and Tõnu Pullerits

Department of Chemical Physics, Lund University, P.O. Box 124, 22100, Lund, Sweden

1. Introduction

In non-linear spectroscopy, an optical signal is created by the interaction of several short laser pulses with a medium. The main goal of such experiments is to relate the out-coming signal to the dynamical processes in the system under study. However, the relation between observables, like time- or frequency spectra, and the underlying processes is often not straightforward. In order to understand and interpret the results, computer simulations are an important tool.

The aim of a simulation is usually to calculate the polarization, which generates the signal in a given experimental situation. In this article we calculate the third order polarization of an excitonic dimer system using response function theory [1]. We assume a weak electron-phonon coupling and determine the time development of coherences and populations using Redfield relaxation theory. A scheme suggested by Kato and Tanimura [2], based on an approach previously applied in non-perturbative calculations [3], was used to extract the polarization components corresponding to the three-pulse photon echo experiment (3PPE).

2. Three Pulse Photon Echo Signals

The response function was evaluated by setting up a two-level dimer system, diagonalzing the site Hamiltonian, and numerically propagating the density matrix using Green functions. A schematic energy level diagram of the excitonic states is given in Fig. 1. To account for static energy disorder, the response was averaged over a number of site configurations. The energies were

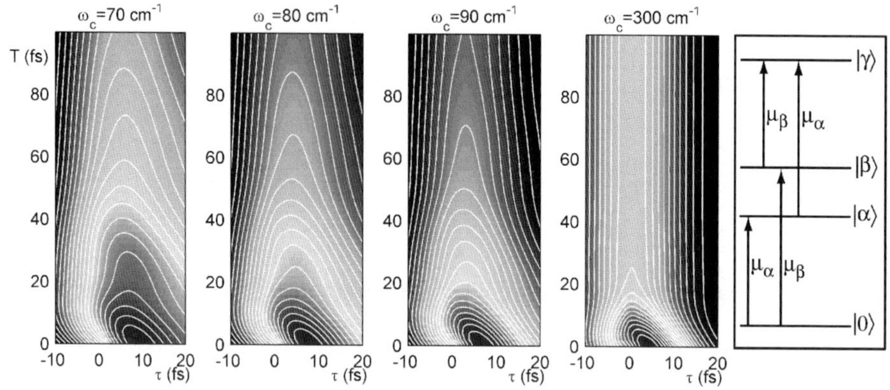

Figure 1: The 3PPE signals from a system where the two-exciton state has been excluded. The signals have been calculated at different cut-off frequencies of the spectral densities. The FWHM of the lasers pulses was set to 15 fs and a pure dephasing parameter of 30 cm^{-1} was used. To the right a schematic energy level diagram of the excitonic states.

set by Monte Carlo sampling a Gaussian distribution of standard deviation σ =300 cm^{-1} centered at 8000 cm^{-1}. The total polarization was calculated by convoluting over the electric field. The standard way of scanning the laser pulses in a three-pulse photon echo experiment was used, see e.g. reference [4]. The total electric field was taken to be a sum of three components, each having a Gaussian pulse envelope function, a carrier frequency, and a wave vector. Finally the polarization components in the four directions $\phi_1=k_1+k_2+k_3$, $\phi_2=k_1-k_2+k_3$, $\phi_3=k_1+k_2-k_3$ and $\phi_4=-k_1+k_2+k_3$ were extracted.

The final 3PPE signal will depend on the system's ability to rephase after three interactions. More precisely, the dephasing of coherences and relaxation of populations will influence the signal. These processes are, to a great extent, determined by the bath modes accessible to the system. In order to investigate the effect of different rates, a spectral density with a cut-off frequency, ω_c, was used. As ω_c increases, the overlap between the spectral density and the energy gap between the exciton states becomes stronger, resulting in faster relaxation and dephasing rates, see Table 1.

To explore the role of the two-exciton state, the 3PPE signals were first calculated for a system where the two-exciton state was excluded. In Fig. 1 the integrated absolute square of the polarization component along the direction $-k_1+k_2+k_3$ is presented. The response function used was averaged over 1000 different site configurations. The most apparent features are that, as ω_c increases, the signal becomes narrower and shifts towards smaller values of τ.

An important observable in 3PPE experiments is the three-pulse photon echo peak shift (3PEPS). It has been shown that information about the energy gap correlation function and the spectral density can be obtained from the peak shift data of a chromophore-solute system [4]. Even though our model does not include a detailed description of the exciton-vibrational coupling, 3PEPS provides a more transparent way of analyzing the 2D-plots in Fig. 1, and some interesting features can be more clearly seen.

Table 1: Relaxation rates at different ω_c averaged over 1000 configurations of site-energies.

ω_c (cm^{-1})	70	80	90	300
$k_{\alpha\to\beta}$ (fs)	1880	916	544	137
$k_{\beta\to\alpha}$ (fs)	92	44	26	5.4

The peak shift of the dimer with and without the two-exciton state is displayed in Fig. 2. The most striking difference between the two results is that the 3PEPS from the system where the two-exciton state has been omitted has a beating structure. The beats appear as a consequence of how the density matrix ρ evolves during the time intervals between interactions with the electric field. Following the first interaction, the non-zero elements of the density matrix are the coherences between ground state and one exciton states $\rho_{\mu 0}$ and $\rho_{0\mu}$ ($\mu = \alpha,\beta$). These propagate with a frequency $\omega_{\mu 0} = \omega_\mu - \omega_0$. The second interaction takes the system into populations $\rho_{\mu\mu}$, one-exciton coherences $\rho_{\mu\nu}$, and coherence between ground state and the two-exciton state $\rho_{0\gamma}$ or $\rho_{\gamma 0}$. The elements $\rho_{\mu\nu}$ oscillate with a frequency $\omega_{\mu\nu}$. After the final interaction, coherences $\rho_{\mu 0}$ and $\rho_{0\mu}$ and, in addition, coherences between one and two exciton states $\rho_{\mu\gamma}$ and $\rho_{\gamma\mu}$ are present. It is the elements $\rho_{\mu\nu}$ which make an

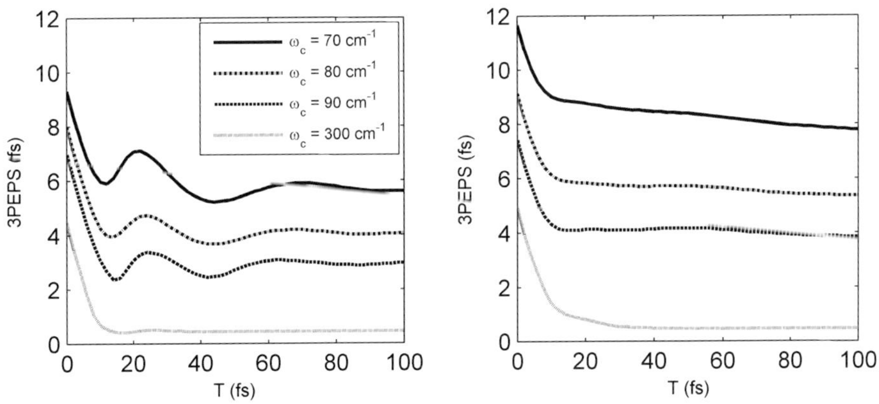

Figure 2: 3PEPS from a dimer with (to the right) and without (to the left) the two-exciton state.

additional contribution to the signal resulting in the beats. This fact is reflected in the response function. The response function at $t_1 = 0$ is shown in Fig. 3. The three-time parameters t_1, t_2, and t_3 are the moments in time of first, second, and third interaction. The longer oscillation, visible in the figure excluding the two-exciton state, originates from coherences between the one-exciton states. These

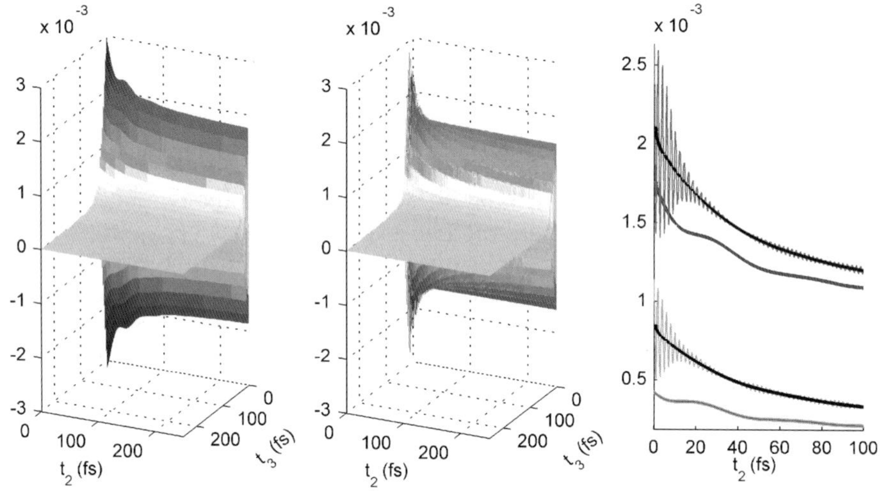

Figure 3: The response function at t_1=0 excluding the two-exciton state (to the left), including (middle) and cuts at 15 and 27 fs (to the right).

oscillations are not present if the two-exciton state is included; this feature of the response function is somewhat obscured by fine oscillations, corresponding to coherences between the ground state and the two- exciton state like $\rho_{0\gamma}$. The absence of the slow modulation is easier to see in the figure showing cuts along the t_2 axis of the two response functions at times 15 and 27 fs. A smooth line, calculated by means of locally weighted linear regression, has been added to the cuts. The smooth line exhibits oscillations if the two-exciton state is ignored and no oscillations if it is taken into account. The oscillations are not present because the two-exciton state introduces additional Liouville pathways, which cancel the contribution from $\rho_{\mu\nu}$ to the final response. If the symmetry of the excitonic dimer were to be broken, by for example a shift of the two-exciton energy level or a change in the dipole moments, beats in the signal will reappear.

Further, the overall downward shift of 3PEPS as the cut-off frequency increases reflects a stronger dephasing due to population transfer.

In conclusion, we have shown that the 3PEPS signal from an excitonic dimer can have some structure even if the detailed picture of the exciton-vibrational coupling is not included. The structure can be related to dephasing mechanisms of coherences. These effects should be considered when analyzing photon-echo experiments of molecular aggregates.

References

1. Shaul Mukamel, Principles of Nonlinear optical spectroscopy, Oxford University Press, New York, 1995.
2. Tsuyoshi Kato and Yoshitaka Tanimura, Chem. Phys. Lett. 341 (2001) 329.
3. Luis Seidner, Gerhard Stock, and Wolfgang Domcke, J. Chem. Phys. 103 (1995) 3998.
4. Taiha Joo, Yiwei Jia, Jae-Yong Yu, Matthew J. Lang, and Graham R. Fleming, J. Chem. Phys. 104 (1996) 6089.

Femtochemistry VII
Fundamental Ultrafast Processes in Chemistry, Physics, and Biology
A.W. Castleman, Jr. and M.L. Kimble

High power synchronized femtosecond Ti: sapphire laser with sub-femtosecond timing jitter

Jinrong Tian[1] Zhiyi Wei[1*], Peng Wang[1], Hainian Han[1], Zhaohua Wang[1], Jie Zhang[1], Tao Yang[2], Jianwei Pan[2]

[1]*Beijing National Laboratory for Condensed Matter Physics, Institute of Physics, Chinese Academy of Sciences, Beijing 100080, China*
[2]*Department of Modern Physics, University of Science and Technology of China, Hefei 230026, China*

1. Introduction

Synchronized femtosecond lasers have been widely used in pump-probe experiments, optical frequency metrology, generation of THz wave, etc. Since the 1990s, synchronized lasers have been achieved by active [1-3] and passive technique [4-6] with at least a timing jitter of 2 fs. The traditional passively-synchronized femtosecond lasers incorporated a single gain medium with a single cavity; thus two independently tunable synchronized femtosecond pulses were obtained. However this scheme suffered severely from gain competition. Furthermore PZT was still necessary to match the cavity length due to the limited mismatch tolerance. In this paper, we describe a new passive synchronization scheme for two independent Ti: sapphire lasers with superior robustness and sub-femtosecond timing jitter. The tolerance of the cavity length mismatch is large enough for us to start the synchronization by hand instead of PZT. The correlation measurement revealed its timing jitter was less than 1 fs. Furthermore, the relative carrier-envelope phase slip between the laser pulses was detected by a beat frequency measurement without frequency conversion, which paved the way for phase-locking the synchronized femtosecond lasers.

2. Setup

The original setup of the synchronized laser is shown in Fig. 1. It comprised two astigmatically compensated femtosecond Ti: sapphire oscillators

(Laser 1: M1~M6, X1, PM1~PM2, OC1; Laser 2: M7~M12, X2, PM3~PM4, OC2). Two
fused silica prism sequences with a tip-to-tip distance of 80 cm were inserted to

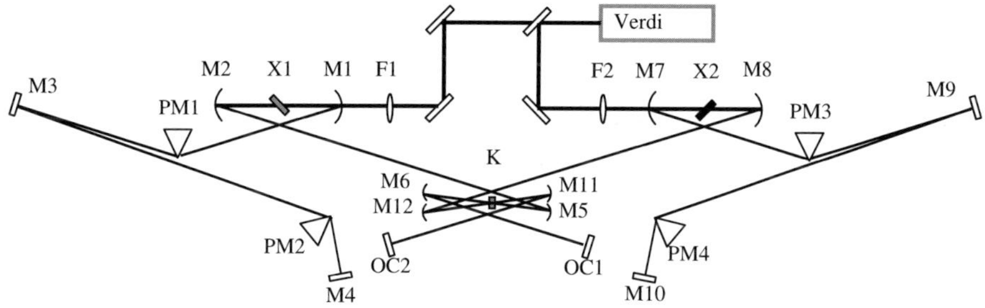

Fig. 1 The original setup of the synchronized femtosecond laser M1, M2, M5~M8, M11, M12: concave mirrors of 10 cm radius of curvature; M3, M4, M9, M10: flat reflectors; PM1~PM4, fused-silica prisms; OC1, OC2, 10% or 20% output couplers; K, Kerr medium.

compensate for the dispersion and provide necessary negative group velocity dispersion. Furthermore, we introduced a coupling cavity (M5~M6, M11~M12) which consisted of two pairs of D-shape concave mirrors with 10-cm-radii curvature and a nonlinear crystal (K). Both lasers were focused into the crystal and crossed inside it, in order to obtain strong cross-phase modulation (XPM) effect for synchronization. One of the end mirrors (M4) was mounted on a translation stage in order to precisely match the cavity length of both lasers. With this configuration, the gain competition was completely eliminated, and stable synchronization was obtained. We further developed a compact design shown in Fig. 2. The size of the whole system is about 800×400 mm^2. The two Ti: sapphire lasers were separately pumped by frequency doubled Nd: YVO4 lasers (Verdi V-10, Coherent Inc.). Under 8 W pump power, the maximum output mode-locked power is 1.3 W. The synchronization could sustain more than 8 hours without external interference. It was noteworthy that we did not use PZT because the synchronization could be obtained by tuning the translation stage by hand slightly, which was due to the strong XPM effect that supported larger tolerance of cavity-length mismatch. The wavelengths of both lasers could be tuned independently from 730 nm to 860 nm, which was desirable in many experiments where dual-wavelength synchronized

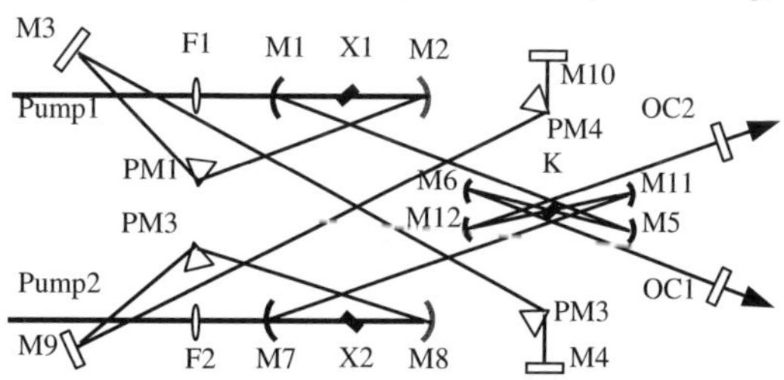

Fig. 2 The compact design of synchronized femtosecond lasers.

femtosecond lasers were necessary.

3. Cross correlation measurement

To estimate the timing jitter, a homemade intensity correlator shown in Fig. 3(a) was used to measure the correlation trace between the two Ti: sapphire lasers. In this configuration one of the lasers passed through a delay line, then collinear (not superimposed) lasers were focused into a 0.1 mm BBO crystal to

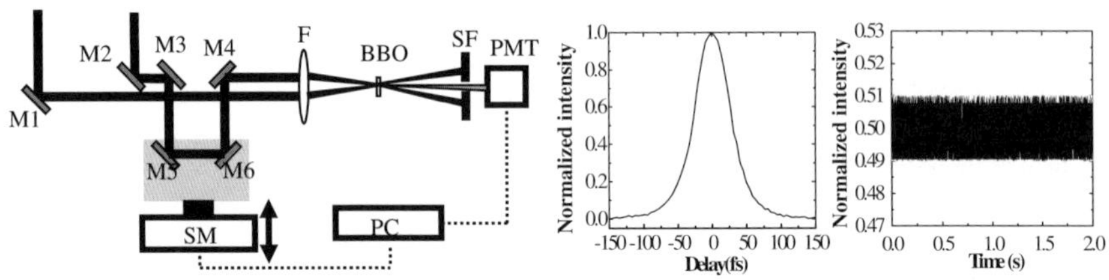

Fig. 3 (a) Schematic of cross correlation traces measurement. M1~M6, gold mirrors; F, 10-cm focal-length lens; SF, slit and filter; PMT, photomultiplier tube; SM, stepping motor; PC, computer. (b) Normalized cross correlation trace of two Ti: sapphire lasers; (c) Intensity fluctuations of the SFG signal at half peak.

generate the sum-frequency signal and the correlation signal was received by a photo-multiplier so that the cross correlation trace was recorded. Fig. 3 (b) shows the typical cross-correlation trace; the FWHM of the correlation trace was 59 fs with a Gaussian assumption. Furthermore, fixing the time delay equal to the FWHM of the correlation trace, we recorded the intensity fluctuation at 1 kHz bandwidth in 2 seconds as shown in Fig. 3 (c), which revealed that the intensity fluctuation was limited to $\pm 1\%$, with an average Root-Mean-Square deviation 0.00653. The corresponding timing jitter was deduced as 0.4 fs. We consider the limiting factor in achieving the lowest timing-jitter was the environmental perturbation.

4. The carrier-envelope phase slip between two synchronized pulses

The carrier-envelope phase slip between both lasers could be determined by the beat frequency measurement. Fig. 4(a) shows the experimental setup. Both lasers were first tuned to center at 790 nm, and after a delay line, two lasers were superimposed spatially after the beam splitter and were dispersed by a holographic grating. The diffracted wavelength component was received by a fast photodiode, which was connected to a spectrum analyzer. When the optical path length was completely compensated, the beat signal would appear. Fig. 4(b) depicts the beat power spectrum collected with the resolution bandwidth of 100 KHz and sweep time of 50 ms, where the heterodyned beat signal δ appeared with a signal-to-noise ratio (SNR) of 41 dB and f_{rep} corresponded to the repetition rate of the laser. The relative carrier envelope offset between two lasers could be determined by $\Delta\Phi = 2\pi\delta/f_{rep}$, in which $\Delta\Phi$ was the carrier phase slip. The high S/N ratio of the observed beat signal will enable us to

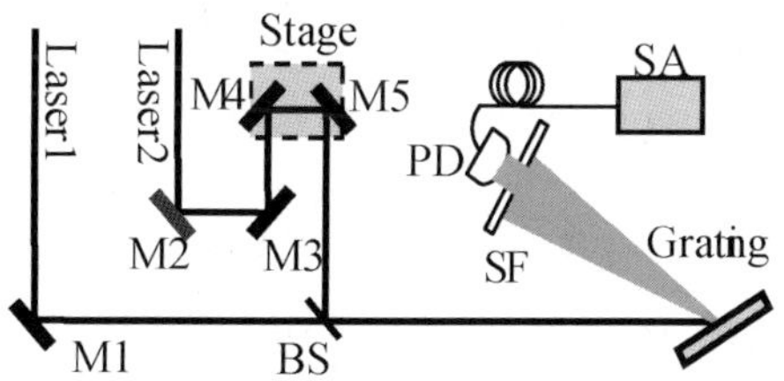

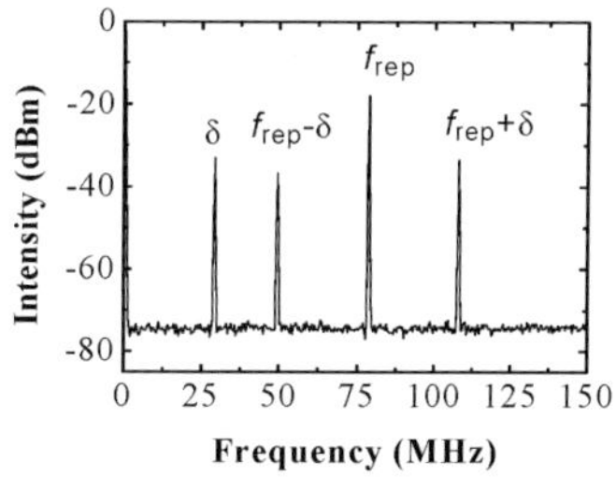

Fig. 4 (a) Schematic of beat-frequency signal measurement. M1~M4, gold mirrors; PD, photodiode; BS, beam splitter; SA, spectrum analyzer; SF, slit and filter. (b) RF power spectrum over a span of 150 MHz. The resolution bandwidth and sweep time were set at 100 kHz and 50 ms, respectively. δ, beat signal; f_{rep}, repetition rate.

realize very robust phase-locking operation with a stabilized phase-locked loop circuit.

5. Conclusion

In summary, we have developed a passively synchronized Ti: sapphire laser. The synchronization operation could be started by hand and the synchronization sustained more than 8 hours with a least timing jitter of 0.4 fs. We believe this laser will provide a reliable tool for many research fields.

Acknowlegdements This work was partly supported by National Science Foundation of China under Grant Nos.60490280, 10227401 and 60225005

References

1. S. A. Crooker, F. D. Betz, J. Levy, and D. D. Awschalom, Rev. Sci. Instrum. 67 (1996) 2068.
2. L.-S. Ma, R. K. Shelton, H. C. Kapteyn, M. M. Murnane, and J. Ye, Phys. Rev. A 64 (2001) 021802.
3. Th. Udem, R. Holzwarth, and T. W. Hänsch, Nature 416 (2002) 233.
4. M. R. X. de Barros and P. C. Becker, Opt. Lett. 18 (1993) 631.
5. A. Leitenstorfer, C. Fürst, and A. Laubereau, Opt. Lett. 20 (1995) 916.
6. Z. Wei, Y. Kobayashi, Z. Zhang, and K. Torizuka, Opt. Lett. 26 (2001) 1806.

Femtochemistry VII
Fundamental Ultrafast Processes in Chemistry, Physics, and Biology
A.W. Castleman, Jr. and M.L. Kimble

Ultrafast vibrational evolution of excited-state coumarin 153 using fs/ps CARS as a transient Raman probe

Benjamin D. Prince, Abhijit Chakraborty, Beth M. Prince, Tse-Ming Hsin, and Hans U. Stauffer

Department of Chemistry, Iowa State University, Ames, Iowa, 50011-3111

1. Introduction

The development of techniques to study molecular vibrations with reasonable frequency resolution in the femtosecond time regime for excited state molecules has opened new avenues of research for the understanding of molecular structure following electronic excitation. Frequency resolution allows one to observe directly the evolution of vibrational modes as a function of some relevant time delay without the need to apply Fourier Transform techniques or to separate complicated beat patterns to determine pairs of involved modes. Here, vibrational evolution upon electronic excitation in coumarin 153 (C153) is observed using a transient Raman probe technique developed in our laboratory that is a variation of time-resolved coherent anti-Stokes Raman spectroscopy (TR-CARS). The technique, named fs/ps CARS, is described in detail elsewhere [1]. C153 has been extensively studied for its ideal solvatochromic properties upon excitation that allow it to be used as a probe for solvent reorganization. However, despite the wealth of work done on the solvation properties, only a few studies have been undertaken to understand the molecular changes of the solute molecule upon electronic excitation.

2. Experimental

A Ti:Sapphire laser system generates pulses of ~50 fs duration centered at 800 nm with a repetition rate of 1 kHz and power of 2.1 mJ/pulse. Part of the 800 nm pulse is used to generate visible pulses at frequencies ω_1 and ω_2 via optical parametric amplification and frequency mixing [1]. The residual 800 nm beam is split to generate a narrowband (ω_3, $\Delta\omega$ ~ 16 cm^{-1}) ps-duration probe pulse centered at 795 nm and a pump pulse (ω_{pump}) near 400 nm via frequency doubling.

The visible fs/ps beams are chosen to be centered at ω_1 = 509 nm (19646 cm^{-1}) and ω_2 = 555 nm (18018 cm^{-1}). After excitation by ω_{pump} ($S_1 \leftarrow S_0$), the ω_2 pulse is resonant with the ground electronic state (stimulated emission, $S_1 \rightarrow S_0$). ω_1 is time overlapped with ω_2 and repopulates the electronic excited state from the transition caused by ω_2 ($S_0 \leftarrow S_1$) (Figure 1).

The three beams associated with the fs/ps CARS probe scheme (ω_1, ω_2, and ω_3) are aligned in a folded BOXCARS arrangement, and the spatially-filtered output signal is focused into a spectrometer (Ocean Optics, USB-2000, ~0.7 nm resolution). This scheme generates vibrationally resolved spectra of Raman active modes while maintaining sub-picosecond temporal resolution associated with time-resolved CARS. The delay of the pump beam, which is aligned to be collinear with the ω_3 pulse, is adjusted via a computer-controlled translation stage. The resultant spectra are fit to Lorentzian lineshapes, and statistical information (intensity, peak frequency, FWHM, *etc.*) is generated and plotted against pump delay.

C153 was used as purchased from Exciton and was dissolved in commercially available solvents for study in these experiments. The solvents include the hydrogen bonding solvents methanol (MeOH) and butanol (BuOH) and non-hydrogen bonding solvent acetonitrile (ACN). The samples were diluted to a concentration of 8 mM. Samples were placed in a 220 μm thick rotating home-built cell to ensure the sample is refreshed between laser shots.

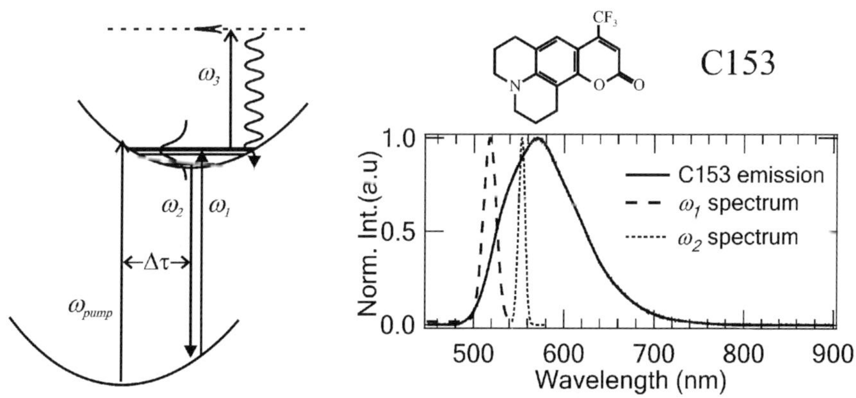

Figure 1: Left: Experimental pulse scheme utilized in this fs/ps CARS experiment. An initial pump beam excites population from $S_0 \leftarrow S_1$, where two broadband time overlapped beams ω_1 and ω_2 prepare population in excited state Raman vibrational modes after some time delay. A final ps narrowband probe beam ω_3 then probes population and a fourth wave is detected. Right: Chemical structure of C153 and emission of C153/MeOH upon 400 nm excitation.

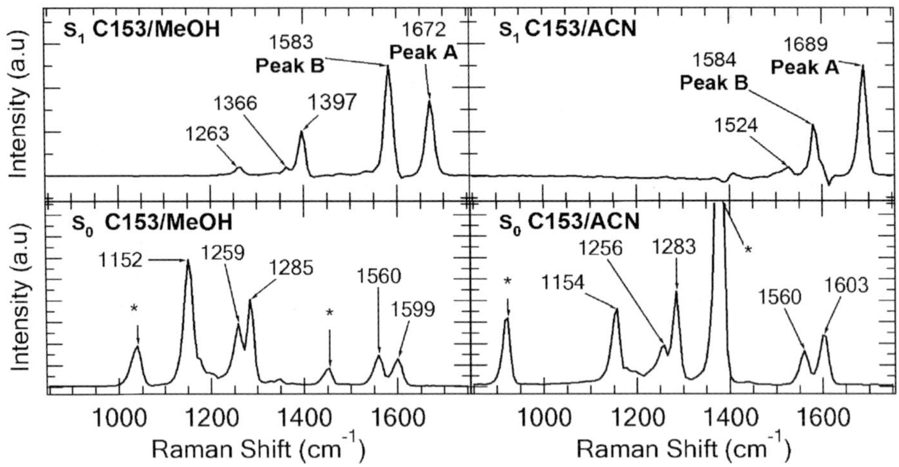

Figure 2: Ground (bottom) and excited state (top) fs/ps CARS spectra of C153 in two solvents: MeOH (left) and ACN (right). Solvents are noted with the "*" symbol. Particular attention is spent on "Peak A" and "Peak B" in this experiment. Excited state spectra are generated with the pump pulse fixed at a long time delay (tens of ps). Note the large frequency difference of "Peak A" in both solvents.

3. Results and Discussion

Figure 2 displays data typical of our experimental setup. The excited state spectra (top graphs) are generated at a fixed excitation pump pulse delay (~30 ps for Figure 2) and have been corrected to remove signal resulting from the ground state (lower graphs). The evolution of the Raman vibrational spectrum from ~1000-2000 cm^{-1} is monitored with the excitation pump delay, and evolution of the modes can be followed within the time resolution (~100 fs) of our experiment. At long delays (~15 ps), the two solvents differ considerably in the peak position of the mode nearest 1700 cm^{-1}. In MeOH, the position is detected at approximately 1671 cm^{-1}, while in ACN the position is near 1690 cm^{-1}. The focus of the results here will be on the peaks denoted as "A" and "B" in the top panels of Figure 2.

The 1584 cm^{-1} frequency (Figure 3) shows some time dependent shift in MeOH, but this shift is either absent in ACN or considerably faster than the time resolution of this setup. To determine whether this time dependent evolution is related to solvent response or to simple molecular evolution, another solvent (butanol) was chosen whose solvation dynamics are considerably slower than the two studied thus far [2]. The frequency evolution in BuOH is similar to that in MeOH but on a considerably slower timescale (3.6 ps vs. ~20 ps). This suggests that this frequency is related to the evolution of a

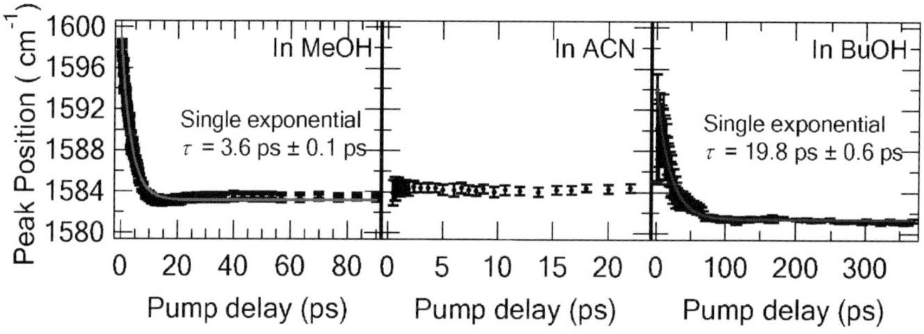

Figure 3: Central peak frequency evolution of "Peak B" (~ 1584 cm^{-1}) in the excited state spectra of C153 in three solvents. The evolution of this mode appears to be solvent dependent rather than purely molecular in nature. The timescales of evolution fit well to single exponentials and appear to be related to slower timescales of solvation.

longer term timescale of solvation. According to previous work by Maroncelli and co-workers, the solvation dynamics of C153 in ACN occurs considerably faster than MeOH, which in turn occurs faster than BuOH [2]. Further studies on the nature of this shift are currently under way in our laboratory.

Previous work on coumarin 102 (C102), similar in structure to C153, has defined the free C=O stretch in the region of 1730-1740 cm^{-1} in the excited state [3]. Thus it is not expected that the excited state mode observed near 1670 cm^{-1} in the alcohols and near 1690 cm^{-1} in the ACN solution is the direct C=O stretch; however, it is potentially a combination mode involving the skeletal motions and a C=O stretch component. Kiefer and coworkers have calculated and assigned the modes of coumarin 152 (C152), which is closely related to C153, in the ground state and compared to experimentally measured Raman spectra in solid and liquid phases [4]. A ground state mode observed at 1603 cm^{-1} contains a C=O stretch component in C152. This agrees with the slight difference observed in the solution phase for this mode in the ground state in our measured C153 spectra (1599 *vs.* 1603 cm^{-1}, Figure 2, bottom panels). Thus, it is expected that in the excited state, the mode near 1690 cm^{-1} has some relationship to a C=O stretch, and further details regarding the behavior of this peak will be addressed elsewhere.

References

1. B. D. Prince, A. Chakraborty, B.M. Prince, and H. U. Stauffer, J. Chem. Phys., in press.
2. M.L. Horng, J.A. Gardecki, A. Papazyan, and M. Maroncelli, J. Phys. Chem., 99 (1995) 17311.
3. E.T.J. Nibbering, F. Tschirschwitz, C. Chudoba, and T. Elsaesser, J. Phys. Chem. A, 104 (2000) 4236.
4. E. Vogel, A. Gbureck, and W. Kiefer, J. Mol. Struct., 550 (2000) 177.

Femtochemistry VII
Fundamental Ultrafast Processes in Chemistry, Physics, and Biology
A.W. Castleman, Jr. and M.L. Kimble

Development of fs/ps CARS as a Sub-100 fs Transient Raman Probe

Benjamin D. Prince,[a] Abhijit Chakraborty,[b] Beth M. Prince,[a] and Hans U. Stauffer[a]

[a]*Department of Chemistry, Iowa State University, Ames, IA 50011-3111*
[b]*Permanent address: Department of Physics, R.K. Mission Vidyamandira, Belur Math, Howrah, 711202, India*

1. Introduction

Recent experimental developments, such as infrared transient absorption and the related terahertz spectroscopy, have made it possible to detect the temporal dynamics of vibrational energy flow in the condensed phase on increasingly fast timescales. Such approaches provide a wealth of information previously unavailable on these short timescales; however, they are often difficult to implement given the challenges of producing and detecting infrared radiation in the vibrational fingerprint spectral region and at lower frequencies. The complementary Raman techniques have not evolved as quickly, although recent development of the femtosecond stimulated Raman scattering (FSRS) technique by Mathies and coworkers has begun to allow spectrally resolved studies of the 600 – 2000 cm^{-1} region of the Raman spectrum with ~100 fs resolution [1]. An alternative approach to this FSRS technique involves the use of time-resolved coherent anti-Stokes Raman scattering (tr-CARS) as a subpicosecond probe of excited state vibrational evolution following electronic excitation. This and related degenerate four wave mixing (DFWM) spectroscopies have been used in several pump-CARS experimental schemes to probe short-time evolution of both solute and solvent following solute electronic excitation. The CARS method typically involves three broadband input pulses: a pump pulse with frequency ω_1, a Stokes pulse with frequency ω_2, and a probe pulse with frequency ω_3. The coherences created by the ω_1 and ω_2 pulses are thus probed by the broadband ω_3 pulse as a function of time delay.

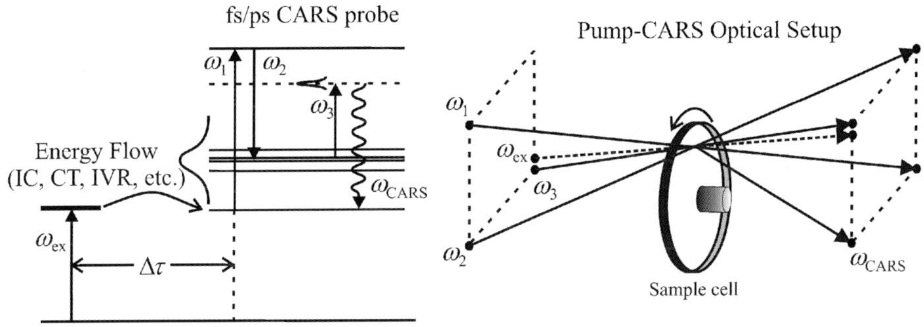

Fig. 1: Left: Schematic of pump-(fs/ps CARS) probe technique. Right: Folded BOXCARS optical configuration for spatial isolation of fs/ps-CARS signal; ω_{ex} alignment is also shown.

Here, we describe a variant of tr-CARS, deemed fs/ps CARS, that allows multiplexed detection of the Raman active modes observable with CARS techniques, while maintaining the ultrafast time resolution necessary to use fs/ps CARS as a simultaneously time- and frequency-resolved probe of vibrational evolution following electronic excitation. This method involves the replacement of the scanning, broadband ω_3 pulse typically associated with tr-CARS detection schemes with a narrowband pulse (Fig. 1), alleviating the need to deduce Raman frequencies from measured quantum beats. Rhodamine 6G (R6G) is used to demonstrate this technique as a Raman probe. Application of this technique to probe the short-time evolution of excited state *trans*-4-dimethylamino-4′-nitrostilbene (*trans*-DANS) following visible excitation is detailed as well.

2. Experimental Setup

The experiments described here have been carried out using a Ti:sapphire oscillator/amplifier system running at a 1 kHz repetition rate, with an output of 2.1 mJ/pulse centered at 795 nm and a duration of 50-60 fs, as determined by autocorrelation [2]. Approximately half is split off to pump an optical parametric amplifier, which produces both signal (ω_s) and idler (ω_i) pulses via parametric downconversion of the input fundamental (ω_f) pulse. These two beams are each subsequently mixed with replicate paths of the fundamental beam to produce visible pulses at frequencies $\omega_1 = \omega_f + \omega_s$ and $\omega_2 = \omega_f + \omega_i$. The bandwidths of these visible pulses are typically 300-600 cm^{-1}. The remainder of the fundamental pulse is again split to produce an excitation pulse (ω_{ex} at 397 nm) via frequency doubling, and the CARS probe pulse, ω_3, which is directed into a 4*f* pulse shaping configuration, equipped with a variable-width slit in the focal plane to allow continuous adjustment of the pulse 3 bandwidth from <16 cm^{-1} to its full spectral bandwidth. All input beams are focused into a motorized rotating cell, allowing sample refreshment after each pulse, using a folded BOXCARS input configuration (Fig. 1). The CARS output, dictated by phase-matching, is spatially filtered and focused into a spectrometer (Ocean Optics USB-2000) with a resolution of ~14 cm^{-1}. The three-color method employed here allows detection of ω_{CARS} in a spectral region (700 - 800 nm, depending on (ω_1 - ω_2) energy), providing improved resolution.

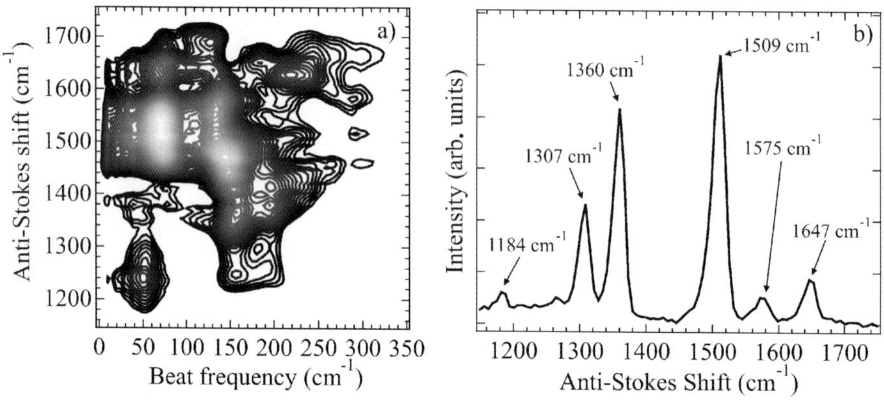

Fig. 2: a) Spectrally-resolved tr-CARS signal of R6G, Fourier transformed to reveal beat frequencies composing typical tr-CARS signal. b) fs/ps CARS spectrum of R6G, directly revealing the frequencies of Raman active modes.

3. Comparison of tr-CARS and fs/ps CARS

A plot of spectrally dispersed tr-CARS signal for ground-state R6G (*i.e.*, no ω_{ex} pulse) reveals several beats associated with the vibrational coherences prepared by pulses 1 and 2 [2]. Following removal of exponentially decaying baseline signal, Fourier transforms were performed on the time-resolved ω_{CARS} signal for each anti-Stokes shift, and the results are shown in Fig. 2a. This contour plot reveals the complexities associated with numerous beat frequencies spaced throughout the spectrally observed range and reinforces the fact that it is difficult to assign vibrational frequencies using the conventional tr-CARS method in molecules with several Raman active modes. In contrast, when we convert to a spectrally narrow probe pulse (~14 cm^{-1}) for fs/ps CARS detection (Fig. 2b), properly timed to remove non-resonant contributions [2], the Raman-active modes are readily observed, and the complexities associated with detection of beat frequencies are removed. The multiplexing associated with this technique allows acquisition in several seconds compared to the minutes to hours required to acquire and analyze the tr-CARS data shown in Fig. 2a. The detected frequencies are in good agreement with previous studies [3].

To demonstate application of this technique as a time-resolved probe, we have initiated studies probing ultrafast excited state dynamics of *trans*-DANS in acetonitrile; DANS is known to undergo several solvent-dependent

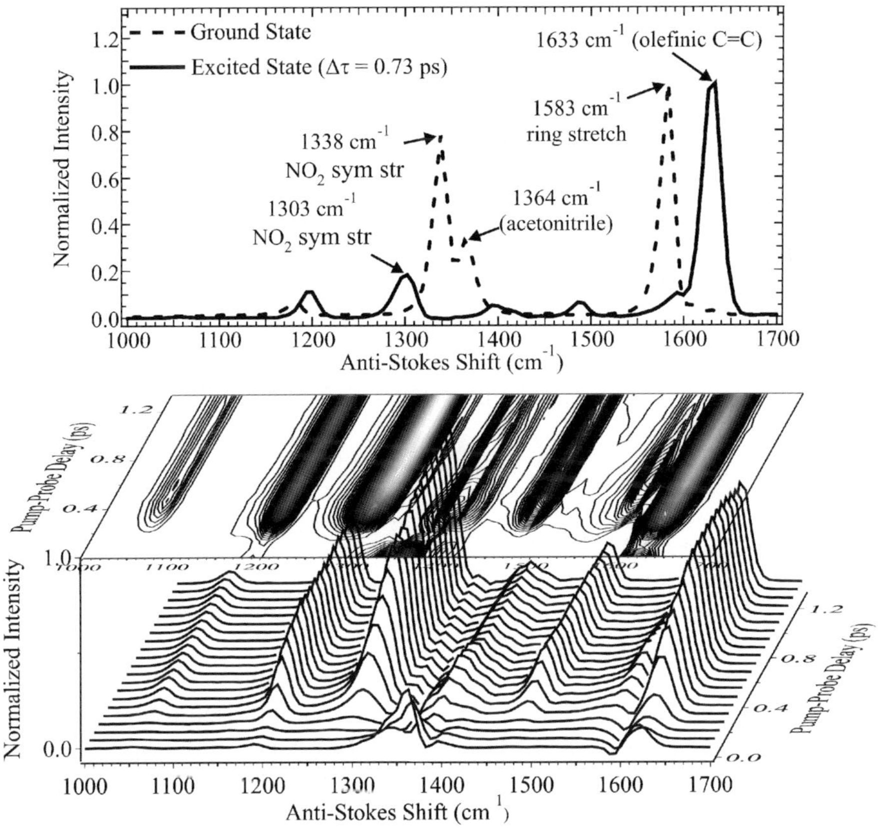

Fig. 3: Top panel: Comparison of ground and excited state fs/ps-CARS spectra of DANS. Bottom panel: Excited state fs/ps-CARS spectra as a function of pump-probe time delay, $\Delta\tau$.

isomerization pathways, including two short-lived twisted intramolecular charge transfer (TICT) isomers, $A^*_{2,4}$ and A^*_5, so named due to the presumed nature of the bonds about which the backbone twisting motion occurs [4]. The observed pump-fs/ps CARS results (Fig. 3) show short-time (~250 fs timescale) evolution of vibrational intensities and frequencies with ~20 cm^{-1} resolution following photoexcitation; this is consistent with the currently accepted kinetic model attributed to photoexcited DANS based on previously observed solvent-dependent dual fluorescence properties [4].

References

1. D. W. McCamant, P. Kukura, S. Yoon, and R. A. Mathies, Rev. Sci. Inst. 75 (2004) 4971.
2. B. D. Prince, A. Chakraborty, B. M. Prince, and H. U. Stauffer, J. Chem. Phys., in press.
3. S. Matsubara and H. Takahashi, Chem. Phys. Lett. 108 (1984) 475.
4. J. Oberlé, E. Abraham, G. Jonusauskas, and C. Rulliére, J. Raman Spect. 31 (2000) 311.

Femtochemistry VII
Fundamental Ultrafast Processes in Chemistry, Physics, and Biology
A.W. Castleman, Jr. and M.L. Kimble

Ultrafast fluorescence parametric amplification for femtosecond molecular dynamics research

Shufeng Wang, Fengming Li, Hong Yang, Qihuang Gong

Physics Department, Peking University , Beijing, China 100871

Abstract

The application of parametric amplification in the field of ultrafast fluorescence dynamics was investigated. A new process was designed in which weak seed signals are amplified up to 10^7 times, and the temporal and spectral characteristics at output are maintained. Compared to widely used OPO/OPA systems, the major issue is linearity in amplification instead of stability of the light source output. Two factors must be taken into account for this purpose: seed intensity and pump energy.

1. Introduction

Optical parametric processes are widely used for tunable light sources in the visible and infrared ranges. Two typical systems are the optical parametric oscillator (OPO) and optical parametric amplifier (OPA) systems. With the development of table-top Ti:sapphire chirped pulse amplification (CPA) laser systems, OPA systems have become widely employed. Their cavity-free design make their alignment quite flexible and make wavelength tuning easy. Broadband output can also be obtained. With post-compression, the broadband outputs generate ultrashort laser pulses. This system is called an optical parametric chirped pulse amplification (OPCPA) system. In 1992, this idea was first introduced by Dubietis *et al.* [1]. In 2000, Kobayashi *et al.*

obtained a broadband spectrum output of 240 nm, which was finally compressed to a 4.7 fs pulse [2]. Also, with this technique, intense laser fields in the terawatt regime were achieved [3,4]. All these achievements indicate that this optical parametric amplification technique is applicable for broadband and weak signal amplification in the spectroscopy field. In 2004, Zhang *et al.* reported the high-efficiency gain of quasi-single-photon amplification; the gain reached 10^8 [5]. Zeng *et al.* used a new method called Seeded Amplification of Colored conical emission (SAC) and successfully amplified signals within a 290 nm range [6]. Liu *et al.* discussed the phase-matching condition for spectrum amplifications [7].

In the field of ultrafast parametric amplification, more efforts have been focused on a broader tunable spectrum range and stable high output power, and these investigations have been quite successful. However, in the field of molecular spectroscopy, such as ultrafast molecular fluorescence dynamics, not much attention has been paid to making use of the benefits of this technique. Currently, there are two main ultrafast fluorescence research techniques: fluorescence up-conversion and the optical Kerr gate (OKG) technique. The former provides true femtosecond temporal resolution, limited by the gate pulse duration, but is accompanied by difficulties in optical alignment and ultraviolet (UV) detection. It also applies single wavelength detection. The whole experimental process covering broad fluorescence spectra takes hours for each sample. The OKG works for all wavelengths in a single scan. However, the nonlinear media of organic solvents, such as CS_2, provide sub-picosecond or picosecond time-resolution because of the slow movement of the nucleus along the polarization of the laser field. For these two methods, the common disadvantage is that only a small portion of both input signals are converted to the detector. This makes the weak signal detection somewhat difficult. Compared to these processes, ultrafast fluorescence parametric amplification (UFPA) provides all-optical weak-signal amplification with a high rate. It can also easily be applied in the IR range. This technique will greatly extend the current research techniques. This work focused on the realization of this fluorescence parametric amplification technique. It was principally proved that this method is effective for spectroscopic investigation.

2. Experimental setup and results

A Spitfire kHz Ti:sapphire laser system from Spectra-Physics was used as the light source with output pulses of 120fs at 800nm and 0.4mJ. After passing a first BBO crystal, frequency doubled (blue) pulses of 400 nm and 70 μJ were generated. These blue pulses were the pump source for the later weak signal amplification. The unconverted fundamentals were used to excite the samples with one-photon or two-photon fluorescence or were frequency doubled as the excitation source to produce weak seed signals. The pump laser pulse and the seed were cross-configured on a second BBO crystal, which was 1 mm in thickness and cut at 24° for type I phase-matching.

With the calculation of the type I phase-matching angle between pump and signal, it was found that the amplified wavelength was very sensitive to this cross-configuration angle. To have proper alignment, colored conical emission (CCE) can be used to successfully reduce the difficulties while increasing the amplification rate dramatically. The CCE is a uniform ring-shaped emission around the pump beam when the pump intensity is higher than the threshold. This ring-shaped pattern contains wavelengths from visible to near infrared at a corresponding radius. The spatial layout of the angle of the output wavelength distribution in CCE represents the phase-matching selection rule of the crystals. Along the cone's surface, if a seed is injected, the corresponding wavelength will satisfy the phase-matching condition. Under certain conditions, a large bandwidth (over 250 nm) overlaps and forms a single orange-colored ring in the CCE. All the seed wavelengths within this bandwidth satisfy the phase-matching condition spontaneously at this angle. A true seed of DCM dye in ethanol, 0.3mM, was injected into the system. A full bandwidth amplified fluorescence signal was obtained.

3. Discussion

3.1. Phase matching

The CCE from the NLO crystal came from the amplification of quantum noise inside. Its output angle satisfied the second-order NLO phase-matching condition. From calculations and experiments, at an angle of 31.5° between pump beam and optical axis, a phase-matching angle from 500 nm to 750 nm appeared within a very small angle distribution. It was proven that the seed input corresponding to this condition was amplified with a full bandwidth.

3.2. Linearity

To experimentally demonstrate the linear amplification dynamic range, supercontinuum white light from a water cell, produced by focused fundamentals, was used as a seed. This seed can be strongly attenuated in intensity. For the pump beam of the parametric process, at high pump intensity above the CCE threshold, the amplification rate increased more dramatically than in non-CCE conditions. At a fixed weak seed intensity, the output increased linearly towards the pump intensity as shown in Fig. 1, left. However, if the pump intensity went too high, the system output became saturated. As shown in the right figure, with variable seed intensity under a certain fixed pump intensity, the input-output linearity was proven with a high amplification rate of about 10^7. However, the stronger seed also was saturated at output, which means the system was suitable for the weak signal detection.

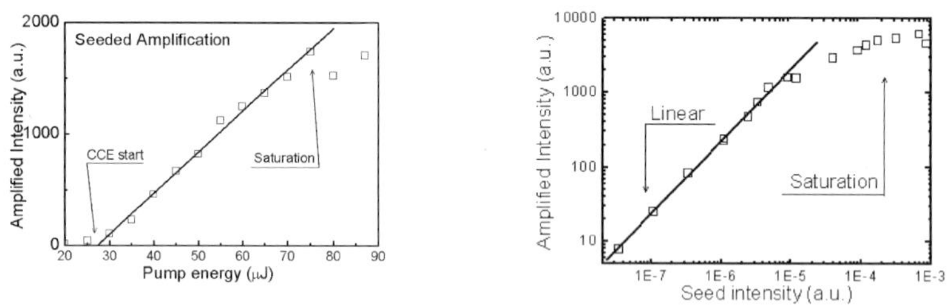

Fig. 1 The saturation effect towards pump energy at fixed seed intensity (left) and seed intensity at a fixed pump energy (right)

4. Conclusion

The initial linearity of the parametric amplification process under CCE conditions was investigated. The amplification rate was higher than 10^7. A potential application in fluorescence detection was demonstrated.

References

1. A. Dubietis, G. Jonusauskas, and A. Piskarskas, Opt. Commun. 88 (1992) 437.

2. T. Kobayashi and A. Shirakawa, J. Lumin. 87-89 (2000) 119.
3. I. N. Ross, J. L.Collier, P. Matousek, C. N. Danson, D. Neely, R. M. Allot, D. A. Pepler, C. Hernandez-Gomez, and K. Osvay, Appl. Opt. 39 (2000) 2422.
4. I. Jovanovic, B. J. Comaskey, C. A. Ebbers, R. A. Bonner, D. M. Pennnington, and R.C. Morse, Appl. Opt. 41 (2002) 2923.
5. J. Y. Zhang, C. K. Lee, J. Y. Huang, and C. L. Pan, Opt. Express 124 (2004) 574.
6. H. P. Zeng, J. Wu, H. Xu, K. Wu, and E. Wu, Phys. Rev. Lett. 9214 (2004) 143903.
7. H. J. Liu, W. Zhao, G. F. Chen, Y. S. Wang, L. J. Yu, C. Ruan, and K. Q. Lu, Chin. Phys. Lett. 211 (2004) 94.

Femtochemistry VII
Fundamental Ultrafast Processes in Chemistry, Physics, and Biology
A.W. Castleman, Jr. and M.L. Kimble

A density-functional study of the structural, electronic, and vibrational properties of Ti_8C_{12} metallocarbohedrynes with relevance to ultrafast time-resolved spectroscopy

M. A. Sobhy[a], J. O. Sofo[b,c] and A. W. Castleman, Jr.[a,b,c]

[a]*Department of Chemistry, The Pennsylvania State University, University Park, Pennsylvania 16802, USA*
[b]*Department of Physics, The Pennsylvania State University, University Park, Pennsylvania 16802, USA*
[c]*Materials Research Institute, The Pennsylvania State University, University Park, Pennsylvania 16802, USA*

1. Introduction

The unique properties of systems with finite size provide a challenging and interesting area of research for the scientific community. Clusters exhibit many intriguing phenomena due to quantum confinement, such as phase transition and collective optical resonance [1]. The time evolution of the excited electronic and vibrational states of an isolated system is determined by its level patterns [2]. In this study, we present the results of spin-polarized density-functional calculations for the structural, electronic, and vibrational properties of metallocarbohedrynes (Met-Cars) [3], of interest for time-resolved spectroscopy.

2. Computational Details

The calculations were performed using density-functional theory (DFT) with a plane wave basis set and the projector augmented-wave (PAW) method as implemented in the Vienna *Ab initio* Simulation Package (VASP) [4]. Exchange and correlation were treated within the generalized gradient approximation (GGA) of Perdew, Burke, and Ernzerhof (PBE) [5]. The clusters were placed in cubic unit cells with a side of 20 Å to reduce the interaction between periodic images to a negligible amount. The geometry optimization was considered to be complete when all the forces acting on each ion were less than 0.01 eV/Å with a convergence of the total energy better than 10^{-4} eV. The vibrational frequencies of the Ti_8C_{12} metallocarbohedryne were calculated using the frozen-vibration method [6]. The *ab initio* molecular-dynamics simulations were initialized with random velocities drawn from a Maxwell-Boltzmann distribution corresponding to 300K and 100K. The simulations were performed in a microcanonical ensemble. The vibrational density of states convoluted with the thermal occupation of the modes were derived from the Fourier transform of the velocity autocorrelation function.

3. Results & Discussion

3.1. Atomic Structure

In this study, we performed geometrical optimization for the seven topological minima of the Ti_8C_{12} metallocarbohedryne [7]. The transition from one topological class to another takes place by the 90° rotation of one or more of the C_2 units at the different faces of a distorted cube constructed from Ti atoms as shown in Figure 1.

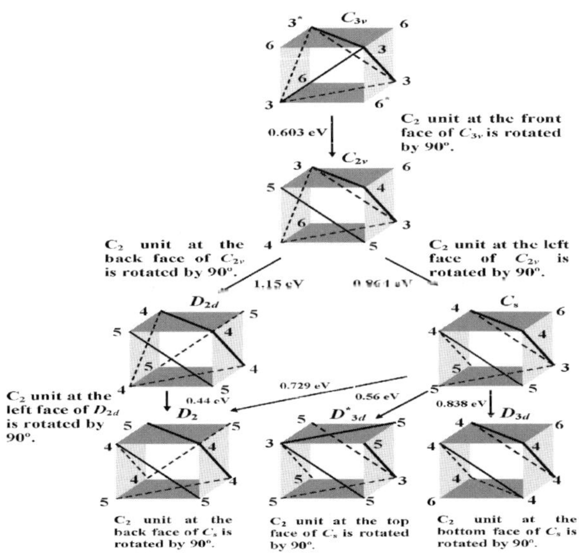

Fig. 1 The different Ti_8C_{12} metallocarbohedryne structures [Ref. 7] and the energy differences between them are shown in the figure. For demonstration, we specify the C_2 units rotated in the figure to obtain one isomer from the other.

The Ti_8C_{12} metallocarbohedryne structures can be divided into four groups depending on the number of the rotated C_2 units. Accordingly, the first group includes the T_d structure and its Jahn-Teller analogue the C_{3v} structure; this group represents the lowest energy or the ground state structures. The second group results from the 90° rotation of one C_2 unit, is represented by the C_{2v} structure, and is approximately 0.6 eV higher in energy than the C_{3v} structure. The third group results from the 90° rotation of each of the C_2 units, in either two opposite or two adjacent faces of the cube made of Ti atoms. The energy of this group ranges from 1.4 to 1.8 eV higher than that of the C_{3v} structure and is represented by the C_s and D_{2d} structures. The last and highest energy group results from the 90° rotation of the C_2 units capping three different faces and is represented by the D_2, D_{3d}, and D^*_{3d} structures, whose energies are 2 to 2.3 eV higher than the C_{3v} structure. Our results show that the ground state, at the level of DFT-PBE calculations for the C_{3v}, D_{3d}, and D^*_{3d} isomers, is a spin triplet. However, in all cases, the difference in energy with respect to the spin singlet state is less than 2 meV/atom. This small energy difference is probably within the systematic error induced by the exchange and correlation potentials and the atomic structure of the cluster is not affected by the electronic spin state.

3.2. Electronic Properties

The electronic density of states (eDOS) plot for the different structures is presented in Figure 2. The C_{3v}, D^*_{3d} and D_{3d} isomers are spin-polarized. Furthur analysis of the partial eDOS shows that, depending on the dimer orientation, carbon atoms and a sub group of the metal atoms form a covalent framework while other metal atoms are bonded to this framework more ionically. This picture is further supported by the charge density of the different structures, where we see that the Ti atoms with higher charge density show less contribution to the covalent bonding of the Ti-C framework.

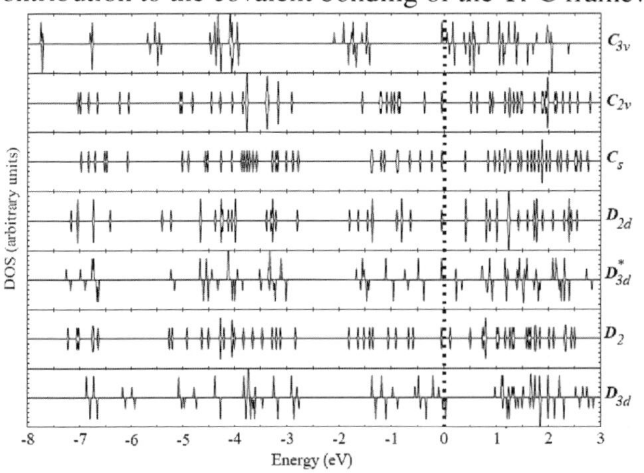

Fig. 2 The electronic density of states plot represents the Kohn-Sham energy spectra obtained from the spin-polarized calculations on the Ti_8C_{12} metallocarbohedryne structures. The vertical dotted line denotes the fermi level.

3.3. Vibrational frequencies

All the structures present a group of high energy vibrations in the region of 1250-1500 cm^{-1}, corresponding to the stretching modes of the carbon dimers, and a low energy group below 700 cm^{-1} corresponding to the vibrational modes of the Ti-C and Ti-Ti bonds. Since the Ti_8C_{12} metallocarbohedryne structures differ in the arrangement of the carbon dimers, the high energy group of vibrations can be considered a signature of the structure [8]. The global similarity of the vibrational density of states suggests that the difference in enthalpy at 0 K is a good indicator of the relative stability of the structures at finite temperatures. The high energy frequencies in the C_{3v} structure form two sets of single and double degenerate modes as a result of the splitting of the three fold degenerate modes in the T_d structure. Also, the D_{3d} and D^*_{3d} structures have single and double degenerate vibrational modes based on the character of their symmetry group [9].

3.4. Molecular Dynamics

The aim of our molecular-dynamics simulations is to study the local stability of the structures and analyze the temperature effects on the vibrational density of states calculated using the frozen-vibration method. The vibrational spectra obtained from the MD simulations at T = 300K and T = 100K show similar features to those obtained from the frozen-vibration method, which confirms the local stability of the structures beyond the harmonic limit explored by the frozen-distortion calculations. Due to the lower frequency resolution of the MD simulations, not all the peaks in the 1200 to 1450 cm^{-1} range are resolved. Since our frozen-vibration calculations have been tested with different distortion sizes, the discrepancy between these two methods can be attributed to the need for longer simulations to decrease the statistical error and resolve the important high energy modes that differentiate between the different isomers.

4. Conclusion

In this study, we employed spin-polarized DFT to investigate the structural build up of the different Ti_8C_{12} metallocarbohedryne clusters. The orientation of the C_2 units relative to the Ti atoms cube was used as the basis for structural classification. The C_{3v} structure was found to be the most stable structure and is 0.6 eV less than the second lower energy isomer. We used the electronic structure of the clusters to rationalize their bonding properties. Finally, we compared the preliminary results of the calculation of the vibrational spectra of the C_{3v} and C_{2v} structures using MD simulations at finite temperatures with those obtained from the frozen-vibration method. An extended version of this work has been published elsewhere [10].

Financial support from the Air Force Office of Scientific Research, Grant no. FA 9550-04-1-0066 is gratefully acknowledged.

References

1. A. W. Castleman, Jr., in Recent Advances in the Chemistry and Physics of Fullerenes and Related Materials, K. M. Kadish and R. Ruoff (Eds.) Electrochemical Society, New Jersey, (1994).
2. A. Tramer, Ch. Jungen, and F. Lahmani, Energy Dissipation in Molecular Systems, Springer, Heidelberg, 2005.
3. B. C. Guo, K. P. Kerns, and A. W. Castleman, Jr., Science 255 (1992) 1411.
4. G. Kresse and J. Hafner, Phys. Rev. B 47 (1993) 558; 49 (1994) 14251; G. Kresse and J. Furthmüller, Comput. Mat. Sci. 6 (1996) 15; Phys. Rev. B 54 (1996) 11169.
5. J. P. Perdew, K. Burke, and M. Ernzerhof, Phys. Rev. Lett. 77 (1996) 3865; 78 (1997) 1396.
6. D. Porezag and M. R. Pederson, Phys. Rev. B 54 (1996) 7830.
7. M. M. Rohmer, M. Benard, C. Henriet, C. Bo, and J. M. Poblet, J. Chem. Soc., Chem. Commun. 15 (1993) 1182.
8. G. K. Gueorguiev and J. M. Pacheco, Phys. Rev. Lett. 88 (2002) 115504.
9. R. L. Carter, Molecular Symmetry and Group Theory, Wiley, New York, 1998.
10. M. A. Sobhy, A. W. Castleman, Jr., and J. O. Sofo, J. Chem. Phys. 123 (2005) 154106.

2. Fundamentals of Reaction Dynamics

Femtochemistry VII
Fundamental Ultrafast Processes in Chemistry, Physics, and Biology
A.W. Castleman, Jr. and M.L. Kimble

Sub 5 fs spectroscopy of polydiacetylene

Takayoshi Kobayashi, Mitsuhiro Ikuta, and Yoshiharu Yuasa

Department of physics, Graduate School of Science, University of Tokyo, 7-3-1 Hongo, Bunkyo-ku, Tokyo, 113-0033, Japan
Tel.: +81 3 5841 4227
FAX: +81 3 5841 4240

E-mail: kobayashi@femto.phys.s.u-tokyo.ac.jp

1. Abstract

We clarified the very early stage of the electronic relaxation in relation with molecular vibration, geometrical relaxation, and internal conversion in one of the most popular one-dimensional conjugated polymers, polydiacetylene.

2. Introduction

This review reports a systematic study of time-resolved measurements of one of the most popular polymers, polydiacetylene. Molecular vibration of several modes in blue-phase polydiacetylene-3- butoxycarbonylmethylurethane (PDA-3BCMU) was real-time observed by a 5-fs pump-probe measurement. The methodologies utilized include (i) "a peak-tracking" method applied to the time-resolved spectrum of the polymer. This provides valuable information of detailed dynamics in the polymer through a one-femtosecond step sequence of electronic absorption and gain spectra and (ii) global fitting after the singular value decomposition of the time trace of the difference absorption spectra induced by pump pulses (iii). The contribution of the vibrational wavepackets in the ground state and in the excited state in the signal was also separated.

By applying these methods, the following results were clarified. The peak of the 0-0 transition from the ground state in which vibrational excitation does not participate oscillates with the frequencies of C-C stretching vibrational modes. The oscillations of energy and intensity of a peak of induced absorption to a higher excited state show different behaviors both in the vibrational amplitude and phase. The results provided us comprehensive information to determine the relative locations of the multidimensional potential hypersurfaces of the ground, the lowest excited, and the higher excited states in the configuration space. The precise analysis of the time evolution of the difference absorption spectra has revealed the thermalization processes due to the energy redistribution among various modes in the polymer chains in the excited and the ground states.

The C=C stretching mode in the ground state starts to oscillate π -out-of-phase with the C≡ C stretching mode. The structure of PDA-3BCMU in the geometrically relaxed state is not pure butatriene-type but mixed with acetylene-type. The frequencies of C=C and C≡ C stretching modes there were determined by the singular value decomposition method to be 1472 ± 6 cm-1 and 2092 ± 6 cm-1, respectively. The double and triple bond stretching frequencies in the ground state are 1463 ± 6 cm-1 and 2083 ± 6 cm-1, respectively.

Even though quantitative values of the parameters mentioned above are expected to vary from one polydiacetylene to the other, the main characteristic features are expected to be common in all blue-phase polydiacetylenes because of their similarity in the electronic spectrum with a typical excitonic peak around 2 eV and phonon side band due to C-C stretching modes.

3. Results and discussion

3.1. Peak tracking analysis

The spectra of excitation and probe pulse laser and absorption of the blue-phase PDA-3BCMU are shown in Figure 1. The vibrational energy levels of C-C double and triple bond stretching are indicated in the graph. The low lying electronic energy states are depicted in the figure together with transitions associated with excitation, internal conversion between the excited states, and induced absorption and relaxation to the ground state. A typical example of experimental data of difference absorption spectra is displayed two-dimensionally in Fig. 2. Several examples of the observed normalized difference transmittance (ΔT (t)/T) traces of the blue-phase PDA-3BCMU sample at several energies of probe photon are shown in Fig. 3. All of the trace curves show signals of a finite size of ΔT (t)/T at negative delay times and sharp and intense peaks around zero probe-delay time. The former is due to the perturbed free induction decay process associated with the third-order nonlinearity induced by the sequential interaction of probe-pump-pump fields modified with molecular vibrations. The latter signals are due to pump-probe coupling induced by the nonlinear process of the pump-probe-pump time ordering. There is another contribution from the interference between the scattered pulses and the probe pulses of which duration are elongated in the polychromator. The details of the vibrational modulation observed at longer delay times than 100 fs are free from the distortions due to the above two mechanisms.

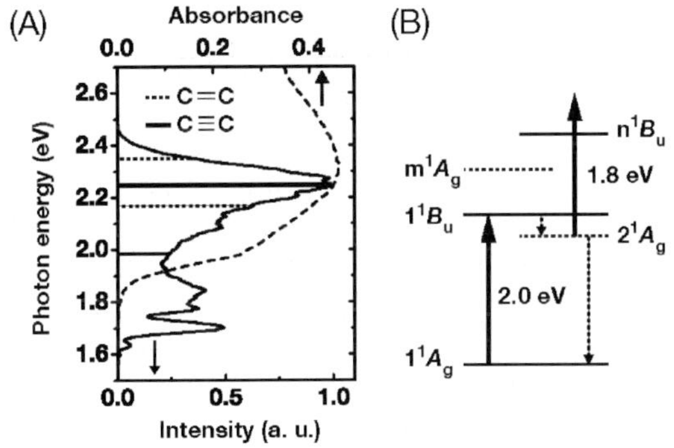

Fig. 1. (A) Spectra of the excitation pulses (solid) and the absorption of PDA (dotted) together with the lines denoting the energy of 1^1B_u state (black line) and its vibrational levels. (B) Schematic energy diagram of the essential electronic states (right figure): ground state (1^1A_g), excited state (1^1B_u), higher excited state (m^1A_g), and geometrically-relaxed state (2^1A_g), from which optical transition is allowed to another excited state (n^1B_u) of PDA relevant to our experiment. The corresponding transition energies are also described. Dashed lines indicate radiationless transitions.

Here, the before mentioned "peak-tracking" analysis developed recently by the authors' group is to be explained. This is to study the peak energies of the transient electronic transition during molecular vibrations. The authors' group applied this method to demonstrate the oscillations of transition peak energies in the blue-phase polydiacetylene (PDA)-3BCMU film [1], where 3BCMU denotes the side-chain of the conjugated polymer. Polymers composed of p-electrons along the one-dimensional carbon-chain [2], including PDAs as a typical example, have large ultrafast electron-lattice interaction [3] and thereby exhibit ultrafast optical response [4]. Nonlinear optical properties in polymers, including polyacetylene and PDAs, can be explained in terms of four essential electronic states [5] and the ground state, relevant to this experiment as shown in Fig. 1B. Pump pulses with the highest peak at 2.25 eV, which also corresponds to the energy of maximum absorbance in the stationary absorption spectrum, predominantly excite the allowed 11Ag → 11Bu transition coupled by C = C stretching (Fig. 1). This creates depleted difference absorption spectra (DAS) due to the population-hole in the ground state. The DAS has a well-defined vibronic structure, including the purely electronic zero-phonon (0-0) peak around 2.0 eV [1,4]. The broadening of the bandwidth of each peak in the bleaching spectra with increasing delay time caused by the thermalization in the ground state indicates that each peak in the bleaching spectra corresponds to a well-defined vibronic exciton-exciton transition. Another clear peak of induced absorption (21Ag → n1Bu) around 1.8 eV [1,4] also appears in the positive DAS (Fig. 2). In the figure, the positive DAS is the downward signal.

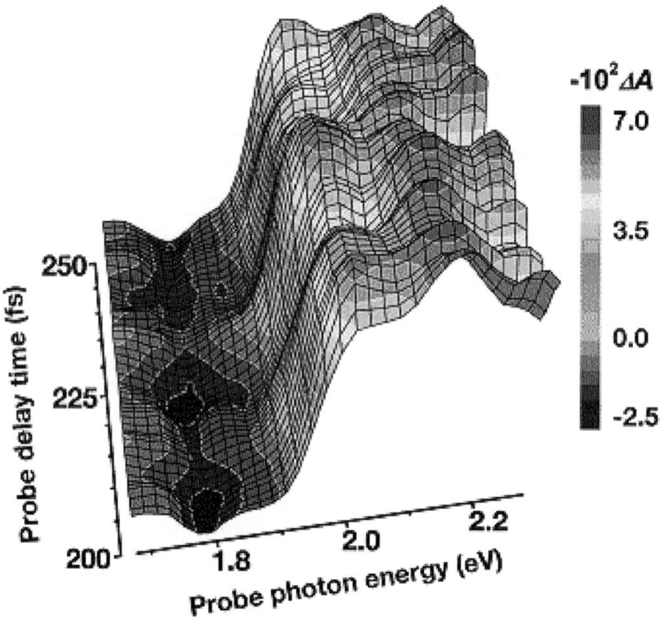

Fig. 2. Time evolution (200–250 fs after excitation) of the sequential difference absorption spectra of PDA-3BCMU with resolution of ≤1 fs (vertical axis) and ≤10 meV (horizontal axis) obtained by the pump-probe measurement. Positive and negative difference absorbances are caused by the induced absorption from 2^1A_g state and the absorption saturation in 1^1A_g state, respectively.

These well-separated features due to the partial "hole burning" enable us to apply the "peak-tracking" method to the transition peaks. This method provides detailed information about the mechanism of the oscillating signals corresponding to all the vibronic transitions separately in DAS. For each vibronic transition, the DAS signal can be analyzed in terms of two contributions: modulations of the transition energy and the transition probability. They appear in DAS as "vertical" and "horizontal" modulation with a molecular vibrational frequency, W, respectively [6]:

$$\Delta A(\omega) \cong \Delta A_0(\omega) + \left(\frac{\delta(\mu^2)}{\mu^2} \Delta A(\omega) + \delta\omega \frac{d\Delta A(\omega)}{d\omega} \right) \cos(\Omega t + \theta) \quad (2)$$

Here $\Delta A_0(\omega)$ is the non-oscillating component of the absorbance change ($\Delta A(\omega)$), and $\delta(\mu^2)/\mu^2$ and $d\omega$ correspond to the probability and energy modulations, respectively. The modulation frequency, Ω, and the initial phase, θ, are corresponding to those of the molecular vibration. The wave-packet motion along the potential curve causes the transition energy modulation [7,8]. Decomposed signal provides the correct time evolutions of transition energies and intensities with phase information and thus enable us to determine the relative locations of multiple PESs in the different electronic states.

The decay of the absorbance change signal size was analyzed by the global fitting method which is to be discussed later in detail to be classified into two components of short-life (60±3 fs) and long-life (890±40 fs)[1,9]. An initial relaxation process of 11Bu → 21Ag (60±3 fs) is observed as the decay of induced emission (IE) (11Bu → 11Ag).

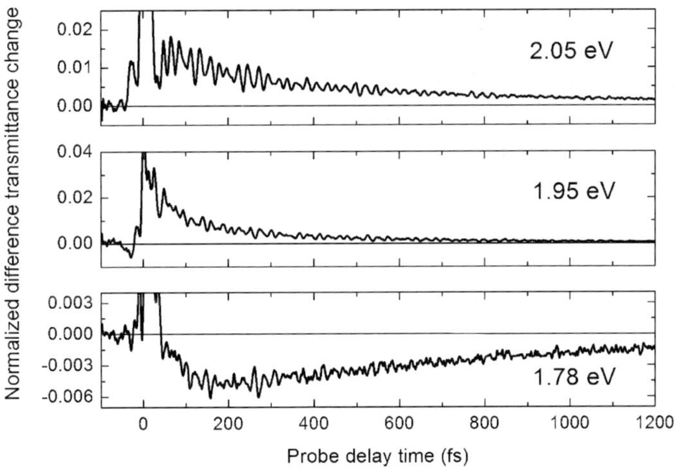

Fig. 3. Pump-probe delay time dependence of the normalized difference transmittance ($\Delta T(t)/T$) on the probe delay time at three probe photon energies.

At a longer delay time than 100 fs, when the relaxation to 21Ag is almost fully completed since the corresponding decay time is about 60fs, the peak-tracking analysis clarified the dynamic behavior of the sequential spectra, composed of the saturation of the stationary absorption (11Ag → 11Bu) and the induced absorption (21Ag → n1Bu), which includes oscillations in the slopes between the peaks. The modulation amplitudes of transition peak energies can be related to the strengths of vibronic couplings involved in the electronic transition. Stabilization energies and Huang-Rhys factors [8] can be obtained in every direction of normal coordinates in the multidimensional configuration space, which provide the displacement of the potential hypersurface minimum of the final state with respect to that of the initial state, of the corresponding transition.

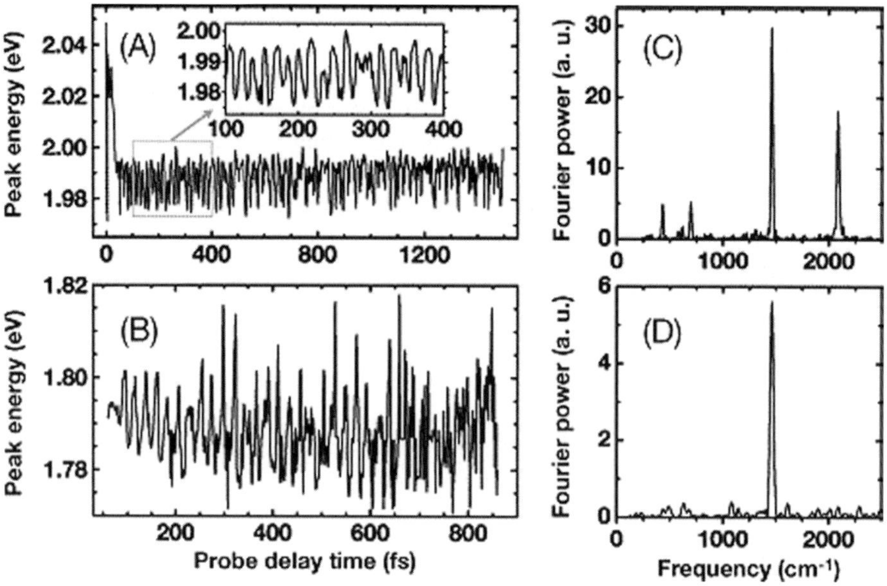

Fig. 4. Dynamic behaviors of transition peaks and their oscillation frequencies. Energy oscillations of the zero-phonon transition peak (A) $1^1A_g \rightarrow 1^1B_u$ (~1.99 eV) and an induced absorption peak (B) $2^1A_g \rightarrow n^1B_u$ (~1.79 eV), and their Fourier power spectra calculated for the data of (C) 100-1500 fs and (D) 100-860 fs, respectively. The inset in (A) is the peak motion with an expanded delay-time scale from 100 to 400 fs.

The excitonic 0-phonon transition peak near 2.0 eV (11Ag → 11Bu) and an IA peak near 1.8 eV (11Ag → 11Bu transition) were tracked in Fig 4. Each peak and its slopes on both sides show energy oscillation following " sine" and " cosine" functions, corresponding to the wave-packet motions in the ground and the excited states [10], respectively. Two intense peaks in Fig. 4(C) represent that two modes of C=C (1460 cm-1) and C≡C (2080 cm-1) stretching are strongly coupled with 11Ag → 11Bu transition. However, the signal intensity of C≡C stretching is dramatically reduced in 21Ag → n1Bu(Fig. 4(D)) with remaining strong peak of C=C stretching mode. This indicates that the potential-minimum shift between 21Ag and n1Bu is largest in the C=C stretching direction, resulting in the impulsive starting of the vibrational mode. Schematic two-dimensional PESs obtained in the C=C and C≡C stretching coordinates are plotted in Fig. 5.

In characterizing the dynamic behavior of the corresponding vibronic transition, we picked up a specific transition among several peaks observed in the spectra. Applying the peak-tracking method, we have decomposed the two-dimensional modulation spectra into the transition-peak energy and the transition-intensity modulations. This method provides a new insight about the vibrational origins relating the vibrational amplitude distributions to the spectral features in DAS and can be used as a general method to determine the multidimensional potential hypersurfaces in molecular systems in the condensed phase, including organic polymers and biomolecules with multiple electronic states experiencing ultrafast dynamics and chemical reactions.

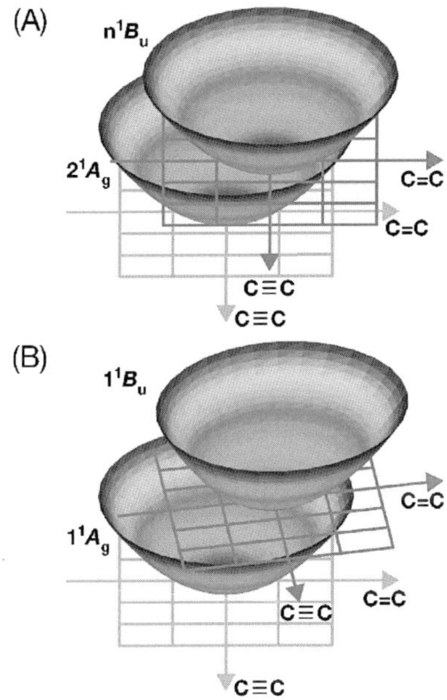

Fig. 5. Schematic potential energy surfaces of the two electronic states involved in (A) $2^1A_g \rightarrow n^1B_u$ and (B) $1^1A_g \rightarrow 1^1B_u$ transitions.

3.2. Bleaching and induced absorption spectra

We smoothed out the oscillating components from DAS by taking sequential convolutions of the data with rectangular functions of 3 fs to 101 fs time widths to investigate the non-oscillating spectral change due to molecular vibration. Then the smoothed signal can be attributed to the population dynamics of the vibrational levels in the electronic states. The time-resolved spectral changes in their shapes reveal that the two processes of the vibrational energy redistribution and the internal conversion process from the 21Ag state to the ground state take place concurrently. The early bleaching spectrum with a vibrational progression suggests that a "hole" is induced in the inhomogeneously broadened absorption spectrum. The transient bleaching spectrum undertakes spectral change due to the vibrational thermalization through the intra- and inter-chain interactions via the low frequency bending and torsional modes of the polymer chains. The initially prepared bleaching spectrum with a smaller inhomogeneity than that of the stationary absorbance relaxes to the thermalized spectrum with a feature similar to that of the stationary spectrum. The recovery of the population in the ground state also refills the "hole" with a relatively large 0-phonon peak in the bleaching spectrum. These thermalization processes cause the spectral blue shift observed in the bleaching (negative DA) region.

3.3. Vibrational thermalization in the ground and the excited states

The non-oscillating spectra also revealed that a positive DA peak due to induced absorption grows up around 1.85-1.90 eV with increasing delay time. Thermalization processes from the nonequilibrium states both in the ground 11Ag and the excited 21Ag states were revealed in the long-term energy shifts of peaks and slopes in DAS. Seventeen transition energies with characteristic features in the spectra were investigated, which included 3 peaks, 4 valleys, 9 slopes, and one zero DA point. Each peak showed a blue or red shift due to thermalization in 11Ag and 21Ag states depending on probe photon energy. The thermalization time, in which population distribution among vibrational levels with distributed quantum numbers in various modes comes to thermal distribution within the main chains is estimated to be about 0.5 ps from the time constant of the spectral shift. The value of the time constant determined here is consistent with our previous papers [11].

The energy shifts of slopes during 1.3 ps from 0.2 ps to 1.5 ps showed that there are two regions with positive and negative energy shifts. The bandwidths of the vibrational peaks in the DAS are broadened with increasing delay time. The results may indicate the spectral diffusion in the 21Ag-excited state. The zero crossing point was around 1.9 eV lower than 0-0 transition energy by about 100 meV, which is consistent with the amount of the Stokes shift in the literature [12].

4. Conclusion

In conclusion, we clarified the very early stage of the electronic relaxation in relation with molecular vibration, geometrical relaxation, and internal conversion in one of the most popular one-dimensional conjugated polymers, polydiacetylene.

The peak tracking analysis revealed the vibronic coupling to the electronic absorption corresponding to the 11Ag → 11Bu transition and 1121Ag → n1Bu transition The peak of 0-0 transition from the ground state in which vibrational excitation does not participate oscillates with the frequencies of C-C stretching vibrational modes. The oscillations of energy and intensity of a peak of induced absorption to a higher excited state show different behaviors both in the vibrational amplitude and phase. The results provided us comprehensive information to determine the relative locations of the multidimensional potential hypersurfaces of the ground, the lowest excited, and the higher excited states in the configuration space. The precise analysis of the time evolution of the difference absorption spectra has revealed the thermalization processes due to the energy redistribution among various modes in the polymer chains in the excited and the ground states.

5. Acknowledgements

The authors would like to thank T. Saito, S. Adachi, K. Nishimura, T. Taneichi and N. Ishii for enlightening discussions. This research is supported partly by the Grant-in-Aid for Specially Promoted Research (#14002003) and partly by the program for the "Promotion of Leading Researches" in Special Coordination Funds for Promoting Science and Technology from the Ministry of Education, Science and Culture.

6. References and Notes

1. T. Kobayashi, M. Yoshizawa, U. Stamm, M. Taiji, and M. Hasegawa, J. Opt. Soc. Am. B 7 (1990) 1558.
2. S. Mukamel, A. Takahashi, H. X. Wang, and G. Chen, Science 266 (1994) 250.
3. W. B. Bosma, S. Mukamel, B. I. Greene, and S. Schmitt- Rink, Phys. Rev. Lett. 68 (1992) 2456.
4. M. Yoshizawa, Y. Hattori, and T. Kobayashi, Phys. Rev. B 47 (1993) 3882.
5. S. Mazumder, D. Guo, and S. N. Dixit, Synth. Met. 57 (1993) 3881.
6. H. Kano, T. Saito, and T. Kobayashi, J. Phys. Chem. B 105 (2001) 413.
7. A. T. N. Kumar, F. Rosca, A. Widom, and P. M. Champion, J. Chem. Phys. 114 (2001) 701.
8. K. Huang and A. Rhys, Proc. R. Soc. London Ser. A 204 (1950) 406.
9. T. Kobayashi, A. Shirakawa, H. Matsuzawa, and H. Nakanishi, Chem. Phys. Lett. 321 (2000) 385.
10. M. Ikuta, Y. Yuasa, T. Kimura, H. Matsuda, and T. Kobayashi, Phys. Rev. B 70 (2004) 214301.
11. M. Yoshizawa, M. Taiji, and T. Kobayashi, IEEE J. Quantum Electron. 25 (1989) 2532.
12. M. Yoshizawa, A. Kubo, and S. Saikan, Phys. Rev. B 60 (1999) 15632.

Femtochemistry VII
Fundamental Ultrafast Processes in Chemistry, Physics, and Biology
A.W. Castleman, Jr. and M.L. Kimble

Dark intermediate states in photochemical *cis-trans* isomerizations and pericyclic reactions

Werner Fuß[a], Sergei A. Trushin[a], Wolfram E. Schmid[a], Constantine Kosmidis[b], Subhasis Panja[c]

[a] *Max-Planck-Institut für Quantenoptik, Hans-Kopfermann-Str. 1, D-85748 Garching, Germany*
[b] *Department of Physics, University of Ioannina, Gr-45110 Ioannina, Greece*
[c] *Present address: Department of Chemistry, Graduate School of Science, Osaka City University, 3-3-138 Sugimoto, Sumiyoshi, 558-8585 Osaka, Japan*

1. Introduction

Photochemical reactions involve large excursions of atoms or groups of them. Crucial parts of the process occur far from the equilibrium (Franck-Condon) geometry, hence in regions on the potential energy surfaces (PES) which are not accessible by vertical excitation from the ground state (S_0), so that they can be called dark. It is the domain of time-resolved spectroscopy to detect and investigate such regions; in a pump-probe experiment the wave packet begins to travel after the pump process, and when it arrives after a delay to remote regions of the PES, the probe laser can interrogate it. For detection it is desirable to be independent of selection rules. For this purpose we use nonresonant ionization, usually at wavelengths of 800–2300 nm. The required probe-laser intensities are typically around 10^{13} W cm^{-2}. The yields of the resulting ions are detected mass-selectively. This means that we record the time-dependent mass spectra. When the molecule relaxes – or the wave packet moves – down along the PES, the ionization energy increases and the ionization probability decreases. But, what is often not recognized, the degree of fragmentation increases at the same time. This is because the total energy is conserved; from the resulting neutral molecule with its excess energy, the probe laser produces a hot ion, which has sufficient time to fragment subsequently. The time constants after some relaxation can therefore only be found in fragments; the examples presented below (see stilbene in particular) are of this kind. The ionization probabilities and fragmentation patterns allow to estimate the relative energies of states. For example, it is always possible to decide whether the wave packet has already arrived at the ground state (of the product or reactant) or is still in an excited state. For methodological details, see e.g. [1].

Correlation diagrams give rise to an important type of dark state, which is two-electron excited. (This gives another reason why it is not accessible by one-photon spectroscopy from S_0. By ionization, however, such states can be easily detected.) A prominent example are pericyclic reactions [2,3]. But *cis-trans* isomerization of polyenes can be described by similar diagrams [4]. In both cases, an S_1 minimum results from avoided crossing of two correlation lines. The reactant S_0 correlates with a two-electron excited state of the product and vice versa (Fig. 1a); two-electron excitation is involved in the correlation, because in chemical reactions the electrons are moved pairwise. This "reactive minimum" ("2A" in Fig. 1) acts as a collection well, collecting the initial population from any "spectroscopic" state ("1B" in Fig. 1), if the two PES intersect [2,3]. From the reactive minimum, the wave packet leaves via a conical intersection (CI) to S_0 (Fig. 1b). The collection well and the CI together form a reaction funnel [2]. The lower cone of the CI is a region of branching between formation of a product (or of products) and return to the reactant, that is, internal conversion. It is obvious in this scheme, how crucial this region is for the reaction, which is far from the Franck-Condon geometry. We demonstrate this importance with three examples (sec. 2-4).

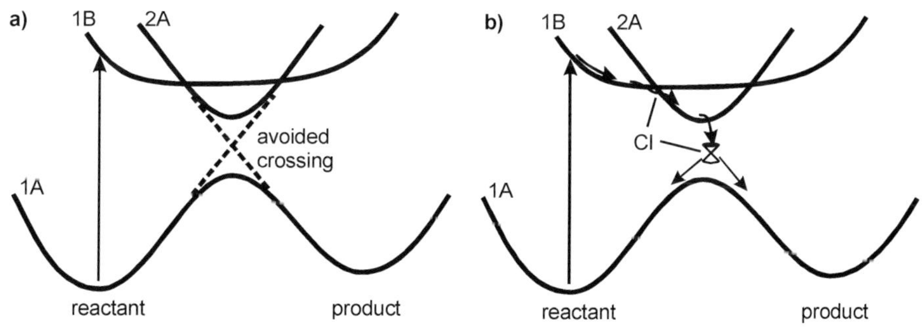

Fig. 1. (a) In *cis-trans* isomerization of polyenes and in photochemical pericyclic reactions, a collection well (2A) and a ground-state barrier are formed by an avoided crossing. The 2A state is two-electron excited and totally symmetric. It is typically crossed by a spectroscopic surface (here: "1B"). (b) The wave packet crosses from surface to surface via conical intersections (CIs), thereby temporarily leaving the drawing plane. Such a path allows ultrafast processes, if there is no barrier.

2. *Cis-trans* isomerization of stilbene follows a Hula-twist path

As a first example we present our results on stilbene. We determined the lifetime of the perpendicular minimum (which is a dark state) of (*cis*-) stilbene and compared it with the corresponding value in a "stiff stilbene", in which aliphatic bridges prevent rotation of the single bonds. Scheme 1 shows the structures and reactions and also explains the paths "one-bond flip" (OBF, in which the ethylenic double bond alone rotates by 180°) and "Hula twist" (in which one adjacent single bond is also turned by 180°, although not necessarily synchronously). A synchronous rotation of θ and φ can either be carried out by holding the central part fixed or by holding the end groups (more or less) in a plane. In the latter case, only an ethylenic CH group (such as the central part of the body in the Hula-twist dance) rotates out of plane, whereas the substituents reorient within the plane. The products of HT and OBF are distinguishable conformers, if stilbene is suitably substituted e.g. at the positions marked with an asterisk.

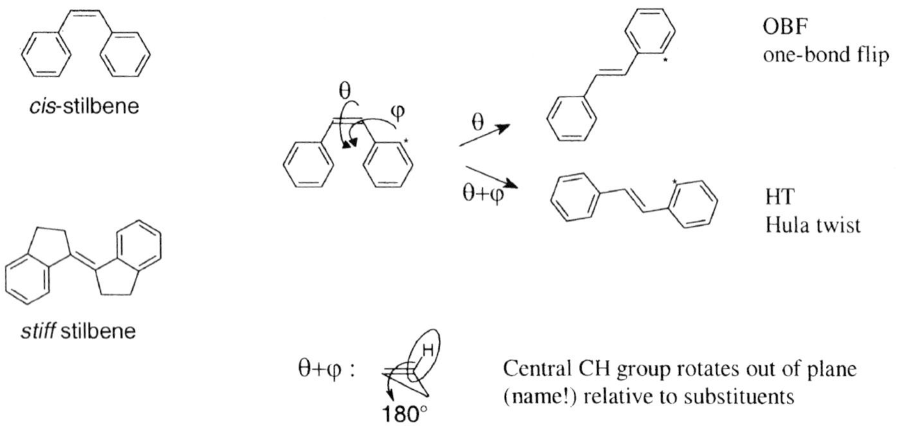

Scheme 1.

The traditional belief has been that *cis-trans* isomerization follows the OBF path. On the other hand, the Olivucci group found by calculation a minimum-energy path that leaves the S_1 minimum of polyenes via a "kink-type CI" [5] with a distortion that was later recognized as Hula twist [6]. Also in stilbene a calculation found an HT-type CI [7]. It would be desirable to have an experimental check of these predictions. The comparison of the two molecules in Scheme 1 seemed suitable. If in fact the most favorable path for stilbene to leave the perpendicular minimum (p*, S_1 minimum with $\theta = 90°$) implies a rotation of φ (by 90° to reach the HT type CI), the lifetime of p* should be much longer for *stiff*-stilbene, where single-bond rotation is blocked.

For this comparison we first determined the lifetime of p* after exciting *cis*-stilbene [8] and found $\tau_{cis} \approx 140$ fs (or 135-160 fs, the exact value depending on an initial step that was not fully resolved [8]). This minimum and its lifetime have been searched in vain for decades, so that it was already called phantom state. So, what was our trick to detect it? For explanation, one should realize that on electronic relaxation down to p* (Fig. 2) the total energy is conserved. Ionization from this p* with its excess energy produces a hot ion, which fragments before it is accelerated (in ≈100 ns) and detected. That is, the p* lifetime cannot be detected from the parent-ion signal (as attempted by previous workers). We deduced it from a fragment ion [8]. The success demonstrates the power of our method.

Then we determined in the same way the p* lifetime for *stiff*-stilbene and found 800 fs [9], 5-6 times longer than in *cis*-stilbene. We concluded that indeed, leaving from p* involves a single-bond twist (φ) in the unhindered stilbenes; that is, the path leads via an HT-type CI. This interpretation could be further supported by the fact that it allowed to explain the relative orientation of *trans*-stilbene photochemically produced from *cis*-stilbene, which was previously observed but not understood (for details, see [9]).

HT was originally suggested as a volume-saving process to explain *cis-trans* isomerization even in constrained environments [10], e.g. in the process of vision or in a matrix. After the first experimental detection of Hula twist [11], many examples were documented for HT in matrixes (see e.g. [12-14]). They were ascribed to the limitation of the available volume in the matrix, assuming that without this constraint the OBF path would be (energetically) preferable. However, our example shows that HT is the lower energy path already in an

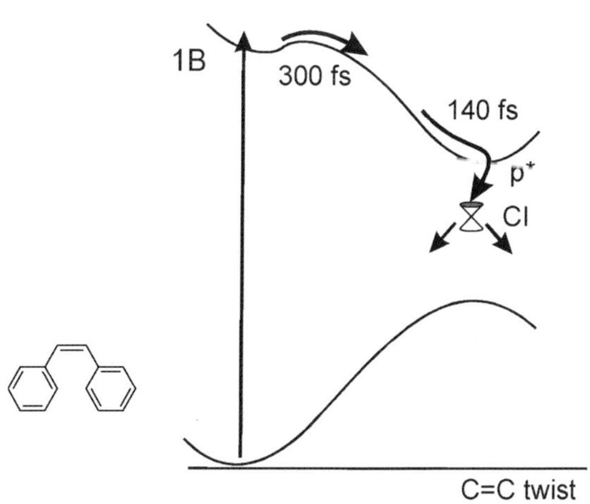

Fig. 2. Relaxation from excited *cis*-stilbene and the lifetime of its perpendicular minimum p* (more details in [8]).

isolated molecule. There must be an internal driving force. This is obviously the slope leading to the conical intersection. (In [6], it was explained in terms of an attractive 1–3 interaction along the axis, about which the CH group rotates in Scheme 1.) According to theory (e.g. [6]), this mechanism is general for nonpolar double-bond systems. The experiment shows that the photochemical path is crucially controlled in regions far from the Franck-Condon geometry and that time-resolved spectroscopy can investigate such features. It was revealed in this way that *cis-trans* isomerization follows a path different from what was believed for about a century.

3. Cyclopentadiene and cycloheptatriene: Even internal conversion happens far from the Franck-Condon region

The main photoreaction of cycloheptatriene is a [1,7]-sigmatropic H migration from the CH_2 group to the adjacent C. It seems that a similar migration ([1,3]-sigmatropic) occurs in cyclopentadiene [15] (Scheme 2). Both are photo-chemically allowed pericyclic reactions, so that we can apply the scheme of Fig. 1. (For simplicity we ignore here the ring closure of both molecules, which leads in very small quantum yields to bi- and tricyclic systems.) We found for both molecules a lifetime of S_1 (i.e. the 2A minimum in Fig. 1) around 70 fs. This is also the time for internal conversion, because the path branches at the S_1/S_0 CI, one branch leading back to the reactant. (This would also be the case, if the molecules were deuterium labeled so that the reactant could be distinguished from the product.) The 70 fs are more than six orders of magni-tude shorter than with benzene (S_1 lifetime ≤100 ns, see [15]). This would be surprising in the traditional approach that describes the internal-

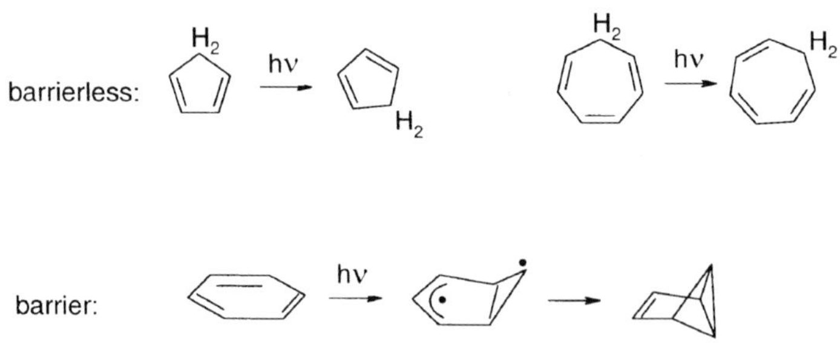

Scheme 2.

conversion rate by Fermi's Golden rule by a product of electronic and vibrational matrix elements (or Franck-Condon factors) and the density of states; a tacit assumption is that the process is vertical. None of these factors is expected to differ by orders of magnitude between these molecules, which are similar in size and electronic structure. In the pathway approach [3], there is an obvious explanation: For the two molecules with the H migrations there is a barrierless path down through the S_1/S_0 CI as in Fig. 1 (for details, see [15,16]), whereas benzene requires activation energy for a ring distortion to reach the corresponding CI (Scheme 2).

We generalize [15]: In particular the fastest internal conversions follow a path with large intermediate excursions of atoms or groups of them. Such paths have their initial part (down to the last CI) in common with an – at least attempted – photochemical reaction.

4. The "conflicting" stereospecificities in electrocyclic ring closure and ring opening

Fig. 1 suggests intermediates (the S_1 minimum and following S_1/S_0 CI) that should be in common for the forward and backward pericyclic reactions. One therefore expects the same product distributions for both. But on irradiation, conjugated dienes form the isomeric cyclobutenes with perfect stereospecificity, whereas the photochemical ring opening of the cyclobutenes is often not stereospecific, as documented by the Leigh group (see e.g. [17], where also different explanations are discussed); that is, the forbidden isomeric diene (ctD in Scheme 3) is also observed. A solution was suggested by Olivucci and coworkers. Their calculations predict that electrocyclic reactions of dienes and their *cis-trans* isomerization can both originate from the same S_1/S_0 CI [18]. This is schematically indicated in Scheme 4b and compared with the previous view (Scheme 4a) that cannot explain the formation of ctD from the cyclobutene.

Scheme 3. Photochemical ring opening of *cis*-disubstituted cyclobutenes (CB) to the allowed products *cis,cis*- and *trans,trans*-diene (ccD and ttD) and the forbidden product *cis,trans*-diene (ctD). The dienes can also be interconverted by photochemical *cis-trans* isomerization.

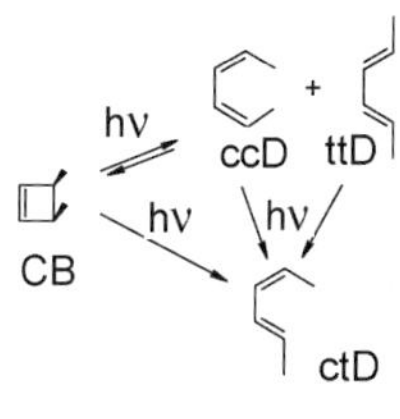

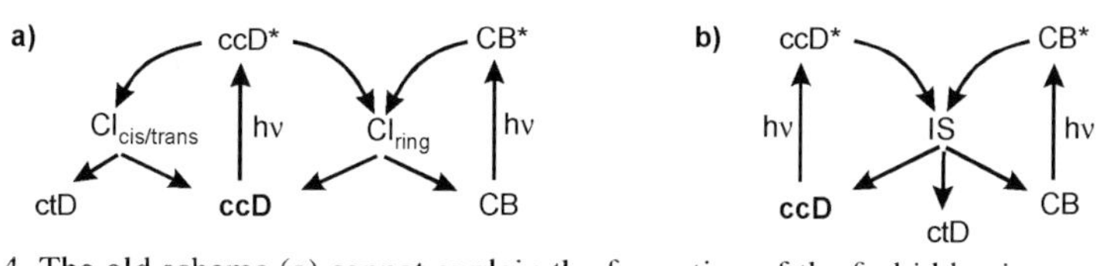

Scheme 4. The old scheme (a) cannot explain the formation of the forbidden isomer ctD observed also from the CB side, in contrast to the Olivucci scheme (b). By "conical intersection" (CI) one usually means a minimum of the intersection space (IS).

Obviously in Scheme 4b, ctD is also accessible from the cyclobutene side. The crucial difference is that Scheme 4a assumes an early branching (e.g.

in the Franck-Condon region) between the *cis-trans* and electrocyclic isomerizations, whereas Scheme 4b postulates a late branching in the S_1/S_0 intersection space (IS). It should in principle be possible to check by pump-probe spectroscopy whether this dark region is common for all reactions as indicated in Scheme 4b.

We investigated the *s-cis*-fixed dienes cyclohepta-1,3-diene (CHD) and cyclo-octa-1,3-diene (COD) and found an earlier branching than predicted, namely at the first conical intersection (Fig. 3) [19]. The *cis-trans* minimum (ct*) was identified by the consecutive reaction ctD → CB, which was detected in the hot S_0; its temperature dependent rate was known before. Actually Fig. 3 is also compatible with the Olivucci calculations [18]. They predict several 2A minima (and conical intersections) with similar molecular geometries. If this is right, the two branches (valleys on the 2A surface) may be both extrapolated to the cyclobutene side (broken lines in Fig. 3), so that also the *cis-trans* channel leading to the forbidden isomer ctD would be accessible from CB. Indeed preliminary experiments with the corresponding CB (bicyclo[4,2,0]octene-7, Fig. 3) show the same time constant (τ_3') as τ_3 for the *cis-trans* channel from the diene side. However, to strengthen the identification of this step, the signal-to-noise ratio should still be improved so that either the consecutive ctD → CB reaction in the hot S_0 or the parallel electrocyclic reaction (from p*, τ_4) can also be detected.

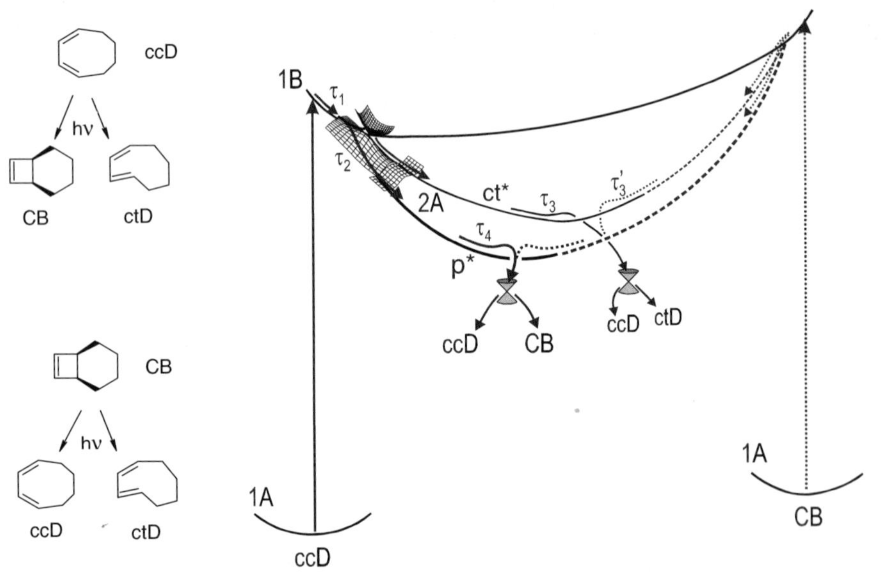

Fig. 3. Potentials and reaction paths for cycloocta[1,3]diene (see the formula scheme) and the corresponding cyclohepta[1,3]diene, as deduced from the time-resolved experiment. The *cis-trans* channel (via ct*) dominates from the side of the diene (ccD), probably also from the CB side. p* is the pericyclic minimum.

The observed difference of stereospecificity seemed to exclude a scheme such as Fig. 1 with a common intermediate for the forward and backward reactions. However, our scheme (Fig. 3) suggests even two common intermediates. The fact that both, including ct*, are accessible also from the CB side explains why CB ring opening is often not stereospecific, whereas ring closure of dienes (where only the p* channel is usually considered) is.

5. Conclusion

The three examples show that crucial parts of the photochemical processes happen far from the Franck-Condon region ("in dark states"). The example of stilbene revealed that *cis-trans* isomerization follows a path (Hula twist) different from what was believed for a century; it is driven by the slope of the PES from the perpendicular minimum towards the conical intersection. Also the traditional view of internal conversion (which invokes Fermi's Golden rule, assuming a vertical process, i.e., no geometry change) must be revised. In particular, if it is ultrafast, the S_1/S_0 transition occurs at the S_1/S_0 CI, which is far from the equilibrium geometry. The problem of stereospecificity of the diene ↔ cyclobutene reactions also seems to find its solution in features of the dark states. Instead of one, there are two common minima and conical intersections for forward and backward reactions.

It is a distinguished goal of time-resolved spectroscopy to investigate the molecules and processes in such remote parts of the PES. The examples also demonstrate that nonresonant ionization with mass-selective detection of the ion yields is a very suitable probing technique for this purpose. The later steps in the process are usually only revealed in the fragment-ion signals (and were therefore overlooked by previous workers who only studied the parent ions). It facilitates assignments to estimate relative energies of the observation windows from the ionization probabilities and the fragmentation patterns. It is also very valuable for assignments, if one can measure a consecutive reaction in the hot ground state whose rate was known before (see the example in sec. 4).

Acknowledgment. We are grateful to W.J. Leigh for stimulating and helpful discussions on the diene/cyclobutene problems and for providing samples of the cyclobutenes. We thank the Deutsche Forschungsgemeinschaft for a grant (FU363/1).

References

1. W. Fuß, W.E. Schmid, and S.A. Trushin, J. Chem. Phys. 112 (2000) 8347.
2. M. Klessinger and J. Michl, Excited states and photochemistry of organic molecules, VCH, New York, 1995.
3. W. Fuß, S. Lochbrunner, A.M. Müller, T. Schikarski, W.E. Schmid, and S.A. Trushin, Chem. Phys. 232 (1998) 161.

4. W. Fuß, Y. Haas, and S. Zilberg, Chem. Phys. 259 (2000) 273.
5. P. Celani, M. Garavelli, S. Ottani, F. Bernardi, M.A. Robb, and M. Olivucci, J. Am. Chem. Soc. 117 (1995) 11584.
6. D. Sampedro Ruiz, A. Cembran, M. Garavelli, M. Olivucci, amd W. Fuß, Photochem. Photobiol. 76 (2002) 622.
7. M.J. Bearpark, F. Bernardi, S. Clifford, M. Olivucci, M.A. Robb, and T. Vreven, J. Phys. Chem. A 101 (1997) 3841.
8. W. Fuß, C. Kosmidis, W.E. Schmid, and S.A. Trushin, Chem. Phys. Lett. 385 (2004) 423.
9. W. Fuß, C. Kosmidis, W.E. Schmid, and S.A. Trushin, Angew. Chem. Int. Ed. 43 (2004) 4178.
10. R.S.H. Liu and A.E. Asato, Proc. Natl. Acad. Sci. USA 82 (1985) 259.
11. A.M. Müller, S. Lochbrunner, W.E. Schmid, and W. Fuß, Angew. Chem. Int. Ed. 37 (1998) 505.
12. R.S.H. Liu and G.S. Hammond, Proc. Natl. Acad. Sci. USA 97 (2000) 11153.
13. R.S.H. Liu, Acc. Chem. Res. 34 (2001) 555.
14. R.S.H. Liu and G.S. Hammond, Chem. Eur. J. 7 (2001) 4536.
15. W. Fuß, W.E. Schmid, and S.A. Trushin, Chem. Phys. 316 (2005) 225.
16. S.A. Trushin, S. Diemer, W. Fuß, K.L. Kompa, and W.E. Schmid, Phys. Chem. Chem. Phys. 1 (1999) 1431.
17. W.J. Leigh, Chem. Rev. 93 (1993) 487.
18. F. Bernardi, M. Olivucci, I.N. Ragazos, and M.A. Robb, J. Am. Chem. Soc. 114 (1992) 2752.
19. W. Fuß, S. Panja, W.E. Schmid, and S.A. Trushin, Mol. Phys. (2005) accepted.

Femtochemistry VII
Fundamental Ultrafast Processes in Chemistry, Physics, and Biology
A.W. Castleman, Jr. and M.L. Kimble

Multi-component fitting of cluster dynamics

K. L. Knappenberger, Jr. and A. W. Castleman, Jr.

Departments of Chemistry and Physics, The Pennsylvania State University, University Park, PA, USA, 16802

1. Introduction

Reaction dynamics in polyatomic systems and cluster assemblies are often complicated by multiple competing pathways. We have developed a method of modeling ultrafast reaction dynamics that permits the measurement of rate coefficients as well as the determination of the branching ratios of such competing processes. This method is based on a general understanding of the rate laws of chemical kinetics and the interaction of ultrafast laser pulses with molecular targets. We have successfully applied this approach to fit pump-probe experiments on sulfur dioxide and sulfur dioxide clusters and were able to determine the influence of solvation on molecular dynamics [1-3]. The model is based on comparing the ratio of the amplitude coefficients of competing pathways and thereby enabling a determination of their branching ratios. This method is described in this paper.

2. Results and Discussion

Experimentally, we populated the electronic Ẽ state of SO_2 by the multiphoton absorption of 312 nm light. Our femtosecond pump-probe experiments revealed that dissociation in SO_2 is mediated by bending of the chromophore, in good agreement with theoretical predictions [4]. We found that cleavage of an S-O bond in SO_2 occurred in 230 fs, resulting in the formation of sulfur monoxide and atomic oxygen. Upon cluster formation, an abrupt change was observed in the dynamics. Indeed, a cluster induced "cage effect" was observed for cluster sizes as small as the sulfur dioxide dimer. This effect was also observed in the mixed dimer of heterogeneous SO_2-H_2O clusters. To fully understand this effect, we developed a fitting model that incorporated the passage of the excited population through a conical intersection. This kinetic model is shown in Figure 1. While the model was constructed to fit the excited state dynamics and energy flow in sulfur dioxide, it is general and can be applied to many systems where there are competitive pathways. The model involves fitting the dynamics to a composite reaction, as shown in Equation 1.

$$A \xrightarrow[k1]{} B \xrightarrow[k2]{} C \qquad (1)$$

The mechanism in Equation 1 depicts the kinetic model shown in Figure 1, where A refers to the excited E electronic state of SO_2, which is the initial population. This population can decay through two pathways depending on coupling to the conical intersection, where B reflects relaxation to a lower-lying bound surface of SO_2 and C reflects a dissociation pathway. Since the formation of B and C is dependent on the consumption of the initial population of A, shown as N_A in Figure 1, the rate of consumption of A must be included in the rate equation. The initial population is the product of the photoabsorption cross-section of the chromophore at a given energy range and the cross-sectional photon density of the interaction region. The rate equation is expressed in Equation 2 below:

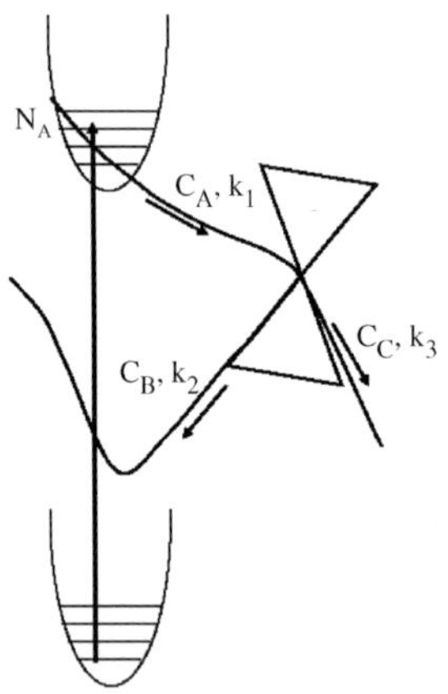

Figure 1. Schemtatic of curve crossings typicallyobserved in polyatomics, such as SO_2. Includes conical intersection and amplitude coefficient of each path.

$$S(t) = G(t)\{C_A(IRL)\exp(-k_1) + C_B(IRL)\exp(-k_2) + C_C(IRL)\exp(-k2)\} \qquad (2)$$

where G(t) is the instrument response function determined by a non-resonant zero marker, and IRL is the integrated rate law of the composite mechanism under study. The terms C_A, C_B, and C_C in equation 2 are referred to as amplitude coefficients for each respective pathway, as illustrated in Figure 1. The formalism shown in Equation 2 is of a general form and intended to show that all components are summed to fit the molecular response function. This is necessary with ion detection, since the dynamics of multiple components appear in the same mass signal. Equation 2 is a versatile function capable of accommodating several kinetic rate laws that encompass a variety of mechanisms. We will expand on this formalism momentarily and consider the pathways relevant to the dissociation processes of SO_2 clusters. It is useful to realize that the pathways represented by amplitude coefficients C_B and C_C are not populated directly in the pump step. Rather, they are populated only by the evolution of the initial population N_A. By taking the ratio of an amplitude coefficient with respect to C_A, one can determine the propensity of the excited population at the conical intersection to follow a specific path. This is possible because we employ a sum of least squares approach to fitting data, i.e. summing all components. By analyzing this ratio as a function of cluster size, the influence of the cluster on population transfer (and energy flow) can be determined.

Considering the kinetic model of Figure 1, pathway C refers to a repulsive surface. Dissociation at this point on the potential surface is an ultrafast process and occurs within the temporal duration of the pulse. No contribution to the parent ion signal is observed from this pathway because a Franck-Condon overlap between the neutral and cationic surfaces cannot exist. As a result, Equation 2 simplifies to:

$$S(t) = G(t)\{A(IRL)\exp(-k_1) + B(IRL)\exp(-k_2)\} \tag{3}$$

where:

$$IRL = \frac{k_1}{k_2 - k_1} \tag{4}$$

At this point it is clear that A reflects the initial population created in the pump step, and B reflects the population that is transferred via the conical intersection to a bound excited state as shown in Figure 1. By monitoring the change in the ratio of the amplitude coefficients, branching ratios can be determined as a function of cluster size. This method has been developed and employed [2,3], where we measured a unique cluster induced cage effect. The technique that we have developed of analyzing the ratio of amplitude coefficients has proved very useful in combination with rate measurements to provide a better understanding of processes associated with cluster dynamics.

The method is first demonstrated with homogeneous $(SO_2)_m$ clusters. Experimentally we observed a sequential increase in the decay rate k_1 as a function of cluster size. Additionally, the pump-probe transient of cluster fragmentation products were clearly formed in the ionic state. When comparing the ratio of the amplitude coefficients C_B/C_A, a significant increase is observed with increasing cluster size. The ratio of C_B/C_A increases by 400 % in going to the dimer and increases by as much as 550% for the tetramer, with respect to the monomer. Correspondingly, a decrease in the decay rate of k_1 was fit along with the increasing ratio of amplitude coefficients. These results are summarized in Table 1.

When studying heterogeneous $(SO_2)_m(H_2O)_n$ clusters, we found that the ratio of amplitude coefficients increased by 250% in going to the mixed dimer and only by 440% when SO_2 was solvated by as many as six water molecules. In going to clusters with as many as eleven water molecules, no value higher than 440% was measured. These results are summarized in Table 2.

Table 1. Homogeneous cluster fitting results

Species	C_1	τ_1 (fs)	C_2	τ_2(ps)	Percent increase (C_2/C_1)
SO_2	7.7	230	0.3	8	-----
$(SO_2)_2$	7.2	323	1.1	3	~400 %
$(SO_2)_3$	4.1	368	0.8	3	~470 %
$(SO_2)_4$	3.8	435	0.8	3	~550%

Table 2. Binary cluster fitting results

Species	C_1	τ_1 (fs)	C_2	τ_2(fs)	% increase (C_2/C_1)
SO_2	7.7	230	0.3	8000	-----
$(SO_2)(H_2O)$	6.5	275	0.6	2000	~250 %
$(SO_2)(H_2O)_3$	4.4	407	0.5	1500	~300 %
$(SO_2)(H_2O)_6$	8.2	485	1.4	1400	~440%

These findings suggest that SO_2 is more efficiently self-solvated than when solvated by water. The cluster "caging" mechanism has been attributed to a steric barrier that is imposed upon the bending chromophore by the cluster environment. Steric hindrance prevents the bending SO_2 from achieveing a linear geometry, and thereby prevents dissociation. The results show that population transfer back to a lower-lying bound well in SO_2 has a higher propensity in homogeneous sulfur dioxide clusters. These findings would not have been possible by considering rate constants alone, but are only revealed by investigating the amplitude coefficients in detail.

Acknowledgements

We gratefully acknowldege financial support from The National Science Foundation, grant number ATM-0089233, and The Department of Energy, grant number FG02-97ER14258.

References:

1. K. L. Knappenberger, Jr. and A. W. Castleman, Jr., J. Phys. Chem. A 108 (2004) 9.
2. K. L. Knappenberger, Jr. and A. W. Castleman, Jr., J. Chem. Phys. 121 (2004) 3540.
3. K. L. Knappenberger and Jr., A. W. Castleman, Jr., J. Chem. Phys. 122 (2005) 154306.
4. K. Kamiya and H. Matsui, Bull. Chem. Soc. Jpn. 64 (1991) 2792.

Femtochemistry VII
Fundamental Ultrafast Processes in Chemistry, Physics, and Biology
A.W. Castleman, Jr. and M.L. Kimble

Photodissociation of SO_2 at 200 nm: Evidence of an anisotropy effect

D. P. Hydutsky, N. J. Bianco, and A. W. Castleman, Jr.

Departments of Chemistry and Physics, Penn State University, University Park, PA, 16802

1. Introduction

Recently our laboratory has taken a great interest in the ultrafast dynamics of SO_2 [1], SO_2 clusters [2-4], and SO_2 clustered with water [4,5]. These experiments investigated both bound and dissociative processes in the monomer and cluster species. Currently our interest lies in sulfur isotope effects in the photodissociation of SO_2 excited by 200 nm pulses to the C/D state. The scientific community is interested in isotope effects in heavy atoms for fundamental reasons [7] as well as for the information inferred about the earth's early atmosphere [8].

A preliminary investigation of SO_2 has also yielded evidence for anisotropy effects. The polarization of the pump and probe beam are found to have a profound effect on our ability to observe the neutral state dynamics of SO fragments that result from the photodissociation of SO_2 with 200 nm light. The evidence for these observations as well as plausible explanations of the results are presented.

2. Experimental

The pump-probe technique, pioneered by Zewail [9-10], was used to examine the temporal evolution of SO_2 excited to the C/D state. The apparatus used to perform these experiments consists of a 10 Hz femtosecond laser system coupled to a home-built time-of-flight mass spectrometer. Neutral room temperature SO_2 is bled into the chamber through a pick-up source where the pump and probe beams ionize the sample for detection in the mass spectrometer. The fundamental wavelength is centered at 790 nm with a

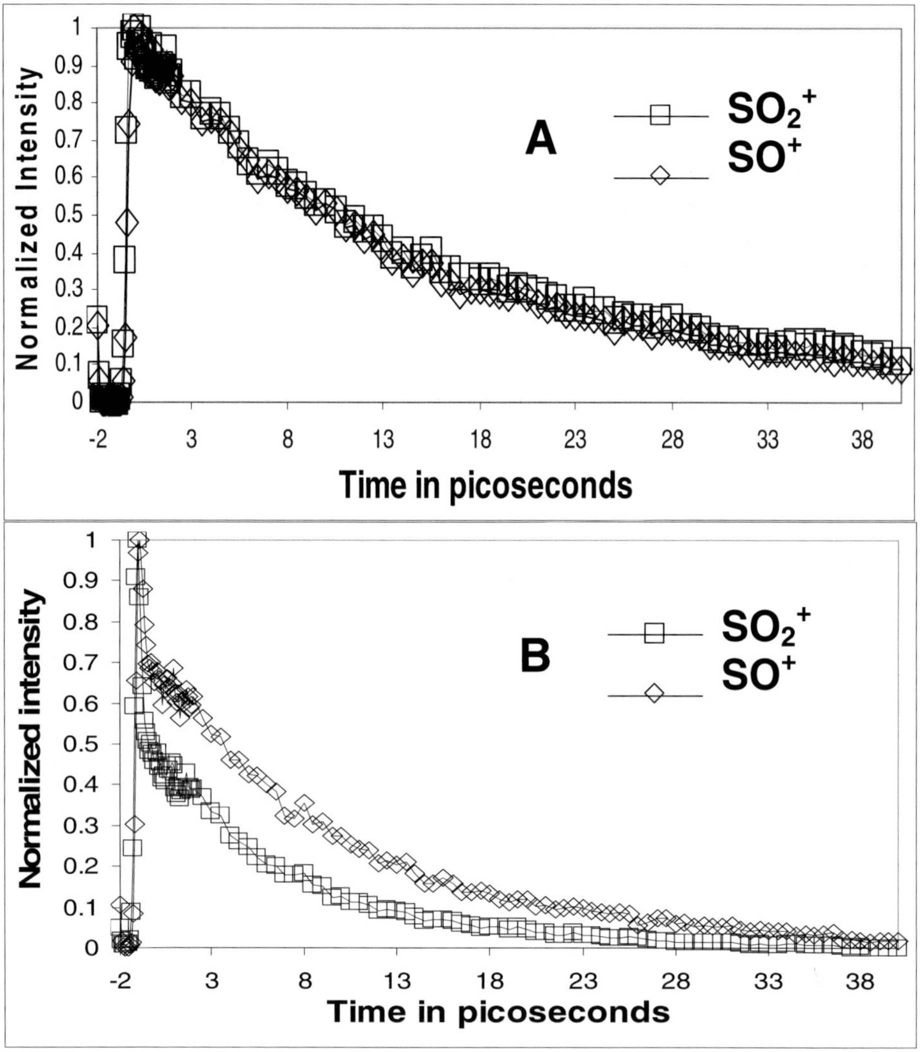

Figure 1. Pump-probe transients of SO_2 with **A**, the pump and probe beams perpendicular and **B**, with the beams parallel.

temporal duration of ~100 fs, and second, third, and fourth harmonics produced by frequency doubling and frequency mixing. For this experiment the fourth harmonic (~200nm) and the second harmonic (~400 nm) are used as the pump and probe beams, respectively. The polarization of the probe beam was changed by altering its routing into the periscopes used in the experiment. Thus, only data for the parallel and perpendicular orientation of the beams are available. All points in pump-probe transients represent mass spectra of 100 shots averaged together.

3. Results

The pump-probe transients shown in Figure 1A represent spectra taken with the pump and probe beams perpendicular to each other while Figure 1B represents spectra where the two beams are parallel to each other. As can be seen in Figure 1A, there is no discernable difference between the intact SO_2^+ species and the ion state fragment SO^+ species. However, in Figure 1B, the SO_2^+ and SO^+ species have similar but distinguishable dynamics, showing that some of the signal recorded results from the dynamics of the neutral SO chromophore. A brief discussion of how this could occur follows.

4. Discussion

The fact that neutral SO is not ionized when the pump and probe pulses are perpendicular but is ionized when the pump and probe pulses are parallel to each other could be giving us information about the orientation of the SO_2 chromophore as it dissociates. In order for the neutral SO fragments to absorb light from the probe pulse, there must be a change in dipole moment. The SO fragment is in line with the polarization of the pump pulse, so the polarization of the probe pulse must be the same in order to detect the neutral fragment. Also, this means the O that dissociates from the molecule is out of the plane of the pump pulse.

5. Conclusion

Our initial investigation into the dynamics of SO_2 excited with 200 nm light has revealed a clear polarization effect for the pump and probe pulse. We can deduce that the dissociating O from the SO_2 chromophore is leaving near-perpendicular to the plane of the pump pulse, from the response of the neutral SO fragment to the polarization of the probe pulse. Further investigation of this system with a phase-sensitive technique such as velocity map imaging could yield more quantitative information on the effect of polarization.

References

1. K. L. Knappenberger and A. W. Castleman, Jr., J. Phys. Chem. A 108 (2004) 9.
2. K. L. Knappenberger and A. W. Castleman, Jr., J. Chem. Phys. 121 (2004) 3540.
3. T. E. Dermota, D. P. Hydutsky, N. J. Bianco, and A. W. Castleman, Jr., in Femtochemistry and Femtobiology, M. M. Martin and J. T. Hynes (Eds.) Cluster Size Effects on Bound Excited States of SO_2, Elsevier, Amsterdam, 2004, p. 25.

4. T. E. Dermota, D. P. Hydutsky, N. J. Bianco, and A. W. Castleman, Jr., J. Phys. Chem. A 109 (2005) 8259.
5. K. L. Knappenberger and A. W. Castleman, Jr., J. Chem. Phys. 122 (2005) 154306.
6. Y. Q. Gao and R. A. Marcus, Science 293 (2001) 259.
7. J. Farquhar, H. M. Bao, and M, Thiemens, Science 289 (2000) 756.
8. A. H. Zewail, Science 242 (1988) 1645.
9. A. H. Zewail, J. Phys. Chem. A 104 (2000) 5660.

Femtochemistry VII
Fundamental Ultrafast Processes in Chemistry, Physics, and Biology
A.W. Castleman, Jr. and M.L. Kimble

Energy Transfer Dynamics in Zn-Porphyrin-Appended Dendrimers

Jane Larsen[a], Ben Brüggemann[a], Tomáš Polívka[b], Joseph Sly[c], Maxwell J. Crossley[c], Villy Sundström[a], and Eva Åkesson[a]

[a]*Department of Chemical Physics, Lund University, Box 124, SE-221 00 Lund, Sweden*
[b]*Institute of Physical Biology, University of South Bohemia, and Institute of Plant Molecular Biology, Czech Academy of Sciences, Czech Republic*
[c]*School of Chemistry, The University of Sydney, NSW 2006 Australia*

1. Introduction

The need for sustainable energy resources has focused a lot of attention to the study of artificial photosynthesis for many years now. The Zn-porphyrin-appended dendrimers presented here are constructed to mimic the natural light-harvesting antenna systems, where sunlight is absorbed and the energy subsequently transferred to the reaction center very efficiently.

1.1. The Zn-Porphyrin-Appended Dendrimers

The dendrimers displayed in Figure 1 are made by attaching monomeric Zn-porphyrins (G0P1) at the periphery of a flexible single-bonded nitrogen-carbon skeleton. The skeleton expands systematically in size from the first generation with 4 Zn-porphyrins (G1P4) up to the fifth generation with 64 Zn-porphyrins (G5P64). In G2P8 and G5P64, the grey circles each symbolize a monomer. A unit of four Zn-porphyrins is referred to as a dendron. The dotted circles behind G5P64 mark the positions of the Zn-porphyrins in the lower generation dendrimers. In addition to G0P1, the dimer (G0P2) is used as a reference.
The steady state absorption and fluorescence spectrum of the monomer is shown in Figure 2A. For G0P2 – G5P64, similar spectra are observed, except for an increase in the extinction coefficient scaling with the number of Zn-porphyrins [1,2]. The main features are the strongly absorbing Soret-band at

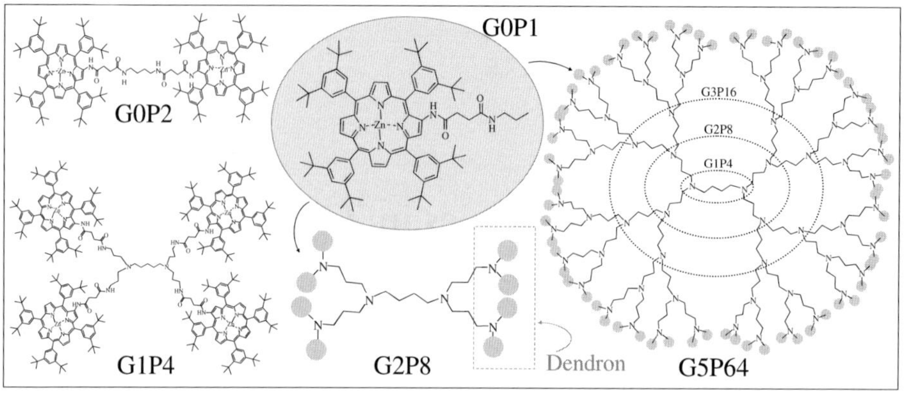

Figure 1. The Zn-porphyrin dendrimers, where the grey circles in G2P8 and G5P64 each mark one monomeric Zn-porphyrin (G0P1) attached as show in G1P4. The dotted circles behind G5P64 mark the position of the monomers in the smaller generation dendrimers. As indicated in G2P8, a unit of four Zn-porphyrins is referred to as a dendron.

429 nm [$S_2 \leftarrow S_0$] and the two weaker vibrational Q-bands observed at 558 and 598 nm [$S_1 \leftarrow S_0$]. Upon excitation in the Soret-band, relaxation to the S_1 state occurs within 1 ps [2], from where fluorescence to the ground state is observed at 602 and 655 nm [$S_1 \rightarrow S_0$]. Intersystem crossing (ISC) from the S_1 state to the triplet manifold is very efficient (84 %), and phosphorescence can be observed above 800 nm [3]. However, since the phosphorescence yield is low and our measurements are performed at wavelengths below 730 nm, phosphorescence contributions to the signals can safely be ignored.

The transient absorption spectra of the monomer measured at 5 ps and 350 ps after excitation in the Soret-band are displayed in Figure 2B. For G0P2 – G5P64, similar transient absorption features are observed [2]. Contributions from ground state bleach and stimulated emission are observed as the three dips between 560 – 660 nm. The peak at 470 nm with a shoulder extending beyond 730 nm is assigned to the S_1 absorption spectrum. An additional feature growing in at ~485 nm can be observed by comparing the 5 ps and the 350 ps spectra. This feature is identified as the T_1 absorption spectrum due to the efficient ISC. The similarities in both the steady state and the transient absorption spectra of the different compounds show that strong interactions between the Zn-porphyrins are not present.

2. Energy Transfer between the Zn-Porphyrins

Energy transfer (ET) between the Zn-porphyrins is studied by two means: (1) fluorescence anisotropy measured at room temperature and at 200 K, and (2) excitation-intensity dependent transient absorption measured at room temperature. In the room temperature fluorescence anisotropy of G0P1, G1P4,

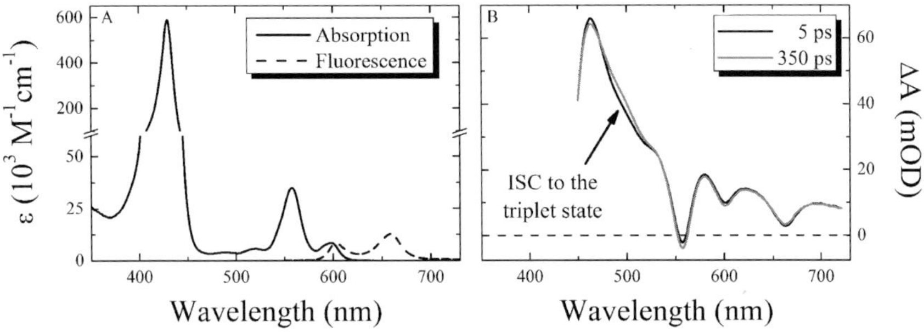

Figure 2. A: The steady state absorption and fluorescence spectra of the monomer marked with the straight and dotted line, respectively. B: The transient absorption spectrum of the monomer at 5 ps (black line) and at 350 ps (grey line) after excitation in the Soret-band.

G3P16, and G5P64, single decay time of ~330 ps is observed for all the compounds [1]. Since only one decay component is present, it is assigned to rotational motion of the Zn-porphyrins. Hence, the individual Zn-porphyrins in the dendrimers are able to rotate almost as freely as the monomer. By cooling the samples down to 200 K, a rotational correlation time of ~1.7 ns is observed in all the compounds, thereby verifying the high degree of structural flexibility within the dendrimers. Moreover, a faster decay component of ~100 ps is observed only in the dendrimers and is thus assigned to ET between the Zn-porphyrins. Employing Förster ET theory shows that the main contribution to the 100 ps ET component originates from ET between the nearest neighboring Zn-porphyrins within a dendron.

To gain further insight into the ET dynamics, excitation-intensity dependent transient absorption measurements were carried out [2]. By increasing the excitation intensity, the probability of simultaneously exciting more than one Zn-porphyrin within the same dendrimer increases. When two Zn-porphyrins are excited and ET between them takes place, one chromophore is de-excited to the ground state and the other is excited to an electronically higher state (S_n) from which it quickly relaxes to the S_1 state again. This ET process and subsequent radiationless decay is called singlet-singlet annihilation. Förster ET theory predicts that the time scale for ET between two excited Zn-porphyrins is between 4 and 8 times faster than the ET observed in fluorescence anisotropy, since the spectral overlap for ET between two excited Zn-porphyrins is larger than the spectral overlap between one excited and one ground state Zn-porphyrin[2]. Figure 3 shows the transient absorption kinetics at different excitation-intensity for G0P2 (the left graph). Similar kinetics are observed in G1P4 – G5P64, but the excitation-intensity required to obtain them depends on the size of the compound – the larger the dendrimer the lower the excitation-intensity. To fit the transient kinetics obtained at different excitation intensities, a rate-equation approach is utilized. This reveals two annihilation ET times of ~18 ps and ~130 ps in G0P2 – G5P64. We assign this to the presence

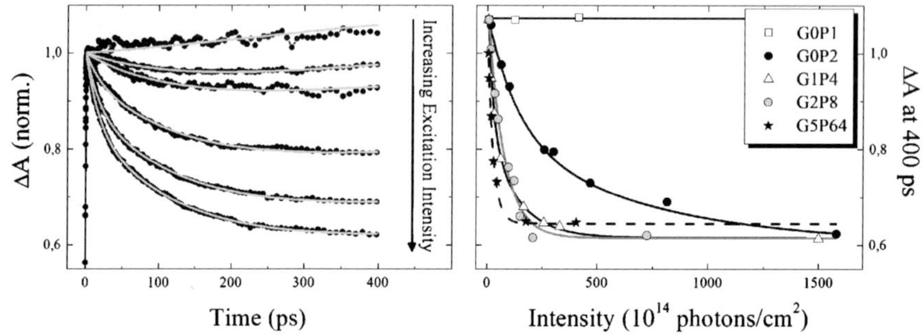

Figure 3. Normalized transient absorption kinetics at different excitation intensities (black circles) along with the fits (grey lines) are shown in the left graph for G0P2 at λ_{probe} = 487 nm. In the right figure, the ΔA value at 400 ps is plotted as a function of the excitation intensity (symbols). The lines are double exponential fit meant only as a guidance for the eye.

of at least two conformer distributions, one related to each of the annihilation ET times [2]. The 18 ps annihilation ET component is five times faster than the 100 ps ET component observed in the fluorescence anisotropy. According to the Förster ET estimates described above, they thus originate from the same structural conformer distribution. Observation of the counterpart to the 130 ps annihilation ET time in the fluorescence anisotropy at 200 K is prohibited by the 1.7 ns rotational correlation time. The excitation-intensity dependence (see Figure 3, the right graph) is directly related to the number of Zn-porphyrins between which annihilation can take place and demonstrates that annihilation can occur with one, three, and six other chromophores in G0P2, G1P4/G2P8, and G5P64, respectively. The structure of G2P8 can therefore be viewed as two separated dendrons, whereas communication between the Zn-porphyrins extends somewhat outside the dendron in G5P64.

3. Conclusion

ET between Zn-porphyrins in the studied dendrimers is clearly observed, and the presence of at least two structural conformer distributions is demonstrated. Furthermore, the number of communicating Zn-porphyrins increases with increasing dendrimer size, though global communication between all the chromophores is not present.

References

1. J. Larsen, J. Andersson, T. Polívka, J. Sly, M. J. Crossley, V. Sundström, and E. Åkesson, Chem. Phys. Lett. 403 (2005) 205
2. J. Larsen, B. Brüggemann, T. Polívka, V. Sundström, E. Åkesson, J. Sly, and M. J. Crossley, J. Phys. Chem. A 109 (2005) 10654

3. J. E. Rogers, K. A. Nguyen, D. C. Hufnagle, D. G. McLean, W. J. Su, K. M. Gossett, A. R. Burke, S. A. Vinogradov, R. Pachter, and P. A. Fleitz, J. Phys. Chem. A 107 (2003) 11331

Femtochemistry VII
Fundamental Ultrafast Processes in Chemistry, Physics, and Biology
A.W. Castleman, Jr. and M.L. Kimble

Vectorial Energy Transfer: Direct Measurement of an Intermediate Transfer Step

Evrim Atas[a], Zhonghua Peng[b], Valeria D. Kleiman[a].

[a]*Department of Chemistry. University of Florida, Gainesville Florida 32611-7200*
[b]*Department. of Chemistry. University of Missouri - Kansas City, Kansas City, MO 64110*

1. Introduction

The search for artificial antennae and light-harvesters has recently focused on hyperbranched macromolecules. Several architectures have been synthesized with chromophores covalently attached to a dendritic backbone [1], or as part of the dendritic backbone itself [2,3]. In a few cases [1,4], vectorial energy transfer has been proposed to occur through a funneling process created by chromophores with an energy mismatch and weakly coupled excited states.

To the best of our knowledge, all evidence of the existence of energy funnel are steady state measurements providing only indirect support for the presence of the cascade. Here, we present direct experimental measurement of energy migration through an intermediate state in a funnel-type dendritic structure.

The presence of vectorial energy transfer is essential to produce artificial light-harvesters with broadband absorption, to facilitate energy transfer through long distances, and to couple chromophores with non-overlapping spectra. Moore, Kopelman, and co-workers were the first to study a phenylethynylene (PE) dendrimer combining PE units of different conjugation lengths [4a]. Klafter and co-workers [4b] suggested that the cascade funneling process is fundamental for the efficient energy transfer observed in these systems. Mukamel and co-workers have investigated theoretically the localization of excitons in PE segments of different length [5], while Thompson *et al.* studied diethynylene phenyl units substituted with H or phenyl groups [6]. Ortiz *et al.* studied the role of Coulombic coupling by combining molecular dynamics simulations with the transition density cube method [7]. Frechet *et al.* [1a] prepared a system consisting of a flexible dendrimer scaffold covalently bonded to two different donors and a perylene-based final acceptor, while Müllen and co-workers [1b] investigated a rigid scaffold with substituted chromophores. In all three systems, efficient energy transfer was observed, and the ladder-type characteristics of the transfer were obtained by indirect measurements based on steady state spectroscopy.

The work presented here focuses on PE dendrimers. The PE units are combined through *ortho*, *meta*, and *para* substitution, leading to architectures with varying conjugation-lengths, which creates an energy gradient [2b, 8].

Direct confirmation of the presence of the gradient can be obtained by following the energy as it migrates through the intermediate steps down the gradient. Measuring the temporal evolution of the fluorescence at the emission wavelength of the intermediate state provides an unequivocal proof of the vectorial nature of the energy transfer in a ladder-type structure.

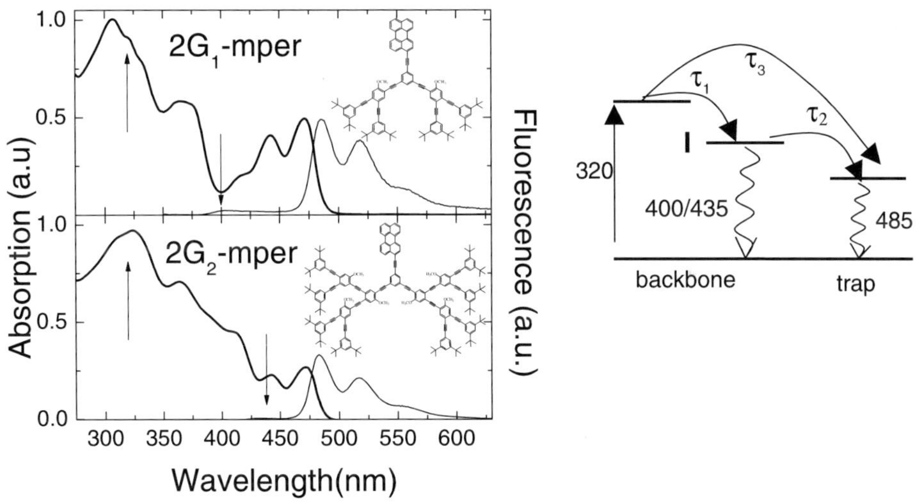

Figure 1: Left: absorption and emission spectra of $2G_1$-mPer and $2G_2$ mPer in CH_2Cl_2. The arrows show the λs used in the up-conversion experiment; right: kinetic model. The intermediate state (I) is detected at 400 nm (gen 1) or 435 nm (gen 2). Emission from trap is at 485 nm.

We investigated dendritic structures based on generation 1 and 2 di-dendrons covalently coupled to an ethynylene-perylene energy sink (Figure 1). For generation 1, the dendrimer backbone (where an OH group replaces the trap, see S1) has its maximum emission at 395 nm with fluorescence quantum yield $\Phi_{fluo} = 0.71$. In the absence of the trap, the generation 2 dendrimer emits with $\Phi_{fluo} = 0.80$, and the emission spectrum has a maximum at 435 nm. Spectra of $2G_1$-mPer and $2G_2$-mPer (Figure 1) show that, at wavelengths shorter than 400 nm or 430 nm, absorption corresponds mostly to the dendrimer backbone and leads to emission from the trap with efficiencies for energy transfer >95% [8].

2. Experiment

Fluorescence up-conversion was used to measure the temporal evolution of the excited state population in a ps time scale. The experiment consists of mixing the samples' fluorescence with a 50 fs gate pulse in a non-linear crystal. The up-converted signal is filtered and detected with a PMT [9]. Time-resolved fluorescence measurements of the dendrimer with and without the final energy sink led us to the 3 state (ladder) kinetic model shown in Figure 1 [9].

3. Results and Discussion

The two time constants involved in the cascade process are τ_1 =260 fs and τ_2 =340 fs for the $2G_1$-mPer and τ_1 =510 fs and τ_2 = 755 fs for $2G_2$-mPer. While the model for $2G_1$-mPer contains only the ladder mechanism, the $2G_2$-mPer data shows that half of the energy transfer process occurs through the ladder process, and the other half proceeds through a direct process (τ_3~440 fs). The relative contributions of these two pathways have a slight dependence on excitation wavelength [9]. Noteworthy in the model is the prediction of an intermediate state (I in Figure 1), consisting of a low-energy state of the backbone. The presence of this state leads to efficient energy transfer after excitation at 320 nm by creating a ladder-type electronic structure, which causes the vectorial energy transfer.

The model predicts that the intermediate step in the ladder-type system corresponds to the lowest-lying excited state of the backbone. Detection of fluorescence at this wavelength (400 nm or 435 nm for the $2G_1$-mPer and $2G_2$-mPer, respectively) provides direct confirmation of the cascade mechanism. As predicted, Figure 2 shows the data obtained after excitation at 320 nm. For both molecules, the fluorescence shows a fast rise time, and a slower decay. If the model is correct, one should be able to simulate the data directly without requiring additional fitting parameters. Using the model obtained in the independent measurements of the trap fluorescence, we simulate the population transfer through this intermediate state as shown in solid line in Figure 2.

The simulation shows excellent agreement with the experimental data obtained from the intermediate state. Note that the only variable in this simulated curve is the normalization constant (F_{max}). Both the decay times and the pre-exponential factors are modeled as:

$$F(t) = \frac{1}{F_{max}}\left(\frac{k_1}{k_2-(k_1+k_3)} \times \left[e^{-(k_1+k_3)t} - e^{-k_2 t}\right]\right) + \text{constant} \; , \tag{1}$$

where $k_1 = (\tau_1)^{-1}$, $k_2 = (\tau_2)^{-1}$, and $k_3 = (\tau_3)^{-1}$ are extracted from the model, and the constant corresponds to the 1.8 ns emission lifetime of $2G_n$-mOH present in the sample as an impurity. Since the quantum yield for energy transfer is > 0.96, the emission from the backbone in the presence of the sink is not observed in the steady state emission spectra, as it can only be detected by using an extremely fast (<6 ps) time window.

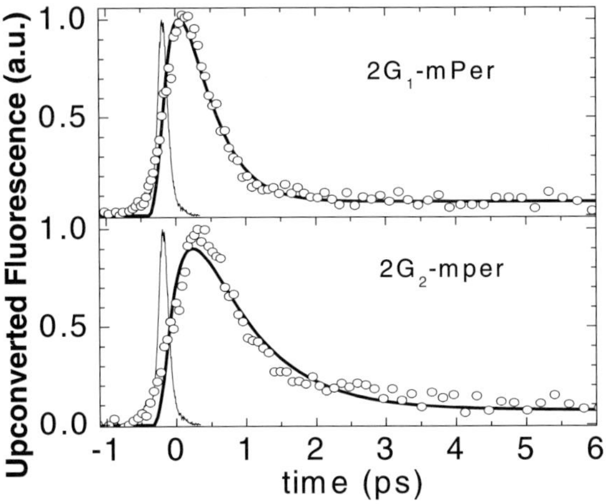

Figure 2: Temporal evolution of the intermediate state population followed by fluorescence up-conversion. The solid line corresponds to a simulation of the population of the excited state. The IRF is also plotted.

In conclusion, we present here the first direct measurement of cascade energy transfer in phenyl ethynylene dendrimers. The presence of gradients makes it possible to use larger dendrimers for light-harvesting, where the intermediate states in the cascade process allow for efficient, unidirectional energy transfer.

Support of this work was provided by NSF (CHE-0239120) and Research Corporation (RI0558).

References

1. a) J. M. Serin, D. Brousmiche, and J. M. J. Frechet, J. Am. Chem. Soc. 124 (2002) 11848; b) T. Weil, E. Reuther, and K. Müllen, Angew. Chem., Int. Ed. 41 (2002) 1900.
2. a) Z. H. Peng, Y. C. Pan, B. B. Xu, and J. H. Zhang, J. Am. Chem. Soc. 122 (2000) 6619; b) Y. C. Pan, M. Lu, Z. H. Peng, and J. S. Melinger, J. Org. Chem. 68 (2003) 6952.
3. Z. F. Xu, M. Kahr, K. L. Walker, C. L. Wilkins, and J. S. Moore, J. Am. Chem. Soc. 116 (1994) 4537.
4. a) M. R. Shortreed, S. F. Swallen, Z. Y. Shi, W. H. Tan, Z. F. Xu, C. Devadoss, J. S. Moore, and R. Kopelman, J. Phys. Chem. B 101 (1997) 6318; b) R. Kopelman, M. Shortreed, Z. Y. Shi, W. H. Tan, Z. F. Xu, J. S. Moore, A. BarHaim, and J. Klafter, Phys. Rev. Lett. 78 (1997) 1239.
5. S. Tretiak, V. Chernyak, and S. J. Mukamel, Phys. Chem. B 102 (1998) 3310.
6. A. L. Thompson, K. M. Gaab, J. J. Xu, C. J. Bardeen, and T. J. Martinez, J. Phys. Chem. A 108 (2004) 671.

7. W. Ortiz, B. P. Krueger, V. D. Kleiman, J. L. Krause, and A. E. Roitberg, J. Phys. Chem. B 109 (2005) 11512.
8. J. S. Melinger, Y. C. Pan, V. D. Kleiman, Z. H. Peng, B. L. Davis, D. McMorrow, and M. Lu, J. Am. Chem. Soc. 124 (2002) 12002.
9. E. Atas, Z. Peng, and V. D. Kleiman, J. Phys. Chem. B 109 (2005) 13553.

Femtochemistry VII
Fundamental Ultrafast Processes in Chemistry, Physics, and Biology
A.W. Castleman, Jr. and M.L. Kimble

Ultrafast Dynamics of Strongly Confined Water

Adriaan M. Dokter, Sander Woutersen, and Huib J. Bakker

FOM Institute for Atomic and Molecular Physics (AMOLF)
Kruislaan 407, 1098 SJ Amsterdam, The Netherlands

1. Introduction

Many biological and chemical processes occur within aqueous environments of nanometer dimensions. The properties of water on this length scale can be studied with so-called reverse micelles: small droplets of water covered by a surfactant and contained in an apolar solvent matrix. Previous linear infrared spectroscopic studies of reverse micelles showed that the nanoconfinement of water leads to a weakening of the hydrogen bonds between the water molecules [1]. Here we use nonlinear infrared spectroscopy to study reverse micelles.

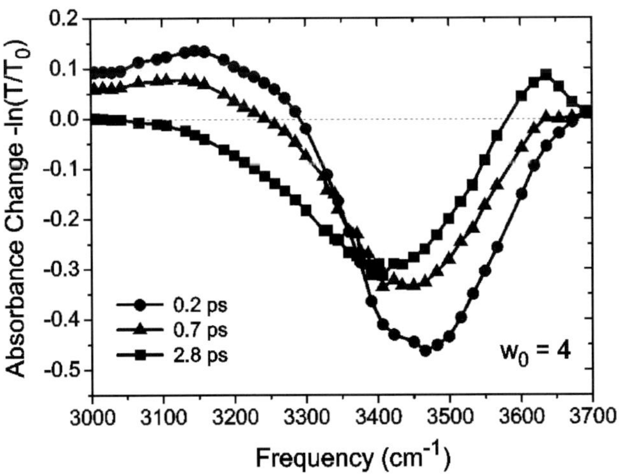

Fig. 1: Transient spectra measured for reverse micelles with w_0=4 at three different delays. The spectra are measured using a pump frequency of 3450 cm^{-1}.

2. Experiment

Reverse micelles were prepared by mixing the surfactant AOT (sodium di-2-ethylhexylsulfosuccinate, Aldrich, 99%), the solvent n-octane (Aldrich, anhydrous grade 99+%), and water H_2O (Aldrich, HPLC grade). The molar water-to-AOT ratio w_0 was set to 1, 2, 4, 7, and 12, creating reverse micelles containing a water pool consisting of a few to ~10.000 water molecules. The radius R in nm of the micelle can be approximated from $R=0.15w_0$, with $w_0=[H_2O]/[AOT]$ [2]. We perform transient absorption experiments using pump and probe pulses that are independently tunable between 2.6 and 3.3 μm (= 3000-3800 cm^{-1}), that have a pulse duration of 150 fs, and energies of 10 and 0.1 μJ, respectively.

3. Results and Discussion

Fig. 1 displays transient spectra for a reverse micelle with $w_0=4$. At early delays, the transient spectra reveal a bleaching at the v=0->1 transition (~3450 cm^{-1}) and an induced absorption at the v=1->2 transition (~3200 cm^{-1}). At longer delays, the transient spectrum takes the form of a thermal difference spectrum: when liquid water is heated the OH-stretch band shifts slightly to the blue and its area is reduced.

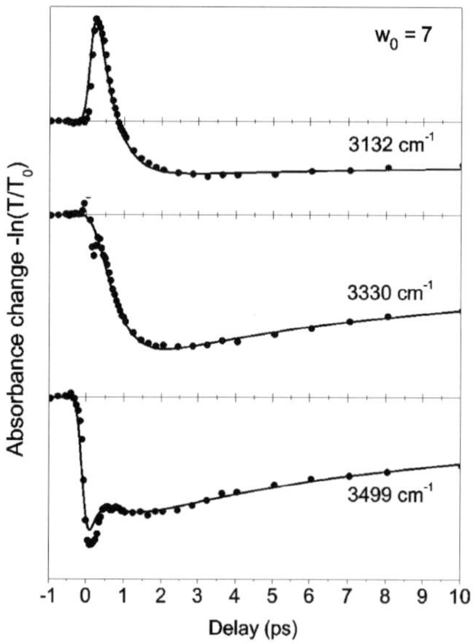

Fig. 2: Delay time scans measured at three different probe frequencies for a reverse micelle with $w_0=7$. The pump frequency was 3450 cm^{-1} for both micelles. The solid curves are obtained from a simultaneous fit of all delay time scans.

In Fig. 2, delay curves are presented. The delay curve in the top panel shows the induced v=1->2 absorption that decays due to the vibrational relaxation of the excited v=1 state. The vibrational relaxation leads to heating of the micelle and thus to an absorption decrease. This absorption decrease decays due to cooling of the micelle. The delay curve in the middle panel shows a rise and decay of a strong absorption decrease. This signal largely represents the effect of heating and cooling of the micelle. The delay curve in the lower panel shows an initial absorption decrease that results from the bleaching of the v=0->1 transition. This signal decays due to the relaxation of the v=1 state. Interestingly, the vibrational relaxation is followed by a renewed rise of an absorption decrease, which means that the thermalization (heating) is delayed with respect to the vibrational relaxation. Hence, we conclude that the relaxation of the v=1 state of the O-H stretch vibration first leads to population of a non-thermal intermediate state, and, only after relaxation of this intermediate state, the energy becomes thermalized over the micelle. We only observe this two-step relaxation mechanism for micelles with w_0=4,7,12, not for micelles with w_0=1 and w_0=2.

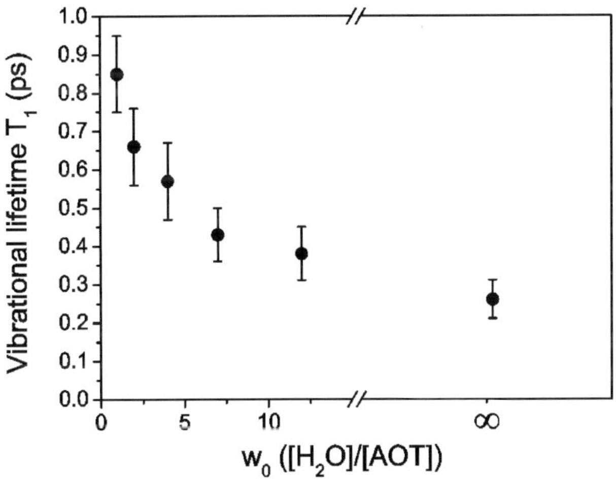

Fig. 3: Vibrational lifetime T_1 of the O-H stretch vibrations as a function of the value of w_0 of the reverse micelle. Also shown is the value of T_1 measured for bulk liquid water [6].

The values of the vibrational lifetime T_1 that follow from modeling the data are shown in Fig. 3. We find that the vibrational lifetime of the O-H stretch vibrations of water strongly increases with decreasing size of the reverse micelles. Interestingly, this increase of T_1 is much stronger than is usually observed for hydrogen-bonded systems [3-5]. Based on the change in frequency of the O-H stretch vibration, the value of T_1 is expected to increase by ~1.5 going from bulk to a micelle with w_0=1. The value of T_1 of bulk liquid water is ~260 fs at 300 K [4], meaning that T_1 is expected to be ~400 fs for a reverse micelle with w_0=1. This value of T_1 is more than two times shorter than the measured value of 0.85±0.1 ps.

The strong increase of T_1 can be explained as follows. When water is confined to nanometer sizes, the hydrogen-bonded network is truncated. As a result, the number of conjugated O-H groups is reduced, leading to a weakening of the hydrogen-bond interactions. This leads to a blueshift of the O-H stretch absorption spectrum and to a change of the relaxation mechanism; for micelles with w_0>=4, the relaxation is observed to result in the population of a non-thermal intermediate state, for micelles with w_0=1, 2, this state is not observed, and the vibrational relaxation is observed to result in direct thermalization. Hence, the nanoconfinement appears to result in the closing of an important relaxation channel at the red side of the O-H stretch absorption band.

The non-thermal intermediate state is likely formed by the overtone of the H-O-H bending mode. This overtone has a frequency of ~3300 cm^{-1} and is thus indeed in resonance with the red wing of the O-H stretch absorption band. This assignment is supported by the observation that the relaxation of the O-H stretch vibrations leads to excitation of the bending mode for bulk liquid water [7], but not for w_0=2 micelles [8]. Recently, the vibrational lifetime of the first excited state of the H-O-H bending mode was measured to be 0.2-0.4 ps [9,10], which is in excellent agreement with the value of τ_{eq} of 0.3±0.1 ps we observe for micelles with w_0=4,7,12.

The Fermi resonance of the O-H stretch vibrations with the overtone of the H-O-H bending mode is a special property of water. For most other systems including HDO and alcohols, the overtone of the bending mode is far out of resonance with the stretch vibration. This explains why the value of T_1 of bulk liquid H_2O is ~3 times shorter than that of bulk liquid HDO:D_2O [5], even though the O-H stretch vibrations of these systems have approximately the same frequency. When water becomes confined to nanometer sizes (w_0=1, 2), the value of T_1 becomes similar to that of HDO:D_2O at the same frequency of the O-H stretch vibration [5] (attained at an elevated temperature). Hence, nanoconfinement of water appears to make this liquid much more similar to other hydrogen-bonded liquids.

References

1. D. S. Venables, K. Huang, and C. A. Schmuttenmaer, J. Phys. Chem. B 105 (2001) 9132.
2. G. Seifert, T. Patzlaff, and H. Graener, Phys. Rev. Lett. 88 (2002) 147402.
3. R. Laenen and A. Thaller, Chem. Phys. Lett. 349 (2001) 442
4. S. Woutersen, U. Emmerichs, and H. J. Bakker, J. Chem. Phys. 107 (1997) 1483.
5. S. Woutersen, U. Emmerichs, H. K. Nienhuys, and H. J. Bakker, Phys. Rev. Lett. 81 (1998) 1106.
6. A. J. Lock and H. J. Bakker, J. Chem. Phys. 117 (2002) 1708.
7. A. Pakoulev, Z. Wang, Y. Pang, and D. D. Dlott, Chem. Phys. Lett. 380 (2003) 404.

8. J. C. Deak, Y. Pang, T. D. Sechler, Z. Wang, and D. D. Dlott, Science 306 (2004) 473.
9. O. F. A. Larsen and S. Woutersen, J. Chem. Phys. 121 (2004) 12143.
10. N. Huse, S. Ashihara, E. T. J. Nibbering, and T. Elsaesser, Chem. Phys. Lett. 404 (2005) 389.

Femtochemistry VII
Fundamental Ultrafast Processes in Chemistry, Physics, and Biology
A.W. Castleman, Jr. and M.L. Kimble

Femtosecond-Domain Dynamical Studies of Energy-Transfer Cassettes

T. G. Kim [a], J. C. Castro [b], A. Loudet [b], G.-S. Jiao [b], K. Burgess [b], R. M. Hochstrasser [a] and M. R. Topp [a]

[a]*Department of Chemistry, University of Pennsylvania, Philadelphia PA 19104, USA*
[b]*Department of Chemistry, Texas A&M University, P.O. Box 30012, College Station, Texas, 77842, USA*

1. Introduction

Energy-transfer labeling cassettes, having two different chromophores in the visible region, are widely used in the fluorescence-aided sequencing of DNA [1-3]. After chemically selective binding of the labeling cassettes to different fragments of DNA followed by gel separation, the mixture is irradiated at a single laser wavelength. The resulting fluorescence in four different wavelength bands provides a means to color-code the fragments, which in turn can identify the terminal bases and provide a method for sequencing. Commercial labeling agents employ a fluorescein derivative as an energy absorber near 500 nm and four different chemically attached rhodamine derivatives to provide fluorescent signals in four resolvable bands ranging from ~520 nm to ~680 nm. The dominant energy-transfer mechanism in such cassettes, where the chromophores are linked by partially saturated chains, and separated by 10-20 Å, is usually modeled according to a Förster energy transfer mechanism (FRET). New applications of fluorescent labels, such as in single-molecule spectroscopy, generate the need for alternative probe molecules, having greater chemical stability and rigid structure.

The work reported here derives from the recent synthesis of energy transfer cassettes, where the absorber and emitter chromophores are rigidly connected via a linker that is internally conjugated but twisted so that it is conjugated to neither the donor nor the acceptor moiety [4, 5]. Such linkers have been used elsewhere to study energy- and electron-transfer reactions in synthetic porphyrin arrays, where the transfer reactions are modeled according to a through-bond scheme [6-8].

The aim of this work was to measure energy transfer rates in a range of rigid cassettes involving structure types, using a single experimental technique. Also, we needed to measure the time-dependent fluorescence anisotropy, to examine the relative alignment of the absorber and emitter transition dipole moments and to determine the extent of polarization of the acceptor fluorescence. The energy transfer rates could be determined by several different approaches, including transient pump-probe methods [4, 9], which we originally used to study the dynamics of the fastest energy transfer cases. Those experiments involved detection via $S_{n>1} \leftarrow S_1$ transient absorption followed by $S_{n>1} \rightarrow S_0$ emission in nine different cassettes involving an anthracene donor and different BODIPY acceptor species. We have now studied many of these samples via fluorescence-upconversion gating methods at a time resolution (fwhm) of <250 fs. Although the time resolution is not quite as good as for pump-probe methods, experiments on the S_1 state provided definitive information about the transient fluorescence anisotropy of both the donor and acceptor moieties.

2. Experiment

Femtosecond-domain second-harmonic pulses from an amplified Ti:sapphire laser delivered an average power ~50 mWatt at ~1 kHz, centerd near 405 nm. These pulses were used directly to excite the anthracene-BODIPY cassettes (Compounds **1** & **8** shown for example). These conditions allowed us to measure both the energy transfer rate and the time evolution of the fluorescence anisotropy for both the donor and acceptor species. Anisotropy measurements of fluorescein-rosamine cassettes (**12**, & **17**) and a commercial dye (Big-ROX) required a tuneable laser source to excite the fluorescein donor near 490 nm, in order to select an optimally polarized transition. Those experiments used a non-collinear optical parametric amplifier (NOPA), built in-house after the design reported by Wilhelm *et al.* [10]. This device was tuned over a range 450-530 nm, with a bandwidth near 20 nm. Careful control of the center wavelength and band shape of the NOPA output allowed us to make ultrafast fluorescence measurements within 30 nm of the peak irradiation wavelength. A detailed experimental description is given in the full paper [11].

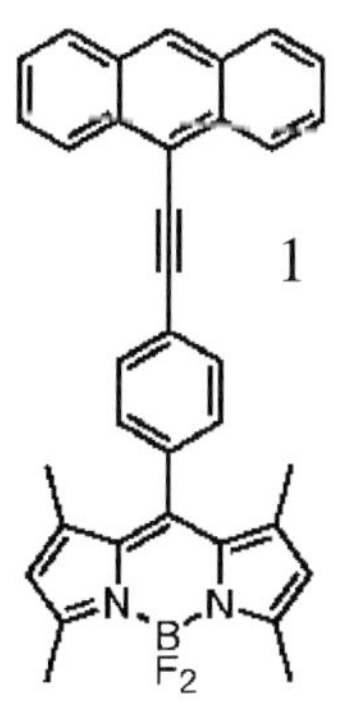

1

3. Results and Discussion

The data for 18 compounds studied are also presented in the full paper [11]. Nine of them (**1** & **8** shown here) have anthracene donors and BODIPY acceptors. Several, such as **1**, have the anthracene $S_1 \leftarrow S_0$ transition moment aligned with the axis of the linker, but perpendicular to the transition moment of the BODIPY acceptor. This was confirmed by the observation of negative anisotropy values for the BODIPY emission (r ~ -0.1) (see Figure 1). Although this transition geometry would strictly have zero Förster overlap for perpendicular moments ($\kappa \rightarrow 0$), the energy transfer times were measured in the range of 300-600 fs. Evidently, a more effective energy transfer mechanism than Förster transfer is involved. Compounds similar to **8** (shown here) revealed even faster transfer rates that were within our experimental time resolution (i.e., <200 fs). This showed that the condition where both the donor and acceptor transition moments were aligned with each other and with the linker axis was particularly favorable for fast energy transfer.

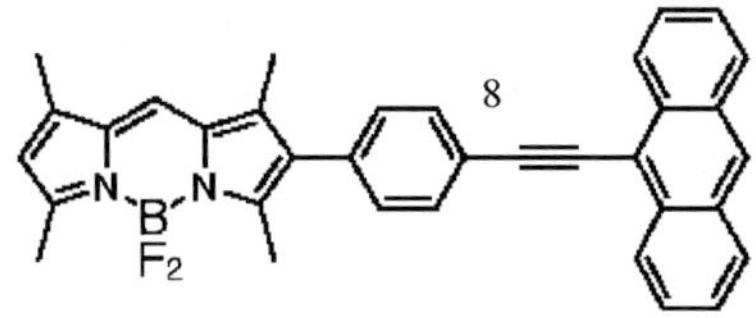

8

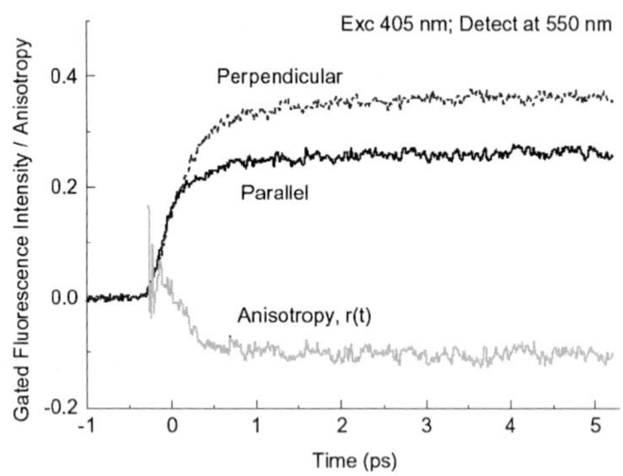

Figure 1: Gated Fluorescence traces for compound **1**, showing complete energy transfer in <1 ps over a distance of >10 Å despite the perpendicular orientation of donor and acceptor transition moments.

The other types of molecules studied (**12** & **17** here) contained a fluorescein donor and a rosamine acceptor. In contrast with the anthracene-BODIPY cassettes, the fluorescein-rosamine types showed ~10× slower energy transfer (i.e., ~6 ps), despite predominantly parallel donor and acceptor transition moments and similar molecular dimensions. An important difference between the two sets of cassettes is that, in a molecule such as **12**, both transition moments are perpendicular to the axis of the linker and are therefore in the least favorable orientation to couple with the principal polarizability axis of the

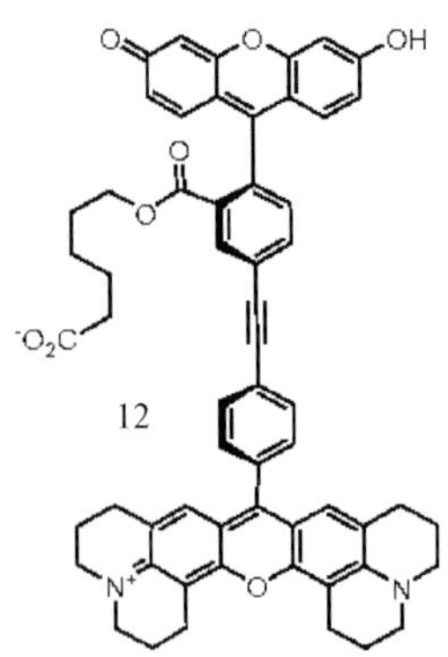

12

linker. On the other hand, anisotropy measurements showed that the alignment of the transition dipoles was predominantly parallel, so that the donor and acceptor moieties are in the optimum configuration for Förster coupling.

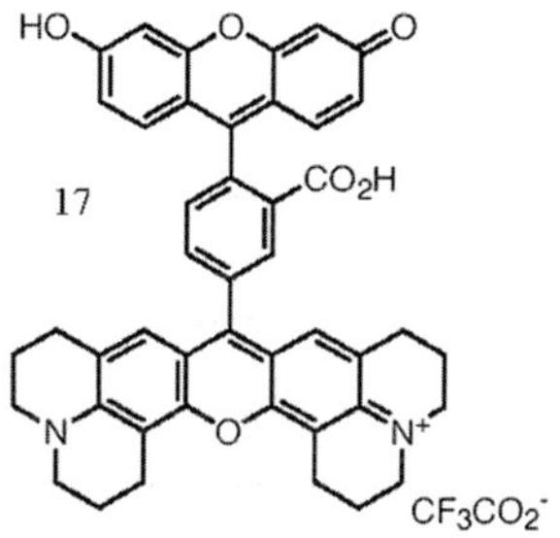

17

The absorption and emission spectra of compounds **12** & **17** are almost the same, although the donor-acceptor spacing was 15 and 9Å, respectively. Yet, removing the phenylethynyl component of the linker (see **17**) caused a 3× *increase* in the energy transfer time (6.3 → 20 ps), despite almost halving the donor-acceptor separation and retaining the parallel transition moments. This slower transfer rate evidently reflects a different transfer mechanism from that which operates in the more widely separated chromophores containing the phenylethynyl linkers. Indeed, without the linker, the energy transfer rate resembled that of the non-rigidly coupled dye cassette Big-ROX (τ~35 ps). In summary, the rigid, conjugated phenylethynyl linkers provide a route for ultrafast energy transfer coupling which also gives rise to polarized acceptor emission.

References

1. J. Ju, C. Ruan, C. W. Fuller, A. N. Glazer, and R. A. Mathies, Proc. Nat. Acad. Sci. 92 (1995) 4347.
2. J. Ju, A. N. Glazer, and R. A. Mathies, Nature Med. 2 (1996) 246.
3. L. G. Lee, S. L. Spurgeon, C. R. Heiner, S. C. Benson, B. B. Rosenblum, S. M. Menchen, R. J. Graham, A. Constantinescu, K. G. Upadhya, and J. M. Cassel, Nucleic Acids Res. 25 (1997) 2816.
4. C.-W. Wan, A. Burghardt, J. Chen, F. Bergström, L. B.-Å. Johansson, M. F. Wolford, T. G. Kim, M. R. Topp, R. M. Hochstrasser, and K. Burgess, Chem. Eur. J. 9 (2003) 4430.
5. G.-S. Jiao, L. H. Thoresen, and K. Burgess, J. Am. Chem. Soc. 125 (2003) 14668.
6. D. Holten, D. F. Bocian, and J. S. Lindsey, Accts. Chem. Res. 35 (2002) 57.
7. S. Speiser, Chem. Rev. 96 (1996) 1953.
8. N. Lokan, M. N. Paddon-Row, T. A. Smith, M. La Rosa, K. P. Ghiggino, and S. Speiser, J. Am. Chem. Soc. 121 (1999) 2917.
9. H. B. Lin and M. R. Topp, Chem. Phys. Lett. 48 (1977) 251.
10. T. Wilhelm, J. Piel, and E. Riedle, Opt. Lett. 22 (1997) 1494.
11. T. G. Kim, J. C. Castro, A. Loudet, G.-S. Jiao, R. M. Hochstrasser, K. Burgess, and M. R. Topp, J. Phys. Chem. A (2006) in press.

Femtochemistry VII
Fundamental Ultrafast Processes in Chemistry, Physics, and Biology
A.W. Castleman, Jr. and M.L. Kimble

Fluorescence of polythiophenes: the effect of intra- and interchain processes on the excited state dynamics

Ignác Bugár,[a] Michal Žitnan,[b] Dušan Velič,[a,b] Gabriel Čík[c]

[a] *International Laser Center, Ilkovičova 3, Bratislava 812 19, Slovakia*
[b] *Departement of Physical and Theoretical Chemistry, Faculty of Natural Sciences of Comenius University, Mlynská dolina CH1, Bratislava 842 15, Slovakia*
[c] *Departement of Environmental Engineering, Faculty of Chemical and Food Technology, Slovak University of Technology, Radlinského 9, Bratislava 812 37, Slovakia*

1. Introduction

Conductive polymers represent a very promising class of materials for application in electronic and optoelectronic devices. After the first demonstration of high conductivity of these polymers [1], intensive investigations of this type of materials started. Naturally, the optical methods play an important role in electronic property analysis of conductive polymers. Doped polythiophene based polymers exhibit high conductivities and are interesting for their high fluorescence quantum yield [2]. Tailoring of thiophene ring by alkyl side chain results in a derivative with advanced conductivity, solubility, and fluorescence properties. Even more, many relations were found between the fluorescence and structural properties of conducting polymers like polyalkylthiophenes (PATs). Macromolecular structure of PATs is very closely connected to the positions of alkyl side chains along the polymer backbone. The regioregularly placed side chains, as head-to-tail, result in the most planar structure with the smaller energy band gap. Irregularities, as head-to-head, of the side chain placement cause conformational defects on the backbone, affecting the electronic-optical properties. Hypsochromic shift by increasing irregularity was also shown for a different type of PATs [3]. Fluorescence properties of poly(3-dodecylthiophene) (PDDT) with different regioregularities were initially investigated in solution [4]. In this work it has been shown that the frequency of structural defects can affect the intrachain exciton hopping on the base of time-resolved study. Studying different types of solution samples, fluorescence spectroscopy proved to be a useful tool for analysis of PDDT intrachain structural properties. In the case of solid state PDDT, it is possible to extend the characterization to the structural determination due to the interchain interaction of polymer chains. The fluorescence techniques might be applicable also for study of interchain structural properties of polythiophenes. The results of steady-state and time-resolved fluorescence spectroscopy will be used to describe the domain structure of solid PDDT.

2. Results and Discussion

PDDT was synthesized by Grignard and oxidative polymerization methods resulting in 61% and 74% head-to-tail dyads, respectively. Solid films were prepared from chloroform solutions by spin coating on 1mm fused silica plates. Fluorescence properties of the prepared samples were studied by using both steady-state and time-resolved techniques. The steady-state fluorescence measurements were performed on a JOBIN YVON Fluorolog 3-11 spectrofluorimeter with simultaneous measurements of fluorescence excitation and emission spectra. Time-resolved fluorescence was performed on a commercial femtosecond laser based up-conversion fluorescence apparatus (CDP). Second harmonics of a Ti:sapphire laser was used to excite the rotating sample, and the fluorescence signal was collected from the opposite site of the sample plate. The sampling up-conversion BBO crystal provides by its simple rotation also frequency resolution of fluorescence. Time resolution of the apparatus is 100 fs with maximum time interval up to 600 ps. All measurements were realized at room temperature. The steady-state absorption and emission spectra of PDDT solution and thin film are shown in Fig.1.

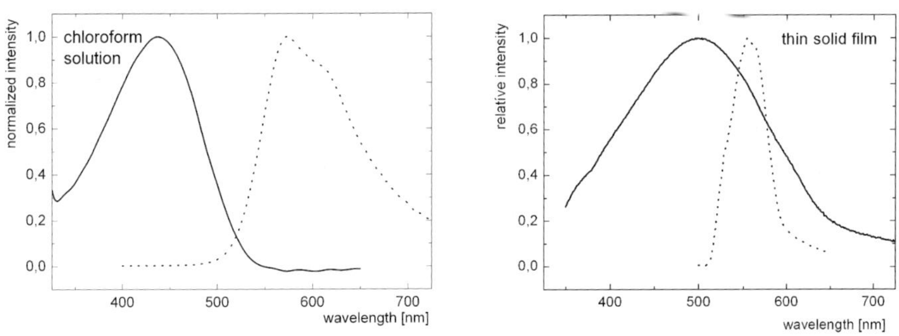

Fig. 1 Steady-state absorption (solid line) and fluorescence (dotted line) spectra of poly(3-dodecylthiophene) in solution and in solid film.

Comparing the absorption spectra, red shift and spectral broadening were observed in the case of the solid sample. The emission spectra were changed in the opposite course, exhibiting a blue shift and spectral narrowing. In the case of polymers, the dominant spectral broadening effect is most probably due to the chain-segment length distribution of the sample. Different segment lengths with non-perturbed conjugations pose different excitation energies, affecting the spectral shape of the absorption and fluorescence bands. Comparing the solution and solid absorption bands, the observed broadening suggests an extension of segment lengths to longer ones. This phenomenon might be caused by an aligning process of the polymer structure, which is not present in solution. The blue shift and spectral narrowing, in the case of the solid film, revealed the fluorescence quantum efficiency is segment length selective. The solid state spectral shape suggests a nonradiative relaxation channel, with higher probability for shorter segments. Segment length selective fluorescence dynamics were confirmed by further steady-state fluorescence studies. Excitation wavelength dependence of the fluorescence band shape was registered in the case of PDDT solid samples with different regioregularity and thickness. All investigated samples exhibited a spectral split enhancement by increasing excitation wavelength. The steady-state fluorescence of the sample with the strongest spectral splitting effect is shown in Fig. 2. At excitation wavelengths longer than 460 nm, two spectral bands occurred with fluorescence minimum at around 580 nm. The time-resolved fluorescence curves were characterized at each wavelength by time τ_e when the initial fluorescence intensity decreased to 1/e. The solution samples exhibited monotonic dependence of τ_e on wavelength in the 40 – 140 ps range. In the case of thin solid films, faster relaxation processes were observed with weaker wavelength dependence in the 20 – 60 ps range. The wavelength dependence graphs of two solutions and solid samples with different regioregularity are shown Fig. 2-right.

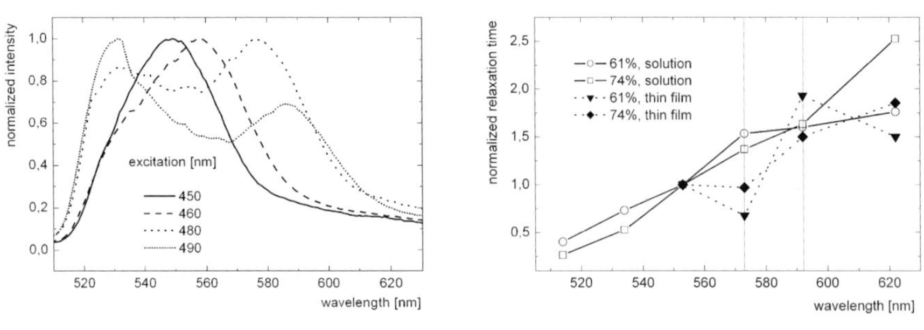

Fig.2. Left: excitation wavelength dependent fluorescence of thin film of PDDT, Right: dependence of fluorescence lifetime on wavelength for samples with different regioregularity.

For better comparison, the τ_e values were normalized for the shortest wavelength of solid samples, 553 nm. The $\tau_e(\lambda)$ dependence of the solid samples exhibited a step change in the 573 – 592 nm ranges, in contrast to the solution samples. An important finding is that the steady-state curves split into two bands in the same wavelength range.

3. Conclusions

On the basis of the steady-state and time-resolved fluorescence spectra, the short time relaxation processes of PDDT are compared in the case of separated molecules in solution and polymeric structures in solid samples. The observed absorption red shift is connected to the enhanced ordering in the solid sample. At the same time, the short wavelength range is not restricted, suggesting that only parts of the polymer chains are involved in the ordering process. The intrachain and interchain exciton hopping process are generally accepted relaxation phenomenon in the case of PATs. In solutions, only the intrachain exciton hopping can be supposed. The observed faster relaxation time of solid PDDT revealed that, in this case, further relaxation channels occur, probably through interchain processes. Both the steady-state and time-resolved results suggest that a segment length selective nonradiative relaxation channel is present in the solid state. According to the observed steady-state spectral splitting and step change of the $\tau_e(\lambda)$ dependence, a concept of domain structures is proposed, which influences the interchain exciton hopping process. The two registered spectral bands with two different relaxation constants originate from two different types of interchain structures. The long wavelength band with τ_e around 60 ps is connected to domains with parallelly ordered chains and the short wavelength band with τ_e around 20 ps to the interdomain space with lower ordering and more broken chains. The system of randomly oriented domains containing ordered polymer structures was observed also by other structural methods. We have shown the fluorescence spectroscopy has a potential to distinguish between the different interchain structures.

Acknowledgement

This work was supported by grants: APVT 298, VEGA 1/0296/03, VEGA 1/2018/05 and VEGA 1/8109/01.

References

1. H. Shirakava and S. Ikeda, Polymer 2 (1971) 231.
2. H.S.O. Chan and S.C. Ng, S.C., Prog. Polym. Sci. 23 (1998) 1167.
3. W.R. Salaneck, Contemp. Phys. 30 (1989) 403.
4. I. Bugar, J. Kovac, K. Matuszna, V. Lukes, and G. Cik, Laser Phys. 14 (2004) 527.

Femtochemistry VII
Fundamental Ultrafast Processes in Chemistry, Physics, and Biology
A.W. Castleman, Jr. and M.L. Kimble

Probing and controlling ultrafast dynamics of a molecular switch

Janne Savolainen, Riccardo Fanciulli, Niels Dijkhuizen and Jennifer L. Herek

FOM Institute for Atomic and Molecular Physics (AMOLF) Kruislaan 407, 1098 SJ Amsterdam, The Netherlands

1. Introduction

Diarylethene derivatives are a promising group of photo- and electrochromic molecular switches for applications to a wide range of optoelectronic devices [1,2]. (See Fig. 1) Upon excitation with UV light, a cyclization reaction occurs to a thermally stable system, resulting in an extended π-conjugation that shifts the absorption towards the red. Cycloreversion (ring opening) is induced by exciting the molecules with visible light. The stability, spectral properties, switching rates, and quantum yields can be passively controlled by the choice of substituents. Our goal is to actively control the switching dynamics using shaped femtosecond laser pulses. Optimization of the ring opening reaction and analysis of the optimal pump pulses will enable us to better understand the switching mechanisms, providing details of the potential energy surfaces and the role of specific vibrational motions.

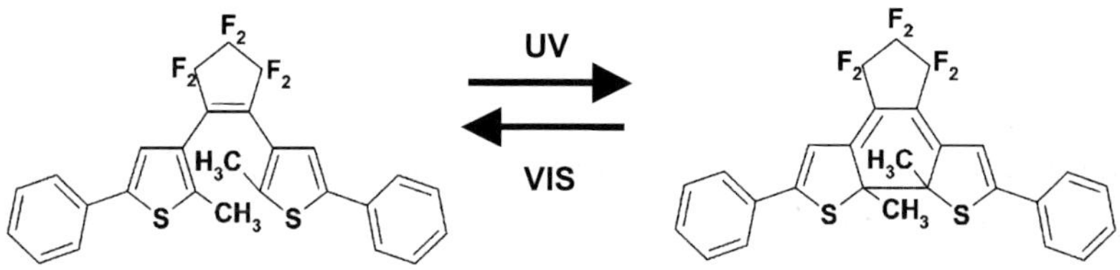

Fig. 1: Photochromic switch bis(2-methyl-5-phenylthiophen-3-yl)hexafluorocyclopentene. By illuminating the open form with UV light, these molecules convert to the closed form. As a result, the π-conjugation extends throughout the molecule producing a new red shifted absorption band in the visible. The ring structure can be broken again by visible photons so that the molecules return to their open form.

2. Experiment

We have developed a setup for adaptive femtosecond pulse shaping with high-resolution phase and amplitude control, which is implemented in a feedback-driven closed-loop experiment [3]. This way, many different pulse shapes are evaluated by the sample and the best pixel voltage patterns of the shaper are selected to reproduce. The loop proceeds until convergence on the target objective is reached and the best pulse shape is found.

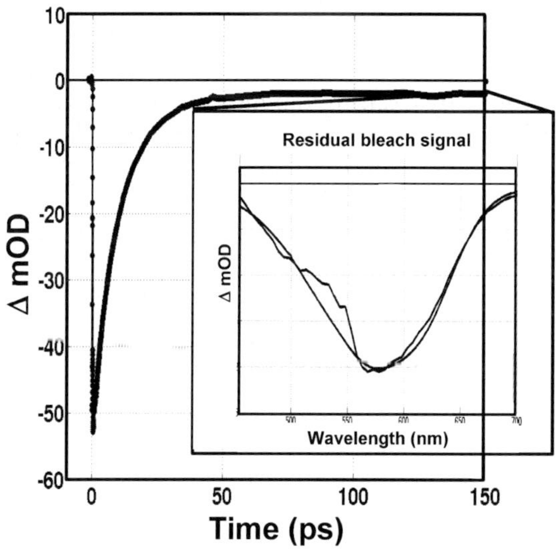

Fig. 2. Recovery dynamics of transient bleach signal from the closed switch. Inset shows corresponding spectrum at 150 ps, overlapped with the (inverted) steady-state absorption spectrum. The discrepancy on the blue edge is due to scatter of the excitation laser.

The sample was excited with originally ~25 fs near-transform-limited laser pulse from a non-collinear optical parametric amplifier (NOPA) centered at 570 nm. The shaping of these pump pulses was done in a pulse shaper that is comprised of two gratings (1200 g/mm), folding mirrors, cylindrical mirrors (f ~540 mm), as well as a 640 pixel double layer spatial light modulator (CRi SLM-1280).

White light continuum, created by focusing a small fraction of the 775 nm output of the Clark CPA2001 laser system to a 2 mm sapphire plate, was used as a probe to record the transient absorption spectrum of the molecule with a broad spectral range (~450-700 nm). A home-built dual-256-diode-array camera operating with a 1 kHz repetition rate was used to detect the white light after the sample. Spectral resolution of the spectrograph was ~1 nm/pixel. Each laser shot was AD-converted, and an average of 2000 shots was used as a measurement point. The 20 different pulse shapes of each generation were averaged 3 times before obtaining the fitness values.

3. Results and Discussion

Pump-probe studies of the reaction dynamics following excitation of the closed switch provide the foundation for the control experiments. The majority of the bleach signal at 570 nm recovers with a time constant of ~10 ps (see Fig. 2). A small residual signal corresponds to the fraction of molecules that undergo switching to the open conformer. This transient signal at time delays >50 ps is ideal for feedback in the optimization experiments aimed at controlling the efficiency of the ring opening reaction, as it correlates directly with yield. However, in the specific switch studied here, the magnitude of this signal is prohibitively small.

We estimate that the efficiency of the ring-opening reaction is no more than 3% (in contrast, the quantum yield of the cyclization reaction is close to 1 [4]. Attempts to optimize this small signal, thereby optimizing the efficiency of ring opening, were unsuccessful due to signal-to-noise limitations on the learning algorithm. The algorithm could not distinguish between the effectiveness of different pulse forms because variations of the fitness values were overshadowed by the noise. Further experiments on other photochromic switches with higher ring-opening efficiency are currently underway.

Nonetheless, by changing the target goal, we succeeded in optimizing another important aspect of the system: the initial population in the excited state. The target in this optimization was to maximize the excited state absorption signal seen at early times (feedback taken at 6 ps), which corresponds to the population in the S_1-electronic state of the closed conformer. In this phase and amplitude optimization, the integrated overlap between the absorption spectrum and the shaped pump spectrum was used as a scaling factor to account for changes in feedback due to differences in the number of absorbed photons.

In Fig. 3, the average fitness value of each generation (comprised of 20 different laser pulse shapes) is shown. The zero-generation point shows the fitness obtained with an unshaped (near transform-limited) pulse. The initial decrease in the first generation reflects the random nature of the created pulse shapes. Gradually, the fitness improves until a 20% increase of the S_1 signal

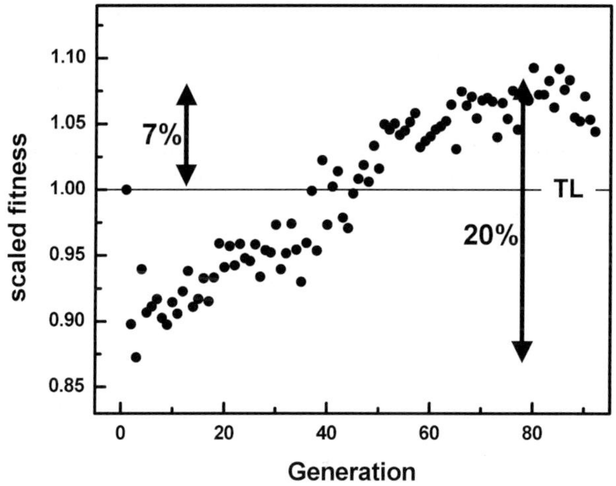

Fig. 3. Learning curve for optimization of population transfer. The data are normalized to the fitness value obtained for excitation with transform-limited pulses (first point).

compared to the initial trial is reached. Relative to the unshaped pulse (first point), the enhancement is 7%.

Analysis of the corresponding pulse shapes can reveal the mechanism of the control interaction. Measured spectrographs (X-FROG traces) at various stages of the optimization reveal complex patterns in time and spectrum. In general, the shaped pulses are not shifted in time and each has approximately the same actinic energy (integral of excitation and absorption profiles). Furthermore, the observed optimization is not simply the result of spreading the pulse energy in time to avoid saturation effects, as confirmed by a series of measurements in which the learning curve is measured by reapplying the best pulse shapes of each generation as a function of excitation intensity. We observe the same effect, even when using the signal of a different spectral band (e.g. the bleach signal) as feedback. More detailed analysis of the pulse shapes, including extraction of the spectral phase, is currently underway in order to establish a better understanding of the underlying mechanism(s) of control over population transfer in this molecular system.

We would like to thank Audrius Pugzlys, Jaap de Jong and Ben Feringa from the University of Groningen, The Netherlands for providing the samples and sharing their expertise and knowledge with us.

References

1. M. Irie, Chem. Rev. 100 (2000) 1683.

2. B. L. Feringa (Ed.), Molecular Switches, Wiley VCH, Weinheim, 2001.
3. H. Rabitz, R. de Vivie-Riedle, M. Motzkus, and K. Kompa, Science 288 (2000) 824.
4. P. R. Hania, R. Telesca, L. N. Lucas, A. Pugzlys, J. van Esch, B. L. Feringa, J. G. Snijders, and K. Duppen, J. Phys. Chem. A 106 (2002) 8498.

3. Solvation Phenomena

Femtochemistry VII
Fundamental Ultrafast Processes in Chemistry, Physics, and Biology
A.W. Castleman, Jr. and M.L. Kimble

Solvation and Photochemical Funnels: Environmental Effects on Conical Intersection Structure and Dynamics

Irene Burghardt[a], Riccardo Spezia[a] and James T. Hynes[a,b]

[a]*Département de Chimie, Ecole Normale Supérieure, 24 rue Lhomond, F-75231 Paris cedex 05, France*
[b]*Department of Chemistry and Biochemistry, University of Colorado, Boulder, CO 80309-0215, USA.*

1. Introduction

It is only relatively recently that it has been appreciated that conical intersections (CIs) play a critical if not dominant role in the passage from photoexcited states to other excited states and – of most interest in the present context – from photoexcited states to the ground electronic state [1]. In particular, CIs serve as electronic 'funnels' allowing often extremely rapid excited state (S_1) → ground state (S_0) nonadiabatic transitions, involving times as short as several tens of femtoseconds. In view of their central role in ultrafast photochemical processes, CIs have been termed 'transition states' for these types of dynamics. Because of their importance, CIs have been the subject of many theoretical (often computational) studies, most commonly focused on isolated molecules, with relatively little effort addressed to CIs for chromophores either in a solution environment or in the protein nanospaces constituting the environment in photobiological situations. In the present brief contribution, we review a few key aspects of our own recent efforts to construct a theoretical approach for the influence of an environment on the approach to and passage through CIs. The interested reader is directed to the original papers [2] for more extensive discussion and referencing.

The influence of an environment on CIs should be most pronounced when there is a significant charge rearrangement in the approach to and passage through a CI and the chromophore is surrounded by an environment containing polar molecules (or polar and charged groups). Strong chromophore-environment electrostatic interactions will then markedly influence both CI structure and dynamics. We have commenced our studies with such a case, detailed further within: the photochemical cis-trans isomerization in protonated Schiff bases (PSBs) (Fig. 1), often taken as models for aspects of the retinal photoisomerization in vision. Our initial choice for the environment of the PSB is a polar solvent, modeled in a dielectric continuum fashion and incorporating nonequilibrium solvation effects. The electronic structure aspects of the problem are described in a Valence Bond approach, which both allows a simple physical interpretation of the charge evolution and a convenient way to include the polar solvent environment. In the following, we outline the theoretical model's essential ingredients and describe the main results for the environment's influence on the CI structure, on paths leading to the CI from the Franck-Condon region, and on the nonadiabatic $S_1 \rightarrow S_0$ transition dynamics.

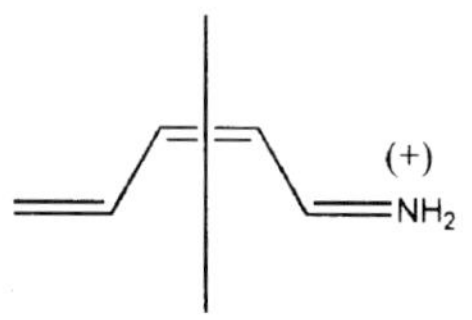

Fig. 1. PSB representative of this study, $C_5H_6NH_2^+$, which has been addressed as a minimal model for retinal isomerization [4].

2. PSB in vacuo

We begin with our description of the isolated PSB molecule and its CI. Fig. 1 displays the PSB molecule used to construct our model from *ab initio* calculations; the vertical line identifies the bond about which photoinduced isomerization is considered to occur. We employ two VB states describing two electronic configurations having fixed charge distributions, corresponding to different locations of the positive excess charge in the PSB. In particular, partitioning the PSB in terms of the left and right sides of the C=C double bond about which isomerization occurs (Fig. 1), we identify first the VB state $|AB\rangle$ with the positive charge on the right side (RHS) containing the N atom (a dot-dot configuration employing the formalism of Bonacic-Koutecky, Koutecky and Michl (BKM) [3]) and second the state $|B^2\rangle$ with the positive charge on the other side (LHS) (the hole-pair configuration in the BKM formalism). This corresponds to a representation in a basis of localized p orbitals that are able to describe a biradicaloid structure at the twisted geometry, a key aspect of the CI [3]. The expansion of the adiabatic ground and excited state wavefunctions in this VB basis is then

$$\Psi_{g,e} = c_{g,e}^{AB}(r,\theta)\psi^{AB}(r,\theta) + c_{g,e}^{B^2}(r,\theta) + \psi^{B^2}(r,\theta) \tag{1}$$

with the geometry-dependent coefficients {c} for the vacuum problem. We have anticipated here that, based on the work of Robb, Olivucci, and coworkers [4], there are two key chromophore coordinates for a first approach to the PSB problem, a stretching-compression coordinate r associated with the C atom skeleton in-plane motion (a bond length alternation (BLA) coordinate), and a torsional coordinate, θ, corresponding to the rotation leading to isomerization. (A more rigorous description is discussed in Ref. [2]). The diabatic model for the coupled energy surfaces then has the form

$$V(r,\theta)=\begin{pmatrix} V^{B^2}(r,\theta) & \gamma(\theta) \\ \gamma(\theta) & V^{AB}(r,\theta) \end{pmatrix} \tag{2}$$

where V^{AB} and V^{B2} are the potential energy functions of the $|AB\rangle$ and $|B^2\rangle$ VB states, respectively, which depend on both of the PSB internal coordinates, and where $\gamma(\theta)$ is the electronic coupling between them depending only on the θ coordinate. Diagonalization of the diabatic matrix eq. 2 gives the ground (S_0) and first excited (S_1) electronic surfaces as

$$V_{g,e}=\frac{V^{B^2}(r,\theta)+V^{AB}(r,\theta)}{2}\mp\frac{1}{2}\sqrt{\left[V^{AB}(r,\theta)-V^{B^2}(r,\theta)\right]+4\gamma(\theta)^2} \tag{3}$$

In the PSB modelling [2], the VB state potential energies depend on the internal coordinates via various contributions. For the |AB> state, $V^{AB}(r,\theta)=V^{AB}{}_{morse}(r)+V^{AB}{}_{shift}(r,\theta)+V^{AB}{}_{\theta}(\theta)$, where $V^{AB}{}_{morse}(r)$ is a Morse function in the stretching-compression coordinate, $V^{AB}{}_{shift}(r,\theta)$ shifts the Morse curve minimum to its appropriate value at $\theta=90°$, and $V^{AB}{}_{\theta}(\theta)$ is the diagonal torsion contribution in the torsion, with a $\cos^2$ dependence. The explicit forms of V^{AB} and V^{B2} are given elsewhere [2]. The V^{B2} potential has the same form as V^{AB}, but with two additional terms, an energy shift, ΔE^{B2}, and a localized Franck-Condon (FC) term, which makes the S_1 surface bound in θ in the FC region (see below). The electronic coupling term, $\gamma(\theta)$, depends simply on the torsional coordinate, $\gamma(\theta)=c_\gamma\cos\theta$. The cosθ dependence arises from a resonance integral proportional to the overlap between the p orbitals on the neighboring centers [3,5]. The vanishing of this coupling at $\theta = 90°$ leads to a CI.

The model eq. 2 was parametrized in order to globally reproduce ground and excited state surfaces of PSB models in vacuum [4,6]. It improves the previous parametrization [2(a)] – which considered only a region close to the S_0/S_1 CI – to account for the FC S_1 surface region. In particular, this surface is now bonding in θ in the FC region, in accord with recent *ab initio* calculations [4], while it becomes non-bonding for larger r values, facilitating the twisting motion towards the CI. Figure 2 shows the resulting ground and excited state surfaces, where the CI is indicated; this will provide a useful reference for the situation in solution.

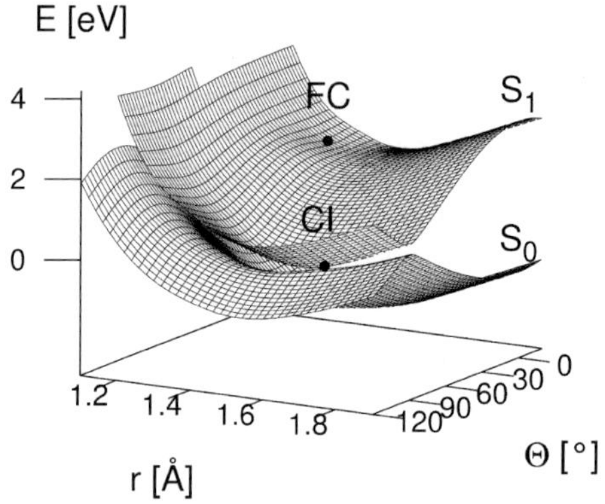

Fig. 2. Ground state (S_0) and excited-state (S_1) adiabatic surfaces for the isolated chromophore. The Franck-Condon (FC) geometry and CI point are marked.

We noted in Sec. 1 that the PSB isomerization is characterized by a charge rearrangment. We have analyzed this in detail [2(a)] via the two VB state description above, with the following qualitative results. The S_0 charge distribution in the ground state cis geometry is approximately that of $|AB\rangle$, with the positive charge on the RHS of the double bond (cf. Fig. 1). The S_1 charge distribution produced in the FC transition is approximately that of $|B^2\rangle$, i.e. the positive charge is now on the LHS of the PBS. Finally, the ground state charge distribution in the trans geometry is again largely that of $|AB\rangle$, with the positive charge now back on the molecular RHS. (These characterizations are in qualitative agreement with the *ab initio* results of Robb, Olivucci, and coworkers [4].) Thus, thinking of the charge rearrangements as charge transfers, there is a charge transfer in the FC transition, followed by a reverse charge transfer in reaching the ground state isomerization product, which as we will see occurs in the CI region.

3. PSB interacting with a polar/polarizable environment

In our model, the solvent is described as a dielectric continuum in terms of a solvent orientational polarization field interacting with the changing electric field generated by the PSB solute. We use the simplified description where the electronic polarization of the solvent is assumed to be instantaneously equilibrated to any solute charge distribution (see Refs. [2] for further details). On the other hand, the solvent orientational polarization is not generally in equilibrium with the solute charge distribution ; the solvent molecules' dipole moments move on a slower time scale than that distribution. This is an important aspect to account for since, as we noted in Sec. 1, the relevant time scales can be as short as a few tens of femtoseconds (fs), with some hundreds of fs being a common time scale. One should note that this important nonequilibrium aspect is not addressed in the often employed equilibrium solvation methodologies for electronic structure in solution.

We express the nonequilibrium orientational polarization field, $\mathbf{P}_{or}(\mathbf{x})$, where $\mathbf{x}$ indicates a field point in the solvent, in terms of the two equilibrium polarization fields, $\mathbf{P}_{or,eq}{}^{AB}(\mathbf{x};r,\theta)$ and $\mathbf{P}_{or,eq}{}^{B2}(\mathbf{x};r,\theta)$, corresponding to the solvent polarization fields in equilibrium with the two VB states describing the PSB charge distribution. Just as for chemical reactions in polar solvents [7], the solvent nonequilibrium polarization field is expressed as a linear combination of these polarizations, $\mathbf{P}_{or}(\mathbf{x};r,\theta)=(1-z)\ \mathbf{P}_{or,eq}{}^{B2}(\mathbf{x};r,\theta)+z\ \mathbf{P}_{or,eq}{}^{AB}(\mathbf{x};r,\theta)$. The solvent coordinate z then identifies the (generally) non-equilibrium solvent orientational polarization state. For small z, the polarization is more similar to that appropriate to the B^2 state, whereas for z close to unity, it is more similar to that appropriate to the AB state. With the previous linear combination of the two equilibrium polarization fields, a free energy representation results which has the same structure as the vacuum eq. (2), with the addition to the diagonal potential energy of each VB state a solvation free energy G_s for that VB state:

$$\begin{aligned} G_s^{AB}(z;r,\theta) &= G_{eq}^{AB}(r,\theta)+\frac{1}{2}k_s z^2 \\ G_s^{B^2}(z;r,\theta) &= G_{eq}^{B^2}(r,\theta)+\frac{1}{2}k_s(1-z)^2 \end{aligned} \tag{6}$$

Here for each VB state, G_{eq} is the solvation free energy when the solvent is equilibrated to that VB state charge distribution, and the parabolic term involving a solvent force constant k_s measures the free energetic cost of a nonequilibrium solvation situation in the presence of that charge distribution, with its extent characterized by the solvent coordinate z. Thus, for example, a z value close to 0 characterizes a solvent polarization close to the appropriate equilibrium value for the |AB> charge distribution but far from the equilibrium value for the $|B^2>$ charge distribution; the situation is reversed for a z value close to 1. When complemented with the kinetic energy associated with the solvent coordinate [2(c)], as well as that associated with the molecular coordinates r and θ, the description has the character of an effective Hamiltonian description for the combined chromophore-solvent system, for which no dissipation is included. The solvent dynamics are in effect described at the inertial level, familiar from solvation dynamics studies [8].

4. CI Structure and Minimum Reaction Path Results

We begin this brief recounting of our principal results with Fig. 3, which shows that, in a solvent environment, the single CI point for the isolated system in Fig. 2 becomes a *seam* of CIs as a function of the solvent coordinate z and the BLA coordinate r, always at the coupling torsional coordinate location θ

= 90° where the coupling between the two VB states (and thus the adiabatic S_1 and S_0 states) vanishes. The degeneracy required for a CI to exist is achieved as a function of the two tuning coordinates r and z by the cancellation of the solute energy and the solvent free energy contributions to the free energy gap between S_1 and S_0 at θ = 90°. Interestingly, this CI seam condition (which is given analytically in Refs [2]) cannot be satisfied for all values of the solvent coordinate. In particular, there is *no* CI for the spectroscopically relevant case where initially the solvent is equilibrated to the ground state S_0 charge distribution and a Franck-Condon (FC) transition to S_1 occurs at the FC geometry. (A refined description of this situation may be found in Ref. [9]). This is a clear indication that the motion in the solvent coordinate is required to reach the CI seam.

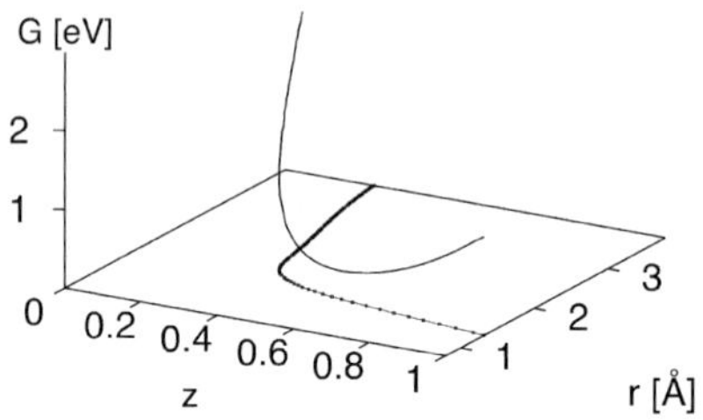

Fig. 3. The CI seam in the (r,z) plane at θ=90° for the combined chromophore plus solvent system described by the potential Eq. 2 in conjunction with the solvent free energies Eq. 6.

We have examined certain aspects of the approach to the CI seam from the FC -accessed S_1 initial condition by calculating the Minimum Energy Path (MEP), the mass-weighted steepest descent path on the G(r, θ, z) S_1 excited state free energy surface. This is a generalization to the excited state of solvent coordinate dependent MEPs for ground state reactions and is also a generalization of the interpretation of Robb, Olivucci, and coworkers [4] for isolated chromophore excited state CI problems to include a solvent coordinate. Although, modulo mass scaling features, the MEP is determined solely by the surface topology and does not bear any necessary relation to the system dynamics, it should reflect certain features of the dynamic process. For perspective, we start with the MEP for the isolated chromophore, whose characteristics are shown in Fig. 4. The behavior is consistent with that found in ab initio studies by Robb, Olivucci, and coworkers [4] ; there is a rapid initial motion in the BLA coordinate as the FC region where the potential in theta is bound is left. The torsional angle initially increases slowly but then follows an accelerating movement towards the CI point θ =90° in the course of the isomerization. Features of the charge rearrangement are illustrated in Fig. 4b in terms of the weight of the $|B^2\rangle$ VB state in the S_1 wavefunction before the CI is reached and in the S_0 wavefunction after the CI (see the wave function decomposition Eq. 1). This portrays the overall evolution of the charge distribution in three stages: (i) adiabatic S_1 charge distribution in the FC region, with a predominant $|B^2\rangle$ character, (ii) polarization of the charge distribution towards an exclusive $|B^2\rangle$ component in the immediate vicinity of the CI (for r less that r^{CI}, the r value at the CI point), and (iii) evolution towards the S_0 trans minimum, predominantly of $|AB\rangle$ character. As we noted in Sec. 2, the net effect of the $S_1 \rightarrow S_0$ transition is a shift of the positive charge from the carbon moiety – which was produced in the initial $S_0 \rightarrow S_1$ FC transition – to the nitrogen-containing end of the PSB. Aspects of the calculated MEP for the PSB in solution are shown in Figs. 6 and 7 for two artificial model solvents, one 'fast' and one 'slow', with respectively solvent masses ~ 0.05 and ~2.5 that of acetonitrile, selected to highlight the influence of the solvent time scale on the MEP. Figure 5 is the solution analogue of Fig. 4a for the isolated PSB. The relative time scales of the torsional coordinate and the solvent coordinate determine at which point the MEP reaches the CI seam.

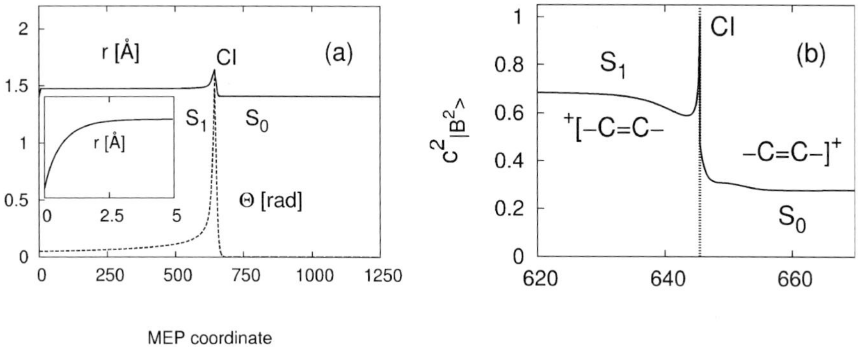

Fig. 4 Minimum energy path for the isolated chromophore for an initial condition at the FC geometry. (a) S_1 evolution of (r,θ) components towards the CI point, and beyond this, S_0 path towards the ground state geometry. (b) Squared eigenvector component, $c^2_{|B2>}$, calculated along the MEP (zoomed on the vicinity of the CI) illustrating a polarization effect at the CI and the net charge transfer S_1 –> S_0.

While the θ motion is qualitatively the final motion in reaching the seam in the light solvent case, motion in z is the final motion in the heavy solvent case (in this latter case, the condition θ = 90° is *not* sufficient as an indicator for reaching the seam). A general feature of these two MEPs is that the CI seam is reached in the sequence corresponding to the time scales of the coordinates in the order: first, the fastest, second, the next fastest, and third, the slowest. It is interesting to observe that this is the *opposite* sequence for reaching a transition state in an activated chemical reaction [10]. This reversal has its origin in the fact that the latter involves an ascent in free energy, while the CI case involves a descent. Figure 6 gives a different representation of the MEP (without the torsional angle) consistent with the discussion just given. The different roles of the solvent coordinate in the MEP for the light and heavy solvent cases is especially to be noted. We have also indicated here the location of the minimum (free) energy CI (MECI); this representation does not address any $S_1 \rightarrow S_0$ transitions, and the path towards the MECI in part evolves along the CI seam.

We next turn to dynamical calculations including such transitions, and we will see how the MEP does reveal some important aspects of the actual dynamics.

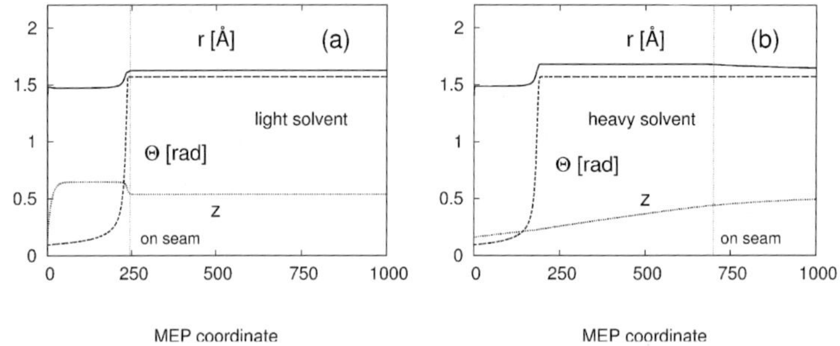

Fig. 5. MEP for the combined chromophore-solvent system, for a light vs. heavy model solvent. For an initial condition at the FC geometry, the evolution of the (r,θ,z) components is showm.

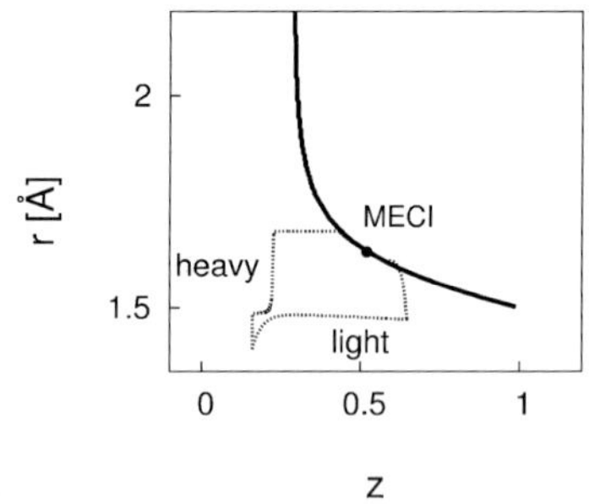

Fig. 6. Projection of the MEP evolution of Fig.5, for light vs. heavy model solvents, onto the $(r,z)_{\theta=90^\circ}$ plane. The CI seam and the minimum energy CI (MECI) are shown ; note that the MECI coincides with the global minimum of the S_1 surface.

5. Dynamical calculations

We have carried out a number of dynamical calculations to explore the influence of the solvent on the passage to and through the CI for the light and heavy solvent cases introduced above. [2(b,c),11]. Here we discuss a few of these results calculated via a multi-configuration time dependent Hartree methodology [12]. Figure 7 shows the time decay of the excited S_1 state population, in which it is seen that the decay is markedly more rapid for the light solvent case. Aspects of the passage from the FC initial configuration in S_1 to the CI seam are shown in Fig. 8, which displays the dynamics in terms of the averages of the BLA and solvent coordinates over the time dependent distributions. Fig. 8a for the heavy solvent case shows that, modulo some oscillations in the BLA coordinate, the approach to the CI seam is quite similar to the image portrayed by the MEP in Fig. 6. On the other hand, the situation is more complex for the light solvent case, Fig. 8b. Initially, there is a certain similarity with the MEP in Fig. 6, in that the dynamical approach to the seam involves significant solvent coordinate motion. However, subsequently there is continuing and very considerable large amplitude solvent motion. This reflects a shortcoming of the present description, namely there is no dissipative component in our effective Hamiltonian description which would 'damp' this extreme solvent motion. Further, although we do not show it here, the torsional angle distributions distribution can continue to spread in time, as opposed to converging to a bimodal distribution with components confined to the S_0 ground state trans and cis well mimima regions. Again, this is a consequence of a lack of dissipation effects, this time, on the torsional coordinate. The inclusion of such dissipative effects is currently underway. Nonetheless, the present results already underline the considerable importance of the solvent dynamics on both the approach to and passage through the CI.

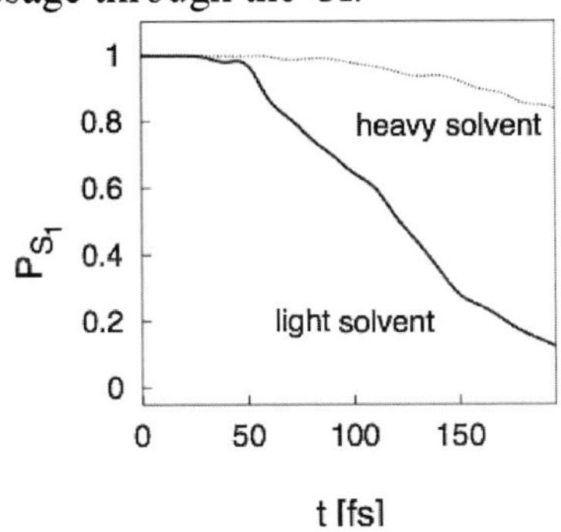

Fig. 7. S_1 adiabatic populations as a function of time, from a quantum dynamical calculation on the (r,θ,z) free energy surface. The population decay depends very markedly on the light or heavy nature of the solvent, which determines the time scale on which the CI seam region is reached.

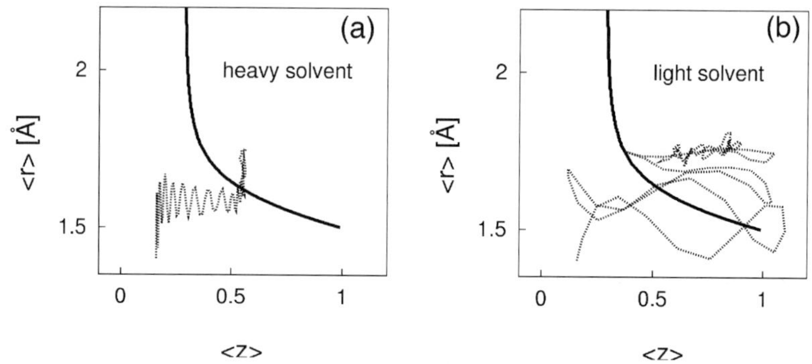

Fig. 8. Time-evolving quantum expectation values of the BLA and solvent coordinates, represented in the (r,z) planc. (a) heavy solvent case, for which the solvent motion is the rate determining step, (b) light solvent case, for which the CI region is reached rapidly. (Note that the distributions in (r,z) remain comparetively compact, so that consideration of expectation values is meaningful.)

6. Concluding Remarks

Here we have provided a brief account of our efforts to construct a theoretical framework for understanding and predicting the important influence of an environment on conical intersection structure and dynamics for ultrafast photoreactions. Much remains to be done, even for the solvent environment case. As just noted, dissipative effects will be incorporated in the description. Further, the continuum description of the solvent will be generalized to a microscopic description. In the photobiological context, a corresponding program is underway, with a first step of treating the local inhomogeneous protein nanospace for the chromophore at a Poisson-Boltzmann level.

This work was supported in part by a CNRS/DFG collaboration project and by NSF grant CHE-0808314. We thank Lorenz Cederbaum, Todd Martínez, Horst Köppel, Evgeniy Gromov, Damien Laage and Monique Martin for valuable discussions.

References

1. (a) W. Domcke and G. Stock, Adv. Chem. Phys. 100 (1997) 59; (b) D.R. Yarkony, J. Chem. Phys. 100 (1996) 18612; (c) W. Schoenlein, L. A. Peteanu, R. A. Mathies, and C. V. Shank, Science 254 (1991) 412; (d) U. K. Genick, S. M. Soltis, P. Kuhn, I. L. Canestrelli, and E. D. Getzoff, Nature 392 (1998) 206.
2. (a) I. Burghardt, L. S. Cederbaum, and J. T. Hynes, Faraday Discuss. Chem. Soc. 127 (2004) 395; (b) I. Burghardt, L. S. Cederbaum, and J. T. Hynes, Comp. Phys. Comm. 168 (2005) 95; (c) R. Spezia, I. Burghardt, and J. T. Hynes, Mol. Phys. in press.

3. V. Bonacic-Koutecky, J. Koutecky, and J. Michl, Angew. Chem. Int. Ed. Engl. 26 (1987) 170.
4. (a) M. Garavelli, P. Celani, F. Bernardi, M. A. Robb, and M. Olivucci, J. Am. Chem. Soc. 119 (1997) 6891; (b) T. Vreven, F. Bernardi, M. Garavelli, M. Olivucci, M. A. Robb, and H. B. Schlegel, J. Am. Chem. Soc. 119 (1997) 12687; (c) M. Garavelli, F. Bernardi, M. Olivucci, T. Vreven, S. Klein, P. Celani, and M. A. Robb, Faraday Discuss. Chem. Soc. 110 (1998) 51; (d) R. Gonzalez-Luque, M. Garavelli, F. Bernardi, M. Merchand, M. A. Robb, and M. Olivucci, Proc. Natl. Acad. Sci. USA 97 (2000) 9379.
5. R. M. Weiss and A. Warshel, J. Am. Chem. Soc. 101 (1979) 6131.
6. R. R. Birge and L. M. Hubbard, J. Am. Chem. Soc. 102 (1980) 2195.
7. (a) S. Lee and J. T. Hynes, J. Chem. Phys. 88 (1988) 6853; (b) H. J. Kim and J. T. Hynes, J. Chem. Phys. 96 (1992) 5088.
8. E. A. Carter and J. T. Hynes, J. Chem. Phys. 34 (1991) 5961.
9. I. Burghardt and J. T. Hynes, submitted to J. Phys. Chem. A.
10. (a) D. Laage, I. Burghardt, T. Sommerfeld, and J. T. Hynes, J. Phys. Chem. A 107 (2003) 11271; (b) I. Burghardt, D. Laage, and J. T. Hynes, J. Phys. Chem. A 107 (2003) 11292.
11. I. Burghardt, R. Spezia, and J. T. Hynes, in preparation.
12. M. H. Beck, A. Jackle, G. A. Worth, and H.-D. Meyer, Physics Reports 324 (2000) 1.

Femtochemistry VII
Fundamental Ultrafast Processes in Chemistry, Physics, and Biology
A.W. Castleman, Jr. and M.L. Kimble

Simulation of resonance Raman spectra of the solvated electron in water and methanol

Stefanie Neumann[a], Wolfgang Eisfeld[a], Andrzej L. Sobolewski[b] and Wolfgang Domcke[a]

[a] *Department of Chemistry, Technical University of Munich, D-85747 Garching, Germany*
[b] *Institute of Physics, Polish Academy of Sciences, PL-02668 Warsaw, Poland*

1. Introduction

The solvated electron in various solvents is a well-known and important species throughout chemistry, physics, and biology (see e.g. [1]) which has been investigated spectroscopically for quite a long time. The optical absorption spectra of the solvated electron in different solvents consist of a single broad and asymmetric band [2, 3].

Very recently, resonance-Raman (RR) spectra of the solvated electron in water and several alcohols have been obtained. Tauber and Mathies and Mizuno and Tahara independently measured the RR spectrum of the hydrated electron [4, 5, 6, 7, 8]. Tauber and Mathies also investigated the RR spectrum of the solvated electron in methanol, ethanol, and *n*-propanol [9]. For the solvated electron in methanol, RR spectra of different H/D substituted isotopomers of the solvent (CD_3OH, CH_3OD, and CD_3OD) have been obtained. The fingerprint lines in the RR spectrum of the hydrated electron are the H_2O intramolecular bending and stretching bands at $\approx$1625 cm^{-1} and $\approx$ 3100 cm^{-1}. For the solvated electron in methanol, the fingerprint lines of the RR spectrum are located at 475 cm^{-1}, 950 – 1020 cm^{-1}, $\approx$ 1330 cm^{-1}, $\approx$1440 cm^{-1}, and $\approx$3000 cm^{-1} [8].

The generally accepted model for the description of the microscopic structure of the solvated electron in liquid water and alcohols is the so-called cavity model in which the electron is supposed to be localized within a cavity of a hydrogen-bonded network of the solvent molecules (see e.g. [1]). An alternative model, the so-called cluster model of the solvated electron, was suggested by Sobolewski and Domcke for the solvated electron in water [10,

11]. It could be shown with restricted open-shell Hartree Fock (ROHF) and density functional theory (DFT) that H_3OW_n (W = H_2O) clusters undergo a spontaneous charge separation, resulting in a solvated H_3O^+ cation and a solvated electron cloud. It was therefore proposed that H_3OW_n clusters can serve as finite-size model systems for the computation of the spectroscopic properties of the hydrated electron. Extending this model of the solvated electron to methanol, the CH_3OH_2 hypervalent radical is the analogue of the H_3O radical. $CH_3OH_2M_n$ (M = CH_3OH) clusters are therefore supposed to represent finite-size models of the solvated electron in methanol. In this article, we report density functional theory (DFT) computational studies of the normal-mode vibrations, vertical excitation energies, oscillator strengths, and RR intensities of H_3OW_9 and $CH_3OH_2M_3$ clusters.

2. Computational methods

2.1. Calculation of RR intensities

The RR intensities have been calculated within the time-dependent picture of RR spectroscopy [12, 13]. Assuming the validity of the Born-Oppenheimer approximation and for short-time dynamics at low-temperatures, the so-called Savin formula [14] gives the relative RR intensities [12, 13]. For a single excited electronic state, the relative intensities of two fundamentals *i* and *j* are given by [14]

$$\frac{I^{(i)}}{I^{(j)}} = \frac{\kappa_i^2}{\kappa_j^2}, \qquad (1)$$

where κ_i, κ_j denote the gradients of the electronic excitation energy with respect to dimensionless normal modes Q_i, Q_j in the electronic ground state.

If several electronic states are resonantly excited, the RR amplitude of a given normal mode is the sum of the RR amplitudes of the individual excited electronic states. Since the dipole moments of different excited states are randomly oriented in complex systems without symmetry, the contributions of the interference terms to the RR intensities cancel to a large extent. The relative RR intensities are therefore given to a good approximation by the incoherent sum of the intensities of the excited states

$$I^{(i)} = \sum_k (f_k \kappa_i^{(k)})^2, \qquad (2)$$

where f_k is the oscillator strength of the excited electronic state *k*, and $\kappa_i^{(k)}$ is the corresponding energy gradient with respect to the normal coordinate Q_i.

2.2. *Electronic-structure calculations*

Equilibrium structures, vibrational frequencies, and normal modes of H_3OW_9 and $CH_3OH_2M_3$ clusters have been calculated using DFT with the Becke-Lee-Yang-Parr (B3LYP) functional [15, 16, 17]. For the excited-state calculations, the time-dependent DFT (TDDFT) method has been employed.

The basis set for the calculations for $CH_3OH_2M_3$ was the 6-31+G** split valence double-ζ Gaussian basis set augmented with polarization functions on all atoms and diffuse functions on the heavy atoms [18]. The basis set has been extended by two additional diffuse sets of functions (one *s* and one *p* set of functions) with an exponent of 0.02 [11] on the oxygen atoms. For the H_3OW_9 calculations, this basis set with additional diffuse functions on the hydrogen atoms has been used.

For the calculations of the electronic ground state of H_3OW_9 and $CH_3OH_2M_3$, the Gaussian03 program package [19] has been used. Vertical excitation energies, oscillator strengths, and the Cartesian gradients of the excited-state potential-energy surfaces (PES) have been calculated at the TDDFT level using the TURBOMOLE program package [20] which provides analytical gradients for excited states.

3. The H_3OW_9 cluster

The H_3OW_9 cluster has C_3 symmetry. Its structure is shown in Fig. 1 together with the singly occupied molecular orbital (SOMO). Fig. 1 reveals the above-mentioned charge separation between the H_3O^+ cation and the unpaired

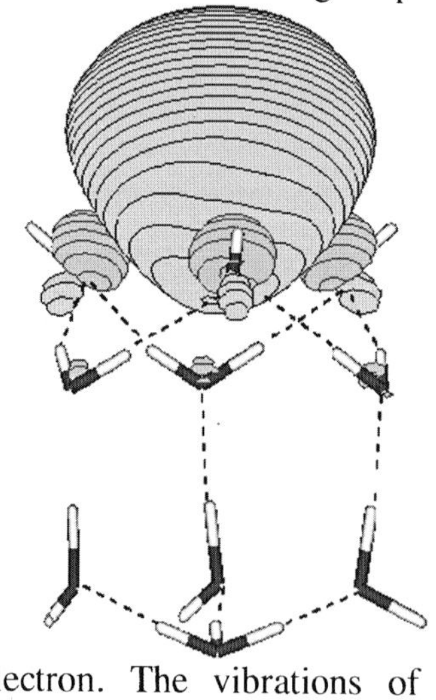

Figure 1: Equilibrium structure and SOMO (electron density cut = 0.03) of H_3OW_9 calculated at the DFT/B3LYP level.

electron. The vibrations of H_3OW_9 can be separated into intermolecular vibrations and intramolecular vibrations of the H_3O group and the water molecules, respectively. For brevity, only the intramolecular vibrations of the

H_3O part and of the water molecules with frequencies in the range 1500 cm^{-1} < ω < 3500 cm^{-1} are considered in the following, since the fingerprint lines of the experimental RR spectrum of the hydrated electron [4, 5, 6, 7, 8] lie within this range.

The vertical electronic excitation spectrum of H_3OW_9 (left-hand side of Fig. 2) has been obtained from vertical excitation energies and oscillator strengths calculated at the DFT/B3LYP level, which have been convoluted with a Gaussian broadening function with a halfwidth of 1.3 eV. The resulting spectrum exhibits a broad and asymmetric band with a maximum at 1.74 eV and an integrated oscillator strength of 0.80. It is quite similar to the experimental absorption spectrum of the hydrated electron (right-hand side of Fig. 2) which also exhibits a broad and asymmetric absorption band with an absorption maximum at 1.79 eV and an oscillator strength of 0.76 [2, 3].

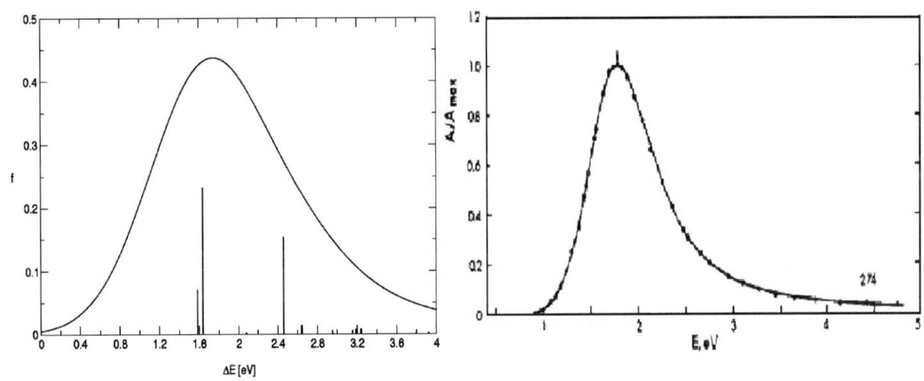

Figure 2: Vertical excitation spectrum of H_3OW_9 (left) and experimental optical absorption spectrum of the hydrated electron, adapted from Ref. [3] (right).

For the calculation of the RR intensities, the states corresponding to the $s \rightarrow p$ transition (the $^2E(p_{x,y})$ and $^2A(p_z)$ excited states) and a further ^2A state, which gains intensity by intensity borrowing, have been included. These are the only excited states with significant oscillator strength.

The simulated RR spectrum of H_3OW_9 is shown on the left-hand side of Fig. 3. It has been obtained from relative RR intensities calculated via Eq. 2 convoluted with a Gaussian broadening function with a FWHM of 19.4 meV for frequencies up to 2250 cm^{-1} and of 46.5 meV FWHM for frequencies larger than 2250 cm^{-1} to account qualitatively for anharmonic couplings and thermal broadening effects. The simulated RR spectrum of H_3OW_9 shows two pronounced peaks, one at about 1700 cm^{-1}, which arises from water bending vibrations, and one at about 3000 cm^{-1}, arising from the OH stretching vibration of the OH bonds which point towards the cloud of the unpaired electron. The frequencies of the OH stretching vibrations are significantly downshifted compared to bulk water. On the right-hand side of Fig. 3, the experimental RR spectrum of the hydrated electron is presented. It exhibits two peaks at about 1625 cm^{-1} and 3100 cm^{-1}.

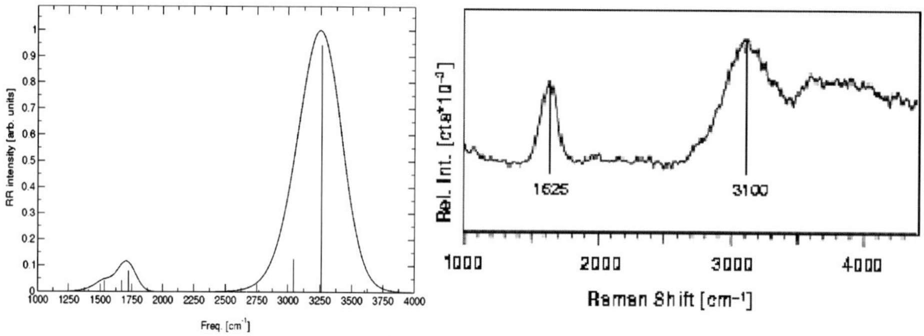

Figure 3: Simulated RR spectrum of H_3OW_9 (left) and experimental RR spectrum of the hydrated electron, adapted from Ref. [7] (right).

Comparing the simulated RR spectrum of H_3OW_9 and the experimental RR spectrum of the hydrated electron, one observes that the simulated spectrum of the cluster is very similar to the experimental spectrum for the hydrated electron and reproduces the important fingerprint lines. The intensity of the peak at 1700 cm^{-1} is too low relative to the peak at 3000 cm^{-1} in the calculated spectrum. This deviation may arise from the finite size of the cluster or may be the consequence of the rather crude estimation of the RR intensities.

4. The $CH_3OH_2M_3$ cluster

The equilibrium structure and the SOMO of the $CH_3OH_2M_3$ cluster are shown in Fig. 4. As for the H_3OW_9 cluster, a charge separation of $CH_3OH_2M_3$ into the $CH_3OH_2^+$ cation and a solvated electron cloud is observed.

The vertical excitation spectrum of $CH_3OH_2M_3$ is shown on the left-hand side of Fig. 5. It has been obtained from vertical excitation energies and oscillator strengths calculated at the DFT/B3LYP level and convoluted with a Gaussian broadening function (FWHM = 1.3 eV). The absorption maximum is 1.76 eV and the integrated oscillator strength 0.84. Compared to the experimental absorption spectrum of the solvated electron in liquid methanol (right-hand side of Fig. 5), the absorption maximum is about 0.2 eV too low, and the oscillator strength is slightly too large. The envelope of the calculated spectrum represents the main features of the experimental absorption spectrum of the solvated electron in liquid methanol reasonably well.

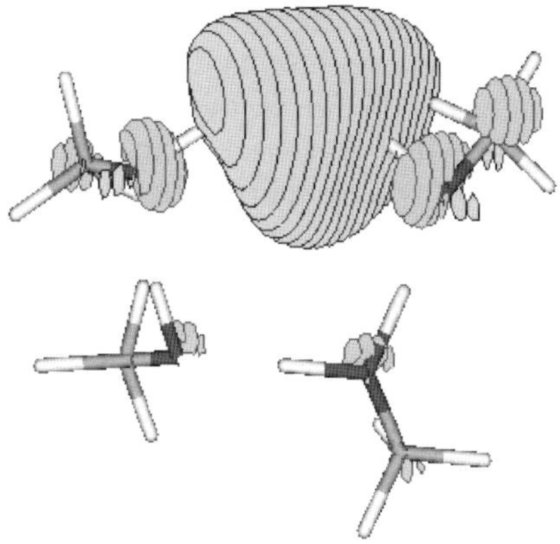

Figure 4: Equilibrium structure and singly occupied molecular orbital (SOMO) (electron density cut = 0.04) of $CH_3OH_2M_3$ calculated at the DFT/B3LYP level

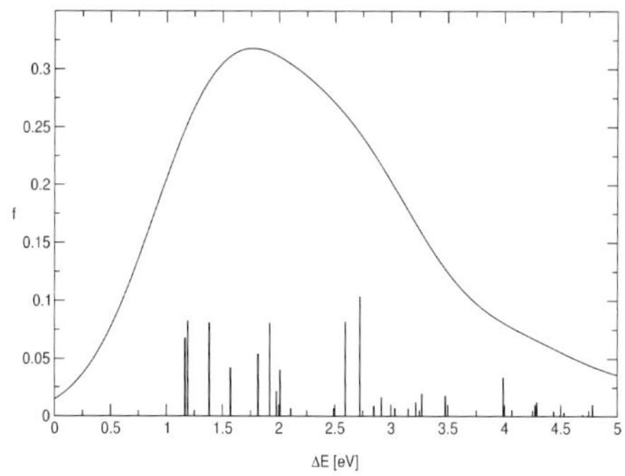

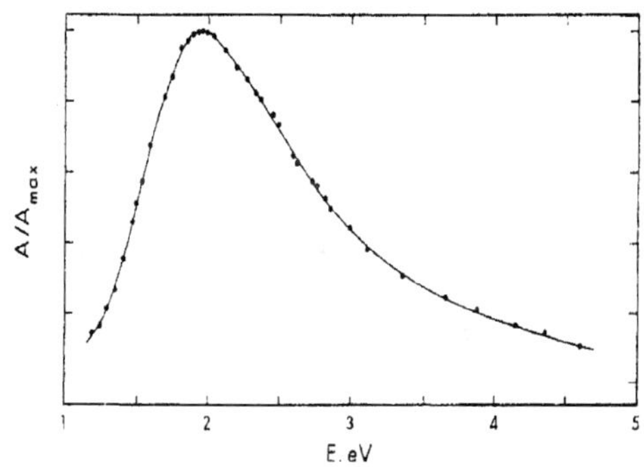

Figure 5: Vertical excitation spectrum of $CH_3OH_2M_3$ (left) and experimental optical absorption spectrum of the solvated electron in methanol, adapted from Ref. [2] (right).

The left-hand side of Fig. 6 shows the simulated RR spectrum of $CH_3OH_2M_3$ which has been obtained from RR intensities calculated according to Eq. 2 convoluted with a Gaussian broadening function with a FWHM of 180 cm^{-1} for frequencies smaller than 2000 cm^{-1} and 571 cm^{-1} for frequencies larger than 2000 cm^{-1} to account qualitatively for anharmonic couplings and thermal broadening effects. The peak at about 500 cm^{-1} arises mainly from an intermolecular vibration. Between 800 and 1800 cm^{-1}, we obtain three peaks which are assigned as the CO stretching, the OH in-plane-bending, and the HOH asymmetric stretching and bending vibrations, respectively. At about 2100 cm^{-1} a peak of low intensity arising from the HOH symmetric stretching vibration exists in the calculated spectrum. The peak with the highest intensity is located at about 3000 cm^{-1} and represents the OH stretching vibration.

On the right-hand side of Fig. 6, the experimental RR spectrum [9] of the solvated electron in methanol is shown. A comparison of the simulated RR spectrum of $CH_3OH_2M_3$ with the experimental RR spectrum of the solvated electron in methanol shows that the cluster spectrum matches the experimental spectrum reasonably well, apart from the different intensities of the four peaks in the frequency range 500 - 1800 cm^{-1}, which are too low. With the exception of the peaks at 500 cm^{-1} and at about 1500 cm^{-1}, which have been assigned as OH torsion and CH_3 deformation in Ref. [9], our assignments agree with those of Tauber, Stuart, and Mathies.

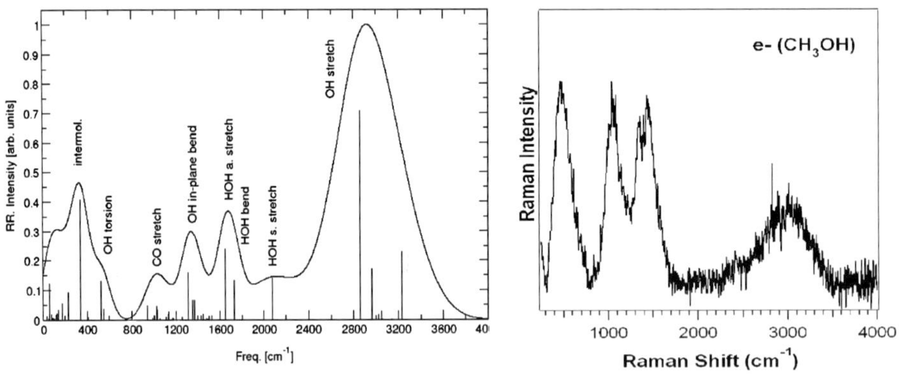

Figure 6: Simulated RR spectrum of $CH_3OH_2M_3$ (left) and experimental RR spectrum of the solvated electron in methanol (with friendly permission of R. Mathies) (right).

Fig. 7 shows the simulated RR spectra for differently H/D substituted $CH_3OH_2M_3$ clusters (left-hand side) and the experimental RR spectra of the solvated electron in correspondingly substituted methanol (right-hand side). For both spectra, it is obvious that the CD_3 substitution does not change the RR spectrum significantly. This indicates that the CH_3 related vibrations do not play a significant role in the RR spectrum of the solvated electron in methanol. The OH $\rightarrow$ OD substitution results in drastic changes in both the cluster spectrum and the experimental spectrum. The OH stretching vibration experiences the expected significant redshift in both spectra. The low-frequency peak exhibits an isotopic shift which would be expected for the OH torsion vibration. It is therefore highly probable that this peak in fact arises from the OH torsion vibration as found for the RR spectra of smaller $CH_3OH_2M_n$ clusters. The fully deuterated spectrum shows the expected effects of the $CH_3 \rightarrow CD_3$ and the OH

→ OD exchanges. The correlations of the peaks in the simulated RR spectra of the deuterated clusters and in the experimental RR spectra are mostly the same. Apart from the assignment of the 500 cm^{-1} peak, the only difference arises for the CH_3 deformation vibration, which does not gain intensity in the calculated spectra.

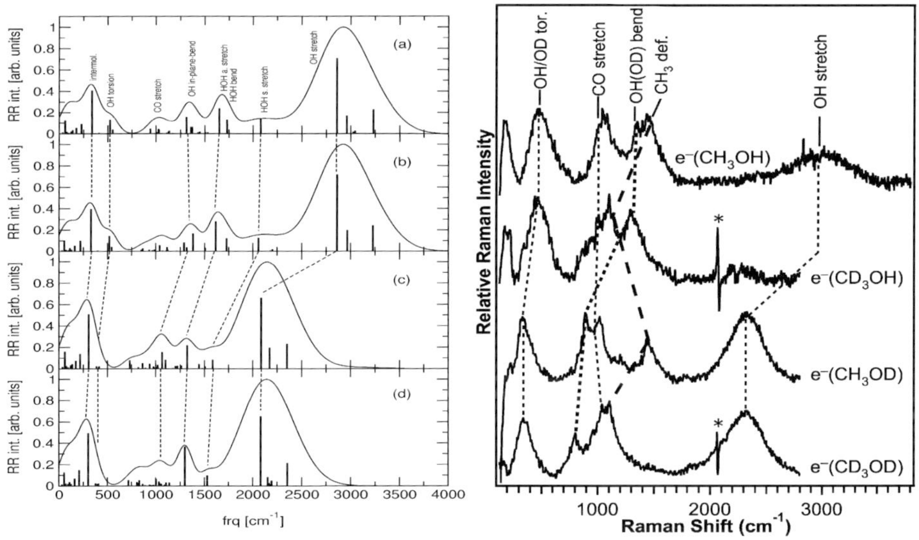

Figure 7: Simulated RR spectra of the H/D substituted $CH_3OH_2M_3$ cluster (left): (a) $CH_3OH_2(CH_3OH)_3$, (b) $CD_3OH_2(CD_3OH)_3$, (c) $CH_3OD_2(CH_3OD)_3$, (d) $CD_3OD_2(CD_3OD)_3$, and experimental RR spectrum of the solvated electron for differently H/D substituted methanol (with friendly permission of R. Mathies) (right).

5. Conclusions

Equilibrium geometries, vertical excitation spectra, and simulated RR spectra of H_3OW_9 and $CH_3OH_2M_3$ have been obtained using (TD)DFT/B3LYP calculations. For the hydronium water cluster as well as the $CH_3OH_2M_3$ cluster, a charge separation between the H_3O^+ / $CH_3OH_2^+$ cation and the unpaired electron is observed, the electron being localized at the opposite end of the cluster.

The envelopes of the vertical excitation spectra of both clusters are broad and asymmetric bands which resemble the experimental absorption bands of the solvated electron in water and methanol, respectively.

The simulated RR spectrum of H_3OW_9 is in good agreement with the experimental RR spectrum of the hydrated electron. In both spectra, two fingerprint lines are observed which arise from the bending and OH stretching vibrations of the water molecules. The simulated RR spectrum of $CH_3OH_2M_3$ matches the experimental RR spectrum of the solvated electron in methanol reasonably well. As the peaks in the simulated RR spectrum of $CH_3OH_2M_3$ arise from several normal modes, the assignment of the peaks in terms of normal modes can only be of a qualitative nature. The behavior of the simulated RR spectra of $CH_3OH_2M_3$ with respect to deuteration is similar to the behavior of the experimental RR spectrum of the solvated electron in correspondingly substituted methanol.

We therefore conclude that H_3OW_n and $CH_3OH_2M_n$ clusters can serve as finite-size model systems for the calculation of spectroscopic properties of the solvated electron in water and methanol, respectively.

References

1. U. Schindewolf, Angew. Chem. Int. Ed. Engl. 17 (1978) 887.
2. F.-Y. Jou and G. R. Freeman, J. Phys. Chem. 81 (1977) 909.
3. F.-Y. Jou and G. R. Freeman, J. Phys. Chem. 83 (1979) 261.
4. M. Mizuno and T. Tahara, J. Phys. Chem. A 105 (20010 8823.
5. M. Mizuno and T. Tahara, J. Phys. Chem. A 107 (2003) 2411.
6. M. Tauber and R. Mathies, J. Phys. Chem. A 105 (2001) 10952.
7. M. J. Tauber and R. A. Mathies, Chem. Phys. Lett. 354 (2002) 518.
8. M. Tauber and R. Mathies, J. Am. Chem. Soc. 125 (2003) 1394.
9. M. Tauber, C. M. Stuart, and R. Mathies, J. Am. Chem. Soc. 126 (2004) 3414.
10. A. L. Sobolewski and W. Domcke, Phys. Chem. Chem. Phys. 4 (2002) 4.
11. A. L. Sobolewski and W. Domcke, J. Phys. Chem. A 106 (2002) 4158.
12. S.-Y. Lee and E. J. Heller, J. Chem. Phys. (1979) 71.
13. E. J. Heller, Acc. Chem. Res. 14 (1981) 368.
14. F. Savin, Opt. Spectrosc. 19 (1965) 555.
15. A. D. Becke, Phys. Rev. A 38 (1988) 3098.
16. C. Lee, W. Yang, and R. G. Parr, Phys. Rev. B 37 (1988) 785.
17. A. D. Becke, J. Chem. Phys. 98 (1993) 5648.
18. R. D. W. J. Hehre and J. Pople, J. Chem. Phys. 56 (1972) 2257.
19. GAUSSIAN 03, Revision B.04, Gaussian Inc. Pittsburgh, PA (2003).
20. R. Ahlrichs, M. Bär, M. Häser, H. Horn, and C. Kölmel, Chem. Phys. Lett. 162 (1989) 165.

Femtochemistry VII
Fundamental Ultrafast Processes in Chemistry, Physics, and Biology
A.W. Castleman, Jr. and M.L. Kimble

Dynamics following charge-transfer to solvent in I^- $(water)_n$ (n=3-28) clusters observed with photoelectron imaging

Aster Kammrath, Jan R. R. Verlet, Graham B. Griffin, Arthur E. Bragg, Daniel M. Neumark[1]

Department of Chemistry, University of California, Berkeley, California 94720, and Chemical Sciences Division, Lawrence Berkeley National Laboratory, Berkeley, California 94720

Abstract

The dynamics of charge-transfer-to-solvent states are studied in I^- $(water)_{n=3-28}$ clusters using time-resolved photoelectron imaging. The photoelectron spectra for clusters with $n \geq 5$ reveal multiple time scales for dynamics subsequent to their electronic excitation. Evolution of the vertical detachment energy (VDE) on two different time-scales is observed. For larger clusters, the faster dynamics suggest isomerization from a cluster with a surface-bound electron to one with more internalized character. Decay of the transient population is commensurate with the growth of very low energy electrons, and is attributed to autodetachment.

1. Introduction

Charge-transfer to solvent (CTTS) states are formed by a variety of anions in polar solvents, an electron ejected from the solute being stabilized by the collective action of the solvent network. First observed in 1928 in aqueous iodide solution [1], they have been extensively studied both theoretically as well as experimentally [2] as a sensitive probe of solvent structure, and as a source of

[1] Corresponding author. Email address: dneumark@berkeley.edu

solvated electrons [3], important in electron transfer studies, radiation chemistry and atmospheric thermodynamics. The analogous electronic excitation in halide-doped water clusters permits observation of the interaction of diffuse electrons with the solvent network [4,5]. Here, we use time-resolved photoelectron (PE) imaging to monitor the dynamics subsequent to excitation of the charge-transfer-to-solvent (CTTS) band in $I^-(water)_{n=3-28}$ clusters.

The bulk-phase experimental work most relevant to the current study was performed by Bradforth and co-workers [6-8] in the late '90s, who used single-photon excitation to excite the lowest CTTS state of aqueous iodide solution, monitoring the subsequent dynamics by ultrafast transient absorption spectroscopy. They interpreted their results based on the model developed by Staib and Borgis [9,10] for the related system of aqueous chloride. Similar work was carried out more recently by Laubereau and co-workers [11] who obtained similar results and found also that solvation and recombination rates are temperature-dependent, slowing as the solution is cooled.

Evidence of cluster precursors to the CTTS state was observed experimentally by Serxner *et al.*[4], in the absorption spectra of $I^-(H_2O)_{n=1-4}$. *Ab initio* studies by Chen and Sheu [12] and Lee and Kim [13] also recover evidence of a precursor state. The relationship between CTTS excitation in clusters and in bulk water was examined by Bradforth and Jungwirth [5].

Lehr *et al.* [14] performed the first experimental time-resolved study of excitation dynamics in the cluster precursor to CTTS states for $I^-(water)_{n=4-6}$ by time-resolved PE spectroscopy with a magnetic bottle photoelectron energy analyzer. They observed a shift to higher vertical detachment energy (VDE) at early pump-probe delays for n=5 and 6, accompanied by an increase in photodetachment signal, which they interpreted as solvent rearrangement resulting in stabilization of the electron in a state localized on the water cluster. Subsequently, the transient exhibited simple exponential decay on a timescale of picoseconds, which was attributed to excited state autodetachment. Similar results were seen for larger iodide-water clusters as well as in clusters with halides complexed to other solvent species [15-17].

Their interpretation of the energetics, however, ignored the role of the iodine atom. Theoretical modeling of the dynamics carried out by Chen and Sheu [18,19], and more recently by Jordan [20], Peslherbe [21], and Kim [22,23] suggests that both the iodine and solvent motion must be taken into account if an accurate picture is to be achieved.

In the current study we probe the dynamics of CTTS excitation in $I^-(water)_{n=3-28}$ by time-resolved PE imaging. We present evidence in support of short-time solvent rearrangement, leading to stabilization of the excess electron, and observe slower dynamics which we attribute to loss of the neutral iodine atom. PE imaging allows observation of low-energy photoelectrons from autodetachment that display depletion dynamics mirroring the growth and decay of pump-probe signal.

We also demonstrate how charge-transfer to solvent may be used to gain insight indirectly to questions about the excess electron in water cluster anions. Recent studies of bare water anion clusters carried out in our lab show two isomers of significantly different VBE [24], with the lower-binding isomer (isomer II) most likely to be a surface-bound isomer [25], while the higher binding one (isomer I) is postulated to have the excess electron in a significantly more internalized structure. Isomer II is formed at a higher pressure of backing gas, usually associated with cooler cluster conditions, suggesting that it is a metastable species trapped in a local energy minimum.

Studies have shown that even in $I^-(H_2O)_n$ clusters, the iodide sits on the surface of the water cluster [26,27]. It is therefore reasonable to assume that the electron undergoing transfer from the iodide to the water cluster will be left in a surface-solvated configuration, suggesting a possible means of generating isomer II water clusters without cooling the cluster to the point where isomerization is no longer possible. The shift to higher VBE observed in all clusters in this study can be correlated (particularly in the larger clusters, n>10) to an isomerization from a surface bound water-cluster anion (isomer II) to the tighter-bound, more internalized isomer I.

2. Experimental

Our experimental apparatus has been described elsewhere [28] and only pertinent details will be given here. Argon at 15-20 psi is passed over a reservoir of CH_3I and bubbled through water, finally undergoing supersonic expansion into vacuum, either through a standard piezo-valve pulsed at 500 Hz (for all H_2O clusters as well as $I^-(D_2O)_{n=5-7,11-28}$) or an Even-Lavie solenoid valve [29] pulsed at 100 Hz (for $I^-(D_2O)_{n=3-10}$). The molecular beam is crossed with an electron beam at 1200 eV (200-400 eV in the case of the Even-Lavie valve). Ions are perpendicularly extracted into a Wiley-Mclaren type mass spectrometer [30]. The cluster of interest is mass-selected and interacts with the laser pump and probe beams and the resulting photoelectrons are collected by velocity map imaging (VMI) [31], with a 70 mm dual multichannel plate coupled to a phosphor screen. Images from the phosphor screen are recorded by a charge-coupled device (CCD) camera.

For small clusters (n=3-10) the 3rd harmonic (265 nm, 4.65 eV, 20 μJ/pulse) of a chirped-pulse amplified Ti:Sapphire femtosecond Oscillator (Clark-MXR, NJA-5, CPA-1000) was used to pump the CTTS precursor state. For the larger clusters, the fundamental (795 nm) and the signal output of an optical parametric amplifier (OPA) at 1200 nm were sum-frequency mixed then doubled, yielding 2 μJ/pulse at 5.1 eV (240 nm). The remaining fundamental light (200 μJ/pulse at 1.56 eV) is used as the probe for all cluster sizes studied and directed onto a computer controlled translation stage which generates the desired pump-probe delays. The beams are collinearly recombined and focused at the interaction region using a 50 cm lens.

Images are four-way symmetrized to account for inhomogeneity in the detector, and the 3-dimensional velocity distribution is reconstructed using the Basis Set EXpansion (BASEX) method [32]. PE spectra are obtained by radial integration of the $\varphi=0$ slice through the reconstructed 3-D distribution, and conversion from velocity to energy space, with a typical resolution of 5%. The reported vertical detachment energies (VDE's) are found by fitting Gaussians to the individual PE spectra.

3. Results

Figure 1 shows the time-resolved PE spectrum of $I^-(D_2O)_7$, typical of clusters of both isotopes in the size range of n=5-7, with clusters of different size ranges displaying differences which are discussed below.

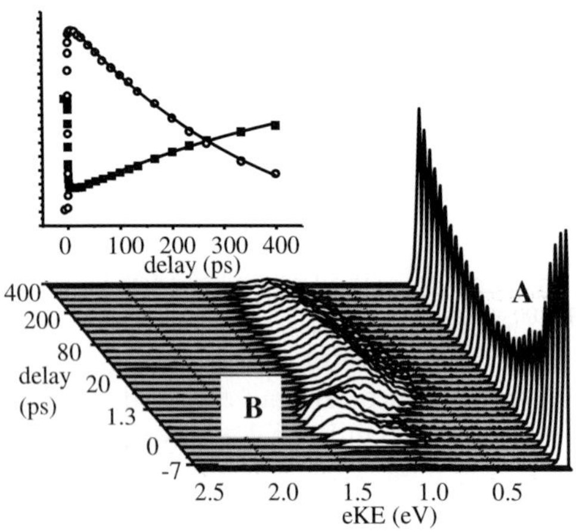

Figure 1) Time-resolved photo-electron spectrum of $I^-(D_2O)_7$. Signal intensity is plotted as a function of eKE on the horizontal axis, with pump-probe delay increasing front to back. The inset shows the integrated intensities of feature A (filled squares) and B (open circles) as a function of pump-probe delay with fits superimposed as solid lines.

Feature B corresponds to detachment through the CTTS precursor state, as previously observed by Davis *et al.*[15]. At early times its integrated intensity, I_{CTTS}, rises sharply with a correspondingly sharp drop in the intensity of feature A, I_A. I_{CTTS} continues to rise slowly over several picoseconds (and I_A continues to fall), long after the end of the pump pulse, so that it cannot be due to an increase in the excited state population, and we agree with Lehr, *et al.*[14] that it indicates an increase in detachment cross-section. At long time the decay of I_{CTTS} is accompanied by recovery of I_A on a similar time-scale.

The short-time dynamics of clusters in the size range n>7 are similar to those observed in clusters of n=5-7. However, for clusters with n=8-10, the long-time decay of I_{CTTS} is no longer smoothly mirrored by the recovery of I_A, which appears to proceed at a slower rate. In the largest clusters studied, n>10, I_{CTTS} decays too slowly for the dynamics to be followed to completion by our existing delay stages, and only the short-time dynamics are studied. Feature A is observed only up to n=15.

The eKE distribution of the CTTS feature also shows time dependence for all clusters considered here. This is illustrated in figure 2, which shows the vertical detachment energy of the CTTS electron (VDE_{CTTS}) as a function of pump-probe delay for $I^-(D_2O)_9$. VDE_{CTTS} is defined as hν – eKE_{max} where eKE_{max} is defined as the eKE at the fitted maximum intensity of feature B. On the rising edge of the cross-correlation of the pump and probe pulses, VDE_{CTTS} decreases sharply (ΔE_1). After time zero, VDE_{CTTS} begins to increase again (ΔE_2) on a somewhat slower timescale, closely matching the slow increase in I_{CTTS}. At long time VDE_{CTTS} shifts again to lower detachment energy (ΔE_3).

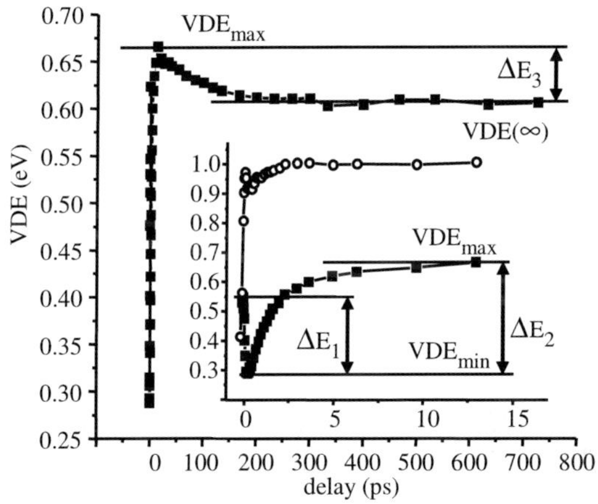

Figure 2) VDE_{CTTS} plotted as a function of pump-probe delay for $I^-(D_2O)_9$. The inset shows details of the short time dynamics, with VDE_{CTTS} shown as filled squares and I_{CTTS} as open circles. The spike in I_{CTTS} at very early delay is due to interference from non-resonant 2-hν detachment.

Figure 3 shows a general comparison of the $I^-(H_2O)_n$ VDE's at minimum and maximum values to the VDE's of the different water cluster isomers observed experimentally. Agreement of VDE_{min} with isomer II and the low-binding isomer of Kim *et al.*[33] is fairly good at most cluster sizes (except in the range n=10-15), consistent with the speculation that this corresponds to an isomer of the water cluster with the electron bound on the surface. VDE_{max} overshoots VDE of isomer I for most of the smaller clusters, but the agreement improves dramatically for the larger clusters, with $VDE_{max}(n)$ progressively approaching VDE of isomer I in the size range n=11-28.

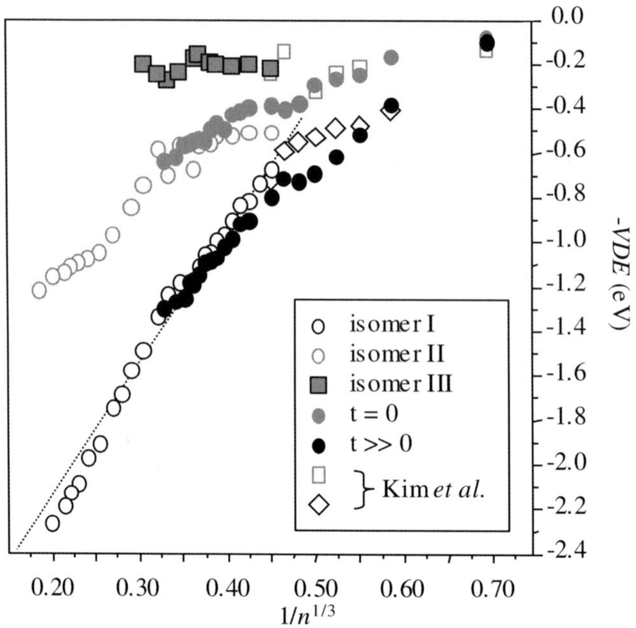

Figure 3) Comparison between VDE_{min} and VDE_{max} of iodide water clusters and VDEs reported for various isomers of bare water clusters. Data for small bare water clusters taken from Ref. [33].

4. Analysis

For all clusters studied, I_{CTTS} reaches a maximum at or shortly after time-zero. It subsequently decays according to one of the following equations:

$$I_{CTTS}(t) = A\exp(-(t-t_0)/\tau_d) \tag{1}$$

$$I_{CTTS}(t) = A_1\exp(-(t-t_0)/\tau_{d1}) + A_2\exp(-(t-t_0)/\tau_{d2}) \tag{2}$$

Here t is the pump-probe delay, t_0 the delay of maximum I_{CTTS} and τ_d are the lifetimes of the signal decay. Selected parameters are shown in Table 1, others are reported elsewhere [34,35]. Uncertainties are determined from the greater of experimental resolution, or spread in results obtained from different data sets. Clusters of n=3-7 show simple exponential decay of the excited population according to Eq. (1), with lifetimes ranging from 600fs for $I^-(H_2O)_3$ to 330ps for $I^-(D_2O)_7$, increasing monotonically with cluster size. For n=5-7 the recovery of I_A proceeds with a lifetime similar to the decay lifetimes determined for I_{CTTS}.

Table 1: Lifetime of CTTS decay (τ_d), magnitude of ΔE_2, time-constants τ_2 and τ_3. ΔE_3=50 meV ±20 meV for all clusters (where observed). Estimated uncertainties are ±15% for τ_d, ±30 meV for ΔE_2, ±30% for τ_2 at n=3,4, ±15% at n=5-10 and ±10% n>10, ±20% for τ_3 in all clusters.

n	τ_d (ps) H_2O/D_2O	ΔE_2 (meV) H_2O/D_2O	τ_2 (fs) H_2O/D_2O	τ_3 (ps) H_2O/D_2O
3	0.59/0.70	30	200	
4	8.2/7.9	150 / 100	200 / 280	
5	56/61	226 / 240	450 / 440	25 / 29
6	136/131	270 / 260	650 / 690	38 / 51
7	324/330	350 / 333	890 / 1110	54 / 73
8		400 / 365	1100 / 1300	86 / 77
9		360 / 350	1180 / 1470	73 / 90
10		290 / 320	1150 / 1430	
11		410	1040 / 1120	
15		490	990 / 1060	
20		580	800 / 850	
25		650	1120	

For clusters with n=8-10, decay of I_{CTTS} proceeds bi-exponentially according to Eq. (2), with τ_{d1} on the order of 100ps and τ_{d2} on the order of several nanoseconds, both increasing monotonically with cluster size. In each case, the faster decay component accounts for ~30% of the population. The recovery of I_A, however, continues to follow a simple exponential with a time-scale close to τ_{d2}. Our delay stages are too short to follow these dynamics to completion, so that the values obtained for τ_{d1} and τ_{d2} cannot be as accurately determined as τ_d for the smaller clusters. Similarly, we were unable to recover lifetimes for the transient in the largest clusters studied, n>10.

VDE_{CTTS} reaches VDE_{min} within the cross-correlation for all clusters studied. It then evolves in two distinct time-regimes according to the equations:

$$VDE(t) = VDE_{\max} - \Delta E_2 \exp(-(t - t_{\min})/\tau_2)) \quad (3)$$

$$VDE(t) = VDE(\infty) + \Delta E_3 \exp(-(t - t_{\max})/\tau_3) \quad (4)$$

Here Eq. (3) applies in the range $t_{min} \leq t \leq t_{max}$, with t_{min} and t_{max} being the times at which the VDE is minimal and maximal, respectively, while Eq. (4) applies for $t \geq t_{max}$. The parameters τ_2 and τ_3 are the time-constants for the energy shifts ΔE_2 and ΔE_3, and all other parameters are as designated in figure 2. Selected parameters are shown in Table 1 (for D_2O clusters when only one value is given), with others reported elsewhere [34,35]. For all clusters, t_{min} is

~100fs. For n=5-9, ΔE_3=50 meV (± 30 meV), while for n>10 it appears to decrease slightly with cluster size, though exact values could not be obtained.

Most of the time-scales and energy shifts show no significant (or consistent) isotope effect. In the range of n=7-10, however, τ_2 for D_2O clusters is consistently 1.2 times as large as τ_2 for the corresponding H_2O clusters. There continues to be ~5% difference in τ_2 for the different isotopes of the larger clusters, with the deuterated clusters displaying the slower time-scale, though this is within the accuracy of the timescales determined.

5. Discussion

5.1. CTTS decay

The PE spectra of $I^-(H_2O)_7$ in Fig. 1 is dominated by the intense feature A near eKE=0, which we assign to resonant excitation to the CTTS state of the cluster followed by autodetachment:

$$I^-(H_2O)_n \xrightarrow{h\nu} [I^-(H_2O)_n]^* \xrightarrow{\tau_d} I\cdot(H_2O)_n + e^-, I + (H_2O)_n + e^- \quad (5)$$

This assignment is supported by the observation that I_A is largely complementary to I_{CTTS}. The time-dependent intensities result from competition between detachment of the excited state by the probe pulse vs. autodetachment from the excited state, as indicated in Eq. (6).

$$\begin{aligned} A^- \xrightarrow{h\nu_{pu}} A^-{}^* &\xrightarrow{\tau_d} A + e^-[eKE \sim 0] \\ &\xrightarrow{t, h\nu_{pr}} A + e^-[eKE = h\nu - VDE] \end{aligned} \quad (6)$$

Hence the time constant τ_d in Eq. (1) reflects the autodetachment lifetime, and we attribute decay of the population excited through the CTTS channel to excited-state autodetachment, as was postulated by Lehr *et al.*[14].

For cluster sizes n=8-10, the decay dynamics are more complicated. We interpret the major decay channel, governed by τ_{d2}, as autodetachment as it is consistent with the slow recovery of I_A observed for these larger clusters. We note that the time constants τ_{d1} for the minor channel are generally quite close to those of τ_3 in Table 1.

5.2. Time-dependent VDE's

The first VDE shift, ΔE_1, is seen in all clusters ($n \geq 5$) considered here and is essentially instantaneous (timescale ~180 fs) at the temporal resolution of our experiment. We speculate that it represents rapid excited state dynamics analogous to the formation of the "contact pair" (I:e$^-$) that occurs in aqueous iodide solution, for which the time scale of formation is 200 fs [8].

As discussed in the introduction, two limiting interpretations of ΔE_2 and ΔE_3 must be considered, described by extensions of the so-called "solvent-driven" and "iodine-driven" models of relaxation in these clusters [21]. The first is that ΔE_2 and ΔE_3 correspond to solvent stabilization of the excess electron and dissociation of the cluster to I + $(H_2O)_n^-$, respectively, with the overall mechanism given by:

$$\begin{aligned} &I^-(H_2O)_n \xrightarrow{h\nu_{pu}} I\cdot(H_2O)_n^- * \xrightarrow{\tau_2,\Delta E_2} \\ &I\cdot(H_2O)_n^- \xrightarrow{\tau_3,\Delta E_3} I+(H_2O)_n^- \end{aligned} \tag{7}$$

Alternatively, ΔE_2 could be from stabilization of the cluster due to the I atom leaving, while ΔE_3 represents subsequent dynamics of the bare (water)$_n^-$ anion:

$$\begin{aligned} &I^-(H_2O)_n \xrightarrow{h\nu_{pu}} I\cdot(H_2O)_n^- * \xrightarrow{\tau_2,\Delta E_2} I+(H_2O)_n^- * \\ &\xrightarrow{\tau_3,\Delta E_3} I+(H_2O)_n^-, I+(H_2O)_{n-1}^- + H_2O \end{aligned} \tag{8}$$

Several factors argue for assigning ΔE_2 to solvent motion. It is strongly size dependent, doubling from n=5 to n=8 and tripling from n=5 to n=20, while the magnitude (650 meV at n=25) seems larger than what might be expected for iodine loss Moreover, if ΔE_2 is due purely to iodine motion, it is difficult to understand the distinct isotope effect in τ_2 for n=7-10 or why VDE_{max} should align so well with the VDE of isomer I of the bare water clusters for n>15.

Conversely, ΔE_3 shows little or no dependence on cluster size in the range of n=5-9. As the VDEs of the bare (water)$_n^-$ in this size range do not increase evenly, or even monotonically, a constant shift seems inconsistent with the loss of a water monomer from a bare water anion cluster, as implied by (8).

Overall, then, "solvent-driven" dynamics appear to be more consistent with the experimental results than "iodine-driven" dynamics. However, ΔE_2 may very well represent combined solvent stabilization and iodine motion (but not dissociation), as was indicated in the simulations on $I^-(H_2O)_3$ by Peslherbe [21].

With this in mind, the excellent agreement of VDE_{min} with the VDE of isomer II of the bare water clusters and VDE_{max} with that of isomer I suggests that ΔE_2 may, in fact, be interpreted as isomerization of the bare (water)$_n^-$ from the weakly bound surface state to the more tightly bound isomer I. In this case, the time-scale of ΔE_2 becomes interesting in light of molecular dynamics simulations done by Barnett *et al.* [36,37] in which they attach a diffuse electron to the surface of $(H_2O)_{256}$. They observe the electron localizes on the surface in the first 150 fs (which could relate to the ΔE_1 we observed), then samples other surface sites for ~1.2 ps, finally penetrating the surface of the cluster (defined as formation of a solvation shell) on a timescale of 1 ps, which is quite similar to

τ_2, especially in the larger clusters. The clusters studied here are considerably smaller than n=256, so care must be used in drawing any conclusions from such a comparison. Nonetheless, the close correspondence of the timescales supports the idea that we are observing internalization of an excess electron originally produced on the surface of the cluster. However, recent theoretical work [25] has questioned whether isomer I really has an internalized electron, so at this point assignment of the observed dynamics to a surface-to-internal conversion must be considered suggestive rather than definitive.

Acknowledgements

This work is supported by the National Science Foundation under Grant no. CHE-0350585. The authors thank Prof. Ori Cheshnovsky for many interesting discussions.

References

1. J. Franck and G. Scheibe, Z. Phys. Chem. 139 (1928) 22.
2. M. J. Blandamer and M. F. Fox, Chem. Rev. 70 (1970) 59.
3. J. Jortner, M. Ottolenghi, and G. Stein, J. Phys. Chem. 68 (1964) 247.
4. D. Serxner, C. E. H. Dessent, and M. A. Johnson, J. Chem. Phys. 105 (1996) 7231.
5. S. E. Bradforth and P. Jungwirth, J. Phys. Chem. A 106 (2002) 1286.
6. J. A. Kloepfer, V. H. Vilchiz, V. A. Lenchenkov, and S. E. Bradforth, Chem. Phys. Lett. 298 (1998) 120.
7. J. A. Kloepfer, V. H. Vilchiz, V. A. Lenchenkov, X. Chen, and S. E. Bradforth, J. Chem. Phys. 117 (2002) 766.
8. J. A. Kloepfer, V. H. Vilchiz, V. A. Lenchenkov, A. C. Germaine, and S. E. Bradforth, J. Chem. Phys. 113 (2000) 6288.
9. A. Staib and D. Borgis, J. Chem. Phys. 104 (1996) 9027.
10. A. Staib and D. Borgis, J. Chem. Phys. 104 (1996) 4776.
11. H. Iglev, A. Trifonov, A. Thaller, I. Buchvarov, T. Fiebig, and A. Laubereau, Chem. Phys. Lett. 403 (2005) 198.
12. H. Chen and W. Sheu, J. Am. Chem. Soc. 122 (2000) 7534.
13. H. M. Lee and K. Kim, J. Chem. Phys. 114 (2001) 4461.
14. L. Lehr, M. T. Zanni, C. Frischkorn, R. Weinkauf, and D. M. Neumark, Science 284 (1999) 635.
15. A. V. Davis, M. T. Zanni, C. Frischkorn, and D. M. Neumark, J. Electron Spectrosc. and Rel. Phenom. 108 (2000) 203.
16. C. Frischkorn, M. T. Zanni, A. V. Davis, and D. M. Neumark, Faraday Disc. (2000) 49.
17. M. T. Zanni, C. Frischkorn, A. V. Davis, and D. M. Neumark, J. Phys. Chem. A 104 (2000) 2527.
18. H. Chen and W. Sheu, Chem. Phys. Lett. 335 (2001) 475.
19. H. Chen and W. Sheu, Chem. Phys. Lett. 353 (2002) 459.
20. F. D. Vila and K. D. Jordan, J. Phys. Chem. A 106 (2002) 1391.
21. Q. K. Timerghazin and G. H. Peslherbe, J. Am. Chem. Soc. 125 (2003) 9904.
22. H. M. Lee and K. Kim, Mol. Phys. 102 (2004) 2485.
23. H. M. Lee, S. B. Suh, and K. Kim, J. Chem. Phys. 119 (2003) 7685.

24. J. R. R. Verlet, A. E. Bragg, A. Kammrath, O. Cheshnovsky, and D. M. Neumark, Science 307 (2005) 93.
25. L. Turi, W. S. Sheu, and P. J. Rossky, Science 309 (2005) 914.
26. O. Cheshnovsky, R. Giniger, G. Markovich, G. Makov, A. Nitzan, and J. Jortner, Journal De Chimie Physique Et De Physico-Chimie Biologique 92 (1995) 397.
27. D. M. Koch and G. H. Peslherbe, Chem. Phys. Lett. 329 (2002) 381.
28. A. E. Bragg, R. Wester, A. V. Davis, A. Kammrath, and D. M. Neumark, Chem. Phys. Lett. 376 (2003) 767.
29. U. Even, J. Jortner, D. Noy, N. Lavie, and C. Cossart-Magos, J. Chem. Phys. 112 (2000) 8068.
30. W. C. Wiley and I. H. McLaren, Rev. Sci. Instrum. 26 (1955) 1150.
31. A. T. J. B. Eppink and D. H. Parker, Rev. Sci. Instrum. 68 (1997) 3477.
32. V. Dribinski, A. Ossadtchi, V. A. Mandelshtam, and H. Reissler, Rev. Sci. Instrum. (2002).
33. J. Kim, I. Becker, O. Cheshnovsky, and M. A. Johnson, Chem. Phys. Lett. 297 (1998) 90.
34. A. Kammrath, J. R. R. Verlet, A. E. Bragg, G. Griffin, and D. M. Neumark, J. Phys. Chem. A in press (in press).
35. J. R. R. Verlet, A. Kammrath, G. Griffin, and D. M. Neumark, J. Chem. Phys. (submitted).
36. R. N. Barnett, U. Landman, and A. Nitzan, J. Chem. Phys. 91 (1989) 5567.
37. R. N. Barnett, U. Landman, and A. Nitzan, Phys. Rev. Lett. 62 (1989) 106.

Femtochemistry VII
Fundamental Ultrafast Processes in Chemistry, Physics, and Biology
A.W. Castleman, Jr. and M.L. Kimble

Ultrafast Photoelectron imaging of the electronic relaxation of a molecule deposited at the surface of an argon cluster.

Lionel Poisson [a], Marc-André Gaveau [a], Eric Gloaguen [a], Jean-Michel Mestdagh [a], Benoît Soep [a] and David H.Parker [b]

[a] *Laboratoire Francis Perrin (CNRS-URA-2453), DSM/DRECAM/ Service des Photons, Atomes et Molécules, C.E.A. Saclay, F-91191 Gif-sur-Yvette Cedex, France*

[b] *Department of Molecular and Laser Physics, University of Nijmegen Toernooiveld 1, 6525ED Nijmegen, The Netherlands*

1. Introduction

Molecules rarely reemit from the initially excited level after the absorption of a photon, and, if the molecule is sufficiently complex, light emission, a slow process, is not the main decay channel for the excited state. The electronic energy is rapidly degraded and thermalized within the ground electronic state of these systems, preventing most of the photochemistry by dissipation of the energy in the medium. These relaxation properties arise from coupling between the electronic configurations of the molecular systems accessible by optical excitation and the other energetically accessible configurations. However, in many molecules, this results in a non-radiative relaxation scheme akin to a tunneling effect that generates a fairly large decay time [1], allowing for photochemistry. Only recently, a direct mechanism has been observed experimentally [2-7] which can allow an extremely fast relaxation in molecules. It involves a crossing between potential energy surfaces [8], named a conical intersection. This intersection or funnel connects directly the relevant excited states and provides a direct descent to the lowest energy surface, guided by the gradients along these surfaces. Nevertheless, discussions [9] show that the passage through a conical intersection may be difficult in a complex system where the descending wave packet can be reflected, lengthening the overall decay. In such case, the relaxation mechanisms can become slow in comparison to molecular movements, if the non-adiabatic coupling between the intersecting surfaces is not sufficient. Given the generally fast decay rates observed in a variety of molecules and typically in ethylene–like compounds [4-7, 10, 11], it is conceivable that, besides the direct connection of the initial and final potential energy surfaces through a conical intersection, a complementary, general mechanism exists and increases the relaxation rate. We have proposed that such a mechanism exists and involves mediating states or surfaces allowing a more direct connection between the relevant excited surfaces [12]. We have identified here these states in the case of a specifically selected ethylenic molecule, tetrakis dimethylaminoethylene, TDMAE, shown in Figure . Briefly, the electronic relaxation from an initially excited valence $\pi\pi*$ state V decays into a Zwitterionic state Z, but this decay is mediated by a diffuse, Rydberg state. This appears from time resolved photoelectron spectroscopy of supersonically cooled TDMAE [12].

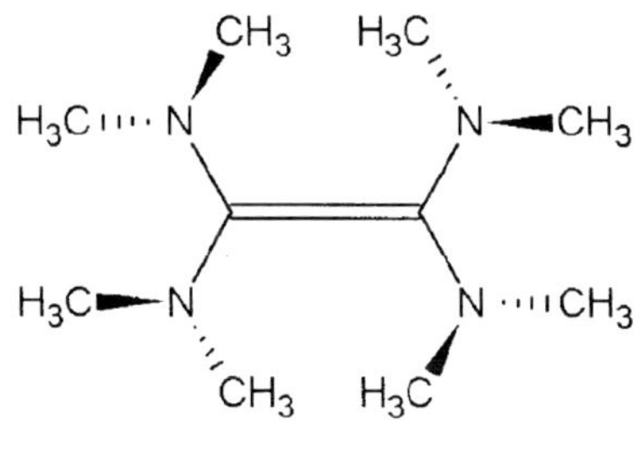

Figure 1: The TDMAE molecule

Since, in general, relaxation is observed in condensed phases, it is essential to conduct the relevant experiments in the presence of a perturbing medium. On the other hand, we wish to keep the potentialities of techniques such as phototelectron spectroscopy to investigate the flow of energy over electronic states. We demonstrate here that this can be achieved at the surface of argon clusters. We compare here the observation of TDMAE, free and deposited at the surface of the argon cluster by a pick up technique. The effect of the cluster and the possibility to record significant photoelectron spectra is thus described here. Indeed, up to now very scanty results have been reported upon the photoelectron spectroscopy in the presence of a medium. Indeed the solvation of halide anions away from the surface of water has been investigated by VUV photoelectron spectroscopy [13, 14]. Here we shall use femtosecond time resolved imaging to resolve the time evolution of the TDMAE $-Ar_n$ system as depicted in Figure 2a.

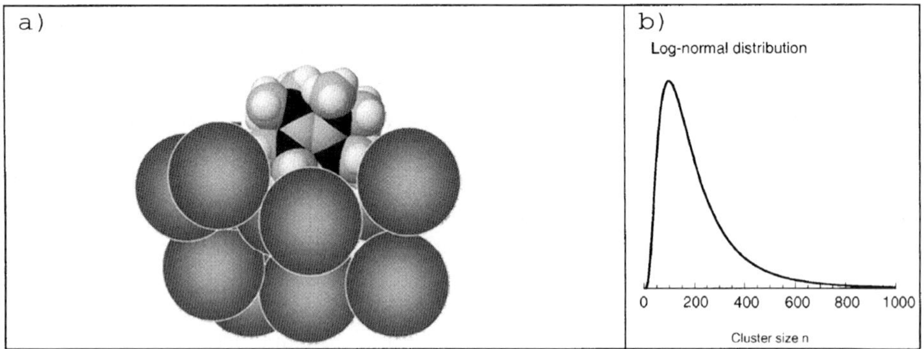

Figure 2: a) schematic, scaled representation, of the TDMAE molecule deposited on an argon cluster after pick-up. The TDMAE molecule almost matches an argon substitutional site. The average size is in general estimated to ~ 100-200 atoms. b) Log normal distribution of argon clusters in the supersonic expansion, maximum at n=100.

2. Experimental methods: the pick-up of TDMAE molecules

Femtosecond pump probe experiments are achieved at the LUCA facility in Saclay on the systems under study. The clusters are excited with 266 nm, typically 2-3 μJ pulses and probed with 400 nm pulses with a 100 fs cross correlation width FWMH. The set-up is described in reference [12] and essentially is used to detect photoions and photon electrons by velocity map imaging.

Argon clusters are generated by a pulsed supersonic expansion of high pressure argon gas (up to 12 bars) through a 100 μm nozzle. Average cluster sizes were calculated evaluated using the empirical relations derived by Hagena [15] and Buck [16]. This results in average argon cluster sizes of c.a. 200. The distribution in sizes is represented in general by a log normal function as shown in Figure 2b.

The TDMAE gas (0.2 mbar) is introduced in the expansion chamber by a leak which is shaped as a rectangle (2*1 mm) and placed close to the nozzle. The impact of the molecule will be captured by clusters of large dimensions (20 Å diameter) and subsequently cooled by evaporation of argon atoms from the cluster surface. If one assumes an energy content of the molecule as ~6000 cm^{-1}, the cooling will evaporate 15 argon atoms. This considers that each argon atom has been evaporated from the surface and had been interacting with 8 atoms with 100 cm^{-1} energy.

However some molecules of TDMAE diffuse in the experiment chamber and are partially cooled. They represent a background that can be substracted from the cluster signal. This is easily achieved by substracting the background signal recorded before the opening of the pulse valve.

At the bottom of Figure 3, the total ion signal has been plotted as a function of the delay between the pulsed valve and the the femtosecond laser. The vertical axis represents to the arrival position of the cluster ions, and the position is proportional to the square root of the mass. Three regions are visible from top to bottom where the masses are increasing downwards. First comes the background TDMAE, constant in time and no velocity; second is TDMAE in the beam. The last broad region corresponds to the extended cluster distribution.

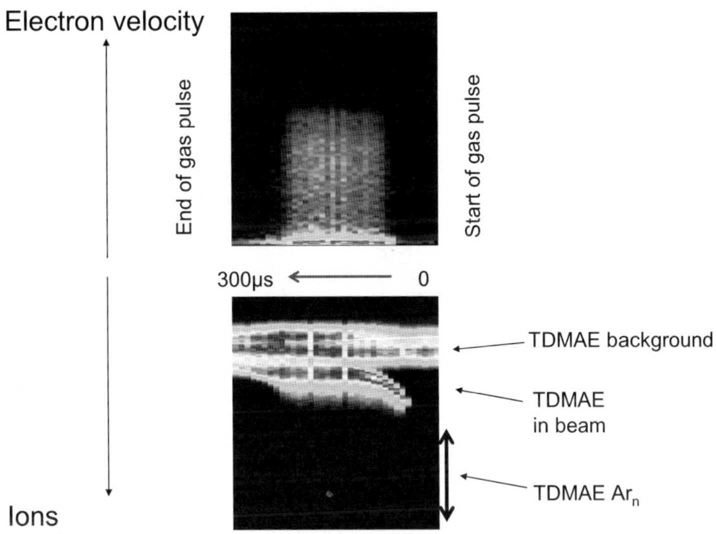

Figure 3: -Top panel 266/400 nm pump/probe photoelectron detection of TDMAE deposited on argon clusters. Horizontal delay between the pulse valve and the ionisation lasers. Vertical kinetic energy release of the electrons. The intensities are in false colors (white is most intense). - Bottom pannel ion detection performed in the same set up with inverse voltages, no selection on masses is applied. The arrow denotes masses of incresing size.

On the other hand if the photoelectron spectrum is represented on the top panel, the emitted photoelectrons arise mostly from TDMAE-Ar_n. This stems from the correspondence between the time dependence of the clusters in the ion imaging and that of the electrons as exagerated by the two white bars. We interpret this as a sign of the lower detectability of cluster ions with masses

>2000 AMU (50 argon atoms), in comparison with their good electron detection.

3. Results

3.1. Time resolved photoelectron spectra

Time resolved images of electron kinetic energy release (KER) are represented in Figure 4, where the angular distribution has been integrated over 2π. The electron energy release is represented as a function of the pump/probe delay. The background substracted spectrum, corresponding to the evolution in the clusters in Figure 4a is compared with that of the background in Figure 4b.

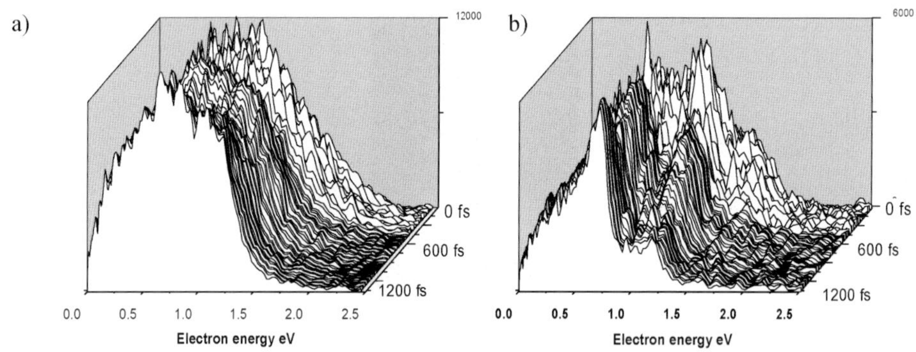

Figure 4: Time resolved photoelectron spectra as a function of the 266 nm/400delay of
-a) TDMAE deposited on an argon cluster,
-b background signal of TDMAE.

In this latter spectrum, the evolution can be clearly described by three bands, A, B, and C, which have been assigned in a previous paper, to respectively, the valence, a Rydberg, and zwitterionic states [12]. The decay is sequential, band A being immediately populated and decaying into B which in turn evolves into C. At first sight, it is difficult to distingish the same evolution on the cluster deposited spectra, which also correspond to a large distribution of sizes, as appears in the log-normal distribution represented in Figure 2b. Nonetheless, if one integrates in time these spectra, as shown in Figure 5, the same band pattern clearly shows up even in cluster deposited spectra; band B and C are visible as maxima and A as a shoulder to B. Returning to Figure 4, the evolution of the large domain corresponding to A, 1.5 eV to 2.5 eV, is apparent on the side of B; at time 0, band B is broad, while at 1200 fs, it descends rapidly on the high energy side.

When the two time integrated spectra (cluster deposited and background) are compared, in addition to the increased widths and the slight shifts of the cluster spectra, another effect is important.

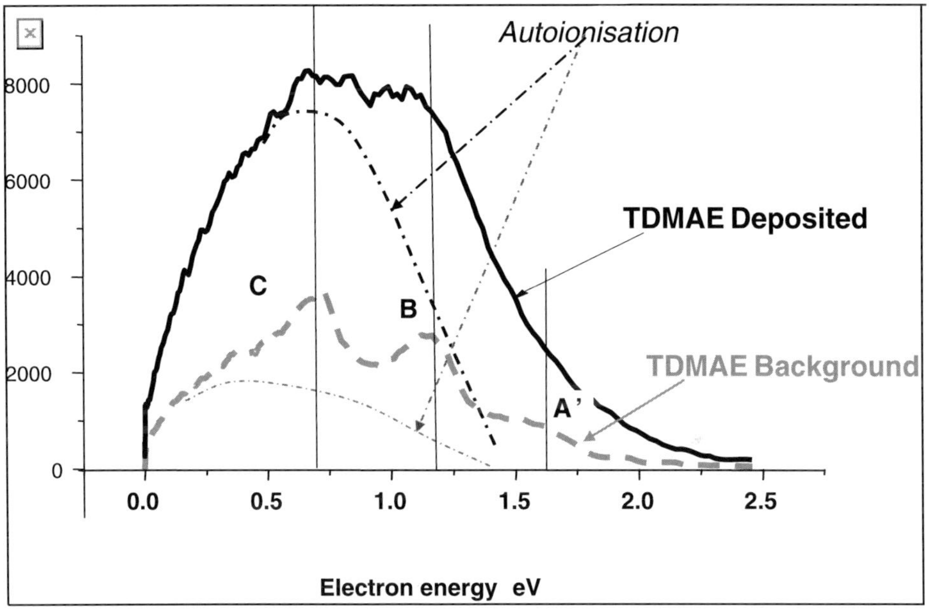

Figure 5: Time integrated photoelectron spectra of the cluster deposited TDMAE *(black)* and the background *(grey)*. Dot dashed curves represent the autoionisation background deduced from the instant rise in the time evolution.

There appears at low electron kinetic energies, at ~0.5 eV, an important backgound which can probably be related to autoionization. Already in the bare molecule, under 266/800 nm pump probe conditions, such a backgound is found in the photoelectron spectrum but to a much lesser extent. It can be related to autoionization and appears as a broad band centered at zero kinetic energy, represented by the dot dash curve in Figure 5. Here a vibrationally hot high lying level is created above the ionization, owing to the different geometries of the ion and the neutral clusters. The large density of van der Waals TDMAE-Ar_n modes will increase the latter autoionization effects. Thus autoionization should generate a broad distribution of low energy electrons. This will be confirmed in the following by monitoring the time evolution of specified energy regions of the time resolved photoelectron spectrum. Indeed we have performed cuts of the spectrum in Figure 4a at different electron energies.

3.2. Time evolution of selected photoelectron bands of the TDMAE-Ar$_n$ cluster

Four slices of the photoelectron spectrum in Figuer 4a have been made in the regions of bands A, B, and C. In region A, represented by the bottom two dot and square curves in Figure 6, an initial decay of the state immediately populated by the laser pulse is found and is in the same time scale as the free molecule, 350 fs viz. 200 fs [12].

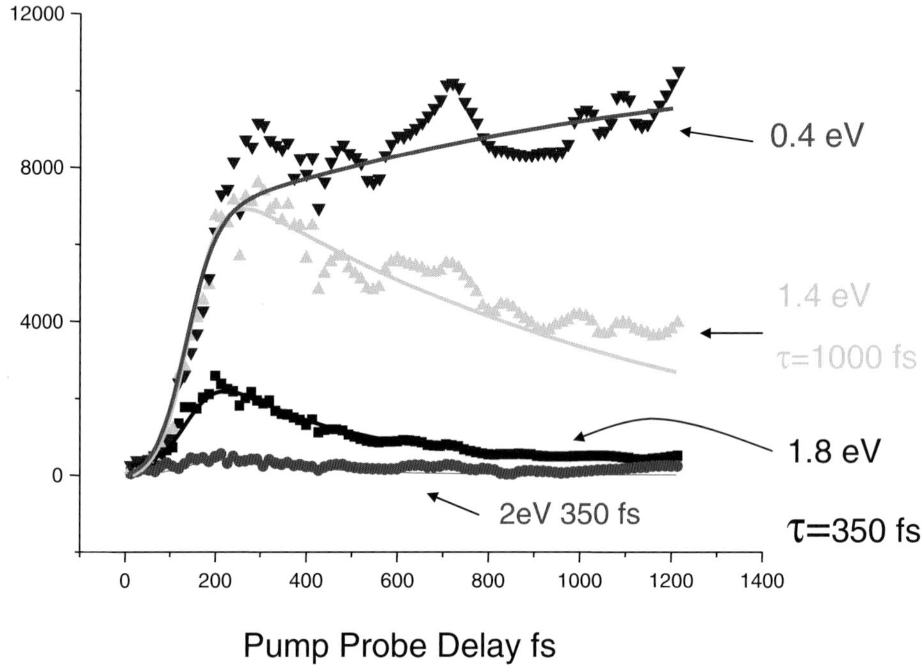

Figure 6: Time evolution of energy slices in the photoelectron spectrum from Figure a, representing the time evolution of bands A, B, C.

The lengthening effect related to clustering has already been observed in TDMAE-argon clusters in a different experiment where the free molecule was co-expanded in an argon beam. There, it can be inferred that TDMAE seeds the argon clustering and is therefore embedded in the cluster. We found a 400 fs decay in that case for interior clusters [17]; this was interpreted as a slowing down of the wavepacket motion by the presence of argon atoms. The effect is almost as important here, indicating an embedment of the TDMAE molecule within the cluster surface as sketched in Figure 2a. The decay of band B is also seen in ~1000 fs and matches an equivalent rise in band C. Since coherence is not lost in the state related to B (a Rydberg state), as wavepacket oscillations remain after a few picoseconds [17], an equivalent mechanism is also at play here.

In addition, from Figure 6, it is clear that a background underlies bands C and B, to a lesser extent. This background has no time dependence aside from its rise within the laser pulse. This supports the hypothesis of a non selective autoionization from the high lying state of the TDMAE-Ar_n cluster above the ionization threshold. This autoionization competes with direct ionization and is more important in the case of cluster deposited photoelectron spectra.

4. Conclusions

We have shown in this report that photoelectron spectroscopy can be applied to the investigation of cluster dynamics. On large clusters, photoelectron spectroscopy proves to be very sensitive. Deposition of large molecule on the argon cluster maintains important information relative to the electronic dynamics occurring in electronic relaxation of the molecules embedded in a medium.

5. Acknowledgements

The authors thank O. Gobert, F.Lepetit and M. Perdrix, who are responsible for developing, maintaining and running the femtosecond laser facility LUCA (Laser Ultra-Court Accordable) of the CEA, DSM/DRECAM. This work is partly supported by the European Community through the PICNIC network (Product Imaging and Correlation: Non-adiabatic Interactions in Chemistry) under contract number HPRN-CT-2002-00183. DHP is also grateful to the European Union for support to access the SLIC/LASERLAB ultrafast laser facility at CEA, Saclay.

References

1. P. M. Felker and A. H. Zewail, Chem. Phys. Lett. 128 (1986) 221.
2. E. W. G. Diau, S. De-Feyter, and A.H. Zewail, J. Chem. Phys. 110 (1999) 9785.

3. Z. Dongping, E. W. G. Diau, T. M. Bernhardt, S. De-Feyter, J. D. Roberts, and A. H. Zewail, Chem. Phys. Lett. 298 (1998) 129.
4. J. M. Mestdagh, J. P. Visticot, M. Elhanine, and B. Soep, J. Chem. Phys. 113 (2000) 237.
5. V. Blanchet, M. Zgierski, T. Seideman, and A. Stolow, Nature 401 (1999) 52.
6. P. Farmanara, V. Stert, and W. Radloff, Chem. Phys. Lett. 288 (1998) 518.
7. W. Fuss, S. Lochbrunner, A. . Muller, T. Schikarski, W. E. Schmid, and S. A. Trushin, Chem. Phys. 232 (1998) 161.
8. E. Teller, J. Phys. Chem. 41 (1937) 109.
9. A.W. Jasper, C. Zhu, S. Nangia, and D. G. Truhlar, Nonadiabatic effects in chemical dynamics, in Faraday Discuss. (2004) 1.
10. S. Pedersen, L. Banares, and A. H. Zewail, J. Chem. Phys. 97 (1992) 8801.
11. D. R. Cyr and C. C. Hayden, J. Chem. Phys. 104 (1996) 771.
12. E. Gloaguen, J. M. Mestdagh, L. Poisson, F. Lepetit, J. P. Visticot, B. Soep, M. Coroiu, A. T. J. B. Eppink, and D. H. Parker, J. Am. Chem. Soc. (2005).
13. B. Winter, R. Weber, I. V. Hertel, M. Faubel, L. Vrbka, and P. Jungwirth, Chem. Phys. Lett. 410 (2005) 222.
14. E. Loginov, D. Rossi, and M. Drabbels, Phys. Rev. Lett. 95 (2005) 163401.
15. O. F. Hagena, Z. Phys. D: At., Mol. Clusters 4 (1987) 291.
16. U. Buck and R. Krohne, J. Chem. Phys. 105 (1996) 5408.
17. S. Sorgues, J. M. Mestdagh, B. Soep, and J. P. Visticot, Chem. Phys. 301 (2004) 225.

Femtochemistry VII
Fundamental Ultrafast Processes in Chemistry, Physics, and Biology
A.W. Castleman, Jr. and M.L. Kimble

Solution Femtochemistry of Two Selected H-bonded Systems

A. Douhal*, J. A. Organero, M. Sanz, L. Tormo and L. Santos

Departamento de Química Física, Sección de Químicas, Facultad del Medio Ambiente, UCLM, Avda. Carlos III, S.N. 4071 Toledo, Spain.

1. Introduction

The design and characterization of molecules which undergo excited state intra- and intermolecular proton transfer reactions continue to engage a great scientific interest because of the unique energy and directional characters that determine the behaviour of many systems with a wide range of applications in life, material sciences and engineering [1-4]. In this contribution, we focus on two systems that exhibit excited state proton-transfer reaction: 2-(2′-hydroxyphenyl)benzoselenazole (HBSE) and 3-hydrxyflavone (3HF) (Scheme). HBSE is a derivative of 2-(2′-hydroxyphenyl)benzothiazole (HBT) which upon electronic excitation rapidly undergoes an intramolecular H-atom (or proton) transfer reaction to give keto tautomer (K) [5,6]. We briefly discuss the results for HBSE in solution using fs detection of its emission. 3HF molecule also undergoes an excited-state intramolecular proton-transfer reaction from its first excited singlet state (S_1) to give a zwitterionic tautomer (Z) in less than 50 fs [8–11]. In hydrogen (H)-bonding media, H-bonded structures to the solvent molecules (Scheme) give normal emission at the blue side of Z emission. Detection of anionic structures at S_0 and S_1 states also has been reported [12-15]. The results described here have been already reported in references [7,15].

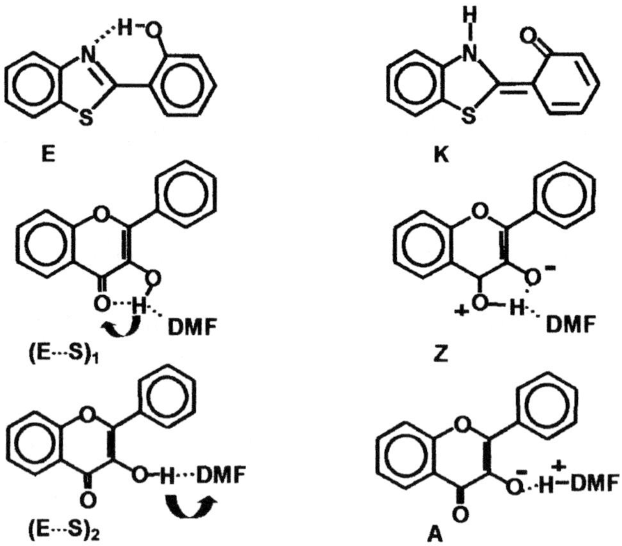

Scheme: Molecular structures of 2-(2′-hydroxyphenyl)benzoselenazole (HBSE) in its enol (E) and keto (K) tautomers and of 3-hydroxyflavonol (3HF) in its solvent-H-bonded forms (E···S)$_{1,2}$, zwiterionic and anionic structures.

2. Discussion and Results

2-(2′-hydroxyphenyl)benzoselenazole. *(HBSE).* Steady-state absorption and emission spectra of HBSE in tetrahydrofuran (THF) show a large Stokes shift (~ 10.000 cm^{-1}) that is interpreted in terms of the formation of a proton transferred tautomer (K) along the OH···N bond, upon a fast electronic redistribution. To obtain information about the involved radiationless processes we used femtosecond up-conversion technique exciting HBSE at 371 nm (without excess of energy) and collecting the emission at different wavelengths in fs regime (Figure 1). At 430 nm the emission decay fits to a bi-exponential function with time constants of 40 fs (98 %) and 230 fs (2%). At 460 nm the emission decay give time constants of 50 fs (94 %), 700 fs (4%) and 20 ps (2%) while at 530 nm are 85 fs (39 %), 700 fs (6%) and 20 ps (55 %). At the red part of the spectrum a rise of 600 fs and a decay of 20 ps were observed (Figure 1A). The first short component changes from 40 ps at 430 nm to 85 fs at 530 nm. We also observed this ultrafast component in HBT in THF (50 fs decay at the blue part of the spectrum).

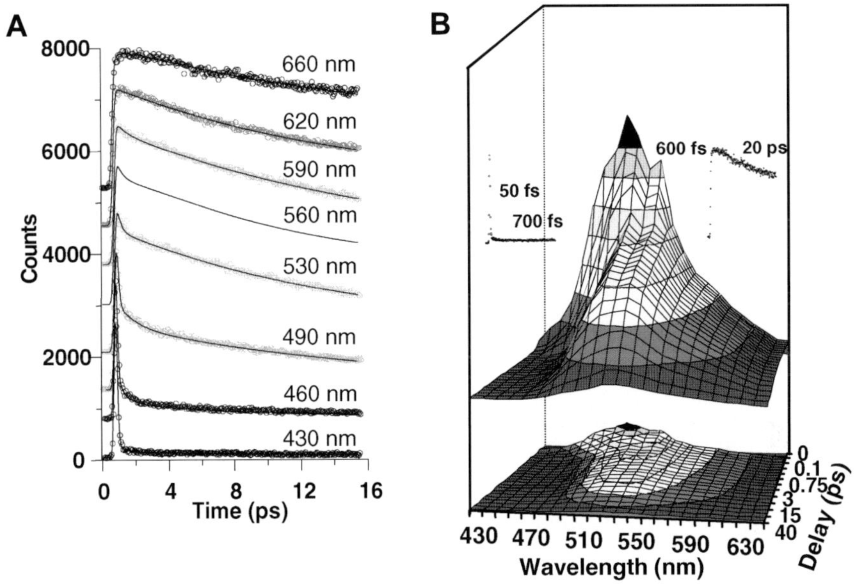

Figure 1: A) fs emission decays of HBSE in THF at different wavelengths. B) fs time-resolved emission spectra excited at 371 nm.

To obtain the origin of this ultrafast component we constructed time-resolved emission spectra of HBSE in THF (Figure 1B). We observed that up to 100 fs the emission spectra are centred at 510 nm, and at 520 nm when we collected the emission at 200 fs. At times longer than 2 ps the emission is centred at 540 nm. The ultrafast red shifts, suggests a high efficiency in the nonradiative processes (quantum yield of HBSE is lower than 10^{-4}). Therefore, the short component in the decays is due to a relaxation of the formed K tautomer populated just after proton-transfer reaction. We did not observe a significant difference in the lifetimes in methyl cyclohexane (a nonpolar solvent). Thus, the assignment of this short component to a solvent polarization can not be the origin. The decreasing of NH···O distance to reach the minimum of energy on the potential energy surfaces is largely implied on the proton transfer motion through the O-H···N coordinate and this distance increases (within ~ 40 fs) when the H atom is transferred to the nitrogen one to obtain a more relaxed K. From the two ultrafast red shifts we can deduce the relaxation frequencies assigned to the relaxation processes of K: 380 and 710 cm^{-1}. These frequencies might be due low frequency modes of in-plane and out-of-plane motions of phenyl and oxazole parts to modulate the N-O distance, and the angle formed by their planes. Finally, the 500–700 fs component is assigned to the intramolecular-vibrational energy redistribution (IVR), and fast vibrational relaxation of K, according to the IR-transient measurements of HBT [16].

3-Hydroxyflavone. 3 HF in the H-bonding solvent DMF gives birth to H-bonded structures to the solvent molecules $(E\cdots S)_{1,2}$ (Scheme). The 0-0 transition of these structures is at ~ 365 nm. Excitation at 340 nm gives a normal emission at the blue side of the emission spectrum (~ 400 nm) and a green emission band at ~ 540 nm due to the zwitterionic forms. However, exciting at 420 nm (absorption band of the anion, A) we only observed the emission band of A at ~ 482 nm. The rates of inter- and intramolecular proton-transfer reactions to give A or Z depend on the nature and strength of the intermolecular H bonds in these species.

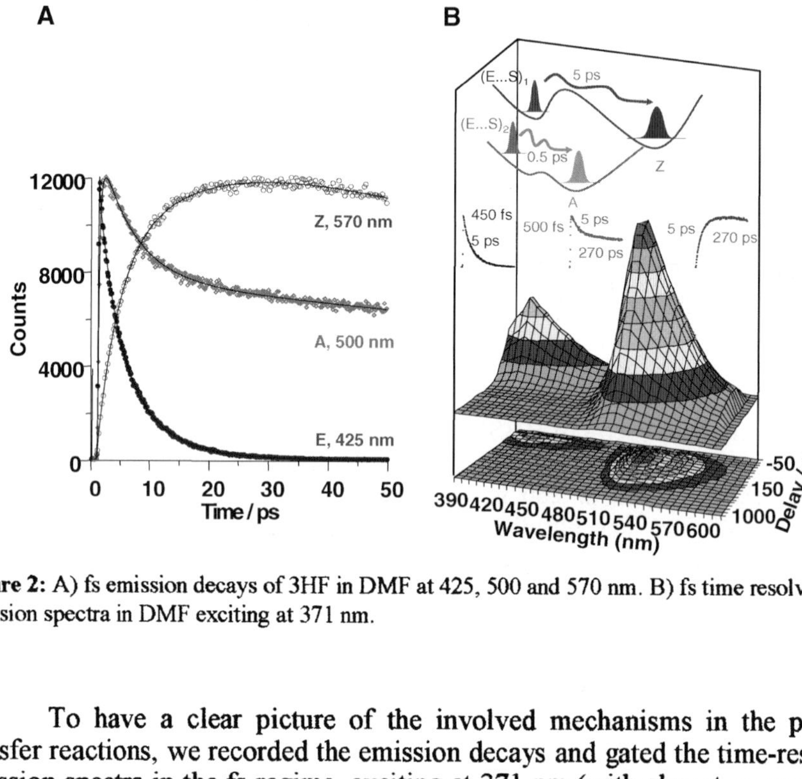

Figure 2: A) fs emission decays of 3HF in DMF at 425, 500 and 570 nm. B) fs time resolved emission spectra in DMF exciting at 371 nm.

To have a clear picture of the involved mechanisms in the proton-transfer reactions, we recorded the emission decays and gated the time-resolved emission spectra in the fs regime, exciting at 371 nm (with almost no excess of energy, Figure 2). Analyzing the emission transients at 425 nm ($(E\cdots S)_{1,2}$ domain) we obtained time constants of 450 fs (19%) and 5.2 ps (81%). The first component is not observed at emission wavelengths beyond 440 nm so we attribute this fs component to the $(E\cdots S)_2$ dynamics giving A. The emission decay observed at 500 nm (emission of A domain) was fitted to a 500-fs rise time (correlated with the time constant of 420 fs observed at 425 nm), and ~5.9 ps (29%) and ~270 ps (38%) decays components assigned to a fast intermolecular proton- transfer reaction in $(E\cdots S)_2$ to give A and the decays of $(E\cdots S)_{1,2}$ and Z structures, respectively. At 560 nm, we observed a 5 ps rising component (48%) and a decaying component of 270 ps (52 %) assigned to an intramolecular proton-transfer reaction in $(E\cdots S)_1$ to give Z and lifetime of Z respectively. We made also a control experiment exciting at 400 nm (where the absorption mainly due to A) and collecting the emission at 500 nm. We observed a bi-exponential decay with time constants of 500 fs and 4.1 ps, assigned to vibrational-energy redistribution/ solvent response and vibrational relaxation in hot A and Z, respectively. No rising component was observed upon direct excitation of A. Furthermore in acetonitrile (solvent with weaker H-bond-accepting ability but with similar polarity of DMF) we observed rise times of ~ 50 fs and ~3.2 ps in the green region assigned to direct (fs regime) and intermolecular H-bond-controlled dynamics of excited-state intramolecular proton-transfer reactions in $(E\cdots S)_1$. Therefore, the time constant for solvent-modulated intramolecular proton transfer in $(E\cdots S)_1$ is shorter for a weaker-H-bonding acceptor solvent (3.2 ps in AC and 5.3 ps in DMF). Further more, the observed intermolecular proton-transfer reaction in $(E\cdots S)_2$ is faster (~500 fs) than the intramolecular proton transfer observed in $(E\cdots S)_1$ (5 ps) in DMF. This is a result of a stronger intramolecular H-bond in the later, relatively to the intermolecular H-bond in the former complex. The later may need a weakening (or a rupture) of the intermolecular H-bond and, probably, a reconfiguration of the solvation shell.

3. Conclusion

Fs-emission experiments on HBSE in solution show that the fast nonradiative processes are related to in-plane and out-of-plane motions in the initially formed K phototautomer. For 3HF in solution, the different observed dynamics of inter and intramolecular proton transfer reaction are affected by H-bonding interactions with the solvent molecule.

Acknowledgment: This work was supported by the MEC and the JCCM through the projects: MAT2002-01829, SAN-04-000-00 and PBI-05-046.

References:

1. A. Douhal, F. Lahmani, and A. H. Zewail, Chem. Phys. 207 (1996) 447.
2. A. Douhal, S. K. Kim, and A. H. Zewail, Nature 378 (1995) 260.
3. E. T. J. Nibbering and T. Elsaesser, Chem. Rev. 104 (4) (2004) 1887.
4. R. Rey, K. B. Moller, and J. T. Hynes, Chem. Rev. 104 (2004) 1915.
5. T. Elsaesser and W. Kaiser, Chem. Phys. Lett. 128 (1986) 231.
6. A. Douhal, F. Amat-Guerri, M. P. Lillo, and A. U. Acuña, J. Photochem. Photobiol. A 78 (1994) 127.
7. M. Sanz and A. Douhal, Chem. Phys. Lett. 401 (2005) 435.

8. A. J. G. Strandjord and P.F. Barbara, J. Phys. Chem. 89 (1985) 2355.
9. M. Itoh and H. Kurokawa, Chem. Phys. Lett. 91 (1982) 487.
10. S. J. Formosinho and L. G. Arnaut, J. Photochem. Photobiol. A: Chem. 75 (1993) 21.
11. S. M. Dennison, J. Guharay, and P. K. Sengupta: Spectrochim. Acta A 55 (1999) 903.
12. M. Itoh, K. Tokumura, Y. Tanimoto, Y. Okada, H. Takeuchi, K. Obi, and I. Tanaka, J. Am. Chem. Soc. 104 (1982) 4146.
13. A. J. G. Strandjord, S. H. Courtney, D. M. Friedrich, and P. F. Barbara, J. Phys. Chem. 87 (1983) 1125.
14. P. T. Dzygan, J. Schmidt, and T. J. Artsma, Chem. Phys. Lett. 127 (1986) 336.
15. A. Douhal, Mikel Sanz, L. Tormo, and J. A. Organero, ChemPhysChem 6 (2005) 419.
16. M. Rini, A. Kummrow, J. Dreyer, E. T. J. Nibbering, and T. Elsaesser, Faraday Discuss. 122 (2002) 27.

Femtochemistry VII
Fundamental Ultrafast Processes in Chemistry, Physics, and Biology
A.W. Castleman, Jr. and M.L. Kimble

Solvent Dependent Conformational Relaxation of *cis*-1,3,5-Hexatriene

D. Ahmasi Harris[a], Roseanne J. Sension[a,b,c]

[a]*Program in Applied Physics,* [b]*Department of Chemistry,* [c]*Department of Physics University of Michigan, Ann Arbor, Michigan 48109 USA*

1. Introduction

Solvent-chromophore interactions play a vital role in chemical reactions by influencing the reactivity of and energy flow in the species involved. In previous studies we have investigate the photochemical ring-opening reaction and ground state conformational relaxation of 1,3-cyclohexadiene (CHD) and the related 7-dehydrocholesterol chromophore responsible for the photoreaction that results in vitamin D formation [1-3]. Ultraviolet excitation of 1,3-cyclohexadiene initiates a ring opening reaction that produces vibrationally hot *cis*-1,3,5-hexatriene (Z-HT) with a quantum yield of 40% [4]. This reaction is shown in Figure 1.

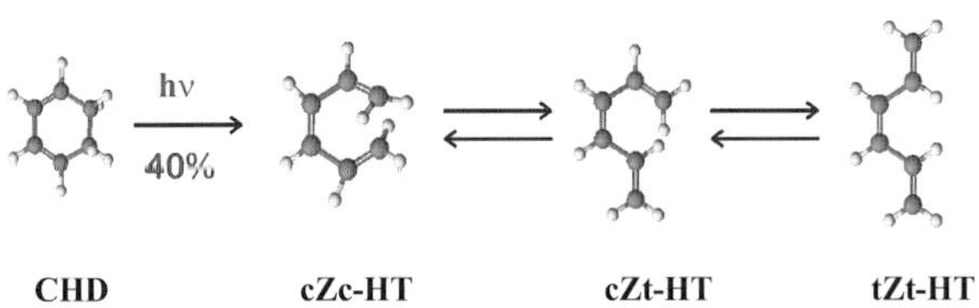

Figure 1. CHD ring opening and conformational relaxation

Upon ground state recovery following photo-initiation of the ring-opening reaction, rapid interconversion among the three conformers of HT produces a dynamic equilibrium mixture of cZc-HT, cZt-HT, and tZt-HT. As the molecules relax, the mixture evolves in composition and deposits energy into the surrounding solvent. Thermal equilibrium with the solvent occurs on a time scale of 10-20 picoseconds, leaving a small portion of cZt-HT molecules trapped for a period »100 ps by a barrier of single bond isomerization to tZt-HT [1-2]. The relevant potential energy surfaces are shown in Figure 2 along with the absorption spectrum of CHD and tZt-HT. We investigate the effect of solvent polarity on the ground state conformational relaxation of vibrationally hot *cis*-1,3,5-hexatriene following the CHD ring opening reaction. We contrast this work with earlier work in which the dynamics of CHD were studied in nonpolar alkane solvents [2]. This work also complements the work done by Lochbrunner *et al.*, where the dynamics of CHD were studied in ethanol [5].

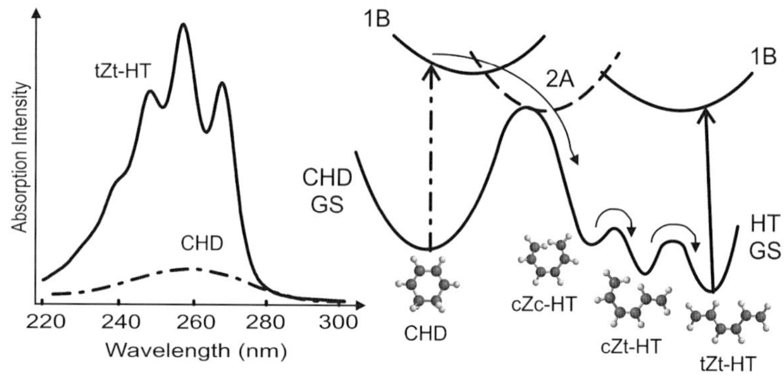

Figure 2. (left) Ground state absorbance of 1,3-cyclohexadiene and 1,3,5-hexatriene, and (right) cartoon of relevant potential energy surfaces involved in CHD ring opening.

2. Experiment

Pump-probe transient absorption experiments were performed using 0.4 mJ laser pulses from a homebuilt 1 kHz titanium-sapphire laser system described in detail elsewhere [2]. 1,3-cyclohexadiene (Aldrich 95%) samples were prepared in solution to have an optical density of 1 at 267 nm for a 1 mm sample thickness, and a pump and flow cell were used to refresh the sample volume. The sample was excited at 267 nm with the output of a 3rd harmonic generation setup followed by a fused silica prism pair for pulse recompression. For kinetic measurements, a tunable ultraviolet probe pulse was generated by taking the second harmonic of a visible optical parametric amplifier (OPA). For spectral measurements, a broadband ultraviolet continuum was generated by frequency doubling white-light continuum produced from focusing infrared laser pulses into flowing ethylene glycol.

Transient absorption spectroscopy was used to investigate CHD in two alcohol solvents, methanol (Aldrich 99%) and 1-propanol (Aldrich 99%), for comparison with data reported earlier for CHD in alkane solvents [2]. Transient absorption kinetics were taken with probe wavelengths in the spectral region 265 nm – 290 nm inclusive. Two photon absorption of temporally overlapped pump and probe pulses in the alcohol solvents gives a cross correlation of ca. 200 fs at the sample. Transient difference spectra were taken at multiple time delays in order to scale the kinetics data for each wavelength. The data is summarized in the left panel of Figure 3.

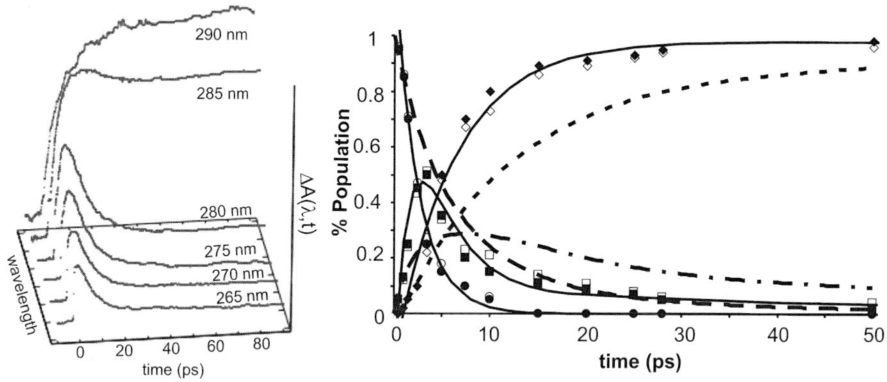

Figure 3. (Left) Scaled transient absorption kinetics of CHD in methanol are shown (similar in propanol). (Right) Calculated hexatriene conformer dynamics in methanol (filled) and propanol (open) are shown: circle(○●) – cZc-HT, square(□■) – cZt-HT, diamond(◊♦) – tZt-HT. Conformer dynamics in alkane solvents are also shown: dashed – cZc-HT, dot-dashed – cZt-HT, dotted – tZt-HT.

3. Results

The transient absorption data details ground state recovery and reveals the conformer dynamics following the ring-opening reaction. The conformer population dynamics, Φ(t), were extracted from the transient absorption data using Equations 1 and 2, where the conformer absorption spectra, A(λ), were calculated previously [2].

$$\Delta A(t,\lambda) = \Phi_{cZc}(t)A_{cZc}(\lambda) + \Phi_{cZt}(t)A_{cZt}(\lambda) + \Phi_{tZt}(t)A_{tZt}(\lambda) \qquad (1)$$

$$\Phi_{cZc}(t) + \Phi_{cZc}(t) + \Phi_{cZc}(t) = constant \qquad (2)$$

The right panel in Figure 3 shows the normalized conformer population dynamics of HT following the CHD ring opening reaction in both alkane and alcohol solvents. The mixture of conformers has a more rapid evolution in the alcohol solvents than in alkane solvents. This is marked both by a faster decay of cZc-HT and a faster growth of cZt-HT. This is also reflected as a larger short time population of the intermediate conformer, cZt-HT. At long times (> 50ps), the trapping of the cZt-HT population is diminished in the alcohol solvents in comparison to the alkane solvents. Similar dynamics were observed in an ethanol solvent by Lochbrunner *et. al.*[5].

4. Discussion

Conventional consideration of the influence of solvent polarity on reaction dynamics requires an excited state dipole moment on a solute molecule that induces a rearrangement of surrounding polar solvent molecules, thus affecting the reaction rates through changing energy barrier heights [6]. This effect is in addition to the potential influence of solvent friction on the reactive molecule [7]. Comparing the conformer dynamics in alkane and alcohol solvents reveals a minimal contribution of solvent viscosity or friction but a considerable dependence on solvent polarity.

For double-bond isomerization, the dipole moment is produced by a "sudden polarization" of the molecule, wherein a molecule may acquire a dipole moment through rearrangement of π-electrons [8]. Considering this source of solute-solvent interactions, the effects of the polar solvent are expected to diminish upon ground state recovery in hexatriene. The reaction coordinate does not involve breaking a double bond but is dominated by steric interactions between hydrogen atoms attached to adjacent σ-bonded carbon atoms. With no π-electrons involved in the ground state single bond isomerization, there is no perceived mechanism for interaction with the polar solvent. In contradiction to this expectation, the polar solvent greatly influences ground state conformer dynamics of hexatriene, increasing single bond isomerization reaction rates in the polar alcohol solvents compared to the nonpolar alkane solvents.

The diminished trapping of the cZt conformer in alcohol solvents points to a change in the activation energy barrier height of ground state single bond isomerization (cZt $\rightarrow$ tZt) in comparision to the alkane solvents. Work is underway to characterize this activation energy barrier height by examining the reaction rate as a function of temperature.

References

1. S. H. Pullen, N. A. Anderson, L. A. Walker, and R. J. Sension, J. Chem. Phys. 108 (1998) 556.
2. N. A. Anderson, S. H. Pullen, L. A.Walker, J. J. Shiang, and R. J. Sension, J. Phys. Chem. A 102 (1998) 10588.
3. N. A. Anderson, J. J. Shiang, and R. J. Sension, J. Phys. Chem. A 103 (1999) 10730.
4. N. G. Minnaard and E. Havinga, Recl. Trav. Chim. Pays-Bas. 92 (1973) 1315.
5. S. Lochbrunner, W. Fuss, W. E. Schmid, and K. L. Kompa, J. Phys. Chem. A 102 (1998) 9334.
6. B. Bagchi, Annu. Rev. Phys. Chem. 40 (1989) 115.
7. P. Hänggi, P. Talkner, and M. Borkovec, Rev. Mod. Phys. 62 (1990) 251.
8. L. Salem, Acc. Chem. Res. 12 (1979) 87.

4. Liquids and Interfaces

Femtochemistry VII
Fundamental Ultrafast Processes in Chemistry, Physics, and Biology
A.W. Castleman, Jr. and M.L. Kimble

Vibrational Echo and Pump-Probe Spectroscopic Studies of the Dynamics of Water Molecules Confined to Nanoscopic Dimensions

Ivan R. Piletic,[a] Howe Siang Tan,[b] David E. Moilanen,[a] Ben Spry,[a] Michael D. Fayer[a]

[a]*Stanford University, Stanford CA 94305, USA*
[b]*Nanyang Technological University, 1 Nanyang Walk 637616, Singapore*

1. Introduction

A question that has a far reaching impact in the fields of surface science, biology, separation chemistry, and reaction dynamics is what happens to the motion of molecules in a condensed phase when boundaries approaching molecular size dimensions are imposed? For instance, the surface of a liquid exhibits surface tension which is a result of anisotropic interactions of molecules at the interface. This conveniently differentiates molecules at the surface from molecules in the bulk. When an interface between two phases is wrapped around on itself to form an enclosed volume (like a sphere) the effects of confinement have to be addressed. Not only do interfacial interactions have to be accounted for in this case but also the effect of the restricted volume which inhibits free motion of all molecules in all directions. A strongly coupled system such as an associated liquid is expected to exhibit slower dynamics since the motion of one molecule affects the dynamical evolution of many others that are all restricted spatially. This strain along with interfacial interactions combine to effect the dynamics of molecules in confined environments. The effects of confinement are prevalent in biological systems since a substantial fraction of water molecules are encapsulated in vesicles and cells as well as in and around proteins.

A system that is typically used as a model for a confined environment such as those encountered in biological and chemical systems is a reverse micelle [1]. A reverse micelle is a micelle in which the nonpolar and polar phases have reversed roles and the orientation of surfactant molecules are inverted so that the head groups point into the enclosed volume containing the polar phase. Reverse micelles provide a convenient way of forming nanoscopic pools containing only a small amount of polar solvent.

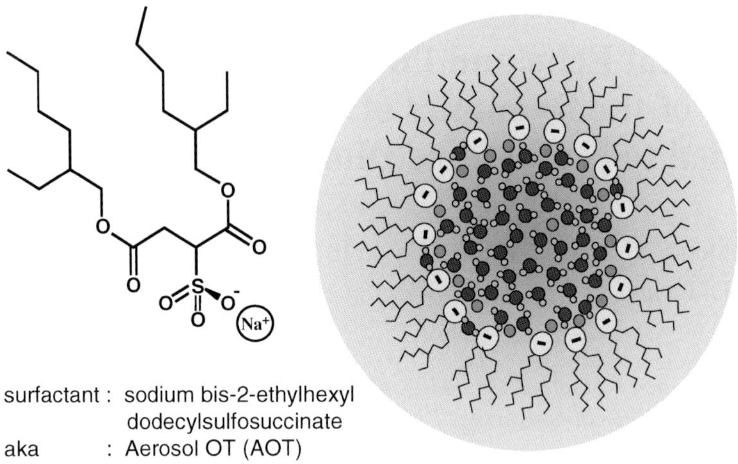

Figure 1. The anionic AOT surfactant and a reverse micelle.

Aerosol OT (AOT) reverse micelles have been the subject of intense research since they readily form under a variety of conditions and their size is easily tunable [2]. Figure 1 displays a schematic of an AOT reverse micelle. The surfactant molecules are anionic with Na^+ counter ions. The size is predominantly determined by the value of the parameter w_0.

$$w_0 = \frac{[H_2O]}{[AOT]} \tag{1}$$

A wide range of reverse micelles was studied whose water pool diameters ranged from 1.7 – 28 nm [2]. This range of reverse micelle sizes encapsulate dozens to hundreds of thousands of water molecules. At what point do the dynamical characteristics of bulk water emerge? This question will be addressed here.

In order to explore the dynamics of molecules in condensed phase systems, ultrafast spectroscopic techniques are required. Molecular vibrations occur on the timescale of tens of femtoseconds while molecular dynamics occur on the femtosecond and picosecond timescales. The coupling of vibrations to environmental changes make ultrafast vibrational spectroscopy ideal for exploring condensed phase dynamics.

Pump-probe and stimulated vibrational echo spectroscopies have the capability of discriminating different types of dynamics that occur in a system. For a vibrational transition, the excited state lifetime is typically on the order of several picoseconds [3,4]. The finite lifetime (T_1) for a spectroscopic transition is one source of line broadening and its contribution can be determined by pump-probe spectroscopy. Energy relaxation processes are important in chemistry since many chemical reactions require the localization of energy in specific bonds [5].

Orientational motions of molecules can also be resolved using pump-probe experimental techniques. In this case, the polarization of the probe pulse is set parallel or perpendicular to the pump pulse. Both signals are used to construct the anisotropy ($r(t)$) which provides a direct measure of the correlation function describing the reorientation of transition dipoles in the sample [6].

Stimulated vibrational echo spectroscopy probes coherences of molecules that decay as a result of irreversible dephasing processes. Fluctuations of the bath that couple to a vibrational transition cause modulations in the absorption frequency. The time evolution of the frequency is called spectral diffusion, and it can be measured with a stimulated vibrational echo experiment. A frequency-frequency correlation function (FFCF) is used to characterize the time evolution of the ensemble of transition frequencies excited by the pulses that span the entire vibrational spectrum [7]. Typically a multi-exponential form is assumed for the FFCF in fitting the data.

$$C(t) = \langle \delta\omega(t)\delta\omega(0) \rangle = \sum_i A_i^2 \exp(t / \tau_i) \tag{2}$$

For water, the FFCF is observed to decay on essentially two timescales: sub-100 fs and 1.5 ps [8]. The two timescales are typically classified as the local and global dynamics. The local dynamics may be ascribed to local translations and librations of water molecules that maintain essentially the same hydrogen bond network structure while the global dynamics represent hydrogen bond network global structural rearrangements that involve the formation and dissociation of hydrogen bonds.

To what extent are these aspects of bulk water behavior different from water confined in reverse micelles? Dielectric relaxation and solvation dynamics experiments suggest dynamics in reverse micelles that are orders of magnitude slower than the dynamics exhibited in bulk water [1]. Ultrafast vibrational spectroscopies provide several advantages over other techniques that make it a more suitable probe of water dynamics. The most important aspect of these experiments is that they directly examine the water dynamics rather than infer the water dynamics by the effect on a probe molecule. Another aspect is that the time resolution is on the order of 50 fs which permits the full range of dynamics to be investigated. All types of water molecules are excited, and the dynamics of water in specific environments within the inhomogeneous distribution can be selected by frequency resolving the signals. Structural fluctuations of the hydrogen bond network of water are strongly correlated to the hydroxyl stretch frequency. The time dependence of vibrational echo signals can be directly related to hydrogen bond network evolution.

2. Experimental Procedures

AOT reverse micelles of various sizes were made by mixing the appropriate amount of 5% HOD in H_2O solution in a stock solution containing AOT dissolved in isooctane. Due to the hygroscopic nature of AOT, the amount of residual water in the stock solution was determined by Karl Fischer titration. All samples were housed in copper cells containing two CaF_2 windows (3 mm thickness) and a variable thickness Teflon spacer. Optical densities of the OD stretch were typically 0.2 – 0.4 for all the samples studied.

The laser system consists of a Ti:Sapphire oscillator that is used to seed a regenerative amplifier pumped by a Q-switched Nd:YLF laser operating at a kilohertz. The output of the regen pumps an optical parametric amplifier (OPA). A BBO crystal is used to generate various near IR wavelengths that are difference frequency mixed in a 0.5mm $AgGaS_2$ crystal. 4 μm mid-IR pulses are generated and subsequently characterized in the time and frequency domains. A monochromator is used to measure the laser power spectrum, and the typical bandwidths achieved lie in the range of 230 – 250 cm^{-1} (FWHM). The time duration of the pulses is measured by crossing three mid-IR beams (used in the vibrational echo experiments) in a nonresonant sample containing CCl_4. A FROG experiment is used to determine the chirp on the pulses. Transform-limited pulses are obtained by adjusting the amounts of CaF_2 and Ge in the mid-IR beam path since they have opposite affects on the 2nd order dispersion of mid-IR frequencies.

Two types of ultrafast vibrational spectroscopies were employed in this study. Pump-probe spectroscopy involves the use of an intense pump pulse followed by a weak probe pulse. The probe pulse polarization is set to 45° relative to the pump, and the resulting pump-probe signal is measured by two detectors with polarizers in front set so that one detector measures the parallel signal and the other detector measures the perpendicular signal. The isotropic and anisotropic signals can be simultaneously resolved using the following equations, respectively [6].

$$P(t) = I_{//} + 2I_{\perp} \tag{3}$$

$$r(t) = \frac{(I_{//} - I_{\perp})}{(I_{//} + 2I_{\perp})} \tag{4}$$

where $P(t)$ is the population decay and $r(t)$ is the orientational anisotropy decay. The signals are subsequently normalized by the probe intensity and processed by gated integrators and lock-in amplifiers and recorded by computer.

Stimulated vibrational echo spectroscopy requires the interaction of three mid-IR pulses with the sample. The three pulses are generated by using coated ZnSe beam splitters and two delay stages which are used for the two inter pulse delays τ and T_w. The three pulses are focused into the sample in a boxcar geometry, and the echo pulse emits in the phase matching direction [7].

$$\overrightarrow{k_{echo}} = -\overrightarrow{k_1} + \overrightarrow{k_2} + \overrightarrow{k_3} \tag{5}$$

The signal is frequency resolved in a monochromator and detected with an InSb detector and processed using a gated integrator and recorded by computer.

3. Results and Discussion

The absorption spectrum of the hydroxyl stretch of water also provides information about the hydrogen bond network since they are strongly coupled. Figure 2 displays the FTIR absorption spectra of bulk water as well as water in

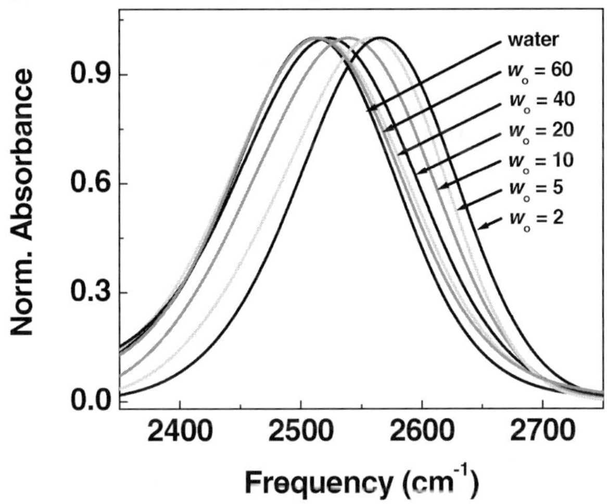

Figure 2. Background subtracted FTIR absorption spectra of 5% HOD in H2O in bulk water as well as water in AOT reverse micelles of various sizes (w_0 = 60 – 2).

AOT reverse micelles ranging in size from w_0 = 60 – 2. A dilute concentration of 5% HOD in H_2O was used to avoid excitation transfer as well as to ensure that the sample absorption was not too high. It is evident from the spectra that the peaks are blue shifting as a function of decreasing reverse micelle size. In bulk water, the position in the inhomogeneous line is rated to the number and strengths of the hydrogen bonds, with stronger hydrogen bonds and more hydrogen bonds being shifted to lower frequency. This is deduced from correlations of hydrogen bond strengths to IR absorption frequencies [9]. At this time, it is unknown what is responsible for the blue shift in the entire spectrum as the size of the water nanopool is made smaller. It is not possible to obtain dynamical information from the lineshapes because they are featureless and their widths depend on a number of broadening mechanisms that are extremely difficult to deconvolute out of the spectra [7]. Nonlinear spectroscopic techniques are required to extract the underlying dynamics.

Polarization resolved pump-probe experiments were conducted to measure the vibrational lifetime (T_1) and the orientational diffusion constant (D_{or}) [6]. Figure 3 displays the isotropic and anisotropic responses of all

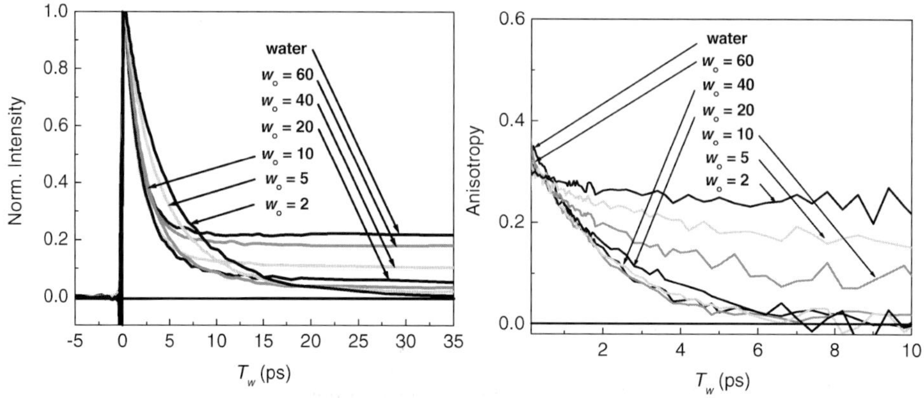

Figure 3. The isotropic (A) and anisotropic (B) signals of water and water confined in AOT reverse micelles detected at the center of the OD stretch absorption line for each sample.

samples under study. The dynamics observed in both plots reveal that the relaxation processes are complex since neither response can be fit with a single exponential. The isotropic signals are complicated by the occurrence of hydrogen bond dissociation during vibrational relaxation. During the experiment, the energy deposited by the pump pulse (at most one photon per micelle) eventually equilibrates and causes the local temperature to increase. The process causes a shift in the equilibrium number of hydrogen bonds to fewer bonds. The resulting positive long-lived signals occur because there is a reduction in the number of molecules that absorb at the detection frequency. The anisotropy decays for the smaller micelles ($w_o \leq 10$; diameter ≤ 4 m) are bi-exponential. The data suggest that the orientational relaxation of water molecules in small reverse micelles is restricted on short time scales to a limited cone of angles, giving rise to the fast exponential. At longer times, full randomization of the orientations occurs, producing the slower exponential.

Vibrational lifetimes for all samples were determined by fitting the data in Figure 3A to an exponential decay at early times. Other data not presented here were taken in the frequency range that probes the increased absorbance of molecules that were excited by the pump pulse. Due to the anharmonicity of the OD stretch mode, the absorbance of these molecules are red shifted by ~160 cm^{-1} from the absorption spectrum maximum. This experiment can measure the lifetime more accurately since the signals are free from ground state dynamics arising from thermal equilibration. The data confirms the vibrational lifetimes displayed in Table 1. The vibrational lifetime is observed to increase by a factor of ~3.5 when comparing bulk water with the $w_o = 2$ reverse micelle. The substantial increase in the vibrational lifetime of various molecular probes has been observed previously [10]. It is difficult to precisely account for the trend observed in the vibrational lifetimes. Fermi's golden rule predicts that either the density of states that the mode relaxes into or the coupling between the initial and final states change for the transition probability (between the excited and ground states) to change. It appears as though both are affected for water in a reverse micelle. Terahertz surface modes have been observed for water in an AOT reverse micelle [11]. Excitation of the head groups provide alternative relaxation routes for water molecules that have been excited in a reverse micelle [12]. The coupling between the states may also change because the nature of the hydrogen bond network varies as a function of reverse micelle size.

Parameters for multi-exponential fits to the anisotropies are also listed in Table 1. τ_{o1} is the fast decay component of the bi-exponential decay, which only occurs for w_o = 10 or less. τ_{o2} is the orientational decay constant for complete orientational randomization.

Table 1. Vibrational lifetimes and anisotropy fit parameters.

Sample	$d_{water\ pool}$ (nm)	T_1 (ps)	τ_{o1} (ps)	τ_{o2} (ps)
water	infinity	1.5	-	2.0
$w_o = 60$	28	1.6	-	2.0
$w_o = 40$	18	1.6	-	2.2
$w_o = 20$	7	1.9	-	2.6
$w_o = 10$	4	2.5	1.6	10
$w_o = 5$	2.4	3.9	1.0	19
$w_o = 2$	1.7	5.0	0.9	78

For large reverse micelles that display single exponential decays, the decays rapidly converge to the response exhibited in bulk water. The bi-exponential decays observed for the small reverse micelles were fit using a restricted orientational diffusion model ('wobbling in a cone') [13]. This model involves fast orientational diffusion within a cone followed by slower, full orientational relaxation. The data provide the cone angles, the diffusion constants for motion within the cones, and the final diffusion constants as a function of the nanopool size. The two processes can be interpreted as a local angular fluctuation of the OD bond followed by global hydrogen bond network rearrangement that permits complete orientational randomization The trend in the long decay constants indicates a longer hydrogen bond network rearrangement time with decreasing reverse micelle size.

Stimulated vibrational echo experiments were conducted to explore the structural evolution of the hydrogen bond network. Vibrational echo decays were obtained at the peak of the absorption spectrum for each size reverse micelle as a function of T_w, the time between the second and third pulses in the vibrational echo pulse sequence. The data are fit using time dependent diagrammatic perturbation theory to obtain the frequency-frequency correlation function (FFCF) for each reverse micelle. The results are compared to the FFCF of water for which the slowest component of the spectral diffusion is 1.5 ps [8]. In bulk water, the slowest time scale dynamics are caused by global structural evolution of the hydrogen bond network that involves making and breaking of hydrogen bonds [8].

To simplify the data display, Figure 4 shows the peak shifts of the vibrational echo decay curves as a function of T_w. As T_w is increased there is

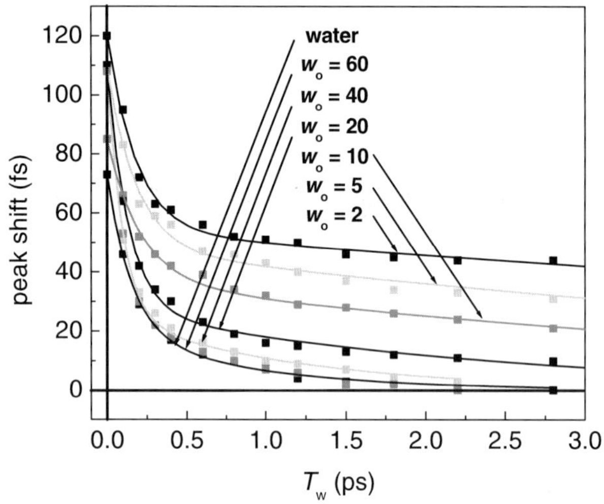

Figure 4. Vibrational echo peak shift data for bulk water and water in AOT reverse micelles.

more time for structural evolution of the hydrogen bond network, which results in more spectral diffusion, and hence, a faster vibrational echo decay with a reduced peak shift. When the peak shift becomes zero, structural evolution is complete in the sense that all configurations have been sampled. The water dynamics occur on a variety of time scales. The greatest difference between bulk water and nanoconfined water occurs on the longest time scale [14]. As the size of the water nanopool is decreased there is a dramatic slowing of the hydrogen bond network evolution. Fitting the data indicates that for the smallest nanopools, the longest time scale component of the water dynamics is ~15 times longer than dynamics in bulk water. Like the orientational relaxation data, the vibrational echo results demonstrate that for water nanopool diameters >7 nm, bulk water characteristics are approached rapidly. It also explains why slower orientational timescales are observed in the smallest reverse micelles. Since global hydrogen bond network rearrangements are suppressed to a greater extent in the smallest reverse micelles, water molecules are restricted from reorienting and consequently exhibit slower orientational relaxation times.

Exploring other confined environments with these techniques will provide increased understanding of the role of water-head group interactions in reverse micelle and other systems. Other notable confining systems include water in non-ionic or cationic reverse micelles, in sol-gel glasses, as well as in Nafion membranes. Since solvents affect chemical reaction dynamics, it is not surprising to observe that there are differences in chemical reactivity of reagents in reverse micelles nanoscopic water pool versus bulk water.

References

1. N. Nandi, K. Bhattacharyya, and B. Bagchi, Chem. Rev. 100 (2000) 2013.
2. M. Zulauf and H. F. Eicke, J. Phys. Chem. 83 (1979) 480.
3. K. J. Gaffney, P. H. Davis, I. R. Piletic, N. E. Levinger, and M. D. Fayer, J. Phys. Chem. A 106 (2002) 12012.
4. T. Steinel, J. B. Asbury, J. R. Zheng, and M. D. Fayer, J. Phys. Chem. A 108 (2004) 10957.
5. F. F. Crim, Science 249 (1990) 1387.
6. A. Tokmakoff, J. Chem. Phys. 105 (1996) 1.
7. S. Mukamel, Principles of Nonlinear Spectroscopy, Oxford University Press, New York, 1995.
8. J. B. Asbury, T. Steinel, C. Stromberg, S. A. Corcelli, C. P. Lawrence, J. L. Skinner, and M. D. Fayer, J. Phys. Chem. A 108 (2004) 1107.
9. P. Schuster, G. Zundel, and C. Sandorfy, The Hydrogen Bond. Recent developments in theory and experiments, North Holland, Amsterdam, 1976.
10. Q. Zhong, A. P. Baronavski, and J. C. Owrutsky, J. Chem. Phys. 119 (2003) 9171.
11. J. E. Boyd, A. Briskman, C. M. Sayes, D. Mittleman, and V. Colvin, J. Phys. Chem. B 106 (2002) 6346.
12. J. C. Deak, Y. S. Pang, T. D. Sechler, Z. H. Wang, and D. D. Dlott, Science 306 (2004) 473.
13. H. S. Tan, I. R. Piletic, and M. D. Fayer, J. Chem. Phys. 122 (2005) 174501.
14. I. R. Piletic, H. S. Tan, and M. D. Fayer, J. Phys. Chem. B (2005) accepted.

Femtochemistry VII
Fundamental Ultrafast Processes in Chemistry, Physics, and Biology
A.W. Castleman, Jr. and M.L. Kimble

Excited-state dynamics of the PYP chromophore in solution. Environment and structure effects

Agathe Espagne, Pascale Changenet-Barret, Jean-Bernard Baudin, Pascal Plaza, Monique M. Martin

UMR CNRS 8640, P.A.S.T.E.U.R. Département de Chimie, Ecole Normale Supérieure 24 rue Lhomond, 75231 Paris Cedex 05, France

1. Introduction

The Photoactive Yellow Protein (PYP) is the photoreceptor responsible for the negative phototaxis of the bacterium *Halorhodospira halophila*. Its chromophore is the deprotonated *trans*-4-hydroxycinnamic acid (Fig. 1), covalently linked to the side chain of the unique cysteine (Cys69) residue by a thioester bond [1,2]. Under blue-light irradiation, PYP undergoes a photocycle that involves *trans-cis* isomerization of the chromophore as the first overall step. Although this step includes still debated femto-picosecond processes, the formation in a few nanoseconds of a relaxed *cis* isomer of the chromophore has been clearly established [3,4]. The *cis* isomer is however not observed in steady-state photolysis of solutions containing isolated derivatives of the chromophore with a thioester-phenyl moiety at the α position of the carbonyl end group (Fig. 1, SC_6H_5 (pCT^-)) [5-7] whereas it is well characterized in the case of analogues with O^- (pCA^{2-}) or NH_2 (pCM^-) as the substituent [5-8]. Our goal is to discriminate the role of the protein environment on the primary molecular steps of the PYP photocycle from that of the chemical structure of the chromophore. For that purpose, we carried out a comparative study of the photoinduced processes in five analogues of the chromophore in solution in several polar solvents by subpicosecond UV-Vis transient absorption spectroscopy. In the present study, special attention was devoted to the impact of the electron-acceptor character of the carbonyl substituent on the coordinates involved in the deactivation pathway, leading or not to the formation of a *cis* product. In addition, the effect of the solvent on the excited-state relaxation is analyzed in order to stress further the existence of different relaxation pathways in these compounds. Hypothetical relaxation mechanisms are discussed.

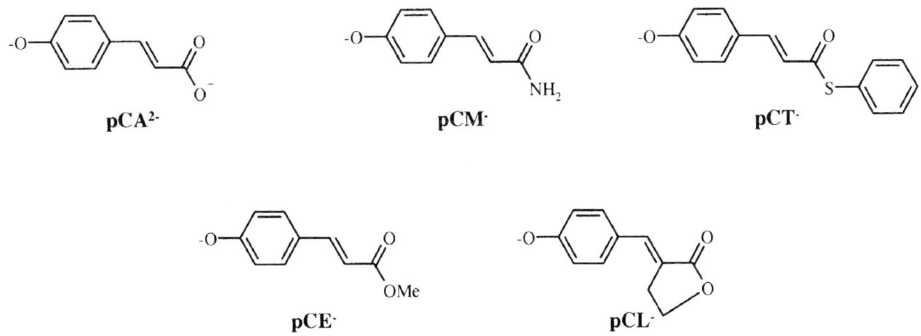

Fig. 1. Chemical structure of the studied deprotonated analogues of the PYP chromophore. The substituent in α of the carbonyl end is varied. O^- : 4-hydroxycinnamic acid (pCA^{2-}); NH_2 : amide derivative (pCM^-); SC_6H_5 : thioester phenyl derivative (pCT^-); OMe : methyl ester derivative (pCE^-); OCH_2- CH_2- : cyclic ester derivative or lactone form (pCL^-)

2. Experimental

2.1. Transient absorption experiments

Pump-probe UV-visible spectroscopy with a white-light continuum probe was carried out by means of an unconventional 500-fs 10-Hz dye laser source essentially described elsewhere [9] (a thinner microcavity, ca. 15 μm, was however used for the generation of subpicosecond pulses). The pump pulses (50 to 100 μJ) were tuned to 355 nm or 430 nm depending on the samples and focused to a ~2 mm diameter spot in the recirculating 1-mm sample cell. Simultaneously, 570 nm pulses were produced and used to generate the white-light probe in a 1-cm water cell. The pump and probe polarizations were set at the magic angle. The differential absorbance spectra were recorded in the 340-700 nm spectral range, through an imaging spectrograph (Jobin Yvon Spex 270M, entrance slit 64 μm), by a CCD camera (Roper Scientific, 1024 x 128). They were averaged over 500 laser shots and corrected from the probe chirp. Kinetics at selected wavelengths were fitted to a sum of exponential functions convoluted by the instrumental response, itself fitted to a ~1.5 ps-FWHM Gaussian function.

2.2. Samples

Trans-4-hydroxycinnamic acid ($pCAH_2$) was purchased from Sigma-Aldrich. *Trans-p*-hydroxycinnamide (pCMH) and *trans-p*-hydroxythiophenyl cinnamate (pCTH) were synthesized following procedures previously described [5,6]. *Trans*-p-hydroxycinnamic methyl ester (pCEH) was synthesized using the Fisher procedure from *trans*-p-hydroxycinnamic acid : methanol as solvent and 2 equivalents of concentrated sulfuric acid. The product was purified by recrystallization in methanol (yield : 90%). The lactone form (pCLH) was synthesized according to the procedure reported by H. Zimmer *et al.* [10]: 4-hydroxybenzaldehyde (6.1 g, 50 mmol) and γ-butyrolactone (8.6 g, 100 mmol) were stirred in benzene (70 mL); sodium methylate was slowly added (15 min). The mixture was warmed (50-60 °C, 4 h), then stirred during 18 h at room temperature. Aqueous sulfuric acid (1 $mol.L^{-1}$) was added until pH ≈ 2, then stirring was continued for 1 h. The crude product was isolated by filtration and purified by recrystallization in water (m.p. : 181°C, yield : 4.25 g, 42%).

Samples of the deprotonated compounds pCA^{2-} and pCM^- in water and alcohols were prepared with 0.05 M KOH (pH > 12). Aqueous samples of pCE^-, pCL^-, and pCT^- were prepared in pH 10.2 buffer solutions (Britton-Robinson or CAPS) to avoid hydrolysis. In alcohols, pCE^-, pCL^-, and pCT^- were dissolved with $5x10^{-4}$ M KOH in order to achieve 95 % deprotonation and sufficient chemical stability [6,7]. For transient absorption measurements, the optical density of the samples was adjusted to 1 per mm. The different solvents used were HPLC grade water and Merk Uvasol methanol, ethanol, and butanol-1. 98%-pure propanol-1, pentanol-1, and octanol-1 were purchased from Merk and 99%-pure decanol-1 from Acros Organics.

3. Results

3.1. Chromophore chemical structure and relaxation path

3.1.1. Time-resolved differential absorption spectra

We previously reported a comparative study of the transient absorption spectroscopy of pCA^{2-}, pCM^-, and pCT^- in aqueous solution [5-7]. pCA^{2-} and pCM^- were found to undergo *trans-cis* photoisomerization without any detectable intermediate, whereas the excited state of pCT^- was found to relax essentially back to the *trans* ground state via a populated intermediate state, characterized by a transient absorption around 450 nm. Typical time-resolved spectra found for pCA^{2-} and pCT^- are recalled in Fig. 2.

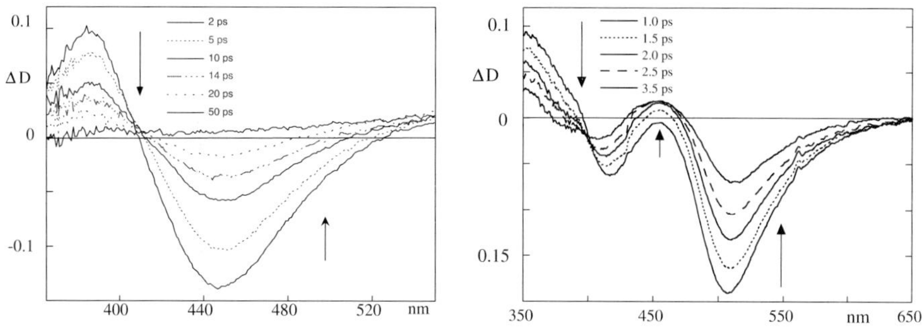

Fig. 2. Comparison of the time-resolved spectra in aqueous solution, of pCA^{2-} (left) which photoisomerizes [8] and of pCT^- (right) which relaxes to the initial ground state via an intermediate absorbing around 450 nm. The ground state absorption spectra are respectively centered around 330 nm (left) and 390 nm (right). The residual broad absorption band found for pCA^{2-} above 420 nm is due to the biphotonic formation of solvated electrons [8].

These results emphasized a major effect of the chemical structure of the chromophore on the light-induced process, which we confirm in the present study. In the first place, we have found here that the linear and cyclic ester derivatives (Fig. 1, pCE^- and pCL^-) do not undergo, as the thioester derivative pCT^-, any measurable *trans-cis* photoisomerization. As a matter of fact, they exhibit time-resolved absorption spectra (Fig. 3) that can be interpreted by the formation of a transient state in the course of the excited-state decay, just like for pCT^- (Fig. 2, right).

It is important to note that the atom in α of the carbonyl group of these esters is oxygen, which has a much larger electron affinity (EA = +141 $kJ.mol^{-1}$) than O^- (EA = -844 $kJ.mol^{-1}$) and N (EA = -7 $kJ.mol^{-1}$). This situation is similar to that of pCT^-, which bears a sulfur atom (EA = +200.4 $kJmol^{-1}$) in α of the carbonyl group. Fig. 3 (upper part) shows, for both esters in water, that an initial absorption between 370 and 400 nm, assigned to the initial excited state, rapidly decays revealing the bleaching band. Simultaneously, an intermediate state absorbing around 400 nm is being produced. A temporary isosbestic point seems to be present in the time-resolved transient absorption spectra around 400 nm, like for pCT^- (Fig. 2, right). A detailed analysis of the kinetics, which will be described elsewhere, shows that in several solvents the decay of the 400 nm transient band is slower than that of the stimulated emission band, although in Fig. 3 (lower part), they seem to be similar. The relaxation mechanism probed for the esters is thus similar to that of the thioester derivative, giving further evidence that the presence of a strong electron-acceptor in α of the carbonyl group is a key element that drives the system towards a metastable intermediate state instead of the *cis* isomer.

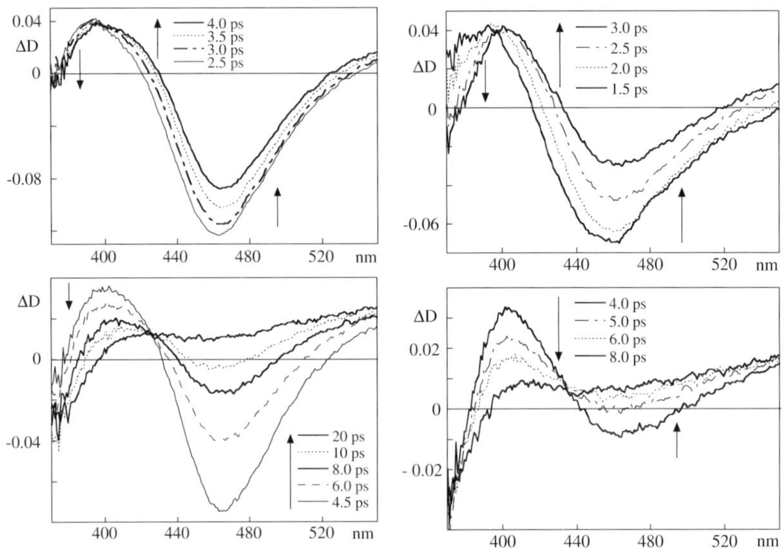

Fig. 3. Time-resolved spectra observed for the linear ester pCE$^-$ (left) and the lactone form pCL$^-$ (right) in aqueous solution. Both compounds relax via a transient state absorbing around 400nm. Their ground state absorption spectra are both centered around 355 nm. The broad residual absorption band above 420 nm is assigned to the biphotonic formation of solvated electrons [5-8].

3.1.2. Excited-state decay kinetics

The excited-state lifetime of the five analogues are compared in Table 1. In our previous reports on pCT$^-$ [5-7], we stressed that the kinetics are complex due to the overlap of several transient bands and to the contribution of solvation dynamics. The solvent reorganization that results from the photoinduced charge shift, from the phenolate group to the ethylenic bond [11], indeed leads to a red shift of the gain band. Experimentally, we nevertheless found that the decay measured at a wavelength slightly red-shifted from the initial maximum of the band is monoexponential. We explain this observation by noting that this maximum is broad and that the shift dynamics, on the explored time-scale, is small at this wavelength. Consequently, the excited-state lifetimes, given in Table 1, were obtained from the fit of the gain decay at this wavelength.

Table 1. Excited-state lifetime of the five studied analogues in aqueous solutions

Analogue	pCA^{2-}	pCM^{-}	pCE^{-}	pCT^{-}	pCL^{-}
τ_{S1} (ps)	10	4	3.0	2.4	1.0

It is seen that the excited-state lifetime decreases when the electron-acceptor character of the carbonyl end group increases. The lactone form is however found to have the shortest lifetime, although the electron affinity of the oxygen atom is smaller than that of the sulfur atom. The electron-acceptor character of the whole lactone group should then be considered, and one can hypothesize it is quite large because charge-transfer reactions in the femtosecond regime were previously reported for other lactones [12]. An alternative explanation might be found in the fact that the internal twisting of the carbonyl group is blocked because of the bridged structure. Possible molecular coordinates involved in the deactivation process will be examined in the discussion section.

3.2. Solvent properties and relaxation dynamics

In order to investigate why the thioester derivative isomerizes within the protein and does not in aqueous solution, we have studied the influence of the solvent properties on the excited-state decay pathway and kinetics. We compare here pCA^{2-} and pCT^{-}, which bear respectively the weaker and the stronger electron-acceptor end group of our series. Fig. 4 shows that the effect of polarity, viscosity, and average solvation time on the excited-state lifetime of these two compounds is different in aqueous and linear-alcoholic solutions. On the top left of the figure, it is seen that the excited-state decay is accelerated when the polarity increases. For a dielectric constant larger than 20, pCT^{-} is slightly more sensitive to the polarity than pCA^{2-}. The especially large value of the S_1 lifetime observed for pCT^{-} in decanol ($\varepsilon = 7$) is thought to be due in part to the large viscosity of this solvent. As a matter of fact, as shown in the lower part of Fig. 4, the excited-state decay of pCT^{-} also exhibits a larger sensitivity to viscosity (η) than pCA^{2-}. A rough analysis of the S_1 lifetime in terms of the power law η^{α} shows that α is twice larger for pCT^{-}. On the top right of the figure, it is seen that the excited-state lifetime increases non-linearly with the average solvation-time $<\tau>$ [13] of the linear alcohols. It is clearly seen that for $<\tau>$ larger than ~20 ps, the S_1 lifetime of pCA^{2-} is not very sensitive to the average solvation time, whereas that of pCT^{-} still increases. Although this comparison suffers from the lack of data for pCT^{-}, it can be considered as a further indication that the two compounds, hence the two classes of compounds they represent, follow different relaxation pathways. Alternatively, supposing that the excited-state relaxation coordinate remains identical, the solvent effects indicate a change in the pathway energy profile.

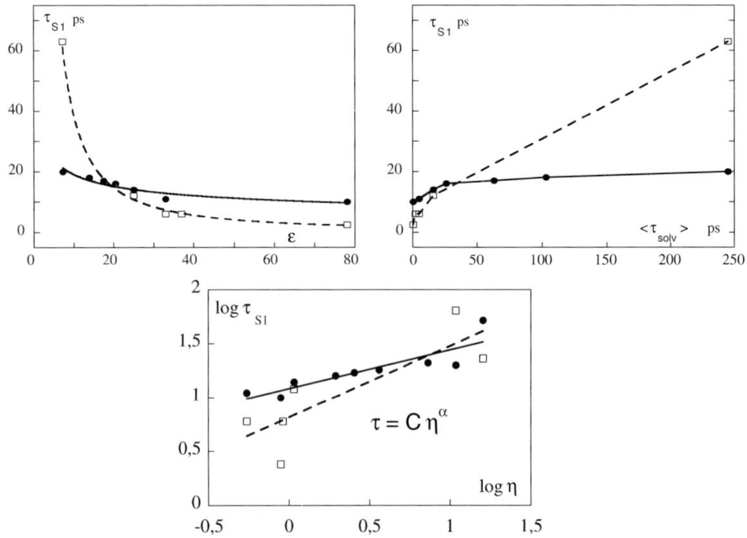

Fig. 4. Variation of the excited-state lifetime τ_{S1} of pCA^{2-} (•) and pCT^- (□) as a function of solvent properties in water and linear alcohols: dielectric constant ε (up left), average solvation time <τ_{solv}> (up right) and viscosity η (down)

4. Discussion

We have seen that when the electron-acceptor character of the end group is weak (pCA^{2-} and pCM^-), *trans-cis* photoisomerization occurs without any detectable intermediate. When it is strong (pCE^-, pCL^- and pCT^-), the excited-state pathway involves a detectable intermediate and does not lead to the *cis* isomer. An interpretation of this difference of behavior is to consider that, in fact, an intermediate state is reached for all molecules. Depending on the electron-acceptor character of the end group, this intermediate is either very short-lived, hence non-observed, and partially leads to the *cis* isomer or it is longer-lived, observed, and only leads back to the *trans* isomer. We will now focus our attention on this intermediate state. The fact that the excited-state decay rate i) increases with the electron-acceptor character of the carbonyl substituent (Table 1) and ii) increases with the solvent polarity is a strong indication that the production of the intermediate state from the initial excited involves a charge transfer. From electrooptical measurements [11] and quantum chemical calculations [14], it is known that optical excitation leads to a dipole moment change corresponding to a shift of the negative charge from the oxygen of the phenolate donor-group to the phenol ring or even to the ethylenic bond.

We thus propose that this charge shift proceeds further in the excited state, giving rise to an even more charge-shifted intermediate state. The driving force of this reaction could be a better localization of the charge, hence a larger solvation energy of this product state.

On the other hand, according to Grote and Hynes' theory [15], the larger constraint exerted by the solvent viscosity on the excited-state decay of pCT^- (as compared to pCA^{2-}) is an indication that its relaxation process involves the crossing of a lower barrier, with a smaller frequency, which implies a larger sensitivity to low-frequency fluctuations of the solvent. This is consistent with the fact that the excited-state lifetime of pCT^- keeps increasing with the average solvation time (which also contains diffusional dielectric responses) when that of pCA^{2-} levels off. This latter effect might also indicate that solvent reorganization is also involved in the barrier crossing of pCT^-. The difference observed in the solvent effects on the excited-state decay of pCA^{2-} and pCT^- is a further indication that these two classes of analogues either have different relaxation pathways or display a noticeable change in the energy profile of their relaxation pathway.

At this stage of the discussion one has to explain why some of our compounds lead to a net *trans-cis* isomerization and the others don't. It is important to bear in mind that the excited-state lifetime of pCT^-, which represents the non-isomerizing compounds, is more sensitive to the solvent polarity than that of pCA^{2-}, which represents the isomerizing compound. One may rationalize those facts in terms of the extent of the photoinduced charge shift in the intermediate state. In the case of analogues with a strong electron-acceptor end group, the charge can be thought to go all the way to the carbonyl oxygen, with a complete reversal of the single and double bond pattern (Fig. 5, top). This analysis also holds for the lactone, the intramolecular bridge of which does not affect the photoinduced change in the bond pattern (Fig. 5, middle). However, the presence of a weak electron-accepting end group, or even a repulsive one in the case of pCA^{2-}, is expected to lead to either a charge shift of a lesser extent (Fig. 5, bottom) or else a complete shift of a partial charge.

In all cases the product state displays a single bond in place of the initial ethylenic bond. We thus think that its geometry tends toward a 90° twist of the ethylenic bond because this rotation enhances the decoupling of the two halves of the molecules and localizes better the negative charge on the carbonyl side. This structure would be comparable to the so-called excited-state *perp* minimum involved in the photoisomerization of stilbene [16] and similarly thought to be the gateway toward a possible *trans-cis* isomerization.

In a first approach, one can hypothesize that reaching the complete 90° twist of the *perp* state requires the concerted motion of other coordinates. For pCA^{2-} and pCM^-, we can see on Fig. 5-bottom that the carbonyl end group can rotate freely. Such a flexibility would also be kept for a complete shift of a partial charge. One can suppose that, in the *perp* geometry, the torsion of the carbonyl end group releases internal constraints, possibly due to steric hindrance and/or electronic repulsion between the substituted carbonyl end group and the phenol ring moiety. This flexibility could thus be thought to stabilize the *perp* state in a 90°-twist configuration, hence to favor isomerization. For pCT^- and the esters, such a flexibility is not available, as shown in Fig. 5-top, and the twist angle of the ethylenic bond in the *perp* state would thus be thought to deviate from 90° in order to avoid the above-mentioned constraints. Consequently, only relaxation down to the initial *trans* ground state would be possible. This twisted excited product would thus be the non-emissive transient state that we observe, the picosecond lifetime of which (not described in this report) would result from a fast internal conversion. The S_1-S_0 energy gap is indeed likely to be small for geometries close to the *perp* configuration.

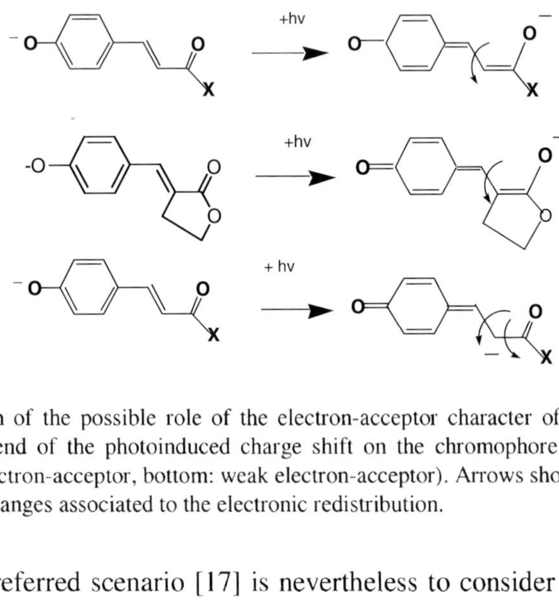

Fig. 5. Illustration of the possible role of the electron-acceptor character of the carbonyl end-group on the extend of the photoinduced charge shift on the chromophore structure (top and middle: strong electron-acceptor, bottom: weak electron-acceptor). Arrows show potential sources for geometrical changes associated to the electronic redistribution.

Our preferred scenario [17] is nevertheless to consider that a fully 90° twisted charge-shifted *perp* state is reached in all cases because both the steric hindrance and electronic repulsion at the *perp* configuration are likely to be negligible. As a result of the large stabilization of this state when the electron-acceptor strength of the carbonyl end group increases, one may propose that an avoided crossing with the ground state occurs at the 90° geometry for pCT^- and the esters only. Such an avoided crossing would lead to a new minimum at the 90° configuration in the ground state. Such a situation has previously been described for the zwitterionic *perp* state of olefins in polar solvents [18]. Due to the destabilization of the *cis* isomer in the ground state, the barrier from this "new" minimum to the *cis* would be higher than that to the initial *trans* configuration, resulting in no net *trans*-to-*cis* isomerization.

Along those lines, one may infer that within the natural PYP photoreceptor, photoisomerization of the thioester chromophore does occur because the expected large photoinduced charge shift is counteracted by the presence of a positive charge on the nearby Arginine52 residue.

5. Conclusion

From ultrafast spectroscopy experiments carried out on five analogues of the Photoactive Yellow Protein chromophore, we have demonstrated that the photoinduced behavior of these compounds is strongly dependent on their electron donor-acceptor structure. Compounds for which the carbonyl group is substituted with a strong electron acceptor, in particular the thioester derivative which models the native chromophore of PYP, do not isomerize in aqueous and alcoholic solutions. Their relaxation pathway involves a new transient, the formation of which is sensitive to the polarity and viscosity of the solvent. Hypothetical relaxation mechanisms are discussed. Our preferred one [17] involves formation of a spectroscopically detectable ground state with a 90° twisted conformation of the ethylenic bond.

References

1. W. D. Hoff, P. Düx, K. Hard, B. Devreese, I. M. Nugteren-Roodzant, W. Crielaard, R. Boelens, R. Kaptein, J. J. van Beeumen, and K. Hellingwerf, Biochem. 33 (1994) 13959.
2. M. Baca, G. E. O. Borgstahl, M. Boissinot, P. M. Burke, D. R. Williams, K. A. Slater, and E. D. Getzoff, Biochem. 33 (1994) 14369.
3. B. Perman, V. Srajer, Z. Ren, T.-Y Teng, C Pradervand, T. Ursby, D. Bourgeois, F. Schotte, M. Wulff, R. Kort, K. J. Hellingwerf, and K. Moffat, Science 279 (1998) 1946.
4. R. Brudler, R. Rammelsberg,, T. T. Woo, E. D. Getzoff, and K. Gerwert, K. Nature Struct. Biol. 8 (2001) 265.
5. P. Changenet-Barret, A. Espagne, N. Katsonis, S. Charier, J-B. Baudin, L. Jullien, P. Plaza, and M. M. Martin, Chem. Phys. Lett. 365 (2002) 285.
6. P. Changenet-Barret, A. Espagne, S. Charier, J-B. Baudin, L. Jullien, P. Plaza, K.J. Hellingwerf, and M.M. Martin, Photochem. Photobiol. Sci. 3 (2004), 823.
7. P. Changenet-Barret, A. Espagne, P. Plaza, K. J. Hellingwerf, and M. M. Martin, New J. Chem. 29 (2005) 527.
8. P. Changenet-Barret, P. Plaza, and M. M. Martin, Chem. Phys. Lett. 336 (2001) 439.
9. N. Dai Hung, P. Plaza, M. M. Martin, and Y. H. Meyer, Appl. Opt. 31 (1992) 7046.
10. H. Zimmer and J. Rothe, J. Org. Chem. 24 (1959) 28.
11. L. L Premvardhan, F. Buda, M. A. van der Horst, D. C. Lührs, K. J. Hellingwerf, and R. van Grondelle, J. Phys. Chem. B. 108 (2004) 5138.
12. T. Bizjak, J. Karpiuk, S. Lochbrunner, and E. Riedle, J.Phys. Chem. A 108 (2004) 10763.
13. M. L. Horng, J. A. Gardecki, A. Papazyan, and M. Maroncelli, J. Phys. Chem. 99 (1995) 17311.
14. G. Groenhof, M. F. Lensink, H. J. C Beredsen, J. G. Snijders, and A. E. Mark, Proteins 48 (2002) 202.
15. R. F. Grote and J. T. Hynes, J. Chem. Phys. 73 (1980) 2715.

16. D. H. Waldeck, Chem. Rev. 91 (1991) 415.
17. A. Espagne, P. Changenet-Barret, P. Plaza, and M. M. Martin, J. Phys. Chem. A 110 (2006) 3393.
18. W. G. Dauben, L. Salem, and N.J. Turro, Acc. Chem. Res. 8 (1975) 41.

Femtochemistry VII
Fundamental Ultrafast Processes in Chemistry, Physics, and Biology
A.W. Castleman, Jr. and M.L. Kimble

Second Harmonic and Sum Frequency of Molecules at Aqueous Interfaces

Eric A. McArthur, Yi Rao, Matthew Comstock, and Kenneth B. Eisenthal

Department of Chemistry, 3000 Broadway, Columbia University, New York, NY 10027

1.Ultrafast Electron Transfer at Organic Liquid/Aqueous Interface

Electron transfer (ET) is one of the most important reactions in bulk [1-5] and interfacial [6-14] chemical and biological systems. In the work reported here the first ultrafast dynamics of ET at a liquid/liquid interface was observed in real time using femtosecond lasers. We have used an earlier approach [15] where the donor molecules serve as one of the solvents comprising the interface to minimize the distance between donor and acceptor molecules, which thereby eliminates translational diffusion as an important precursor to electron transfer. The interface selectivity of second harmonic generation (SHG) makes it possible to observe interfacial equilibrium and dynamic processes, concerning ET, without interference from the bulk media forming the interface [16].

In this presentation the time resolved ultrafast excited state ET at the dimethylaniline (DMA)/aqueous interface is presented. Photoexcited coumarin 314 (C314) and the DMA solvent molecules, respectively serve as acceptor and donor molecules. The coumarin dye was selected because of its strong SHG signal at twice the fundamental probe frequency and because of our extensive knowledge of its equilibrium and dynamic processes at aqueous interfaces (solvation and rotation).

The time dependent changes of the C314 ground and excited state populations were probed with a 140fs pulse at 832nm following the photoexcitation of C314 (1) with a 120fs laser pulse at 424nm. By selecting 832nm the SHG generated is resonant with C314, and consequently the detected SHG is dominated by ground and excited state C314, i.e. the contributions of interfacial DMA and water to the SHG signal are small and furthermore are essentially constant in time because of their very high interfacial populations. In its excited state the electron affinity of C314 is sufficient to effect an electron transfer from a ground state DMA molecule (2), thereby producing the radical anion $C314^{\bullet-}$ and the radical cation $DMA^{\bullet+}$. Subsequent to this ET step, the back electron transfer (BET) from $C314^{\bullet-}$ to $DMA^{\bullet+}$ yields the neutral ground state C314 and DMA molecules (3).

$$(1) \quad C314 + \hbar\omega \longrightarrow C314^{*}$$

$$(2) \quad C314^{*} + DMA \xrightarrow{ET} C314^{\bullet-} + DMA^{\bullet+}$$

$$(3) \quad C314^{\bullet-} + DMA^{\bullet+} \xrightarrow{BET} C314 + DMA$$

In our experiments C314 is adsorbed to the DMA/water interface where its interface population was determined to be ~10^{13} C314/cm^2, from interfacial tension measurements [17].

Dynamics of the excited state C314 are characterized by a biexponential fit which is convoluted to include instrument response time (Figure 1). The observed ultrafast decrease in the second harmonic field, $E_{2\omega}$, at the time of photoexcitation, t=0, is due to the bleaching of ground state C314 molecules. In addition, the amplitude of $E_{2\omega}$ is reduced because the phase of the initially resonant excited state, C314*, is opposite to that of ground state, C314, which results in a cancellation and therefore a further reduction in the SHG signal.

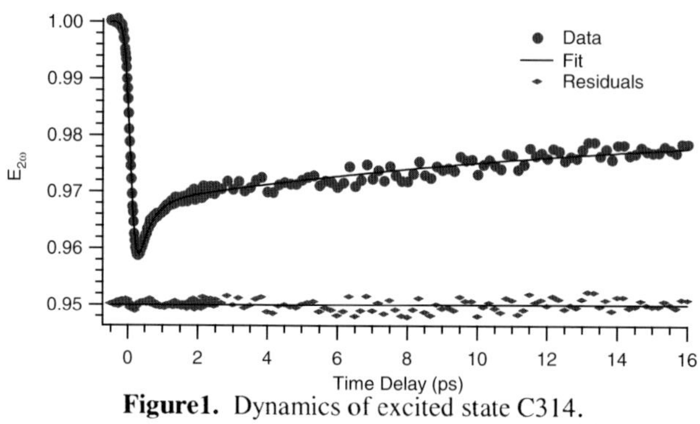

Figure1. Dynamics of excited state C314.

The extracted times are τ_1 = 362±60fs with an amplitude of 0.53 and τ_2 = 14±2ps with an amplitude of 0.47. The 362fs component is attributed to the excited state C314* solvation dynamics at the DMA/aqueous interface, this solvation occurs in a time scale similar to C314* solvation at an air/aqueous interface [18], average of 850±70fs, and similar to C314* solvation at a neutral surfactant (steric acid) monolayer/aqeuous interface, 400±60fs. It should be noted that the SHG signal increases as solvation of C314* occurs because the solvation of C314* results in a Stokes shift. Because the energy difference between the solvated C314* with its ground state is no longer equal to the probe wavelength, it follows that the contribution of the C314* to the SHG signal is reduced. Since the C314* contribution to the SH signal is reduced, the net cancellation is diminished, and as a result, the signal increases as seen in Figure 1. The longer time component of 14ps (Figure 1) is assigned to ET from DMA to excited state C314*.

To confirm that the 14ps component is due to ET dynamics and not the dynamics of other processes, we probed the time evolution of the radical cation $DMA^{\bullet+}$. As given in Eq.2 the transfer of an electron from DMA to C314* generates the $DMA^{\bullet+}$ radical cation. The formation of $DMA^{\bullet+}$ thus tracks the electron transfer process. To selectively follow the $DMA^{\bullet+}$ dynamics, we use a probe wavelength of 904nm, which generates an SHG signal that is resonant with the ground to lowest excited state transition of $DMA^{\bullet+}$ giving a large resonant enhancement of the SHG signal. In this way the SHG signal is dominated by $DMA^{\bullet+}$ with C314*, C314, DMA, and water contributing to a weak background signal. A confirmation of the $DMA^{\bullet+}$ dominance in the SHG signal is the absence in Figure 2 of the initial bleach due to C314 seen in Figure 1. Figure 2 shows the time evolution of the $DMA^{\bullet+}$ radical cation.

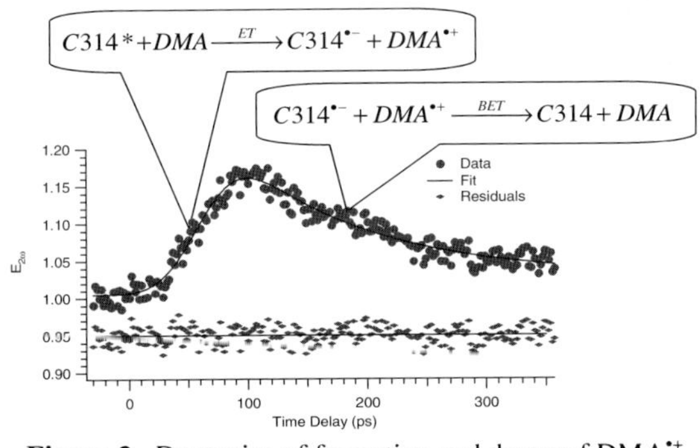

Figure 2. Dynamics of formation and decay of $DMA^{\bullet+}$.

The formation time, 16±3ps, of the $DMA^{\bullet+}$ is conclusive evidence of the ET dynamics, in agreement with the 14±2ps obtained from the coumarin dynamics. The back electron transfer time constant is 163±4ps.

Although there have been no studies of electron transfer from photoexcited C314 to DMA in the bulk liquid DMA, there have been a number of studies of other coumarins in bulk DMA [1-3], most of which show a biexponential decay of the photoexcited coumarin. The two decay times in the bulk DMA studies with C153, which has a reduction potential very close to that of C314 (10mV difference), are 3ps and 25ps. Our results are not very different from the bulk DMA studies. Considerable work on ET at the liquid/lqiuid interface has been performed using electrochemical and photocurrent methods. Comparision to these studies are difficult since these measurements yield heterogenous rate constants in units of cm/s. Converting to units of 1/s requires knowledge of an effective length [13,14] over which the process occurs which is somewhat ambiguous. Further studies to investigate these and other issues, including comparison with theoretical models, are underway.

2. Absolute Orientation of Molecules at Interfaces

Knowledge of molecular orientation at surfaces and interfaces is of widespread interest to scientists because of its relevance to chemical reactions, long- and short-range molecular interactions, phase transitions, and the electronic and vibrational states of interfacial molecules [19-21]. A method to determine the absolute orientation of molecules at liquid interfaces by Sum Frequency Generation (SFG) is reported.

The basic idea of the method reported here is to measure the orientations of two SFG active chromophores with respect to the surface normal to obtain the absolute orientation defined by both symmetry axes and the molecular plane with respect to the surface normal. Combining these measured values with the known orientations of the chromophores with respect to each other in the molecule, together with a rotation matrix operation formulation, we obtain the absolute orientation of the interfacial molecule. Generally the alignment of the molecule at the interface, i.e. which part of the molecule projects towards which bulk phase, is known. If not, phase measurements will provide the up vs. down alignment. We chose m-tolunitrile, which is a planar molecule having two distinct vibrational chromophores, as a proof of method example to demonstrate how to obtain the absolute orientation of the molecule at the interface. It should be noted that this method to determine the absolute orientation of an interfacial molecule requires two distinct local vibrational modes that have a fixed relative orientation. Many molecules satisfy this restriction.

The sum frequency intensity is proportional to the square of the sum over all macroscopic sum frequency susceptibility terms separated into a non-resonant part $\chi^{(2),NR}$ and a resonant part $\chi^{(2),R}(\omega)$

$$I^{SF}(\omega) \propto \left| \chi^{(2),NR} + \sum_{q}^{n} \chi^{(2),R}(\omega, q) \right|^2 ; \chi^{(2),R}(\omega, q) \propto \frac{A_q}{\omega_{IR} - \omega_q + i\Gamma_q} \quad (1)$$

where A_q contains the product of the Raman and infrared matrix elements of the q-th normal mode, ω_q denotes the resonant frequency of the q-th vibrational normal mode, ω_{ir} is the frequency of the IR light, and ω_q is the spectral width of the vibrational state. In the case of dielectric interfaces, the contribution from the non-resonant parts are small compared to the resonant terms of interest.

We approximate the macroscopic sum frequency susceptibility tensors $\chi_{IJK}^{(2),R}(\omega)$ in the laboratory coordinates (X; Y; Z) (I,J,K being the three coordinates) with the microscopic hyperpolarizability tensor elements $\beta_{ijk}^{(2)}$ in the molecular coordinates (x; y; z) (i,j,k being the three coordinates) yielding,

$$\chi_{IJK}^{(2)} = N_s \left\langle (\hat{I} \cdot \hat{i})(\hat{J} \cdot \hat{j})(\hat{K} \cdot \hat{k}) \right\rangle \beta_{ijk}^{(2)} = N_s \left\langle \hat{R}_{I \cdot i} \hat{R}_{J \cdot j} \hat{R}_{K \cdot k} \right\rangle \beta_{ijk}^{(2)} \quad (2)$$

where Ns is the number density of the interface moiety under investigation, $\hat{R}$ is the element of the rotation transformation matrix from the molecular frame to the laboratory frame, and $\langle \ \rangle$ denotes the ensemble average over the orientation of the interfacial molecules of interest.

In the theory of vibrational SFG with a single resonance at the IR frequency, the tensor elements of the hyperpolarizability are related to the IR and Raman properties of the vibrational mode and are given by the partial derivative of the Raman polarizability tensor and the IR transition dipole [22] moment of the q-th mode with respect to the normal coordinate of the same mode [23]. It is given by the following equation,

$$\beta_{ijk,q}^{(2)} = \frac{1}{2\varepsilon_o \omega_q} \frac{\frac{\partial \alpha_{ij}}{\partial Q_q} \frac{\partial \mu_k}{\partial Q_q}}{\omega_{ir} - \omega_q - i\Gamma_q} \quad (3)$$

where $\frac{\partial \alpha_{ij}}{\partial Q_q}$ is the derivative of the Raman polarizability element and $\frac{\partial \mu_k}{\partial Q_q}$ is the derivative of the transition dipole moment for the q-th normal mode. A new approach to the measurement of vibrational spectra using SFG (broadband sum frequency generation) has been developed and has proven to be successful [22]. It entails combining a spectrally broad IR pulse, which is characteristic of femtosecond IR pulses, with a spectrally narrow visible pulse [22]. In the frequency-resolved SFG experiment, the bandwidth of the incident IR pulse is much larger than that of the vibrational resonance. The frequency resolved sum frequency signals cover several hundred wavenumbers, which makes it possible to more rapidly obtain vibrational spectra of interest. The spectral resolution results from the spectrally narrow ($10cm^{-1}$) picosecond visible laser pulse.

An 800nm regeneratively amplified Ti:sapphire system (Spitfire, Spectra Physics) seeded with a MaiTai 80MHz, 80fs oscillator, at a one kilohertz repetition frequency was employed in the present experiments. A reflected 20% output from the regenerative amplifier was used to produce a picosecond 800nm pulse. A signal generated by the input of the 800nm femtosecond pulse into an OPA is in the range of 1.1-1.6μm and the idler is in the range of 1.6-2.8μm. By difference 5 frequency generation in a 1mm AgGaS2 crystal, tunable infrared light between 3μm and 8.5μm with a bandwidth of about 150 cm^{-1} (FWHM) was generated.

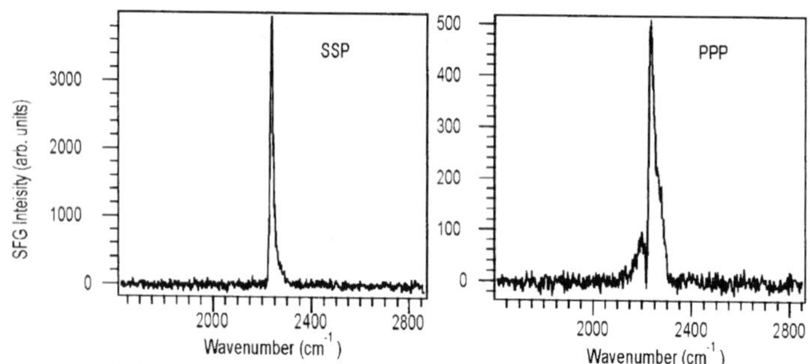

Figure 3: The nitrile (-CN) SSP and PPP sum frequency spectra of 6mM m-tolunitrile with visible light at 800nm and the IR centered at 4.7μm.

Figure 3 shows the SFG spectra of a 6mM m-tolunitrile aqueous solution taken at two different polarization combinations SSP and PPP with the visible beam at 800nm and the IR centered at 4.7um with a bandwidth of $140cm^{-1}$. The polarization S is perpendicular to the plane of incidence, and the polarization P is in the plane of incidence. The first letter gives the polarization of the SFG signal, the middle letter gives the polarization of the incident visible light, and the last letter gives the polarization of the incident IR light. From fits of the SFG spectrum using two Lorentzian functions, the peak at 2234 cm^{-1} in the SSP polarization spectrum was assigned to the symmetric stretching mode of -CN,with a shoulder at 2270cm^{-1} , which becomes more prominent in the PPP spectrum. The assignment of the other features will be published at a later time.

There are three non-vanishing elements for the $C_{\infty v}$ symmetry of -CN, namely, $\beta_{xxz} = \beta_{yyz} = r\beta_{zzz}$, where r is β_{xxz}/β_{zzz}. The subscripts refer to coordinates in the molecular frame. From Eq.3 we see that r = $\alpha_{xx}^{Raman} = \alpha_{zz}^{Raman}$, i.e. the ratio of the Raman perpendicular polarizability element to the Raman parallel polarizability element of the -CN bond. From Raman polarization measurements two possible ratios are obtained, which one is correct cannot be determined from the Raman measurements alone. However, as has been shown in previous work, a determination of the relative sign of $\chi_{xxz}^{(2)}$ and $\chi_{xzx}^{(2)}$ from SFG measurements can be used to identify the correct value of the polarizability ratio r, which is 0.26 [24,25]. To obtain the orientation of the -CN chromophore, null angle measurements of the nitrile stretch mode were performed. The measured intensity plotted against the polarization angle of the analyzer is shown in Figure 4. The null angle Ω_{CN} can be fitted from Figure 4 [26-28], from which we found that the orientational angle μ1 of the nitrile group is 53:0 ± 1:3° with respect to the surface normal .

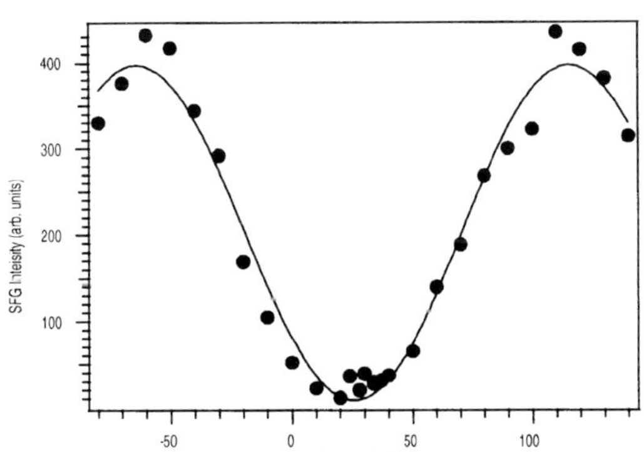

Figure 4: SFG intensity of the nitrile (-CN) 2234cm^{-1} peak plotted against the polarization angle of the ananlyzer, the solid line is the calculated fit.

Figure 5 shows the SFG spectra of m-tolunitrile at the air/aqueous interface taken at two different polarization combinations SSP and PPP with the visible wavelength at 800nm and the IR wavelength centered at 3.4μm with a bandwidth of $150cm^{-1}$. The PPP polarization combination spectrum shows a strong peak centered at $2961cm^{-1}$, which was assigned to the antisymmetric stretch [29,30]. The strong peak at $2945cm^{-1}$ in the SSP SFG spectrum is assigned to the symmetric stretch of the methyl group [29,30].

For the symmetric stretch mode, two non-vanishing elements of the hyperpolarizabilities exist, namely, $\beta_{xxz} = \beta_{yyz} = r\beta_{zzz}$, where the ratio r value of 2.3 was used by combining the Raman depolarization of the methyl group in acetonitrile with the SFG measurements of the relative sign of the $\chi^{(2)}_{XXZ}$ and $\chi^{(2)}_{XZX}$ susceptibility elements [24]. Using the null angle method, the calculated orientational angle of the C_3 axis of the methyl group is 23:0 ± 1:6°.

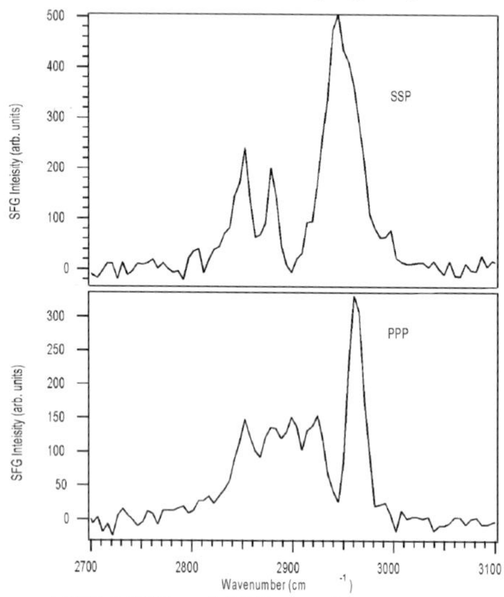

Figure 5: The methyl group ($-CH_3$) SSP and PPP sum frequency spectra of a 6mM m-tolunitrile aqueous solution with the visible light at 800nm and the IR centered at 3.4μm.

Now that we have obtained the orientational angle μ_1 of the nitrile group and the orientational angle μ_2 of the methyl group, we can determine the absolute orientation of the m-tolunitrile molecule at the air/aqueous interface.[31] To extract the remaining angle needed to determine the absolute orientation of m-tolunitrile, we use a rotation matrix method; we can calculate the angle ' of the normal to the molecular plane e_x and the surface normal Z, which is also the angle between the molecular plane and the surface plane to be 70°. See Figure 6.

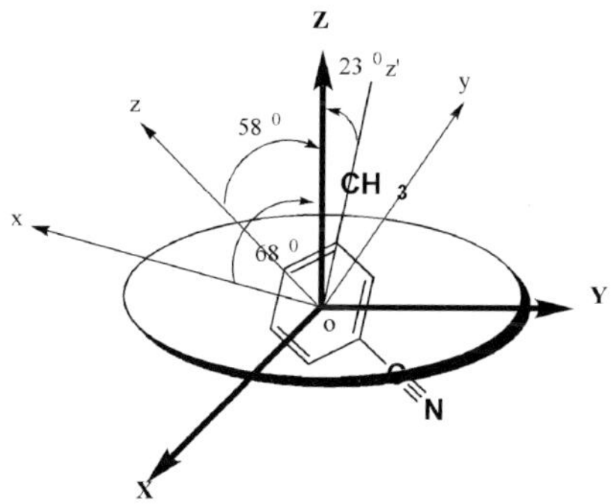

Figure 6: The –CN axis is along the molecular z axis and at an angle of 53° with respect to the surface normal $\hat{Z}$, the C3 axis of the methyl group is at an angle of 23° with respect to the surface normal Z and the normal to the molecular plane is along the molecular x axis and at an angle of 70° with respect to the surface normal.

Acknowledgment

This work was supported by the National Science foundation and the Chemical Sciences, Geo-sciences and Biosciences Division, Office of Basic Energy Sciences, Office of Science, US Department of Energy. Y. Rao thanks Dr. Feng Wang, Dr. Na Ji and Daohua Song for beneficial discussions. The authors also thank Professor Hong-fei Wang for his valuable discussions.

References

1. E. W. Castner, D. Kennedy, and R. J. Cave, J. Phys. Chem. A 104 (2000) 2869.
2. H. Pal, Y. Nagasawa, K. Tominaga, and K. Yoshihara, J. Phys. Chem. 100 (1996) 11964.
3. B. Akhremitchev, C. F. Wang, and G. C. Walker, Laser Chemistry 19 (1999) 403.
4. R. A. Marcus and N. Sutin, Biochim. Biophys. Acta 811 (1985) 265.
5. M. D. Newton, Chem. Rev. 91 (1991) 767.
6. H. L. Tavernier, A. V. Barzykin, M. Tachiya, and M. D. Fayer, J. Phys. Chem. B 102 (1998) 6078.

7. K. L. Kott, D. A. Higgins, R. J. McMahon, and R. M. Corn, J. Am. Chem. Soc. 115 (1993) 5342.
8. S. Amemiya, Z. F. Ding, J. F. Zhou, and A. J. Bard, J. Electroanalytical Chem. 483 (2000) 7.
9. R. Lahtinen, D. J. Fermin, K. Kontturi, and H. H. Girault, J. Electroanalytical Chem. 483 (2000) 81.
10. N. Eugster, D. J. Fermin, and H. H. Girault, J. Phys. Chem. B 106 (2002) 3428.
11. Y. B. Zu, F. R. F. Fan, and A. J. Bard, J. Phys. Chem. B 103 (1999) 6272.
12. D. J. Fermin, Z. F. Ding, H. D. Duong, P. F. Brevet, and H. H. Girault, J. Phys. Chem. B 102 (1998) 10334.
13. R. A. Marcus, J. Chem. Phys. 94 (1990) 4152.
14. B. B. Smith, J. W. Halley, and A. Nozik, J. Chem. Phys. 205 (1996) 245.
15. T. J. Chuang and K. B. Eisenthal, J. Chem. Phys. 59 (1973) 2140.
16. K. B. Eisenthal, Chem. Rev. 96 (1996) 1343.
17. A. Yeung, T. Dabros, and J. Masliyah, J. Colloid Interface Science 208 (1998) 241.
18. A. V. Benderskii and K. B. Eisenthal, J. Phys. Chem. B 104 (2000) 11723.
19. A. W. Adamson, Physical chemistry of surfaces, 5th ed., Wiley, New York, 1990.
20. Y. R. Shen, Ann. Rev. Phys. Chem. 40 (1989) 327.
21. L. T. Richter, T. P. Petralli-Mallow, and J. C. Stephenson, Opt. Lett. 23 (1998) 1594.
22. R. Superfine, J. Y. Huang, and Y. R. Shen, Chem. Phys. Lett. 172 (1990) 303.
23. D. Zhang, J. Gutow, and K. B. Eisenthal, J. Phys. Chem. 98 (1994) 13729.
24. S. Y. Yakovenko, E. Maiwald, A. Wurflinger, and J. Pelzl, Liquid Crystals 26 (1999) 23.
25. R. Lu, W. Gan, and H. F. Wang, Chinese Science Bulletin 48 (2003) 2183.
26. J. M. Hicks, K. Kemnitz, and K. B. Eisenthal, J. Phys. Chem. 90 (1986) 560.
27. R. Lu, W. Gan, B. H. Wu, H. Chen, and H. F. Wang, J. Phys. Chem. B 108 (2004) 7297.
28. E. L. Hommel and H. C. Allen, Analyst 128 (2003) 750.
29. G. R. Bell, Z. X. Li, C. D. Bain, P. Fischer, and D. C. Duffy, J. Phys. Chem. B 102 (1998) 9461.
30. H. Goldstein, Classical mechanics, 2d ed., Addison-Wesley Pub. Co., Reading, Mass., 1980.

Femtochemistry VII
Fundamental Ultrafast Processes in Chemistry, Physics, and Biology
A.W. Castleman, Jr. and M.L. Kimble

Solvation Dynamics in Ionic Liquids, Results from ps and fs Emission Spectroscopy

Sergei Arzhantsev[a], Hui Jin[a], Gary A. Baker[b], Naoki Ito[c], and Mark Maroncelli[a]

[a]Department of Chemistry, The Pennsylvania State University, 104 Chemistry Building, University Park, PA 16803, USA
[b]Chemical Science Division, Oak Ridge National Laboratory, P.O. Box 2008 Oak Ridge, TN 37831-6110 USA
[c]Current Address: Department of Chemistry and Biochemistry, Arizona State University,Tempe, AZ 85287-1604, USA

1. Introduction

Ionic liquids are a new class of room temperature solvents which are attracting considerable attention as potential "green" replacements for conventional organic solvents and for the advantages these "designer solvents" hold in a number of applications [2]. Our interest in ionic liquids lies in exploring their potentially unique solvating characteristics, specifically how energetic and dynamical aspects of solvation are different in ionic liquids compared to conventional dipolar solvents and how these differences affect elementary chemical processes like electron transfer.

The present report summarizes our recent studies of "polar solvation dynamics" [3] {Horng, 1995 #27}, the time-dependent response of a solvent to an electrical perturbation of a solute, in ionic liquids. These dynamics are measured by exciting a probe fluorophore with a short laser pulse, observing its emission spectrum evolve in time as the solvent surroundings relax. To characterize the solvation response we use the normalized spectral response function,

$$S_\nu(t) = \frac{\nu(t) - \nu(\infty)}{\nu(0) - \nu(\infty)} \tag{1}$$

where $\nu(t)$ is some measure of the frequency of the spectrum at time t such as the peak or first moment of the spectrum. We employ two techniques to obtain the requisite time-resolved emission spectra, picosecond time-correlated single photon counting (TCSPC) and femtosecond Kerr-gated emission spectroscopy (KGE). The subsequent discussion will be somewhat artificially divided into two sections according to technique.

For later reference, the probes and ionic liquids studied here are illustrated in Fig. 1.

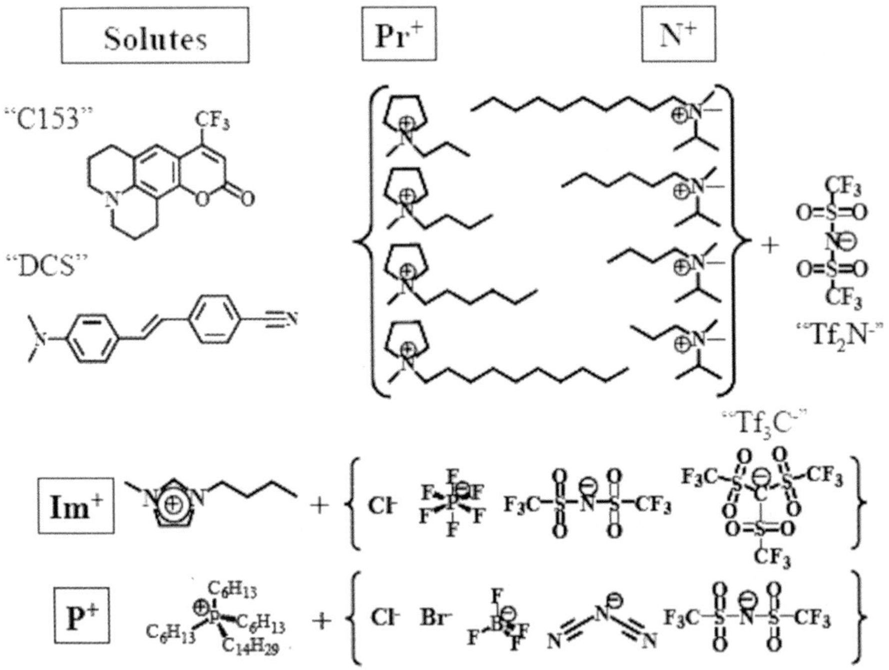

Figure 1: Ionic liquids studied here consist mainly of two homologous series of pyrrolidinium "Pr+" and ammonium "N+" liquids all with the *bis*(trifluorosulfonyl)imide anion "Tf2N-", a series consisting of 1-butyl-3-methylimidazolium "Im+" with four different anions, and another anion series based on a large phosphonium cation "P+". (Two miscellaneous liquids that fall into the N+ and Im+ classes are not depicted here.)

2. TCSPC Measurements

Several research groups have used time-correlated single photon counting or related picosecond methods with instrumental response times between 20–80 ps to monitor solvation dynamics in a range of ionic liquids4-16. Although there is some dependence of the observed dynamics on choice of probe molecule [5,6,9] (and considerable variability in the results reported by different laboratories [9,13]), a reasonably consistent picture is beginning to emerge from these studies. The spectral response functions $S_\nu(t)$ observed can usually be represented by double or stretched exponential functions of time (but see below). This response typically occurs over the hundreds of picoseconds to nanoseconds time ranges. The times required to achieve solvent equilibration are considerably slower in ionic liquids than in most conventional polar solvents near room temperature, wherein solvent relaxation is complete in 0.1-10 ps [3,17]. The observation of much slower dynamics in ionic liquids is not surprising given the very high viscosities of ionic liquids (20-2000 cP at room temperature in the ILs we have examined).

In contrast to the well known dielectric continuum models of solvation which are capable of providing reliable first estimates of solvation times in dipolar solvents [3,18,19], there are no corresponding models for calibrating expectations in ionic liquids. Whereas the dominant solvation mechanism in dipolar fluids is reorienation of solvent molecules [20], one expects the primary mechanism in ionic liquids to involve translation of ions. Thus, it is reasonable to suppose that there should be some link between solvation times and such properties as viscosity and the translational diffusion constants of the ionic components, but the details are unclear.

In an effort to uncover such relationships, we have undertaken a survey of solvation dynamics in 19 different ionic liquids using the solvation probe C153. Figure 1 catalogues the liquids involved, which include many of the most common types of cations and anions currently in use. This collection contains ions whose mass, volume, and shape span wide ranges.

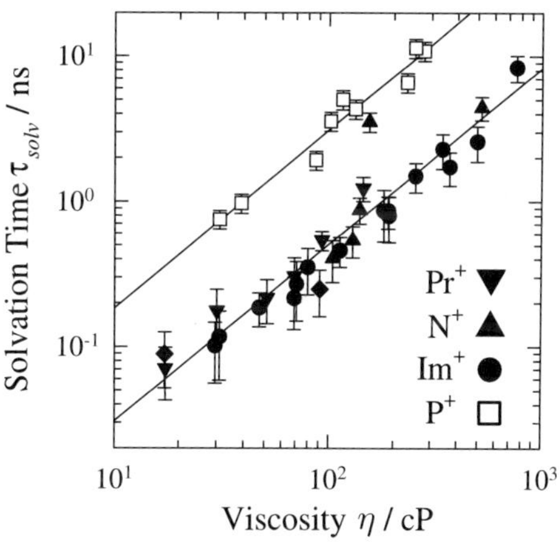

Figure 2: Observed solvation times [1] of C153 plotted against solvent viscosity.

Solvation times of C153 measured in these ionic liquids are plotted as a function of solvent viscosity in Fig. 2. In some cases multiple temperatures, within the range 263-343 K, are plotted for a single liquid. In previous work [7-9] we have shown that solvation times track viscosity as temperature is changed within a single solvent. This is not unexpected. What is more interesting is the fact that the data in Figure 2 show that, to a reasonable approximation, solvation times in many distinct ionic liquids share a *common* relationships with solvent viscosity. In fact, with the exception of a single point, all of the data appear to break up cleanly into two groupings. One grouping consists of the P^+ series, all liquids comprised of the large phosphonium cation $PC_{32}H_{68}^+$ and five assorted anions (see Fig. 1), and the other grouping consists of all of the remaining liquids. However, the correlation lines shown in Fig. 2 are of the form $\tau_{solv} \cong C\eta^p$ with p = 1.2 rather than unity in both cases. In series of liquids consisting of a single cation and a variety of anions, the P^+ series being the primary example [10], we find that the choice of anion has a substantial effect on the solvation time. For example, there is a six-fold variation in the solvation times among the P^+ liquids at 25 °C. However, these variations appear to be largely accounted for by the effect that the anion has on the liquid's viscosity. As is clear from Fig. 2, the same cannot be said of the identity of the cation. On average, the same viscosity produces a roughly six times slower solvation response in the P^+ liquids compared to the remaining (Pr^+, N^+, and Im^+) series.

Some feature(s) of the P^+ cation must distinguish the dynamics the P^+ liquids from those of the remaining cations. The most obvious distinguishing feature is the large size of $PC_{32}H_{68}^+$ -- its van der Waals volume and mass are twice as large as the next largest cation studied. If we attempt to take the cation size into account in a simple empirical manner, we find that all of the data collected here can be correlated to within a mean unsigned error of 26% using the empirical form

$$\tau_{solv} / \mathrm{ns} = \frac{\eta / \mathrm{cP}}{T / \mathrm{K}} \{-15.0 + 4.81(R_+ / \text{Å})\} \tag{2}$$

where R_+ is the estimated van der Waals radius of the cat on. The interpretation of such a correlation is not obvious.

A more suggestive correlation is obtained by relating the observed viscosity and ion size dependences to the translational diffusion of the cation. NMR measurements of ion diffusion constants by Watanabe and coworkers [21-23] have shown that cation diffusion constants in a range of ionic liquids similar to those studied here follow a modified Stokes-Einstein type law,

$$D_+ = f_+ \frac{k_B T}{6\pi\eta} \tag{3}$$

where f_+ is an empirical correction, $f_+ = 0.53 \pm .06$. (An analogous relation holds for anion diffusion with $f_- = 0.74 \pm .09$.) To put diffusion constants on a comparable basis to τ_{solv}, we consider "cation diffusion times" defined by:

$$t_+ \equiv \frac{R_+^2}{6D_+} \cong \frac{\eta R_+^3}{f_+ k_B T} \tag{4}$$

t_+ is the time required for a cation to move a root mean square distance equal to its radius, assuming that its motion is purely diffusive [24]. Figure 3 shows a plot of τ_{solv} versus cation diffusion times estimated using the second equality in Eq. 4. This correlation again represents the data to within a mean unsigned error of 26%.

What meaning can be ascribed to the correlations shown in Figs. 2 and 3 and Eq. 2? It seems safe to assume that the proportionality between solvation time and solvent viscosity (or η/T) implies that the underlying dynamics are in some way related to structural rearrangements of the solvent25. As already mentioned, it is reasonable to expect that translation of ions is the primary mode of relaxation, although contributions from ion reorientations cannot be ruled out. The apparent role of cation size and the correlation shown in Fig. 3 implicate cation translational motion as being a controlling factor in the solvation dynamics. Why the cation? One possibility is that cations dominate the solvation energetics of C153. Some support for this idea comes from studies of solvation dynamics in dilute ionic solutions [26], where it was observed that probes such as C153 are much more sensitive to the presence of small cations (Li^+, Na^+, K^+, etc.) than they are to anions. But given the size of the cations involved here, this explanation seems unlikely. A more plausible explanation is that the cations appear to control the solvation dynamics because they are almost always the larger of the two ions in the liquids we have examined. If, as one might expect, solvation involves collective motion of both ionic species, it is not unreasonable to suppose that the slowest components of these collective dynamics are linked to the size of the largest species involved, here the cation. (The solvation times considered here are integral times and therefore most heavily weight the slowest components of the response.) Finally, it should be noted that Fig. 3 shows τ_{solv} to be even smaller than t_+, the time required for a cation to diffuse and r.m.s. distance equal to its radius. In most cases the estimated extent of anion motion during a time τ_{solv} is also not much greater than its own radius. Thus, the data indicate that the amount of ion motion involved in the solvation act is relatively small.

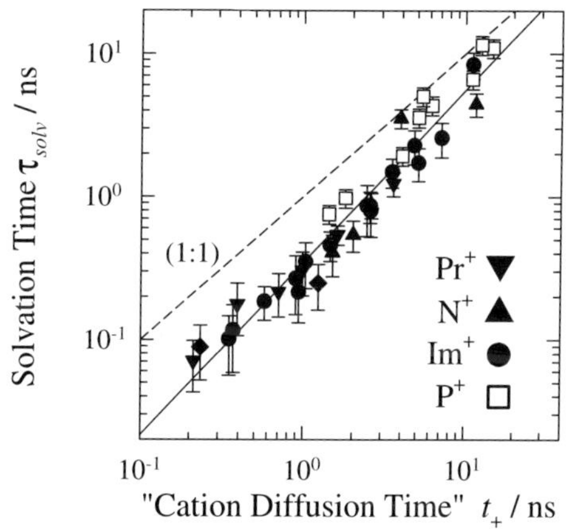

Figure 3: Solvation times [1] of C153 versus estimates of the "cation diffusion time" (Eq. 4).

3. Kerr-Gated Emission Measurements

An important fact about picosecond experiments of the sort described above is that the extent of the spectral relaxation actually observed is often considerably smaller than what is expected. In our experiments, which employ a TCSPC system with a 25 ps response (FWHM), we find that up to half of the dynamics is too fast (<5 ps) to be registered by this method. The solvation times described in the previous section have been approximately corrected for the fast dynamics not observed in the picosecond experiments [1]. But, in order to learn more about the nature of this "missing dynamics", we have also observed timedependent Stokes shifts using the femtosecond Kerr-gated emission (KGE) spectroscopy. The spectrometer and analysis methods we employ have been described in detail in Ref. [27]. The instrument has a response time of 450 fs (FWHM), and after deconvolution it is capable of resolving spectral response times down to the 100 fs range.

An example of the sort of data we obtain is shown in Figs. 4 and 5. Here and in the following discussion, data are obtained using the solute DCS [28] (Fig. 1) rather than C153 because its shorter lifetime is advantageous for the KGE experiment. In the ionic liquid illustrated in Figs. 4 and 5 the TCSPC method captures the majority of the spectral relaxation (~70%). The "missing" dynamics occur in the .1-10 ps time window and are easily captured by the KGE experiment. As shown in Fig. 5, we only collect KGE data to 200 ps after which time the TCSPC experiment provides much better data. To construct the complete spectral response function in these ionic liquids we therefore combine the data from the two experiments simply by splicing together curves like those in Fig. 5 where they intersect (~20 ps).

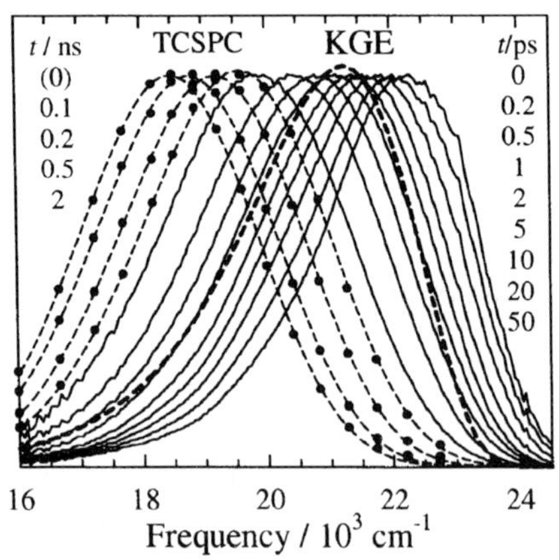

Figure 4: Time-resolved spectra of DCS in $[Im_{41}^+][Tf_2N^-]$ at 25 °C. Points denote TCSPC data and dashed curves log-normal fits to these data at the times indicated. (The heavy dashed curve shows the fitted t=0 spectrum from TCSPC.) The solid curves are the KGE data.

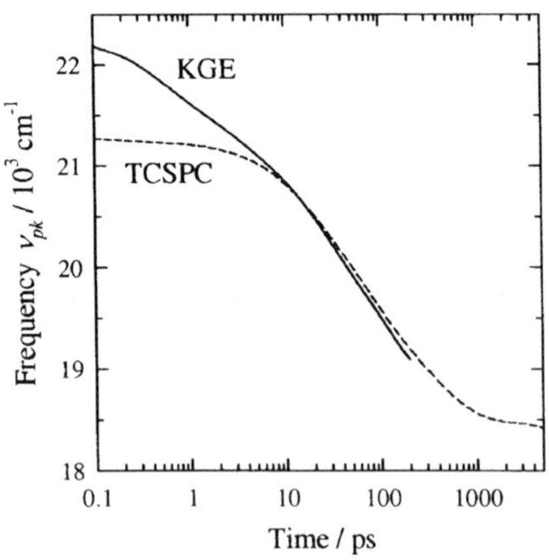

Figure 5: Evolution of the peak frequency of the spectra of DCS in $[Im_{41}^+][Tf_2N^-]$ (25 °C) shown in Fig. 4.

To date we have used this combination of KGE + TCSPC spectroscopies to measure the complete solvation response of DCS in three liquids having the common cation 3-butyl-1-methylimidazolium "Im_{41}^+" and three anions of varying size, PF_6^-, Tf_2N^-, and Tf_3C^- (Fig.1) [29]. Figures 6 and 7 show the spectral response functions $S_\nu(t)$ (Eq. 1) observed at 25 °C. Very similar results are observed at 40 °C[29] .

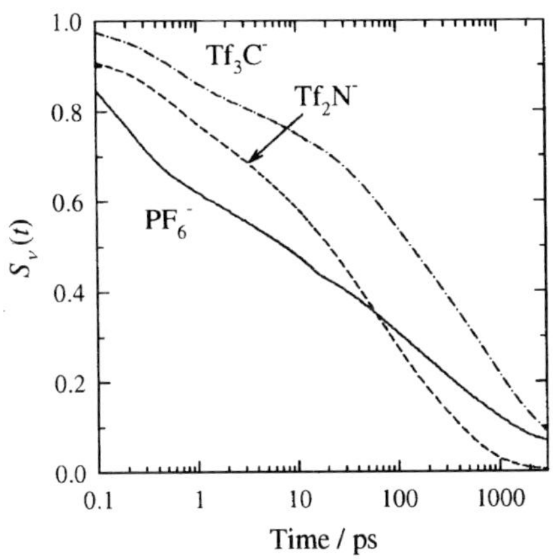

Figure 6: Spectral response functions measured with the probe DCS in three [Im$_{41}^+$] ionic liquids at 25 °C.

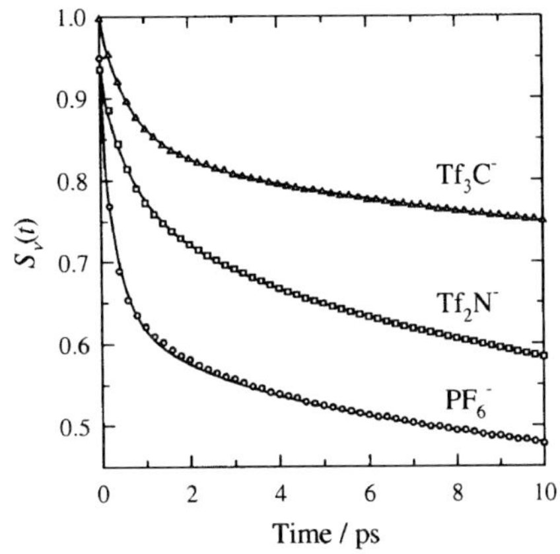

Figure 7: Short time behavior of the spectral response functions from Fig. 5. Points are (thinned) experimental data and the curves are fits to Eq. 5.

Figure 6 highlights the unusually broad distribution of relaxation times associated with solvation in ionic liquids. Significant relaxation takes place over times stretching from 100 fs to 3 ns, i.e. over more than 4 decades in time. Whereas the spectral response functions we had previously measured only with TCSPC could be fit adequately using a single stretched exponential function [7-10], the data in Fig. 6 require an additional fast component to capture the data to within uncertainties:

$$S_\nu(t) = f\exp(-t/\tau_1) + (1-f)\exp\{-(t/\tau_2)^\beta\} \qquad (5)$$

The stretching exponents observed are in the range $0.3 \leq \beta \leq 0.5$, again indicating a remarkably broad distribution of relaxation times in these systems. It should be mentioned that we place no particular interpretation on the functional form used in Eq. 5 -- it is merely a convenient representation of the experimental data. Other functions such as a multi-exponential sum would also suffice, and functions motivated by the predictions of mode coupling theory [30] might provide additional insights, but for now we chose to represent the data with the fewest fitting parameters possible.

The early time dynamics of $Sv(t)$ are better displayed in Fig. 7. From this figure it is clear that the amplitude f and perhaps the time scale τ_1 associated with the faster portion of the solvation response varies systematically with anion. The smaller (lighter) the anion the more prominent is the fast component. Such behavior is at least qualitatively similar to the behavior recently found in recent simulations of solvation dynamics in ionic liquids [31-34]. These observations contrast with the results of the previous section, where we found that, beyond its effect on viscosity, the identity of the anion did not seem to play a clear role in determining the (integral) solvation times τ_{solv}. Where the identity of the cation did seem to have a distinct effect on τ_{solv} was in cases where the cation was unusually large. We conjectured that here translation motions of the largest species dictate the slowest part of the solvation response and thus the solvation times. In an analagous fashion, one might argue that the smallest ion should determine the fastest part of the solvation response. Consistent with this idea, the fast response in $[Im_{41}^+][PF_6^-]$ is much more pronounced than in the other two solvents, whose constituents are all considerably larger than PF_6^-.

However, it should also be noted that even the long-time dynamics of $[Im_{41}^+][PF_6^-]$ are considerably different from the dynamics in the other two liquids. Although the solvation times track viscosities in roughly the expected manner, the dynamics in $[Im_{41}^+][PF_6^-]$ are clearly dissimilar to those in the other two liquids, which can be at least approximately related by a time scaling. Thus, when one looks beyond only the solvation times, it is clear that the identity of both species comprising the ionic liquid is important for determining the nature of the solvation response. The correlations established in the previous section are therefore only a starting point for understanding these systems.

4. Summary and Conclusions

In this work we have summarized much of the data currently available on solvation dynamics in ionic liquids. The data show that complete solvent equilibration is much slower (~1 ns) in these high-viscosity solvents than it is in most conventional solvents (~1 ps). Attempts to correlate solvation times with viscosities and ion diffusion constants suggest that translational diffusion over small distances is the likely mechanism of solvation in ionic liquids. Despite the slow overall response in these systems, fs KGE measurements also reveal significant relaxation occurring in the ps and sub-ps range. In some cases the response may be clearly bimodal, but in most cases it appears that solvation in ionic liquids is characterized by a remarkably broad distribution of relaxation times stretching over the range of 100 fs – 10 ns. This broad distribution of times together with a number of other experimental observations, such as nonexponential probe [8-10] and ion reorientation times [30], non-Arrhenius dependence of viscosity and other transport properties on temperature [21-23], and heterogeneous emission [35], all suggest the "glassy" character of dynamics in ionic liquids. However, we are only at the beginning stages of characterizing these dynamics. Much more experimental work, together with help from computer simulations and theory, will be needed before a satisfying understanding of the dynamics of these interesting new liquids will be achieved.

Acknowledgements
This work was supported by funds from the Office of Basic Energy Sciences of the US Department of Energy.

Notes and References

1. The solvation times plotted in Figs. 2 and 3 are integral times of $S_v(t)$ which account in a crude manner for the fact that in many cases we estimate that only a fraction f_{obs} ($0.5 \leq f_{obs} \leq 1$) of the total dynamic Stokes shift is observed due to insufficient time resolution in the TCSPC experiments. We do so by plotting data at values $\tau_{solv} = <\tau>_{obs} (1 + f_{obs}) / 2$ and the error bars are (at least) $\pm <\tau>_{obs}(1 - f_{obs})/2$, where $<\tau>_{obs}$ is the integral time associated with the observed dynamic Stokes shift.
2. R. D. Rogers and K. R. Seddon (Eds.) Ionic Liquids IIIA, B: Fundamentals, Progress, Challenges, and Opportunities, Vol. 901, 902, American Chemical Society, Washigton, DC, 2005.
3. M. Maroncelli, J. Mol. Liq. 57 (1993) 1.
4. R. Karmakar and A. Samanta, J. Phys. Chem. A 106 (2002) 4447.
5. R. Karmakar and A. Samanta, J. Phys. Chem. A 106 (2002) 6670.
6. R. Karmakar and A. Samanta, J. Phys. Chem. A 107 (2003) 7340.
7. J. A. Ingram, R. S. Moog, N. Ito, R. Biswas, and M. Maroncelli, J. Phys. Chem. B 107 (2003) 5926.
8. S. Arzhantsev, N. Ito, M. Heitz, and M. Maroncelli, Chem. Phys. Lett. 381 (2003) 278.
9. N. Ito, S. Arzhantsev, and M. Maroncelli, Chem. Phys. Lett. 396(2004) 83.
10. N. Ito, S. Arzhantsev, M. Heitz, and M. Maroncelli, J. Phys. Chem. B 108 (2004) 5771.
11. S. N. Baker, G. A. Baker, C. A. Munson, F. C., E. J. Bukowski, A. N. Cartwright, and F. V. Bright, Ind. Eng. Chem. Res. 42 (2003) 6457.
12. D. Chakrabarty, P. Hazra, A. Chakraborty, D. Seth, and N. Sarkar, Chem. Phys. Lett. 381 (2003) 697.
13. P. K. Chowdhury, M. Halder, L. Sanders, T. Calhoun, J. L. Anderson, D. W. Armstrong, X. Song, and J. W. Petrich, J. Phys. Chem. B 108 (2004) 10245.
14. S. Saha, P. K. Mandal, and A. Samanta, Phys. Chem. Chem. Phys. 6 (2004) 3106.
15. J. F. Wishart, S. I. Lall-Ramnarine, R. Raju, A. Scumpia, S. Bellevue, R. Ragbir, and R. Engel, Rad. Phys. Chem. 72 (2004) 99.
16. P. K. Mandal and A. Samanta, J. Phys. Chem. B 109 (2005) 15172.
17. M. L. Horng, J. A. Gardecki, A. Papazyan, and M. Maroncelli, J. Phys. Chem. 99 (1995) 17311.
18. M. Maroncelli, J. Chem. Phys. 106 (1997) 1545.
19. X. Song and D. Chandler, J. Chem. Phys. 108 (1998) 2594.
20. B. M. Ladanyi and M. Maroncelli, J. Chem. Phys. 109 (1998) 3204.
21. A. Noda, K. Hayamizu, and M. Watanabe, J. Phys. Chem. B 105 (2001) 4603.
22. H. Tokuda, K. Hayamizu, K. Ishii, M. A. B. H. Susan, and M. Watanabe, J. Phys. Chem. B 108 (2004) 16593.
23. H. Tokuda, K. Hayamizu, K. Ishii, M. A. B. H. Susan, and M. Watanabe, J. Phys. Chem. B 109 (2005) 6103.
24. The assumption of purely diffusive motion $< R_2 (t) > = 6Dt$ may seem inappropriate for distances as small as $R=R_+$. However, examination of the $< R^2 (t) >$ behavior observed in simulations of ionic liquids [36,37] indicates that the values of t_+ estimated in this way should in fact be reasonably accurate.
25. One might also consider the solute rotation and translation, which will also be proportional to η/T. We have measured rotation times of C153 in many of these solvents and find that (L=2) rotation times are about 10-fold slower than τ_{solv}, except in the P^+ solvents, where they are comparable. We do not therefore expect solute reorientation to play much of a role in most solvents whereas it might hasten the relaxation in P^+ solvents. We have no direct information about solute translation.

26. C. F. Chapman and M. Maroncelli, J. Phys. Chem. 95 (1991) 9095.
27. S. Arzhantsev and M. Maroncelli, Appl. Spectrosc. 59 (2005) 206.
28. S. Arzhantsev, K. Zachariasse, and M. Maroncelli, J. Phys. Chem. B (2006).
29. S. Arzhantsev, J. Hui, N. Ito, and M. Maroncelli, Chem. Phys. Lett. (2006).
30. H. Cang, J. Li, and M. D. Fayer, J. Chem. Phys. 119 (2003) 13017.
31. Y. Shim, M. Y. Choi, and H. Kim, J. Chem. Phys. 122 (2004) 044511.
32. Y. Shim, J. Duan, M. Y. Choi, and H. J. Kim, J. Chem. Phys. 119 (2003) 6411.
33. M. N. Kobrak and V. Znamenskiy, Chem. Phys. Lett. 395 (2004) 127.
34. C. J. Margulis, Mol. Phys. 102 (2004) 829.
35. P. K. Mandal, M. Sarkar, and A. Samanta, J. Phys. Chem. A 108 (2004) 9048.
36. M. G. Del Popolo and G. Voth, J. Phys. Chem. B 108 (2004) 1744.
37. Z. Su and M. Maroncelli, unpublished results (2005).

Femtochemistry VII
Fundamental Ultrafast Processes in Chemistry, Physics, and Biology
A.W. Castleman, Jr. and M.L. Kimble

Femtosecond relaxation dynamics of solvated electrons from liquid metal-ammonia solutions: Temperature jump versus local density jump

Jörg Lindner, Peter Vöhringer

[a]*Rheinische Friedrich-Wilhelms-Universität, Institut für Physikalische und Theoretische Chemie, Wegelerstraße 12, 53115 Bonn, Germany*

1. Introduction

Localization phenomena of excess electrons in condensed phases has been the subject of extensive femtosecond (fs) time-resolved experiments in the past. The primary focus was on elucidating the relaxation dynamics of solvated electrons in liquid water where their preparation can be accomplished by laser-induced detachment either from the solvent or from suitable solutes [1-4].

Recently, it became also possible to study the femtosecond relaxation dynamics of equilibrated solvated electrons in liquid ammonia solutions [5]. However, rather than producing the excess charges flash-photolytically, the electrons were synthesized chemically by preparing metal-ammonia solutions which are well known for their characteristic blue color. Whereas flash photolysis is connected with a poorly understood mechanism for electron formation, the chemical preparation allows for full control over the state variables of the sample like e.g. pressure, p, temperature, T, or even the electron concentration.

The fs pump probe experiments on both the solvated electron in bulk water (i.e. the "hydrated electron") and the solvated electron from metal-ammonia solutions (i.e. the "ammoniated" electron) yield a spectro-temporal response that can be characterized by the following common features: Immediately after photo-excitation, a transient bleach can be observed in the spectral region of the stationary absorption spectrum of the solvated electron. This bleach is accompanied by a transient absorption that is strongly red-shifted with respect to the linear resonance of the equilibrated excess charge. Both bleach and absorption component to the dynamic spectrum seem to be highly correlated because their decay occurs on identical time scales. Whereas in liquid water, decay constants of 300 fs and 1 ps were reported [3], the dynamics in ammonia could be described by time constants of 150 fs and 2 ps [5]. In both systems, the faster of the two exponential components displayed the larger amplitude.

Such a fs spectral response in bulk liquids is in principle consistent with two distinct relaxation scenarios: (i) an adiabatic mechanism in which the absorption/bleach decays reflect processes associated with the excited state of the electron such as dynamic solvation, and (ii) a non-adiabatic mechanism in which the same response reflects ground-state relaxation processes preceded by an ultrafast internal conversion from the excited to the ground electronic state.

Here, we discuss two different non-adiabatic relaxation models to interpret the experimental fs-spectroscopy of solvated electrons in liquid ammonia, namely a laser-induced local temperature-jump model and a laser-induced local density-jump model. As opposed to the adiabatic relaxation picture, an *a priori* knowledge of the frequency dependent absorption cross section for transitions from the excited electronic state to higher lying states (such as continuum states) is neither required for the T-jump nor for the local ρ-jump model. In fact, both models rely on a minimum number of free adjustable parameters to simulate the experimental transients and require no input other than the shape of thermalized linear electronic resonance of the solvated electron and its dependence on the thermodynamic state variables of the fluid, i.e. its temperature or its density.

2. The linear electronic resonance

The absorption spectrum of solvated electrons in liquid ammonia solutions [5] at a temperature of -50°C and under its own saturation pressure is centered around 6840 cm^{-1} and has a full spectral width at half maximum of 3260 cm^{-1}. The line shape is highly asymmetric, as evidenced by the inability of conventional symmetrical line shapes such as Gaussians or Lorentzians to fit the data. The resonance is often believed to originate from transitions between hydrogen-like states of the electron, which are localized in a quasi-spherical cavity formed by the nearby solvent molecules, i.e. the spectrum is due to a transition of the type "1s"→"2p". The origin of the asymmetry as well as the line broadening mechanisms that give rise to the enormous spectral width remain as yet unclear. Deviations from a non-spherical cavity will lift the p-state degeneracy thereby contributing substantially to the spectral width. The time-scale of the dynamical fluctuations of the molecular cavity and the concomitant stochastic scrambling of the Bohr frequencies of the three individual 1s-2p-components are believed to be of the order of a few femtosecond as deduced from recent fs photon echo studies on the hydrated electron by Wiersma and coworkers [4].

In contrast to this cavity-picture of the electron, recent *ab initio* quantum chemical as well as time-dependent density functional calculation on finite size aqueous clusters by Domcke and coworkers suggest that the hydrated hydronium radical $H_3O(H_2O)_n$ may actually be the chemical species which carries the spectroscopic properties of the solvated electron in bulk water. Similarly, one might conclude that ammoniated ammonium clusters $NH_4(NH_3)_n$ are responsible for the characteristic spectrum of metal ammonia solutions [6].

The absorption spectrum of the ammoniated electron is a strong function of the temperature, T. Indeed, by increasing T while continuously following the saturation pressure of the solution, the electronic resonance shifts markedly to lower frequencies [5] according to an apparent temperature coefficient of

$$\left(\frac{\partial \tilde{\nu}_{max}}{\partial T}\right)_{Sat} = 17\frac{cm^{-1}}{K}. \quad (1)$$

Experimental spectra along the gas-liquid coexistence curve are available in the interval 203 K $\leq T \leq$ 298 K where the liquid density, ρ, decreases by 20 % from ~0.73 g/ml to ~0.6 g/ml. This yields an apparent density coefficient of

$$\left(\frac{\partial \tilde{\nu}_{max}}{\partial \rho}\right)_{Sat} = 225\frac{cm^{-1}}{mol\ dm^{-3}}. \quad (2)$$

Therefore, the pure expansion of the liquid in response to a temperature increase cannot be neglected. The two individual effects on the spectral position, namely its T-dependence at constant ρ and its ρ-dependence at constant T, can be disentangled in the following fashion. According to Farhataziz *et al.* [7], the electronic resonance shifts with increasing liquid density according to an isothermal density coefficient of

$$\left(\frac{\partial \tilde{\nu}_{max}}{\partial \rho}\right)_{T} = 159\frac{cm^{-1}}{mol dm^{3}}, \quad (3)$$

when the temperature is held constant at 296 K. A detailed re-evaluation of pressure-dependent absorption spectra from Vogelsang and Schindewolf [8] yields a coefficient of 173 cm^{-1} mol^{-1} dm^{-3} for a temperature of 272 K and a value of 156 cm^{-1} mol^{-1} dm^{-3} at 214 K. Therefore, it is reasonable to assume that the isothermal density coefficient is independent of the temperature.

Eqn. (3) lets us retrieve the isochoric temperature dependence of the spectral position provided we have an expression for the T-dependence of the density of liquid ammonia along its gas-liquid coexistence curve. In the interval 210 $\leq T \leq$ 300 K, i.e. well below the critical point, this dependence can be linearly approximated with sufficient accuracy by

$$\left(\frac{\partial \rho}{\partial T}\right)_{Sat} = -0.0771 \frac{mol}{dm^3\,K}. \tag{4}$$

It can easily be shown that under these circumstances, the isochoric temperature dependence of the spectral position is given by

$$\left(\frac{\partial \tilde{\nu}_{max}}{\partial T}\right)_{\rho} = \left\{\left(\frac{\partial \tilde{\nu}_{max}}{\partial \rho}\right)_{Sat} - \left(\frac{\partial \tilde{\nu}_{max}}{\partial \rho}\right)_{T}\right\} \times \left(\frac{\partial \rho}{\partial T}\right)_{Sat} = -5.3 \frac{cm^{-1}}{K}. \tag{5}$$

This analysis shows that only one third of the apparent red shift of the absorption spectrum along the saturation curve can possibly be a true temperature effect. The remainder must simply be due to the pure expansion of the liquid and the high sensitivity of the electronic resonance to the solvent density.

3. The dynamical response of the electronic resonance

We have recently shown that the transient optical response of the ammoniated electron to femtosecond excitation can be simulated quantitatively with a dynamically evolving Kubo line shape,

$$\varepsilon(\tilde{\nu},\tau) = \mathrm{Im}\left\{1 - 2\pi i c\,\tilde{\nu} \int_0^{\infty} dt\; e^{2\pi i c \tilde{\nu} t}\; \psi(t,\tau)\right\} \tag{6}$$

where $\Psi(t,\tau)$ is the relaxation function, c is the speed of light, and the frequency, $\tilde{\nu}$, is in units of wavenumber. The underlying stochastic modulation of the "1s-2p"-Bohr frequency is assumed to be a Gauss-Markov-process, i.e. the energy gap correlation function decays exponentially with a rate, γ, and is characterized by an amplitude, Δ^2, corresponding to the mean-squared magnitude of the frequency fluctuations. The resultant relaxation function is given by

$$\psi(t,\tau) = \exp\left[-\frac{\Delta^2}{\gamma^2}\left(e^{-\gamma t} + \gamma t - 1\right)\right] \cos\left[2\pi c \tilde{\nu}_r(\tau) t\right]. \tag{7}$$

It acquires a pump-probe delay dependence only through the resonance frequency of the "1s"→"2p" transition according to

$$\tilde{\nu}_r(\tau) - \tilde{\nu}_r(-\infty) = \Delta\tilde{\nu}_r \exp(-\tau/\tau_{rel}) \tag{8}$$

while the line broadening parameters are kept constant.

The pump laser-induced frequency jump, $\Delta\tilde{\nu}_r$, from the stationary resonance, $\tilde{\nu}_r(-\infty)$,was found to be roughly –3600 cm^{-1}, whereas the dominant relaxation time was 150 fs. These values translate into a jump of the frequency of maximal absorbance of 3330 cm^{-1}. Figure 1 displays a dynamical pump-probe simulation according to Eqns. (6) through (8) that quantitatively satisfies our previously published experimental data recorded at –52°C and saturation pressure [5].

To reiterate, the pump pulse comes in at time t = 0 and excites the 1s-2p transition of the solvated electron thereby leaving a pronounced differential bleaching signal (black contours) in the spectral region of the linear absorption spectrum. The simulation assumes further that the residence time of the electron in its excited state is well below the experimental time resolution and that, after excitation, the electron finds itself in its 1s-type ground state but far from equilibrium with respect to the surrounding solvent cage. It is this non-equilibrated electronic ground-state which gives rise to a resonance that is strongly red-shifted with respect to the linear resonance. As a consequence, the differential pump-probe spectrum displays a transient absorption in the far-red wing of the linear resonance and beyond. The decay of the transient absorption and the simultaneous bleach recovery reflects a dynamic blue-shift of the resonance back to its original spectral position according to Eq. 8. It is brought about exclusively by the ensuing relaxation of the electronic ground state back to equilibrium.

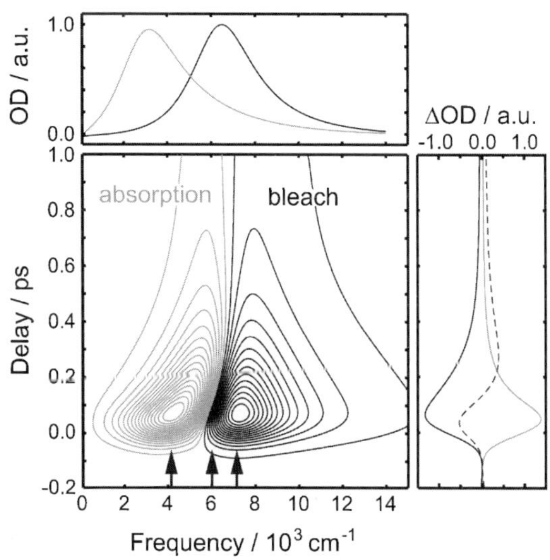

Figure 1: Top panel: Calculated line shapes of the linear electronic resonance of the solvated electron in liquid ammonia before (black) and after (gray) an infinitely fast mid-IR excitation. Middle panel: Simulation of the spectro-temporal response of the electrons following excitation with 120-fs mid-IR pulses. The vertical arrows indicate the probe frequencies of representative femtosecond transients displayed in the right panel.

4. T-jump versus ρ-jump

From the time-dependent Kubo line-shape we can extract the time-dependent resonance frequency of maximal absorbance of the solvated electrons following optical excitation and subsequent non-adiabatic 2p-to-1s transition. This quantity is shown in Figure 2A. Here, the question arises: What is the molecular nature of the relaxation dynamics that are responsible for such a time-dependent resonance frequency?

We have previously attributed the dynamic blue shift from Figure 2A to electron cooling [5]. In this temperature-jump scenario, the non-adiabatic transition effectively converts the photon energy into kinetic energy of the solvent cage. Provided that, at each pump-probe delay, this kinetic energy is randomly distributed in a canonical fashion over all nuclear degrees of freedom of the solvent cage that couple to the electronic transition dipole, the resonance frequency can be taken as an indicator of the local temperature of the solvent surrounding the excess charge. In the past we have used the temperature dependence of the linear absorption spectrum of the solution under its own saturation pressure to convert the time-dependent resonance frequency into a time-dependent temperature. Such a procedure implies that, while the solvent shell cools, it contracts according to the temperature dependence of the liquid solvent density.

Using Eqn. (1), the time-dependent resonance frequency is then mapped onto the temperature scale denoted T_{Sat} in Figure 2A. The magnitude of the laser-induced T-jump amounts to roughly 195 K. Therefore, using the molecular heat capacity of ammonia of 6.7 cm^{-1} / K, this temperature jump equals a jump of the internal molecular energy of the solvent particles surrounding the electron of 1313 cm^{-1}. Given the energy of the pump photon of 7815 cm-1 used in the experiments of Ref. [5], this energy jump implies that on average the electron is coordinated to six nearest neighbor ammonia molecules. Note that such a six-fold coordination is highly reminiscent of the hydrated electron which is believed to be trapped in an octahedral array of water particles [4]. However, using the equivalent interpretation of cavity cooling but now keeping volume of the solvated electron cavity constant, the laser-induced T-jump and, therefore, the increase of the molecular internal energy are roughly three times as large as before. Following again the above lines of arguments, an isochoric laser-induced temperature-jump scenario of the solvent cage is consequently consistent with an effective coordination number of only two, thereby supporting the chemical species notion of the solvated electron supported by the quantum chemical calculations of Ref. [6].

So far, we have neglected that the optical excitation results inevitably in a quasi-instantaneous formation of a "bubble" due to the enhanced overlap between the solvent shell particles and the spatially more demanding 2p electron cloud (see Figure 2B). Such a mechanism was previously discussed by Chergui and coworkers for the ultrafast relaxation dynamics of optically prepared Rydberg-states of atomic and molecular impurities in cryogenic noble gas matrices [9].

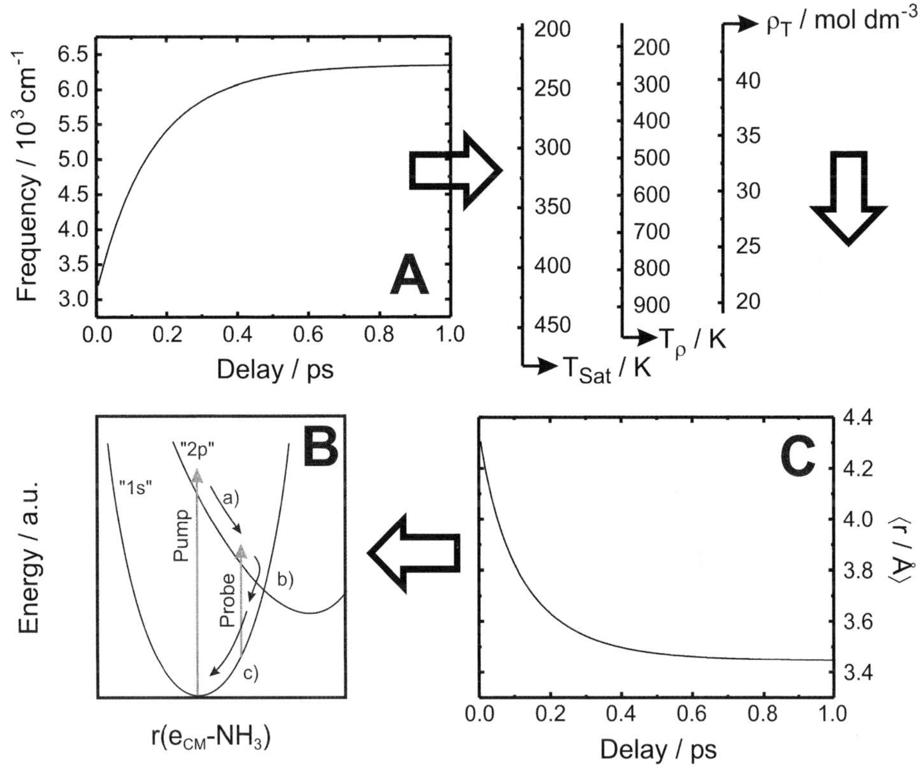

Figure 2: A:Time-dependent resonance frequency of the ammoniated electron obtained from the model simulations. The frequency scale can be converted into a time-dependent solvent temperature using Eqn. (1) or (5) or into a time-dependent solvent density using Eqn. (3). The latter can be translated into a mean solvent interparticle distance thereby relating the spectral position of the electronic resonance with the average solvent geometry around the electron cloud.

In the case of the ammoniated electron, the expansion of the solvent cage due to the Pauli repulsion between the excess charge and the electrons of the solvent particles is likely to be the motion which guides the system towards the conical intersection between the 1s and the 2p electronic states (see arrow a in Figure 2B). Upon the non-adiabatic transition, which occurs within the experimental time resolution (arrow b), the newly formed 1s-ground state experiences a solvent shell which is now in a non-equilibrium geometry with respect to its own charge density. Optical Franck-Condon transitions from such a geometry will be strongly red-shifted as compared to the linear absorption spectrum of the solvated electron. With the electron back in the ground state, a "bubble" contraction will set in, thereby driving the resonance back to its original spectral position in the blue (arrow c).

This interpretation can be put on a more quantitative footing by using the isothermal density dependence of the linear absorption spectrum as given by Eqn. 5 and assuming that the transient solvated electron spectrum accurately reports the solvent density in the immediate vicinity of the excess charge. The resulting local density scale is denoted ρ_T in Figure 2A. We can see that the local density decreases by almost a factor of two in response to the sudden preparation of the 2p-state. The pump-induced dilatation is then followed by a slower contraction of the solvent with the characteristic time of 150 fs.

Assuming that at thermal equilibrium the average distance between molecules forming the solvent cavity is identical to the intermolecular distance in the bulk, we can convert the time dependent resonance frequency into a time-dependent "bulk-equivalent" interparticle distance within the cage via the isothermal density dependence of the resonance frequency (see Figure 2C). Interestingly, the equilibrium intermolecular distance of 3.43 Å in the bulk is roughly equal to the cavity radius of the ammoniated electron of 3.34 Å as determined from dilatometry measurements [10]. This coincidence makes liquid NH_3 a perfectly size-adapted environment for the localization of an excess charge simply by replacing an ammonia molecule by the excess electron (cavity notion) or by attaching it to a fully solvated ammonia molecule (chemical species notion).

5. Conclusions

In summary, we have presented two distinct non-adiabatic relaxation scenarios for interpreting the fs-spectral response of solvated electrons from metal-ammonia solutions, namely, the local T-jump and the local ρ-jump models. Depending upon what model is preferred, the time-dependent solvated electron resonance merely serves as a local thermometer or a local densitometer of its own solvent cavity following the ultrafast non-adiabatic transition.

References

1. A. Hertwig, H. Hippler, A. N. Unterreiner, and P. Vöhringer, Ber. Bunsenges. Phys. Chem. 102 (1998) 805.
2. C. L. Thomsen, D. Madsen, S. R. Keiding, J. Thøgersen, and O. Christiansen, J. Chem. Phys. 110 (1999) 3453.
3. J. C. Alfano, P. K. Walhout, Y. Kimura, and P. F. Barbara, J. Chem. Phys. 98 (1993) 5996.

4. A. Baltuška, M. F. Emde, M. S. Pshenichnikov, and D. A. Wiersma, J. Phys. Chem. A 103 (1999) 10065..
5. A.-N. Unterreiner, J. Lindner, and P. Vöhringer, in Femtochemistry and Femtobiology, World Scientific, Singapore, 2002, p. 261-269.
6. S. Neumann, W. Eisfeld, A. L. Sobolewski, and W. Domcke, Phys. Chem. Chem. Phys. 6 (2004) 5297.
7. Farhataziz, L. M. Perkey, and R. R. Hentz, J. Chem. Phys. 60 (1974) 4383.
8. R. Vogelsang and U. Schindewolf, Ber. Bunsenges. Phys. Chem. 75 (1971) 651.
9. F. Vigliotti, L. Bonacina, and M. Chergui, J. Chem. Phys. 116 (2002) 4553.
10. S. R. Gunn and L. G. Green, J. Chem. Phys. 36 (1962) 363.

Femtochemistry VII
Fundamental Ultrafast Processes in Chemistry, Physics, and Biology
A.W. Castleman, Jr. and M.L. Kimble

Multiplex Electronic Sum Frequency Generation Spectroscopy of Dye Molecules at the Air/Water Interface

Shoichi Yamaguchi, Kentaro Sekiguchi, Tahei Tahara

Molecular Spectroscopy Laboratory, RIKEN (The Institute of Physical and Chemical Research), 2-1 Hirosawa, Wako, Saitama 351-0198, Japan

1. Introduction

Second-order nonlinear spectroscopy is known as a powerful tool to study the structure and dynamics of molecules at interfaces that are very important in fundamental science and industrial application [1]. Vibrational spectra of interfacial molecules are measured by IR-visible (vis) sum frequency generation (SFG) spectroscopy that has been widely utilized since the first demonstration by Shen and co-workers [2]. The sensitivity of the IR-vis SFG spectroscopy has been improved to a great extent by using broadband IR pulses and multichannel detectors, and high-quality IR-vis SFG spectra can be measured nowadays. On the other hand, electronic $\chi^{(2)}$ (second-order nonlinear susceptibility) data of interfacial molecules are measured by the resonance second harmonic generation (SHG) method. In the spectral measurements, the SHG light intensity is recorded point by point with scanning the fundamental excitation wavelength, and this technique has not been changed since Heinz *et al.* first reported the SHG spectra of dye molecules on silica [3]. Because interface-specific electronic spectra are indispensable for the studies of the static and dynamic properties of molecules at interfaces, it is highly desirable to obtain precise electronic $\chi^{(2)}$ spectra. However, it is difficult to measure high-quality electronic spectra of interfacial molecules by the conventional scanning method. Last year, we reported a new multiplex SFG technique that utilized the white light continuum to measure interface-specific electronic $|\chi^{(2)}|^2$ spectra with an unprecedentedly high signal-to-noise ratio (*S/N*) [4]. Herein we describe the method and its application to the studies of static and dynamic properties of dye molecules adsorbed at the air/water interface.

2. Experimental

Figure 1 shows the experimental setup of the electronic $\chi^{(2)}$ multiplex SFG spectroscopy. The ω_1 and ω_2 pulses were noncollinearly focused onto the same spot at the air/water interface of a sample solution at an incident angle about 45° from normal. When the ω_1 and ω_2 pulses were temporally overlapped, the sum frequency (SF) $\omega_1 + \omega_2$ was generated at the interface. The SF light was collected and focused onto the entrance slit of a single polychromator. The spectrally dispersed SF light was detected by a CCD. We normalized the SF spectrum of the sample by that of a quartz plate to obtain the electronic $|\chi^{(2)}|^2$ spectrum. By introducing the pump (ω_p) pulse to photoexcite the sample, femtosecond time-resolved $|\chi^{(2)}|^2$ spectra were also measured. Pump-induced changes of the $|\chi^{(2)}|^2$ spectrum (hereafter called $\Delta|\chi^{(2)}|^2$ spectrum) were recorded at each time delay to investigate the dynamics at the interface.

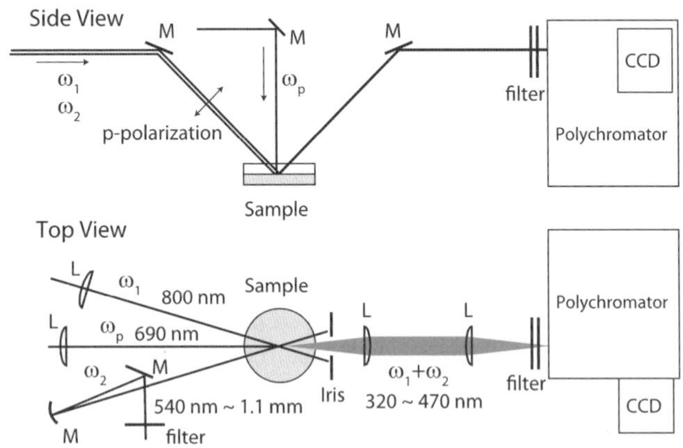

Figure 1. Experimental setup for electronic $\chi^{(2)}$ spectroscopy based on multiplex SFG. Labels L and M denote lens and mirrors, respectively. Some optical elements shown in the side (or top) view are not shown in the top (or side) view for clarity.

3. Results and Discussion

3.1. Steady-State Electronic $|\chi^{(2)}|^2$ Spectra

Figure 2a shows the electronic $|\chi^{(2)}|^2$ spectrum of a dye oxazine 750 (OX750) at the air/water interface measured with the ω_1 wavelength of 800 nm. The bottom and top horizontal axes stand for wavelength corresponding to $\omega_1 + \omega_2$ and ω_2, respectively. The bulk UV-visible (vis) absorption spectrum of OX750 in water is shown for comparison below and above the $|\chi^{(2)}|^2$ spectrum along the bottom and top horizontal axes, respectively. The comparison leads us to readily recognize that the prominent $|\chi^{(2)}|^2$ bands at 341 nm and 366 nm are resonantly enhanced by both the one-photon transitions (600 – 700 nm) and the broad two-photon transitions (320 – 400 nm). The $|\chi^{(2)}|^2$ signals around 400 – 465 nm are not resonant with the one-photon transitions but are only two-photon-resonantly enhanced. To more quantitatively investigate the resonance enhancement spectral feature, we measured the $|\chi^{(2)}|^2$ spectrum with different ω_1. Figure 2b shows the electronic $|\chi^{(2)}|^2$ spectrum of OX750 measured with the ω_1 wavelength of 740 nm. As the ω_1 wavelength becomes shorter, the $|\chi^{(2)}|^2$ band at 366 nm that is clearly seen in the spectrum taken with the ω_1 wavelength of 800 nm becomes smaller. The ω_1 dependence of the $|\chi^{(2)}|^2$ spectrum is systematically explained by a global fitting analysis based on the general expression of $\chi^{(2)}$ [4]. The $|\chi^{(2)}|^2$ spectra are successfully reproduced by the fitting curves shown as dotted lines in Figure 2. From this analysis, we can determine one- and two-photon resonant excited states of OX750 at the interface.

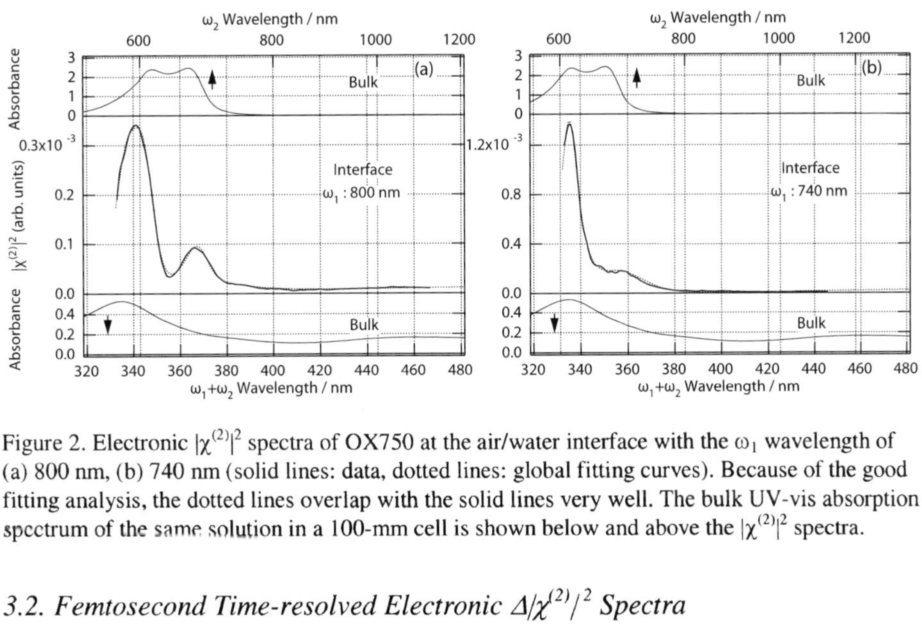

Figure 2. Electronic $|\chi^{(2)}|^2$ spectra of OX750 at the air/water interface with the ω_1 wavelength of (a) 800 nm, (b) 740 nm (solid lines: data, dotted lines: global fitting curves). Because of the good fitting analysis, the dotted lines overlap with the solid lines very well. The bulk UV-vis absorption spectrum of the same solution in a 100-mm cell is shown below and above the $|\chi^{(2)}|^2$ spectra.

3.2. Femtosecond Time-resolved Electronic $\Delta|\chi^{(2)}|^2$ Spectra

Figure 3 shows the steady-state $|\chi^{(2)}|^2$ and femtosecond time-resolved $\Delta|\chi^{(2)}|^2$ spectra of a dye rhodamine 800 (R800) at the air/water interface. It is noted that the $\Delta|\chi^{(2)}|^2$ spectra, which have comparable *S/N* to conventional transient absorption spectra, are realized for the first time. Around 370 – 390 nm, large negative signals appear just after the photoexcitation. The negative signals are ascribed to the pump-induced decrease of the ground-state population of R800. Actually, the band shape of the negative signals at 0 – 0.2 ps is in good agreement with the steady-state $|\chi^{(2)}|^2$ spectrum of R800 apart from the sign. As the negative signals around 370 nm partly recover with the time constant of 0.3 ps, positive signals around 360 – 365 nm rise up with the same time constant. The positive signals are probably assigned to the lowest excited singlet (S_1) state of R800. Full interpretation of these data will be reported elsewhere. With the present time-resolved electronic SFG method, we can investigate the femtosecond interfacial dynamics as thoroughly as we do with the transient absorption spectroscopy for the bulk dynamics.

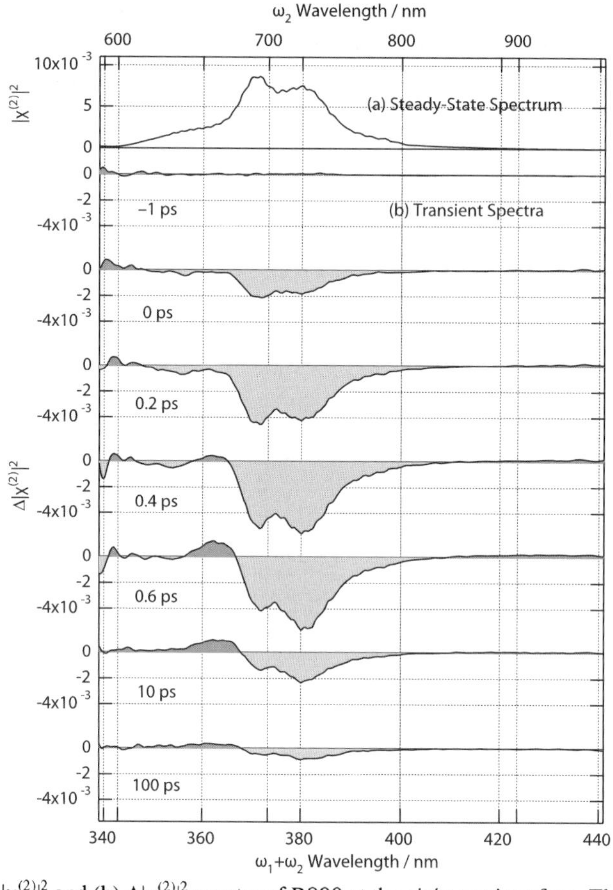

Figure 3. (a) $|\chi^{(2)}|^2$ and (b) $\Delta|\chi^{(2)}|^2$ spectra of R800 at the air/water interface. The ω_1 wavelength was 800 nm. The pump wavelength for the $\Delta|\chi^{(2)}|^2$ spectra was 690 nm.

References

1. C. T. Williams and D. A. Beattie, Surf. Sci. 500 (2002) 545.
2. X. D. Zhu, H. Suhr, and Y. R. Shen, Phys. Rev. B 35 (1987) 3047.
3. T. F. Heinz, C. K. Chen, D. Ricard, and Y. R. Shen, Phys. Rev. Lett. 48 (1982) 478.
4. S. Yamaguchi and T. Tahara, J. Phys. Chem. B 108 (2004) 19079.

Femtochemistry VII
Fundamental Ultrafast Processes in Chemistry, Physics, and Biology
A.W. Castleman, Jr. and M.L. Kimble

Photodetachment of aqueous iodide: variation of the excitation energy and solution temperature

H. Iglev,[a] A. Trifonov,[b] I. Buchvarov,[c] T. Fiebig[b] and A. Laubereau[a]

[a]*Physik-Department E11, TU München, D-85748 Garching, Germany*
[b]*E. F. Merkert Chemistry Center, Boston College, Chestnut Hill, 02467 Massachusetts*
[c]*Faculty of Physics, University of Sofia, 1164 Sofia, Bulgaria*

1. Introduction

The important role of aqueous electrolytes in biological and chemical systems makes them a subject of significant scientific interest. In the last decade, particular attention has been paid to the electron photodetachment of halide anions in solutions by excitation in their charge-transfer-to-solvent (CTTS) sta-tes [1, 2]. Molecular-dynamics (MD) simulations predict two formation channels of the solvated electron after population of a high-lying halide CTTS-state [3]. The first channel with a small quantum yield directly leads to the fully solvated electron via a fast nonadiabatic transition. The dominant second channel is characterized by a fast relaxation within the CTTS states to the lowest charge transfer (LCT) state and subsequent adiabatic separation of the electron from the parent atom. An intermediate was predicted consisting of a halogen atom, an electron and a water molecule in between [3, 4]. The lifetime of this complex is governed by diffusion-controlled dissociation in the presence of an attraction potential between the geminate particles [4].

Femtosecond studies of the process [5, 6] support the microscopic picture developed by the MD simulations. Experimental information on two important aspects however is still missing: the dissociation of the atom-electron pair and the predicted branching of the initial relaxation pathway after excitation into higher-lying CTTS states. Here we report on the temperature dependent relaxation dynamics after excitation into the first (5.13 eV) and second (6.4 eV) CTTS band of aqueous iodide. Probing is carried out in the visible and near-infrared.

2. Experimental

Our fs spectrometer with different UV pump pulses is based on a commercial Ti:sapphire amplifier system yielding 150-fs pulses of 800 μJ and 1 kHz repeti–tion rate at 775 nm. To generate a 242-nm pump pulse, the output of a first two-stage non-collinear optical parametric amplifier (NOPA) is tuned to 484 nm and frequency-doubled, providing pulses with energy of 1 μJ and duration of 100 fs. Excitation radiation at 194 nm (0.7 μJ and 500 fs) is achieved by fourth harmonic generation of a portion of the laser pulse. For probing purposes, a second NOPA generates 80-fs pulses at 1140 nm that produce a white light continuum by focusing into a rotating CaF_2 plate. The applicable probing range extends from 400 to 1100 nm. The pulse duration of the different spectral components is 100±10 fs. The relative orientations of the polarization planes of pump and probe beams are set to the magic angle (54.7°) for rotation free observations. The probe and reference spectra are detected independently and simultaneously behind the sample, using a monochromator and a CCD array.

The sample cell, equipped with 1-mm windows of UV grade fused silica, has a thickness of 250 μm and is filled with circulating solutions of 2.5 M NaI in de-ionized water. The experiments are performed at 25, 50, and 75°C.

3. Results and Discussion

Examples for the measured transient absorption spectra at 25°C after excitation at 242 nm are shown in Figs. 1a and b in ΔmOD. A small absorption centered at 1000 nm is already seen at delay time $t_D = 0$. This spectral feature grows within 1 ps, while the absorption narrows notably and shifts to shorter wavelengths. The changes in the band shape terminate at $t_D = 1.5$ ps. The finding shows that the initial relaxation dynamics are accompanied by a continuous reorganization of the electron solvation cavity. The agreement with the absorption of the fully solvated electron shows that at least the first solvation shell of the separated electron is already fully established at this time. The measurements for larger delay times up to 1 ns (see Fig. 1b) indicate that the changes in the absorption band shape have terminated at $t_D = 2$ ps. The subsequent relaxation is assigned to population decay, i.e. geminate recombination of the electron with the iodine.

Corresponding measurements at temperatures of 50 and 75°C are carried out, obtaining a similar time evolution of the spectra. Generally, the transient absorption spectra measured at higher temperatures are found to be slightly broadened and red-shifted as compared to the 25°C measurements. A strong temperature effect on the measured relaxation dynamics is observed. This finding suggests that the long-time dynamics are dominated by diffusion-controlled geminate recombination.

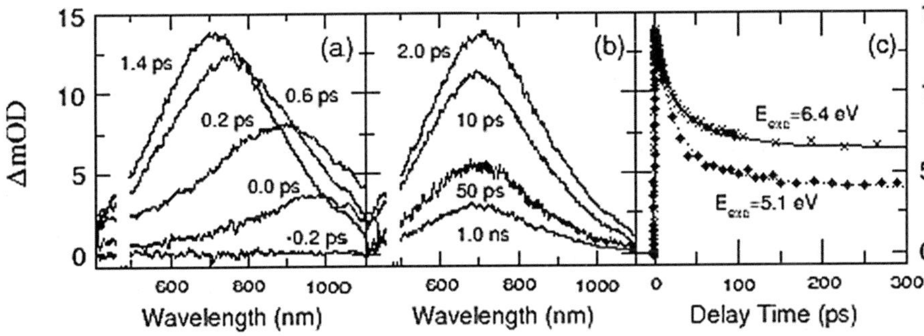

Figure 1. Time-resolved data measured in 2.5 mM aqueous NaI solution at 25°C. (a) and (b) Transient absorption spectra measured after excitation at 5.1 eV; different delay times indicated in the inset. (c) Signal transients taken at 725 nm for excitation at 5.1 eV (dotted line, filled points) and 6.4 eV (solid line, crosses).

The physical picture developed from the experimental data presented above is as follows. The initially excited LCT state shows a broad absorption with a maximum around 1000 nm at all three bulk temperatures. The released electron escapes the anion solvation shell with a time constant $t_1 = 220$ to 180 fs going from 25°C to 75°C. The separation of the excess electron from the iodine atom and the assembling of a new solvation configuration require $t_2 = 700$ fs to 540 fs in the mentioned temperature interval. The process is connected with a strong blue shift of the induced absorption (see Fig. 1a) and is accounted for by a separate intermediate assigned to the iodine-electron complex in a transient solvation structure. The extracted average spectrum of this precursor is centered at 825 nm and is only weakly affected by the temperature variation in the investigated range. Subsequently, an equilibrated hydrated iodine-electron pair is formed. This complex is stabilized by an attractive interaction potential with depth of $850\pm100\ cm^{-1}$ as derived from observed temperature dependence. The full half-width of the potential well decreases from 260 pm at 25°C to 200 pm at 75°C. The absorption band of the hydrated pair cannot be distinguished from that of the solvated electron. The long-time dynamics indicated by an absorption decrease are found to be strongly temperature-dependent and explained by a competition of diffusion-controlled pair dissociation and geminate recombina tion (with subsequent conversion to the ground-state anion). Our data show that the ratio of fully solvated electrons 1 ns after the optical excitation to atom-electron pairs at $t_D = 1.5$ ps increases from $\eta = 0.25$ to 0.53 going from 25°C to 75°C, respectively. In other words the formation of hydrated electrons is more effective at higher temperature.

In order to study the predicted branching of the initial relaxation pathways, we studied the electron photodetachment dynamics after excitation in the second CTTS state of aqueous iodide at 6.4 eV using pump pulses at 194 nm. The signal transient measured at 725 nm for 25°C is presented in Fig. 1c (crosses, solid line). The corresponding result for 5.1 eV (242 nm pump) is also shown (points, dotted line). The maximal absorbance changes measured for pumping at 6.4 eV are normalized to the value for 5.1 eV for a direct comparison. The probe wavelength of 725 nm is close to the absorption band maxima. Thus the dynamics measured at t_D>2 ps are mainly due to population decay of the excess electrons, while spectral shifts are of minor importance. Fig. 1c indicates a strong increase of the electron survival probability for 6.4 eV compared to the 5.1 eV experiment. The data analysis reveals that initial excitation of higher-lying CTTS states enhances the probability for electron separation from the iodine, while the relaxation channel discussed above is not significantly affected, i.e. evidence for the direct nonadiabatic channel is obtained. The quantum yield of the direct channel was measured to be 0.3±0.05 after excitation into the second CTTS state. The phenomenon could explain the strong variation in the relaxation dynamics reported in femtosecond photodetachment experiments using different excitation energies.

4. Summary

We report on the photodetachment of electrons in a 2.5 mM aqueous iodide solution at temperature values of 25, 50, and 75°C. After one-photon excitation by a 100 fs-pulse at 242 nm, the dynamics are studied with 100 fs probe radia–tion in the spectral range of 400 – 1100 nm. The initially excited lowest charge transfer state shows a broad excited state absorption in the NIR with a maximum around 1000 nm. The excess electron escapes the anion solvation shell within 200 fs. The further separation of the electron from the iodine atom and the assembling of a new solvation configuration around the atom-electron pair requires 700 to 540 fs. This process is accounted for by equilibration of the intermediate. The subsequent slower dynamics indicated by an absorption decrease are found to be strongly temperature-dependent and explained by a competition of pair dissociation and geminate recombination. The relaxation dynamics are described by diffusion of the particles in the presence of an attractive interaction potential with depth of 850±100 cm^{-1}.

We observe that an initial excitation of higher-lying CTTS states opens a new channel for direct nonadiabatic electron separation from the iodine. The quantum yield of the additional direct channel was measured to be η=0.3±0.05 after excitation into the second CTTS state. The observed energy dependence of the relaxation dynamics is the first experimental verification of the theoretically predicted branching of the relaxation pathway after excitation of high-lying CTTS states of aqueous halides.

References

1. M. Assel, R. Laenen, and A. Laubereau, Chem. Phys. Lett. 289 (1998) 267.
2. H. Iglev, R. Laenen, and A. Laubereau, Chem. Phys. Lett. 389 (2004) 427.

3. W.-S. Sheu and P. J. Rossky, J. Phys. Chem. 100 (1996) 1295.
4. A. Staib and D. Borgis, J. Chem. Phys. 104 (1996) 9027.
5. J. A. Klopfer *et al.*, Chem. Phys. Lett. 298 (1998) 120; J. Chem. Phys. 117 (2002) 766.
6. H. Iglev *et al.*, Chem. Phys. Lett. 403 (2005) 198 ; Proc. SPIE 5826 (2005) 212

Femtochemistry VII
Fundamental Ultrafast Processes in Chemistry, Physics, and Biology
A.W. Castleman, Jr. and M.L. Kimble

A combined femtosecond fluorescence and TD-DFT study of uracil derivatives in aqueous solution

Thomas Gustavsson[a], Elodie Lazzarotto[a], Dimitra Markovitsi[a], Ákos Bányász[a], Roberto Improta[b], G. Scalmani[c]

[a]*Laboratoire Francis Perrin, CEA/DSM/DRECAM/SPAM - CNRS URA 2453, CEA Saclay, F-91191 Gif-sur-Yvette, France*
[b]*Dipartimento di Chimica, Universita Federico II, Complesso Universitario Monte S. Angelo, Via Cintia, I-80126 Napoli, and Istituto Biostrutture e Bioimmagini /CNR, V. Mezzocannone 6 –80134 Napoli, Italy*
[c]*Gaussian, Inc., 340 Quinnipiac St. Bldg 40, Wallingford, CT 06492, USA*

1. Introduction

DNA nucleobases (as well as the corresponding nucleosides and nucleotides) are characterized by extraordinarily short excited states lifetimes [1], suggesting the existence of very fast non-radiative decay channels from the excited to the ground state. The fluorescence upconversion technique has been applied with success in the study of the excited states of these molecules [2-6], showing that their decays in aqueous solution are faster than 1 ps. In parallel, new theoretical methods have been developed that allow a correct description of the excited state, including solvent effects [7].

Uracil and the closely related thymine (5-methyluracil) are the simplest nucleobases present in RNA and in DNA, respectively. It is therefore of fundamental interest to understand the processes involved in their excited state deactivation. In this communication we present a combined experimental and theoretical characterization of the static and dynamical behavior of the excited states of uracil and ten of its derivatives in aqueous solution by means of time-resolved fluorescence and TD-DFT quantum chemical calculations [8].

2. Femtosecond fluorescence decays

Fluorescence decays were recorded for ~2.5×10^{-3} mol/dm^{-3} aqueous solutions of the eleven uracils, uracil (U), 1-methyluracil (1MU), 3-methyluracil (3MU), thymine (5-methyluracil) (T), 6-methyluracil (6MU), 1,3-dimethyluracil (1,3DMU), 1-methylthymine (1MT), 5-trifluoromethyluracil (trifluorothymine) (TFT), 5-fluorouracil (5FU), 5-fluoro-1,3-dimethyluracil (5F1,3DMU), and 5-chlorouracil (5ClU). The excitation wavelength was 267 nm (the third harmonic of a Ti:S laser), and the detection of fluorescence was at 330 nm. Total fluorescence intensities were constructed from separate recordings at parallel and perpendicular excitation. As an example, the resulting fluorescence decay curves for uracil, thymine, and 5-fluorouracil are shown in Figure 1. Also shown in this figure is the apparatus function (330 fs fwhm at 330 nm).

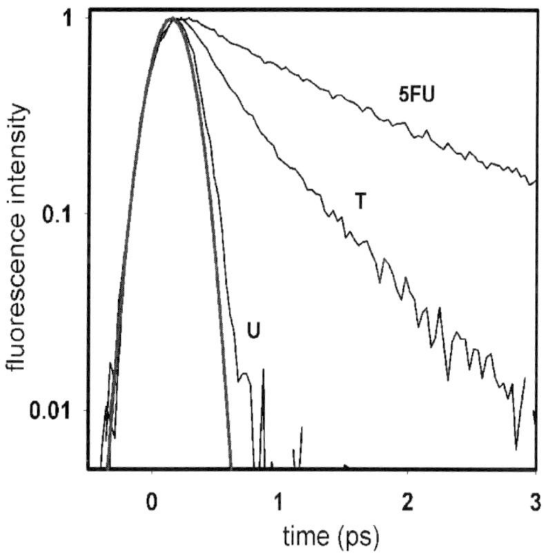

Figure 1. Fluorescence decays (on a semi-log scale) at 330 nm after excitation at 267 nm of uracil (U), thymine (T) and 5-fluorouracil (5FU) in room temperature aqueous solutions (~2.5×10^{-3} M). Also shown is the 330 fs (fwhm) Gaussian apparatus function.

A thorough comparative analysis of the fluorescence decays of the eleven derivatives reveals that uracil, 1MU, 3MU, 6MU, and 1,3DMU all possess very rapid decays ($<\tau>$ ~100 fs). The only position that significantly affects the excited state lifetime is C5. A bi-exponential fit of the fluorescence lifetimes of the 5-substituted uracils gives one ultrafast component varying between 140 and 700 fs and one much slower component varying between 330 fs and 1.7 ps. In particular, 5-fluorination makes the excited state lifetime substantially longer, both for uracil and 1,3DMU, but the fluorescence lifetime of 5FU is longer than that of 5F1,3DMU.

3. TD-DFT calculations

Excited state energies were calculated by the TD-DFT method using PBE0 hybrid functionals [7]. The solvent contribution was accounted for by a hybrid model. In the first solvent shell, 4 water molecules were included on the quantum level; the bulk contribution was calculated using the Polarizable Continuum Model (PCM). It is found that the geometry of the first solvation shell strongly depends on the electronic state. Analytical TD/PBE0 geometry optimizations in aqueous solution allowed accurate computations of the fluorescence spectra. Interestingly, at this level of calculations, the first electronically excited state is the $\pi\pi^*$ state for all compounds; the $n\pi^*$ state is situated at a higher energy. Comparison with steady-state fluorescence spectra show that an acceptable quantitative agreement with experiments and the right ordering of the eleven compounds is only obtained using the hybrid solvent model.

4. S_1/S_0 conical intersection

Two-dimensional Potential Energy Surfaces (PES) were computed at the PCM/TD-PBE0/6-31G(d) level for the three representative compounds uracil, thymine, and 5-fluorouracil in aqueous solution. The two coordinates are (i) the out of plane motion of the C5 substituent (ϕ, improper dihedral) and (ii) a "collective" coordinate x, defined as a linear interpolation between the remaining internal coordinates of the minimum on S_1 and the S_1/S_0 conical intersection (CI). The CI was located at the CASSCF 8/8 level. As an example, the resulting excited state PES for uracil is shown in Figure 2.

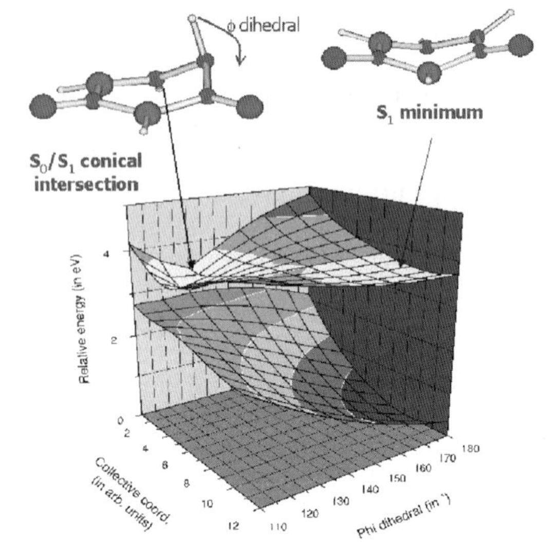

Figure 2. S_0 and S_1 potential energy surfaces (in eV) in the region connecting the minimum of the π/π^* state and the S_0/S_1 conical intersection for uracil.

The key motion to reach the S_1/S_0 conical intersection is the pyramidalization at C5, while an out of plane motion (improper dihedral angle ϕ) brings the C5 substituent towards a 'pseudo perpendicular' arrangement with respect to the molecular plane. The energy barrier height between the S_1 minimum and the CI are increasing in the order uracil (0.1 eV) < thymine (0.2 eV) < 5-fluorouracil (0.5-1 eV). This is in line with the experimentally determined excited state lifetimes.

Indeed, the calculated PES shows that the reaction path leading from the local minimum (Min-S_1) of S_1 to the CI is consistent with a very fast excited state decay. One must note the Min-S_1 is already more stable than the initially excited Franck-Condon state by ~ 1 eV (~ 8×10^3 cm^{-1}).

Once Min-S_1 is reached, the kinetic energy of the wave packet should be enough to easily overcome the small energy barrier towards the CI, since it also has the right inertia towards the out of plane motion.

5. Conclusions

This joint experimental/theoretical study provides several interesting indications on the excited state behavior of uracil derivatives in aqueous solution. Experiments show that the excited state lifetimes of all the compounds examined, except 5-fluoro uracil, are dominated by a component shorter than 250 fs. This is in line with earlier studies of other DNA bases. Calculations indicate that the relevant state for discussing the absorption and fluorescence spectra of uracil derivatives in aqueous solution is the $\pi\pi^*$. The key steps in the ultrafast internal conversion are the pyramidalization of C5 and the out of plane motion of C5 substituent, which explains why the excited state lifetime depends mainly on the nature of the C5 substituent. Though a dynamical study is necessary to unambiguously assess this point, the shape of the potential energy surface of the S_1 electronic states seems consistent with a very fast non-radiative decay. The S_1 minimum is indeed more stable than the Franck-Condon state by ~1eV, and thus the wave packet may easily pass the barrier towards the CI. The relative barrier height on the path towards the CI provided by TD-DFT calculations are consistent with the different excited state lifetimes of uracil, thymine, and 5-fluorouracil observed experimentally. This result strongly supports the reliability of the general picture of the ultrafast ground state recovery of excited uracils in aqueous solution. Further work on uracil-like and other nucleobases in different solvents is under way.

References

1. C. E. Crespo-Hernandez, B. Cohen, P. M. Hare, and B. Kohler, Chem. Rev. 104 (2004) 1977.
2. J. Peon and A. H. Zewail, Chem. Phys. Lett. 348 (2001) 255.
3. T. Gustavsson, A. Sharonov, and D. Markovitsi, Chem. Phys. Lett. 351 (2002) 195.
4. T. Gustavsson, A. Sharonov, D. Onidas, and D. Markovitsi, Chem. Phys. Lett. 356 (2002) 49.
5. D. Onidas, D. Markovitsi, S. Marguet, A. Sharonov, and T. Gustavsson, J. Phys. Chem. B 106 (2002) 11367.
6. A. Sharonov, T. Gustavsson, V. Carré, E. Renault, and D. Markovitsi, Chem. Phys. Lett. 380 (2003) 173.
7. R. Improta and V. Barone, J. Am. Chem. Soc. 126 (2004) 14320.
8. T. Gustavsson, A. Banyasz, E. Lazzarotto, D. Markovitsi, G. Scalmani, M. J. Frisch, V. Barone, and R. Improta, J. Am. Chem. Soc. 128 (2006) 607.

5. Aggregates/Particles/Surfaces

Femtochemistry VII
Fundamental Ultrafast Processes in Chemistry, Physics, and Biology
A.W. Castleman, Jr. and M.L. Kimble

Ultrafast electronic and lattice processes of plasmonic nanoparticles of different shape

Wei Qian, Wenyu Huang, Qusai Darugar, Mostafa A. El-Sayed

Laser Dynamics Laboratory, School of Chemistry and Biochemistry, Georgia Institute of Technology, 770 State Street, Atlanta, GA 30332-0400, USA

Abstract

Ultrafast spectroscopic studies of the electron phonon relaxation processes in gold and copper nanoparitcles of different sizes and shapes have been carried out and compared. The coherent lattice vibrations of prismatic gold nanoarrays in the sub-cm^{-1} range are determined optically from their effect on the optical frequency of the electron surface plasmon absorption. The mechanism of photothermal transformation of the prismatic nanoparticles to spheres is studied and discussed.

1. Introduction

Over the last decade, studies of plasmonic nanoparticles (e.g. gold and silver) have increased significantly as a consequence of their potential use in a variety of applications, including biosensing, imaging, and long-range energy transfer waveguides [1-4]. In this paper, we mainly concentrate on the studies of the ultrafast electronic and lattice properties of these nanoparticles.

Section 2 describes the preparation of the periodic prismatic and spherical gold nanoparticle arrays by using the nanosphere lithography technique [5]. The femtosecond transient absorption experimental setup we used is also described in this section. In section 3, we discuss the effect of gold nanoparticle shape and copper nanoparticle size on electron-phonon relaxation. Section 4 focuses on the ultrafast laser-induced coherent lattice oscillation in prismatic-shaped gold and silver nanoparticles assembled in monolayer periodic arrays and the effect of interparticle coupling on the period of coherent lattice oscillation [6]. Experimental results on femtosecond laser photothermal shape transformation of prismatic-shaped gold nanoparticles are discussed in section 5, and a summary of conclusions of the work presented in this paper is given in section 6.

2. Experimental Section

2.1. Preparation of plasmonic nanoparticles with different shapes

Periodic arrays of gold and silver nanoparticles of different shapes are prepared by the nanosphere lithography (NSL) method [5]. The commercially available monodisperse polystyrene (PS) sphere suspensions (Duke Scientific Corp.) were used to form a self-assembled monolayer on quartz slides (Technical Glass Products, Inc.). The quartz slides are pre-cleaned in piranha solution (3:1 H_2SO_4:30% H_2O_2) at 80 °C for 1.5 h and then sonicated in a solution of 5:1:1 H_2O:NH_4OH:30%H_2O_2 for one hour. To form a close packed single layer polystyrene sphere mask, five microliters of the PS sphere suspension was dropped on the quartz slides, which were then tilted to distribute the suspension evenly over the substrate. The substrate was slowly immersed in distilled water to remove excess PS spheres from its surface. One drop of 2% dodecylsodiumsulfate surfactant was then added to the water to separate the excess floating PS spheres from the substrate, and the substrate was taken out slowly from a clean area of the water in order to prevent additional PS spheres deposition. After the evaporation of the water in ambient conditions, the PS spheres form a monolayer that can cover several square millimeters on the substrate. The mask was then mounted in a thermal evaporator (Denton DV-502A) to deposit 50-nm-thick gold or silver (99.999%, Alfa Aesar) into the voids between the spheres. After deposition, the PS spheres were removed by sonication in absolute ethanol. The gold or silver deposited into the voids between the PS spheres is shown to form a periodic array of prismatic nanoparticles.

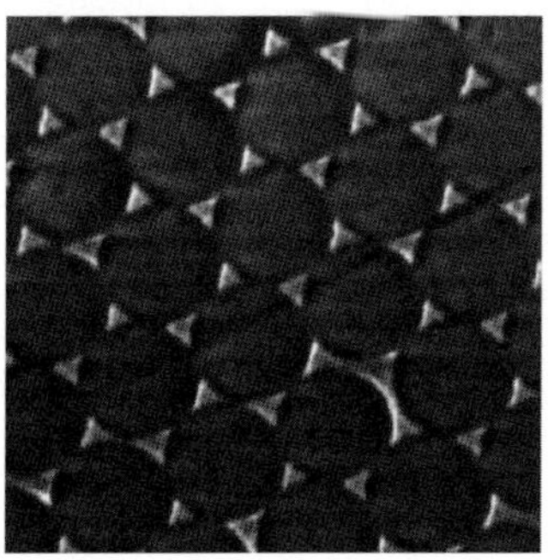

(A)

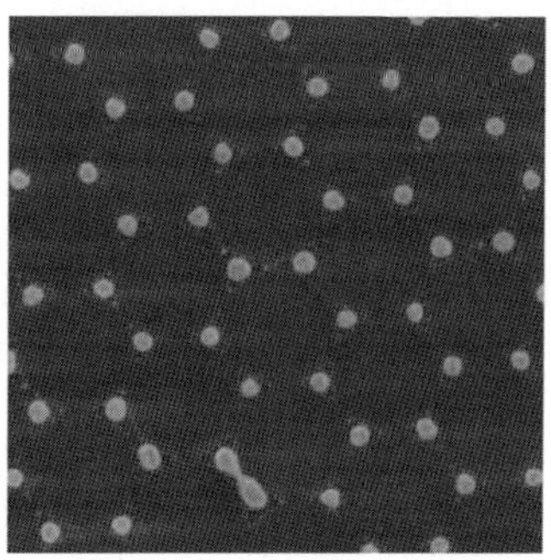

(B)

Figure 1. SEM images of the prismatic nanoparticle before (A) and after (B) thermal annealing at 1000 °C for 1.5 min.

Figure 1A shows a typical SEM image of the arrays formed. The periodic arrays of spherical gold nanoparticles are formed by thermal annealing of the original prismatic gold nanoparticle array sample in an oven at 1000 °C for 1.5 minutes (Figure 1B). The SEM image shows that, after thermal annealing, the prismatic gold nanoparticles are reshaped to spheres but keep the same periodic hexagonal geometry.

The copper nanoparticles in ethanol solution used in this paper are prepared by photochemically reducing a commercially available precursor bis(2,4-pentanedionato)copper(II), $[Cu(acac)_2]$ with UV light as proposed by Giuffrida *et al.* [7]. In a typical copper nanoparticle synthesis, $5\times10^{-4}\,M$ $Cu(acac)_2$ in deoxygenated ethanol is irradiated at 254 nm at an intensity of 5 mW/cm^2 under anaerobic conditions. Growth of copper nanoparticles is monitored by the appearance of an absorption peak around 570 nm, which is characteristic of the surface plasmon band of copper nanoparticles. In order to obtain stable copper nanoparticles, poly(N-vinylpyrrolidone) was added to the solution before irradiating with UV light. Two copper nanoparticle samples, Cu1 and Cu2, with different sizes were prepared for the experiments.

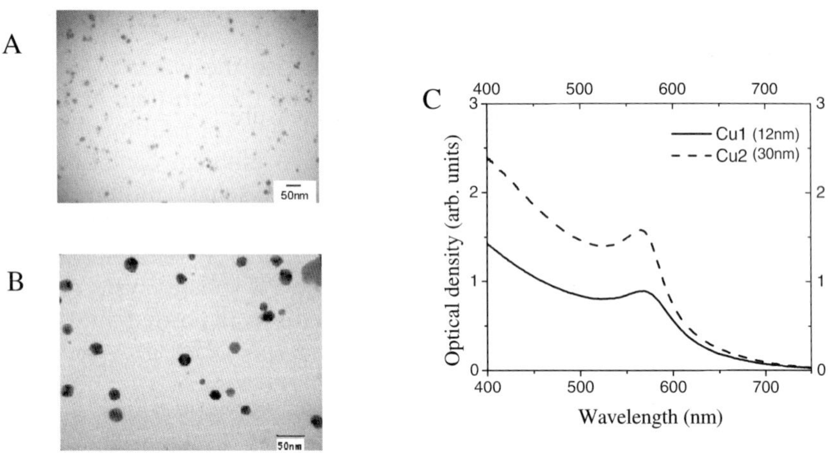

Figure 2. Transmission electron microscope images of (A) Copper 12nm and (B) Copper 30nm particles. The average diameters of these nanoparticles were determined by measuring sizes of roughly 200 particles. In (B), we also observed nonspherically shaped copper nanoparticles. (C) Absorption spectra of copper nanoparticles in ethanol solution. The band around 570 nm is characteristic of the surface plasmon absorption of copper nanoparticles.

The morphology and size of the linear absorption spectra of copper nanoparticles examined by using a JEOL 100CX II TEM and a Shimadzu UV-

3101-PC spectrophotometer are shown in Figure 2A-C. The average diameters of Cu1 and Cu2, determined by averaging the sizes of roughly 200 particles, were 12 nm and 30 nm, respectively. The surface plasmon absorption band is clearly seen in both samples with a slight peak shift from 568.5 nm for Cu1 to 565.0 nm for Cu2.

2.2. The femtosecond pump probe transient absorption system

Time-resolved transient absorption experiments performed with a femtosecond laser system have been described elsewhere [8]. Briefly, an amplified Ti-Sapphire laser system (Clark MXR CPA 1000) is pumped by a frequency-doubled Nd:Vanadate laser (Coherent Verdi). This generates laser pulses of 100 fs duration (fwhm) with energy of 1 mJ at 800 nm and a repetition rate of 1 kHz. The femtosecond laser pulses of 800 nm are frequency doubled in a 1 mm BBO crystal to generate 400 nm pump pulses. A small portion of the fundamental pulse (about 40 μj) was focused on a 1 mm sapphire plate to generate a femtosecond white light continuum, which was used as the probe. The pump and probe beams were focused and overlapped at the sample with a diameter of about 250 μm. The induced changes of the sample absorption were probed at the peak of the surface plasmon absorption by measuring the differential transmission ($\Delta T / T$) as a function of the temporal delay between the pump and probe. The high stability of our femtosecond laser system and phase lock technique permit high sensitivity measurements with a noise level for ($\Delta T / T$) in the 10^{-4} range. The cross-correlation between the pump beam and probe beam had a FWHM of about 180 fs at the sample position. All the experiments reported below were carried out at room temperature with pump laser power used within 0.01-0.3uJ/pulse.

3. Electron-phonon Relaxation

One of the most fundamental aspects of the studies on plasmonic nanoparticles is the investigation of the properties of the electron-phonon interactions in these nanoparticles [9-11]. This is because electron-phonon interactions are essential in addressing many fundamental questions, such as thermal and electrical conductivity. With the advent of ultrafast laser spectroscopy, time-resolved transmission and reflection experiments have been widely used for studying electron-phonon relaxation dynamics in plasmonic nanoparticles [12-17]. Extensive work has been carried out on studying the effects of changing the nanoparticle size and shape on their electron-phonon relaxation dynamics. It was thought that when the plasmonic nanoparticle's radius is smaller than the mean free path of the conduction electrons (40-50 nm in gold and silver), the measured electron-phonon relaxation rate would increase with decreasing nanoparticle size because of the additional contribution of the electron-surface phonon scattering to the electron energy relaxation processes.

However, Link *et al.* [15] in our group observed nearly the same electron-phonon relaxation time for 9 nm and 48 nm diameter gold nanopartices in aqueous solutions, being 1.6 ps and 1.7 ps, respectively. Similarly, for smaller gold nanoparticles, Ahmadi *et al.* [13,14] did not observe size dependence of the electron-phonon relaxation dynamics for 1.9, 2.6, and 3.2 nm diameter thiol-passivated gold nanocrystals compared to the measured relaxation dynamics for 30 nm colloidal gold nanoparticles prepared by the citrate method. In addition, Hartland's group [12,16,17] also found no change in the electron dynamics between 10 to 50 nm diameter silver nanoparticles and between 4.6 and 120 nm diameter gold nanoparticles in aqueous solutions. The electron energy relaxation rates and the electron-phonon coupling constants measured in both gold and silver nanoparticles were found to be similar to their corresponding bulk values [16,18]. The reason for the observed size independence of the electron phonon relaxation is due to the relatively small contributions of surface phonon scattering to the total electron-phonon coupling.

In the front part of this section (*3.1*), we discuss the effect of changing the copper nanoparticle size on the electron surface phonon relaxation. In the second part (*3.2*), we examine the effect of changing the shape of gold nanoparticles.

Quantum-kinetic calculation [18,19] indicates that the contribution of the surface phonon modes to the total electron-phonon coupling is determined by the ratio of the free electron density to the metal density (n/ρ), which is proportional to the valence of the metal divided by its atomic mass. For gold nanoparticles with a diameter as small as 2 nm, calculations [18] show that the contribution of the electron surface phonon scattering to the total electron-phonon scattering processes is less than 10%. However, copper nanoparticles may offer a better system to observe the size dependent electronic coupling to the surface phonons because the values of (n/ρ) for gold, silver, and copper are $3.06\times10^{21}\,g^{-1}$, $5.57\times10^{21}\,g^{-1}$, and $9.46\times10^{21}\,g^{-1}$, respectively.

3.1. Effect of the size of copper nanoparticle on the electron phonon relaxation

As we mentioned above, because of its larger ratio of free electron density to the metal density (n/ρ), copper nanoparticles offer a better system to observe the size dependent electronic energy relaxation due to the coupling to the surface phonons. Two copper nanoparticle samples, Cu1 and Cu2, with average diameters about 12 nm and 30 nm, respectively, were synthesized for this purpose. Pump power dependent transient bleach experiments have been shown to be a powerful technique for measuring the electron-phonon coupling constant, g [16,17]. A reliable value of g can be obtained by extrapolating the lifetime measured at various pump powers to zero, where the intercept is given by $\gamma T_0/g$ (T_0 is the ambient temperature and γ is electronic constant, which is $96.6 Jm^{-3}K^{-2}$ for bulk copper metal). The normalized traces of the transient bleach signals with the pump at 400 nm and the probe at 568 nm (the peak of surface plasmon absorption of copper nanoparticles) for Cu1 (12 nm) and Cu2 (30 nm) are shown in Figure 3A and 3B, respectively. The insets show the linear dependence of electronic relaxation lifetimes on pump pulse energies.

Extrapolating the lifetimes to zero pump pulse energy, the intercepts for Cu1 and Cu2 are 0.37 ± 0.04 ps and 0.51 ± 0.05 ps, respectively. The electron-phonon coupling constants of Cu1 and Cu2 are $(7.8 \pm 0.4) \times 10^{16} Wm^{-3}K^{-1}$ and $(5.7 \pm 0.3) \times 10^{16} Wm^{-3}K^{-1}$, respectively. Both of these electron phonon coupling constants are larger than the bulk value for copper, which is $4.8 \times 10^{16} Wm^{-3}K^{-1}$ [20,21], and become higher as the size of the nanoparticle is decreased. The observed size dependent electronic energy relaxation is due to the contribution from surface phonons as well as the energy transfer to surface defects of copper nanoparticles [22].

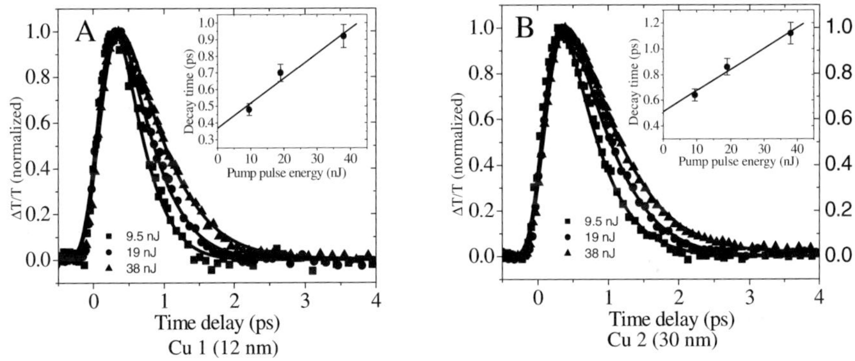

Figure 3. Transient bleach signal with pump at 400 nm and probe at 568 nm (the peak of surface plasmon absorption of copper nanoparticles) of (A) Cu1 (12 nm) and (B) Cu2 (30 nm). The linear fits of relaxation time versus the pump pulse energy are shown in the inserts.

3.2. Effect of changing the shape of gold nanoparticles on their electron phonon relaxation

The effect of gold nanoparticle shape on electron-phonon relaxation processes is studied by comparing the electronic relaxation dynamics acquired from femtosecond transient bleach experiments. The normalized traces of the transient bleach signals with the pump at 400 nm and the probe at 560 nm (corresponding to surface plasmon absorption of spherical gold nanoparticle arrays) and at 850 nm (corresponding to in-plane dipolar transition of prismatic gold nanoparticle arrays) are shown in Figure 4A and 4B, respectively. The electron-phonon coupling constant g for spherical and prismatic gold nanoparticle arrays obtained from the signal shown in the Figure 4A & 4B are

$(2.2 \pm 0.3) \times 10^{16} Wm^{-3}K^{-1}$ and $(2.6 \pm 0.5) \times 10^{16} Wm^{-3}K^{-1}$, respectively ($\gamma = 96.6 Jm^{-3}K^{-2}$ for bulk gold metal). Within the experimental error, the electron-phonon coupling constants are the same for prismatic and spherical gold nanoparticles and do not show any shape dependence. It is not surprising that the electron-phonon coupling constants are close to the value of bulk gold because the sizes of both the prismatic gold nanopaticles (bisector 104 nm) and the spherical gold nanoparticles (diameter 84 nm) are much larger than the mean free path of the conduction electrons in gold (40-50 nm).

Our observation of shape independence of the electron-phonon coupling constant for gold nanoparticles is consistent with previous studies carried out by Link *et al.* in our group [23]. These workers have studied the electron-phonon relaxation dynamics of colloidal gold nanorods encapsulated in micelles with an average aspect ratio of 3.8. They found that the transient bleach of gold nanorods at both transverse (520 nm) and longitudinal (750 nm) plasmon absorption recovered with the same time constant as that of gold nanospheres prepared by photothermal reshaping of the rods and gold nanospheres synthesized chemically by the citrate reduction method. These results suggest that the changes in both the gold nanoparticle shape (facets) and the mode of the surface plasmon oscillation do not have an effect on the electron-phonon relaxation dynamics [10,23].

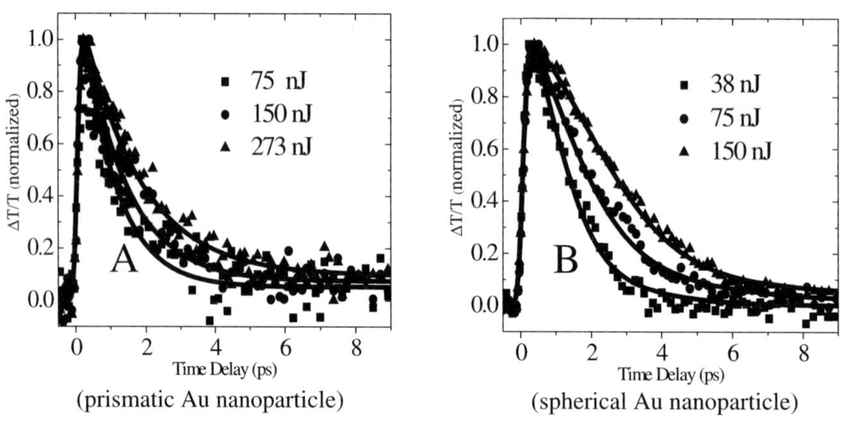

Figure 4. The normalized traces of the pump power dependent transient bleach signals with pump at 400 nm of (A) gold nanoprism (of 104 nm bisector) arrays probed at 560 nm and (B) spherical gold nanoparticle (of 84 nm diameter) arrays probed at 850 nm.

4. Coherent lattice oscillation

Using ultrafast laser pulses to excite the electronic surface plasmon oscillation, rapid electron phonon relaxation leads to a coherent lattice oscillation. This leads to a periodic changing of the nanoparticle's electron density of the nanoparticles. This causes the surface plasmon absorption spectrum to change periodically with the lattice vibration period, thus allowing one to determine the period optically. The great advantage of this technique is that it could study vibration modes with very low frequency, which could not be observed in Raman scattering experiments due to the overlapping intense Rayleigh scattering signal. The possibility of modulating the reflected and transmitted light with the coherent lattice oscillation gives plasmonic nanoparticles potential applications as a high frequency (megahertz) light modulator. Because of the significance in both science and application, the studies of coherent lattice oscillations in plasmonic nanopartilces have attracted much attention.

Ultrafast laser-induced coherent lattice oscillations have been observed for plasmonic nanoparticles of different shapes and in different environments [6,16,24-26]. The frequency of the lattice oscillation is inversely proportional to the dimension of the nanoparitcles. The prismatic-shaped gold and silver nanoparticles assembled in monolayer periodic arrays using the NSL method have well-defined particles size, shape, and interparticle separation. This provides good opportunity for us to study the effect of coupling among nanoparticles on the period of coherent lattice oscillation. Optically detected lattice phonon oscillations induced in 60.5 nm and 103.7 nm prismatic gold nanoparticle monolayer arrays with a 100 fs laser pulse at 400 nm and monitored near the absorption maximum of each nanoparticle are shown in Figure 5A and 5B, respectively. The oscillation is fitted with a damped cosine function and two exponential decays resulting from the electron-phonon and phonon-phonon relaxation processes. The oscillation periods are found to be 39 ± 2 and 73 ± 2 ps for 60.5 and 103.7 nm prismatic gold nanoparticles, respectively. The dependence of the coherent lattice oscillation periods on the nanoparticle size of prismatic silver and gold monolayer arrays are shown in Figure 6A and 6B.

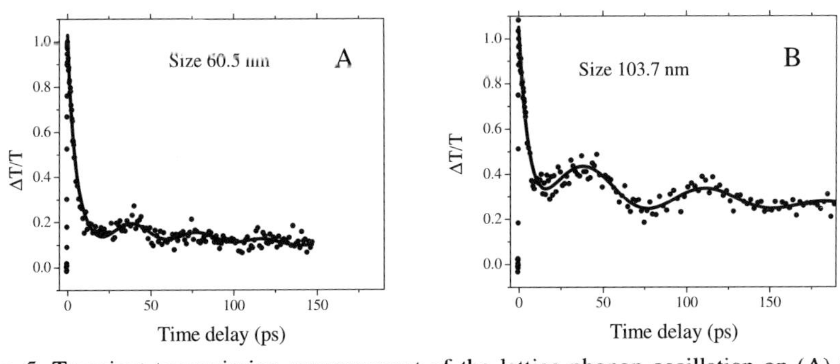

Figure 5. Transient transmission measurement of the lattice phonon oscillation on (A) 60.5 nm and (B) 103.7 nm prismatic gold nanoparticle arrays. The oscillation is fitted to a damped cosine function and two exponential decays to account for the electron-phonon and phonon-phonon relaxation processes. The oscillation periods are 39 ± 2 and 73 ± 2 ps for 60.5 and 103.7 nm prismatic gold nanoparticles, respectively.

In Figure 6, we show the observed dependence of the oscillation periods on size [6]. While the observed dependence is in good agreement with the theoretical single-particle calculation for silver prismatic nanoparticle arrays, there is an obvious deviation for the gold prismatic nanoparticle arrays which increases with the size of the particle. These observations have been explained [6] by using a dipole exciton model of coupling between the sharper prismatic gold nanoparticles in the array. For silver nanoparticle arrays, the interparticle coupling is negligible because of the more-rounded tips which are due to the lower melting temperature and the possibility of surface oxidation of the silver nanoparticles.

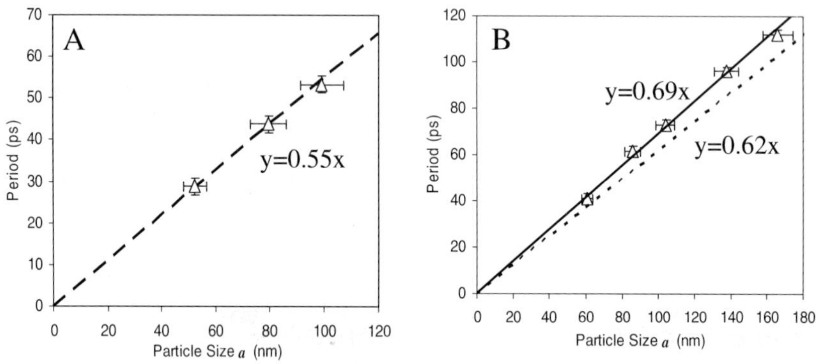

Figure 6. The dependence of the coherent phonon oscillation period on the particle size for prismatic silver (A) and gold (B) monolayer nanoparticle arrays. The dashed line in (A) and (B) is the calculated linear relation between the period and the particle size calculated from $\tau = 2a / V_l$ based on one-dimensional standing wave model, where V_l is the longitudinal sound velocity in bulk silver (or gold) and a is the particle bisector.

5. Photothermal shape transformation

The irradiation experiments with 400 nm femtosecond laser pulses were performed on a sample made with 0.45 μm polystyrene spheres by the NSL technique [5,27]. From SEM measurements, the bisector of the as-deposited array nanoparticles is 103.7 ± 6.8 nm, and their thickness is 30 nm. In Figure 7a, the black spectrum shows that the prismatic array nanoparticles have a strong absorption band with a maximum beyond 850 nm before laser irradiation, which is due to in-plane dipolar transitions [6]. The weak plasmon resonance band around 570 nm results from the in-plane quadrupole resonance transition [6]. There is also a dip around 500 nm which is the result of the gold film [28]. The absorption spectrum increases below 500 nm, which is due to the absorption of the interband transition (from the Au 5*d* band to empty 6*sp* bands above the Fermi level [16]).

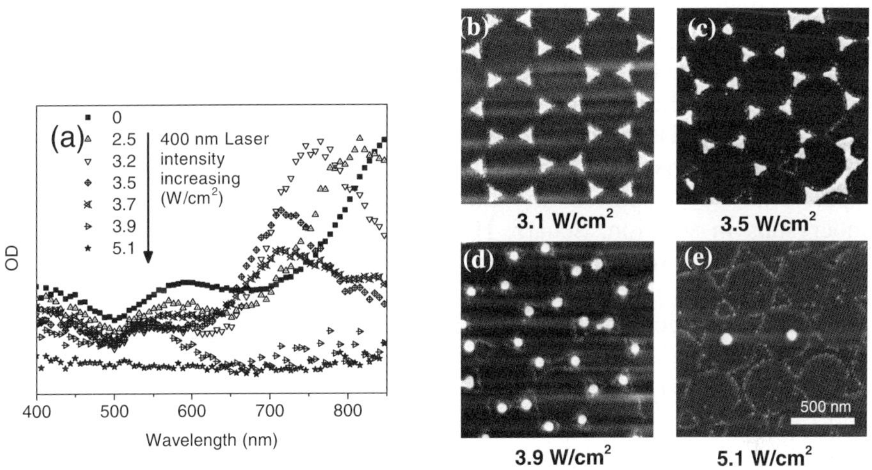

Figure 7. (a) Micro-absorption spectra and (b-e) the corresponding SEM images of a monolayer array of prismatic gold nanoparticles before and after irradiation with 400 nm femtosecond laser at different intensities. Assuming the laser pulse has a Gaussian energy distribution, the laser power density increases as we move from the edge to the centre of the laser irradiation profile. The numbers in (a) correspond to different power densities (W/cm^2) of the examined region. From the assumed Gaussian laser energy distribution profile, the calculated laser densities for the SEM images are: (b) 3.1 (c) 3.5, (d) 3.9, and (e) 5.1 W/cm^2.

Figure 7a shows that the in-plane dipolar transition bands blue shift as the power density of the laser irradiation increases from 0 to 3.2 W/cm^2. This is due to the decrease in the nanoparticle size as a result of annealing and/or atomic evaporation. Similar phenomena have been observed for thermal annealing [29] and solvent treatment [30] of the Ag prismatic array nanoparticles. Figure 7b is an SEM image of the sample after irradiation by 400 nm femtosecond laser at 3.1 W/cm^2, which clearly shows the rounding of the nanoparticle tips and the decrease in the nanoparticle size. Two reasons can explain this observation [31].

The first is the dewetting properties of gold on a quartz surface, which will force the particle to retract at the edge after melting. The second reason is a lower melting temperature at the tips due to the size-dependent surface melting temperature [5,32]. Thus, the tips of the prismatic nanoparticles are initially melted and retracted to form rounded tips.

As the power density increases further to 3.5 W/cm^2, the intensity of the absorption bands decreases due to the loss of the gold material as the particles become smaller. Quite interestingly, we were able to find tripod shaped particles when we measured the SEM image in the same region (Figure 7c) [31]. Since the most thermodynamically favored particle shape is a spherical shape that has a lower surface energy than any other shape, the formation of the gold tripodal nanoparticles is quite unusual. Apparently, the gold nanoparticle is not totally melted, otherwise it would be transformed from prismatic directly to the spherical shape (Figure 7d). It is obvious that the tripodal shape has lower surface energy than the prismatic shape.

A possible explanation of the forming of the tripod shaped nanoparticle by femtosecond laser irradiation is lattice rearrangement [33] of the as-deposited gold prismatic array nanoparticles to form a metastable tripodal shape. This is supported by the successful synthesis of gold tripodal nanocrystals in solution. Schatz, Hupp, and co-workers [34] showed that the gold tripod can be formed within a mixture solution of sodium citrate and 30% H_2O_2. The citrate method is one of the most well known methods for the synthesis of spherical gold nanoparticles [35]. In this method, citrate acts not only as a reducing agent to reduce $HAuCl_4$ but also as the capping material to stabilize the formed gold nanoparticles. With the assistance of the H_2O_2 at room temperature, the gold tripod can be formed with 40-50% yield [34]. With the same citrate as the capping agent, both the gold spherical and tripodal nanoparticles can be synthesized. This means the tripod shape is a metastable shape and can coexist with the spherical shape. This supports our conclusion that the tripodal shape is a metastable state of gold nanoparticles on their way to the most stable spherical shape as shown in Figure 8. With appropriate laser irradiation energy density (~ 3.5 W/cm^2 in this case), the prismatic array nanoparticles can be transformed to the tripod shape, while the energy is not sufficient to change the particle further to the spherical shape.

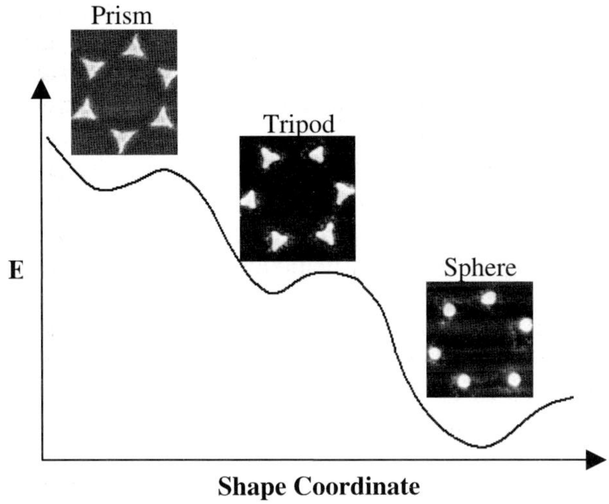

Figure 8. A proposed energy potential curve of nanoparticle arrays of different shapes. The surface potential decreases as the original prismatic nanoparticle changes to a tripodal shape and finally to a spherical shape as the laser energy input increases. The tripodal shape is believed to be a metastable configuration during the transformation (the gold tripod nanoparticles were also synthesized in solution) [34,36].

As the power density of the laser irradiation reaches 3.9 W/cm^2, we found that the dipolar absorption band at 850 nm completely disappears, while an absorption band at 540 nm appears. Figure 7d is the SEM image taken in the same region. It shows that all of the original prismatic nanoparticles change to a spherical shape with the diameter of the particles around 70 nm. The formation of the spherical shape suggests that the absorbed irradiation energy is sufficient to transform all the particles to the most thermodynamically stable shape.

The absorption band around 540 nm disappears when the power density of the laser irradiation is increased above 4.5 W/cm^2. Figure 7e is the SEM image measured in this area, which shows that most of the nanoparticles have disappeared. This is due to the ablation of the gold array nanoparticles by the high laser power density.

6. Conclusions

The effect of gold nanoparticle shape on electron-phonon relaxation has been studied. The observation of shape independence of the electron-phonon coupling constant for gold nanoparticles is consistent with the previous study, and the electron-phonon coupling constant is close to the value of bulk gold. For copper nanoparticles with diameters smaller than the bulk electron mean free path of copper, we find that the electron-phonon coupling constant is size dependent due to the interaction between electrons and surface phonon modes.

By using ultrafast laser spectroscopy and the optical detection technique, we have studied the coherent lattice vibrations of both silver and gold prismatic nanoparticles with very low frequencies. For silver, the observed size dependent oscillation period is in good agreement with the theoretical calculation. However, for gold there exists obvious deviation. This deviation is explained by the dipole exciton model of coupling between the sharper prismatic gold nanoparticles in the array. In addition, we show that femtosecond laser pulses at 400 nm could cause a continuous shape transformation from a prismatic shape to a tip rounded triangle, then to a tripodal shape, and finally to a spherical shape as the laser pulse energy increases. The tip rounded nanoprisms formed because of the dewetting properties of gold on quartz and the low melting temperature at the tips. The formation of the tripodal nanoparticles is because this shape is proposed to be a metastable configuration between the prismatic shape and the spherical shape.

Acknowledgment

The authors thank Mr. Christopher Tabor for his careful reading of the manuscript. This work is supported by the Division of Material Research of the National Science Foundation (Grant 0138391).

Reference

1. M. A. El-Sayed, Acc. Chem. Res. 34 (2001) 257.
2. J. C. Riboh, A. J. Haes, A. D. McFarland, C. R. Yonzon, and R. P. Van Duyne, J. Phys. Chem. B 107 (2003) 1772.
3. J. J. Storhoff, R. Elghanian, R. C. Mucic, C. A. Mirkin, and R. L. Letsinger, J. Am. Chem. Soc. 120 (1998) 1959.
4. M. Quinten, A. Leitner, J. R. Krenn, and F. R. Aussenegg, Optics Lett. 23 (1998) 1331.
5. J. C. Hulteen and R. P. Vanduyne, J. Vac. Sci. Tech. A - Vac. Surf. Films 13 (1995) 1553.
6. W. Huang, W. Qian, and M. A. El-Sayed, J. Phys. Chem. B (2005) accepted.
7. S. Giuffrida, G. G. Condorelli, L. L. Costanzo, I. L. Fragala, G. Ventimiglia, and G. Vecchio, Chem. Mat. 16 (2004) 1260.
8. S. L. Logunov, V. V. Volkov, M. Braun, and M. A. El-Sayed, Proc. Nat. Acad. Sci. USA 98 (2001) 8475.
9. S. Link and M. A. El-Sayed, J. Phys. Chem. B 103 (1999) 8410.
10. S. Link and M. A. El-Sayed, Int. Rev. Phys. Chem. 19 (2000) 409.
11. S. Link and M. A. El-Sayed, Ann. Rev. Phys. Chem. 54 (2003) 331.
12. J. H. Hodak, I. Martini, and G. V. Hartland, J. Phys. Chem. B 102 (1998) 6958.
13. T. S. Ahmadi, S. L. Logunov, and M. A. El-Sayed, J. Phys. Chem. 100 (1996) 8053.
14. S. L. Logunov, T. S. Ahmadi, M. A. El-Sayed, J. T. Khoury, and R. L. Whetten, J. Phys. Chem. B 101 (1997) 3713.
15. S. Link, C. Burda, Z. L. Wang, and M. A. El-Sayed, J. Chem. Phys. 111 (1999) 1255.
16. J. H. Hodak, A. Henglein, and G. V. Hartland, J. Chem. Phys. 111 (1999) 8613.
17. J. Hodak, I. Martini, and G. V. Hartland, Chem. Phys. Lett. 284 (1998) 135.
18. J. H. Hodak, A. Henglein, and G. V. Hartland, J. Chem. Phys. 112 (2000) 5942.

19. M. Nisoli, S. Stagira, S. DeSilvestri, A. Stella, P. Tognini, P. Cheyssac, and R. Kofman, Phys. Rev. Lett. 78 (1997) 3575.
20. S. D. Brorson, A. Kazeroonian, J. S. Moodera, D. W. Face, T. K. Cheng, E. P. Ippen, M. S. Dresselhaus, and G. Dresselhaus, Phys. Rev. Lett. 64 (1990) 2172.
21. T. Q. Qiu and C. L. Tien, Int. J. Heat Mass Transfer 35 (1992) 719.
22. Q. Darugar, W. Qian, and M. A. El-Sayed, J. of Phys. Chem. B, submitted.
23. S. Link, C. Burda, M. B. Mohamed, B. Nikoobakht, and M. A. El-Sayed, Phys. Rev. B 61 (2000) 6086.
24. N. Del Fatti, C. Voisin, F. Chevy, F. Vallee, and C. Flytzanis, J. Chem. Phys. 110 (1999) 11484.
25. M. Perner, S. Gresillon, J. Marz, G. von Plessen, J. Feldmann, J. Porstendorfer, K. J. Berg, and G. Berg, Phys. Rev. Lett. 85 (2000) 792.
26. G. V. Hartland, M. Hu, O. Wilson, P. Mulvaney, and J. E. Sader, J. Phys. Chem. B 106 (2002) 743.
27. A. Kosiorek, W. Kandulski, P. Chudzinski, K. Kempa, and M. Giersig, Nano Lett. 4 (2004) 1359.
28. S. Norrman, T. Andersson, C. G. Granqvist, and O. Hunderi, Phys. Rev. B 18 (1978) 674.
29. T. R. Jensen, M. D. Malinsky, C. L. Haynes, and R. P. Van Duyne, J. Phys. Chem. B 104 (2000) 10549.
30. M. D. Malinsky, K. L. Kelly, G. C. Schatz, and R. P. Van Duyne, J. Am. Chem. Soc. 123 (2001) 1471.
31. W. Huang, W. Qian, and M. A. El-Sayed, J. Appl. Phys. (2005) accepted.
32. J. C. Hulteen, D. A. Treichel, M. T. Smith, M. L. Duval, T. R. Jensen, and R. P. Van Duyne, J. Phys. Chem. B 103 (1999) 3854.
33. Z. L. Wang, R. P. Gao, B. Nikoobakht, and M. A. El-Sayed, J. Phys. Chem. B 104 (2000) 5417.
34. E. Hao, R. C. Bailey, G. C. Schatz, J. T. Hupp, and S. Li, Nano Lett. 4 (2004) 327.
35. J. Turkevich, P. C. Stevenson, and J. Hillier, Discuss. Farad. Soc. 11 (1951) 55.
36. S. H. Chen, Z. L. Wang, J. Ballato, S. H. Foulger, and D. L. Carroll, J. Am. Chem. Soc. 125 (2003) 16186.

Femtochemistry VII
Fundamental Ultrafast Processes in Chemistry, Physics, and Biology
A.W. Castleman, Jr. and M.L. Kimble

Relaxation of photoexcited states in mass-selected Ag_n^- and $Ag_nO_2^-$ clusters

Marco Niemietz, Markus Engelke, Young Dok Kim and Gerd Ganteför

Department of Physics, University of Konstanz, 78457 Konstanz, Germany

1. Introduction

The relaxation of excited electronic states is different for nanoparticles and clusters with respect to isolated atoms and bulk materials. In bulk metals a single electron photoexcited up to a few eV above the Fermi energy decays very rapidly dissipating its energy among the surrounding metal electrons. This results in extremely short lifetimes for such excited states in metals [1]. In contrast, the lifetimes of excited states in single atoms are of the order of nanoseconds, because Auger-like decay channels are not allowed due to the low density of states. Relaxation can occur only by photon emission, which is a slow process. This gives rise to the question how the electron dynamics behave in small nanoclusters?

Another possible decay channel of an excited electronic state is desorption of a monomer or a molecule, which may be bound to the cluster. This process can occur either direct, if an anti-binding state of the chemisorption bond is excited, or thermally. In the latter case the energy of the excited state is first dissipated within the cluster resulting in a vibrational excitation which then induces an unimolecular decay of the molecule or thermal desorption.

Accordingly, we can roughly distinguish between four different decay channels: photoemission corresponding to nanosecond lifetimes, thermalization through electron-phonon interaction yielding lifetimes of the order of picoseconds, Auger-like decays with extremely short lifetimes in the order of 10 fs - 100 fs, and finally photodissociation, which may also have a short lifetime in the case of a direct process. Which decay channel takes place depends on the electronic structure of the cluster and the photon energy.

Such processes can be studied using time-resolved photoelectron spectroscopy of mass-selected cluster anions [2-10]. This method has been pioneered by D. Neumark's group studying the photodissociation [2] of I_2^- and related systems [3,4]. The electronic relaxation of excited states in transition metal clusters as small as Ni_3^- occurs on time scales comparable to the corresponding bulk materials, which is explained by the already high density of d-derived states enabling fast Auger-like relaxation processes [5-7]. The photodesorption of CO from a Au_2^- cluster has also been studied using the same method, but the process turned out to be thermal and real-time observation of the changing electronic structure during the desorption process was not possible [8]. More recently, Cheshnovsky et al. studied the electronic relaxation in Hg_n^- clusters [9,10], which have a band gap similar to bulk semiconductors. An electron excited within the conduction band relaxes in a stepwise process by electron-phonon interaction. It approaches the lower edge and may recombine with a corresponding hole in the valence band.

Here, we present new results on bare Ag_n^- clusters continuing our earlier work on Au_n^- and Al_n^- clusters [11-13]. The electronic structure of Ag_n^- cluster [14] is much simpler than the one of open d-shell transition metal clusters and can be understood using the electronic shell model [15]. Figure 1 displays photoelectron spectra of Ag_7^- and Ag_8^- with the shell configurations $1s^2 1p^6$ and $1s^2 1p^6 1d$, respectively. The observed features can be assigned to emission from the occupied 1p and 1d shells. As expected in a shell model, for the larger cluster Ag_8^- the same features are observed as for the smaller cluster Ag_7^- but shifted towards higher binding energies. The 9th electron in Ag_8^- has to occupy the next shell (1d), which gives rise to the small peak at low binding

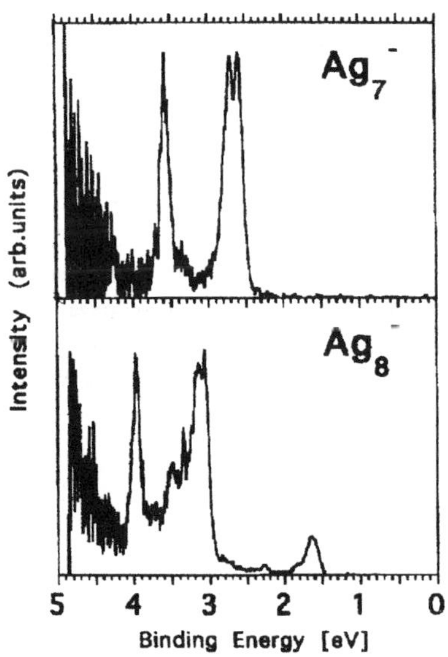

Figure 1. Photoelectron spectra of Ag_7^- and Ag_8^- recorded with a photon energy of 5.0 eV. The two features visible in the spectrum of Ag_7^- can be assigned to emission from the 1p shell. Similar peaks in the spectrum of Ag_8^- are shifted slightly towards higher binding energy. The small feature at low binding energy corresponds to the 9th electron in the 1d shell.

energy (1.6 eV). For the magic cluster Ag_7^- we found a rather long-living excited state which might indicate that this cluster, having a low density of states, behaves more atom-like rather than bulk-like. For Ag_{19}^- we were able to observe a step-like relaxation within the 1f-derived manifold of states indicating a semiconductor-like behavior. However, these are first results, and our preliminary interpretation needs further backup from calculations, which are underway. Finally, we studied the relaxation of excited states in $Ag_2O_2^-$ and $Ag_8O_2^-$. For these species direct photodissociation seems to be the most likely decay channel. The spectra reveal the signatures of these excited states and their lifetimes.

2. Experimental set up

The experimental set-up has been described in detail elsewhere [16,17]. Briefly, Ag cluster anions are produced by means of a pulsed arc cluster ion source (PACIS) and mass-selected using a time-of-flight (ToF) mass spectrometer. A selected bunch of cluster anions is irradiated by two laser pulses generated in a femtosecond laser system. The cluster temperature is estimated to be room temperature [18], and the energy resolution in our photoemission spectra is about 0.1 eV. For the spectroscopy of the excited states of the bare Ag_n^- clusters, the first pulse (pump pulse) has a photon energy of 1.55 eV and the second pulse (probe pulse) a photon energy of 3.1 eV. The pump pulse results in photoexcitation of a certain fraction of the anions in the bunch, and the probe pulse detaches electrons from a fraction of the anions in the same bunch. The photon flux of both pulses has to be optimized to maximize the number of the species hit by a single photon of each pulse and to minimize the number of the species hit by two or more. Figure 2 displays a

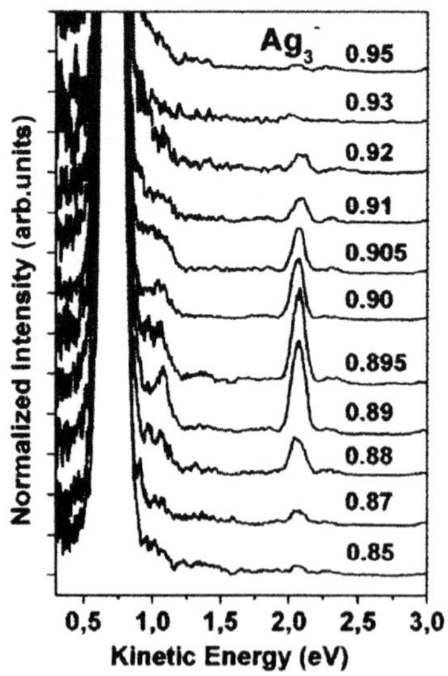

Figure 2. Pump-probe photoelectron spectra of Ag_3^-. The large feature at low kinetic energy corresponds to slow electrons detached by photons of the probe pulse, which has a photon energy higher than the electron affinity. The interaction with both pulses (pump + probe) gives rise to a small feature at higher kinetic energy. The delay between the pulses is given in mm difference of the optical path length. From such spectra the zero delay and the time resolution are determined.

series of pump-probe photoelectron spectra of Ag_3^- plotted versus the kinetic energy of the electrons. Ag_3^- has no excited state in the energy range covered here and only a non-resonant 2-photon signal is observed. Since this signal appears only if the two laser pulses overlap, the zero point and the time resolution of about 120 fs can be determined. $Ag_nO_2^-$ species are generated by introducing molecular oxygen [19] into the extender using a pulsed valve for adjusting the amount of O_2.

3. Results and Discussion

Ag_7^- has 8 electrons, the 1p shell is filled, and there is a pronounced gap between the highest occupied and the lowest unoccupied molecular orbital (HOMO-LUMO gap). This gap roughly corresponds to the energetic spacing between the 1s and 1p electronic shells, which can be estimated from the photoelectron spectra of Ag_8^- to be about 1.4 eV [14]. In the single particle picture an excitation with an energy of 1.55 eV corresponds to the promotion of a 1p electron into the empty 1d shell. Since this is the lowest possible excited state, Auger-like processes are not allowed, similar to the case of the atom. We expect long lifetimes and relaxation either through photon emission or thermalization.

Figure 3 displays a series of pump-probe photoelectron spectra of Ag_7^-. We found only a weak pump-probe signal, and for a better visibility, the displayed spectra are difference spectra. The electron signal which exhibits no delay dependence has been subtracted. This constant signal has been determined by recording a spectrum at a large delay where no further change of the electron signal is observed. This background spectrum is a simple sum of the contribution of the two electron signals from the pump and the probe laser

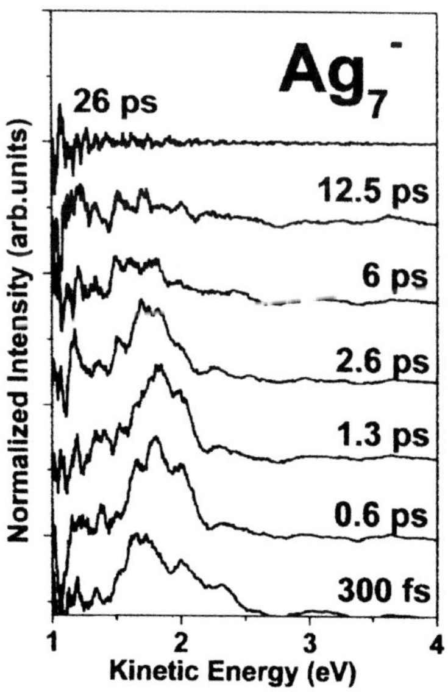

Figure 3. The delay dependent signal of the pump-probe photoelectron spectra of Ag_7^-. The pump photon energy is 1.55 eV and the probe energy 3.1eV. The peak at a kinetic energy of about 1.8eV can be assigned to an excitation of an electron in the uppermost occupied 1p orbital into the empty 1d orbital (see Fig. 1). From this orbital the electron is detached resulting in a kinetic energy of $-2.6 + 1.55 + 3.1 = 2.05$ eV. This rough estimate neglects all relaxation effects, but still the agreement with the measured kinetic energy is very good. The lifetime extracted from the delay dependence of the intensity is 5.5 ± 1 ps.

pulses alone. The reason for the low signal level of the pump-probe signal might be that the photon energy of the pump pulse is fixed at 1.55eV and might not fit exactly the energy of the 1p → 1d excited state of Ag_7^-, resulting in a small cross section. However, this signal indicates the existence of an excited state with a lifetime of 5.5 ± 1 ps, which is much shorter than in case of isolated atoms. We think that thermalization, i.e. relaxation by electron-phonon interaction, is the process responsible for this decay. Typical relaxation rates for this mechanism are in the order of 1 ps. Since the HOMO-LUMO gap of Ag_7^- is much larger than the energies of the phonon modes of this cluster, relaxation must involve many phonons, which might explain the somewhat increased lifetime. A long-living excited state has also been found for the neutral magic cluster Ag_8 in two-photon ionization experiments [20,21].

Ag_{19}^- has the electronic shell configuration $1s^2 1p^6 1d^{10} 2s^2$. It is a closed shell cluster with 20 electrons. The size of the 2s - 1f gap can be estimated from the photoelectron spectra of Ag_{20}^- to be about 0.7 eV [14]. The gap size is smaller than in Ag_7^- since the density of states generally increases with increasing cluster size. The gaps will vanish for larger clusters, marking the transition to metallic properties. Since in Ag_{19}^- the gap size is smaller, phonon assisted relaxation might be faster although Auger-like processes are still suppressed at low excitation energies due to the lack of low lying empty states.

Fig ure 4 displays pump-probe photoelectron spectra of Ag_{19}^-. As in the case of $Ag_7^-$the signal which is not delay-dependent has been subtracted. The integral time-dependent signal vanishes with a time constant of 1.5 ± 0.5 ps, which is faster than in the case of Ag_7^- due to the smaller gap size. However, there is an interesting change in the peak shape and position; although the signal

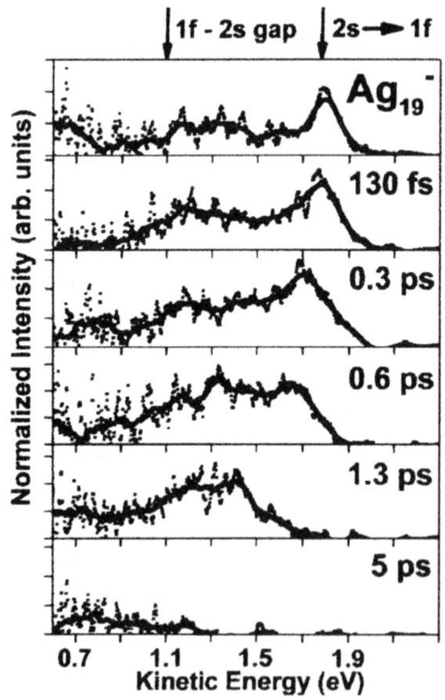

Figure 4. Pump-Probe photoelectron spectra of Ag_{19}^-. The non-time dependent signal has been subtracted. At zero delay a narrow feature appears at a kinetic energy corresponding to the excitation of 2s electron into one level of the 1f manifold of states. There, the electron relaxes stepwise down to the lower edge of the 1f states. There the process stops and the feature seems to vanish analogous to the case of Ag_7^-. This behavior reminds of a semiconductor.

to noise ratio does not allow a reliable quantitative analysis, at zero delay a rather narrow peak at high kinetic energy appears. This feature broadens with increasing delay time towards lower binding energy until it reaches a kinetic energy of 1.1 eV. Here, it narrows again and finally vanishes. This reminds of the behavior of an excited electron within the manifold of states of the conduction band in a bulk semiconductor, and a similar behavior has also been observed recently in Hg_n^- clusters [9,10]. There is one tempting interpretation of these observations: the pump photon with an energy of 1.55 eV excites a 2s electron to a level within the 1f-derived manifold of states. Then the electron relaxes stepwise via electron-phonon interaction down to the lowest-lying 1f-level resulting in a broadening of the photoelectron signal and a shift towards lower binding energy. At the lower edge of the 1f-manifold of states this process stops and the peak disappears. This behavior indicates an electron dynamics similar to a semiconductor in Ag clusters of this size.

In bare Ag_n^- cluster anions we did not observe photoinduced fragmentation on the timescales accessible in our experiment (up to 5 ns). This is different for clusters with inbuilt weak bonds like metal clusters with an attached molecule. Even numbered Ag_n^- and Au_n^- clusters readily chemisorb O_2 molecules [19,22,23]. The even-odd alternation can be understand in terms of a simple charge transfer picture [23]: the bond between the metal cluster anion and the O_2 molecule is formed by charge transfer from the cluster to the O_2. The easier the transfer, the faster the reaction goes. Au_n^- and Ag_n^- clusters exhibit a pronounced even-odd alternation of the electron affinities with the clusters having an odd number of electrons (= even number of metal atoms) having low electron affinities [15]. Figure 5 displays a section of a mass spectrum of Ag_n^- clusters reacted with O_2. The conditions (temperature, partial pressures, reaction time) are chosen in a manner that almost all Ag_8^- anions chemisorb a single O_2 molecule, while the neighboring species Ag_7^- and Ag_9^- remain unreacted [19].

Using the relatively low photon energy of 1.55 eV for excitation we could observe no pronounced pump-probe signal. However, if the order of the pump and the probe laser pulses is reversed, a very strong-pump probe signal

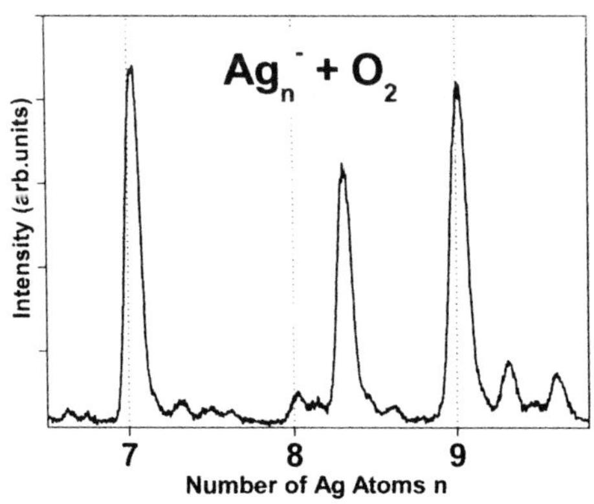

Figure 5. Section of a time-of-flight mass spectrum of Ag_n^- clusters reacted with O_2 molecules. Dotted lines mark the positions of the unreacted clusters. The reactivity of the even numbered cluster anions is much higher than for the odd numbered. At the conditions chosen here almost all Ag_8^- chemisorb a single O_2 molecule while the Ag_7^- and Ag_9^- clusters remain almost unreacted [19].

appears. An intense peak at a kinetic energy of about 1eV is observed, which exhibits a pronounced variation of shape and lifetime with changing the cluster size n. It seems that all the $Ag_nO_2^-$ clusters we studied so far have an excited state reachable with a pump-energy of 3.1 eV. Unfortunately, in this inverse ordering of our pump-probe set up the probe energy is only 1.55 eV and does not allow any deeper insight into the dynamics of this state except for the case of Ag_2^-. The bare Ag_2^- has an extremely low electron affinity of 1.1 eV [14] and even with a 1.55 eV photon, an electron can be detached and would appear at a kinetic energy of 0.45 eV. Therefore, if the excited state of $Ag_2O_2^-$ decays by O_2 desorption, a peak at 0.45 eV kinetic energy should appear for sufficiently long pump-probe delays.

Figure 6 displays a series of pump-probe photoelectron spectra of $Ag_2O_2^-$. The strong feature observed at zero delay can be assigned to the excited state of $Ag_2O_2^-$, which decays on a time scale comparable to our experimental resolution. A smaller peak appears on the same time scale, which is right at the expected position of photoelectrons from Ag_2^-. We tentatively assign this series of spectra to the following process:

$$Ag_2O_2^- + \text{photon (3.1 eV)} \rightarrow Ag_2O_2^{-*} \rightarrow Ag_2^- + O_2$$

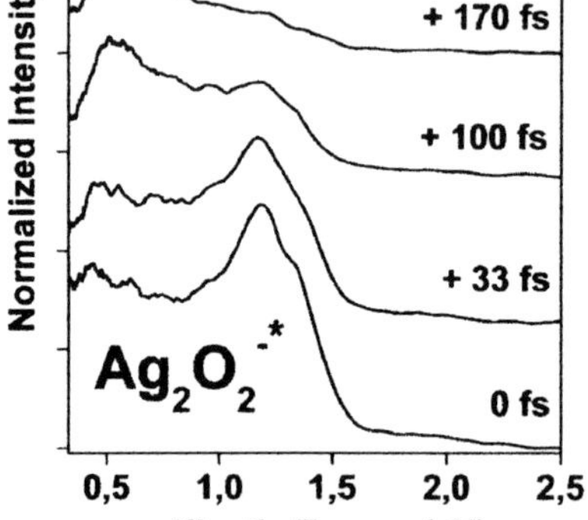

Figure 6. Series of pump-probe photoelectron spectra, recorded using a probe pulse of 1.55 eV after photoexcitation of $Ag_2O_2^-$ with 3.1 eV. At negative delays this ordering is reversed and almost no electron signal (time dependent and non-time dependent) is observed. There is only a weak signal from single photon detachment from $Ag_2O_2^-$ by the 3.1 eV laser pulse. The peak at 1.1eV is assigned to an excited state of $Ag_2O_2^-$, which has an extremely short lifetime (< 120 fs) comparable to our experimental resolution. The appearing peak at 0.5 eV can be assigned to photodetachment from Ag_2^- with 1.55 eV photons.

This assumption bases on the correct assignment of the peak at 0.5eV kinetic energy to photodetachment from Ag_2^-. Further experiments using a higher photon energy for the probe pulse are necessary to support this first tentative assignment. If the assignment is correct, the desorption process most likely is direct. If it was thermal, the excited state would thermalize on a different time scale than the subsequent thermal evaporation of the O_2 molecule. In Figure 6 the appearance of the peak assigned to Ag_2^- is accompanied by the disappearance of the excited state feature, indicating a direct process. However, at the moment we cannot exclude a thermal desorption as fast as the relaxation or even faster. Studies with improved time resolution will reveal more details and may even show the transient electronic structure during the reorganization of the orbitals.

Recently, Wöste *et al.* [24] conducted a NeNePo experiment on the same species $Ag_2O_2^-$. In this type of experiment, the pump photon serves to photodetach an electron, neutralizing the anion. The probe photon ionizes the neutral species. The intensity of the generated ions is recorded and plotted versus the pump-probe delay. For the first step a photon energy close to our pump-energy is used, and the neutral clusters are ionized with a photon energy of about 3 eV. Of course this photon energy is smaller than the ionization potential and, therefore, multiphoton processes must be involved in the second step. The NeNePo signal exhibits an oscillatory behavior which is assigned to the vibration of the neutral Ag_2 from which the O_2 dissociates. According to their analysis, the O_2 desorbs in the neutral ground state because the neutral species Ag_2O_2 is unbound. In the NeNePo experiment photodesorption takes place after the detachment, while in our tentative analysis it is the other way around. Future studies, especially using a higher photon energy of the pump pulse, will certainly clarify the understanding of these processes.

Finally, Figure 7 displays a similar set of time-resolved photoelectron spectra for $Ag_8O_2^-$ (see also Fig. 5). The spectrum shows a broad feature which vanishes with increasing delay time with a time constant comparable to the case of the dimer. The shape and position of the feature are similar in both cases and it might be possible that a similar process occurs. However, the electron affinity [14] of Ag_8^- is higher than for the dimer, and therefore the low photon energy of the probe pulse makes it impossible to detach electrons from Ag_8^-. It is also possible that the excited state of $Ag_8O_2^-$ relaxes by other decay channels. However, since the position, the decay time, and even the finestructure of the features are similar, the decay mechanism might be the same. Future studies will clarify this. A detailed analysis of the finestructure, which might be due to vibrational excitations, can also shed some light on the nature of this excited state.

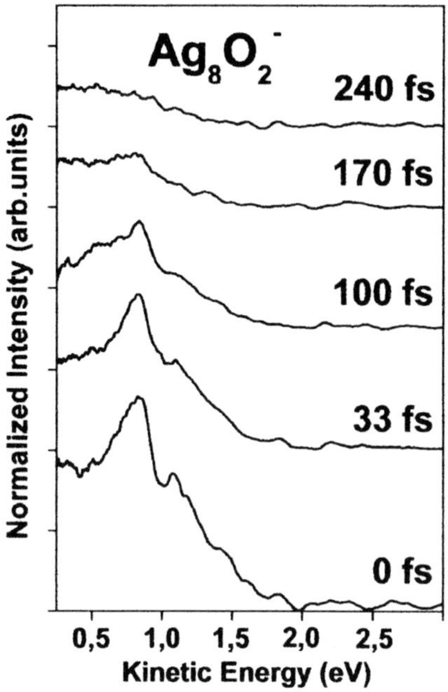

Figure 7. Series of pump-probe photoelectron spectra of $Ag_8O_2^-$ with pump- and probe-energies of 3.1 eV and 1.55 eV, respectively. The broad feature centered around 0.8eV kinetic energy can be assigned to an excited state of $Ag_8O_2^-$ with a lifetime of less than 120 fs.

4. Conclusion

We present time resolved photoelectron spectra of bare Ag_n^- and reacted $Ag_nO_2^-$ cluster anions. Electronic relaxation in Ag_7^- takes place on a time scale much longer than in the corresponding bulk, which is to be expected because of the large HOMO-LUMO gap of 1.4 eV of this cluster. Relaxation via auger-like processes is not possible, and most likely the excited state decays by electron-phonon interaction. In the case of Ag_{19}^- we observe a shift of the energetic position of the excited electron, similar to electrons excited into the conduction band of a semiconductor. Although these data are incomplete and a quantitative analysis is difficult because of the limited signal to noise ratio, the data indicate that certain small Ag_n^- clusters exhibit an atomic-like electron dynamics, while certain medium sized species show a semiconductor-like relaxation. For even larger clusters, Auger-like processes should become the most prominent relaxation process, resulting in very fast decay similar to bulk metals.

If a single O_2 molecule is bound to the cluster, electronic relaxation is different. For excitation with 1.55 eV photons we find no pump-probe signal. Using a higher photon energy of 3.1 eV, a pronounced signal marks an excited state of these species. In the case of $Ag_2O_2^-$, this state seems to decay via photodesorption into Ag_2^- + O_2. For $Ag_8O_2^-$ a similar pump-probe signal is measured. Our data support the idea that, for these particles, an excited state exists for photoexcitation with 3.1 eV, which decays by direct photodesorption. A detailed analysis using a higher photon energy of the probe pulse should reveal the details of the desorption process and its cluster size dependence. It would be interesting to directly observe the breaking of the chemical bonds in such a cluster system.

5. Acknowledgement

We acknowledge the support by the Center of Applied Photonics (CAP) of the University of Konstanz.

References

1. M. Wolf, Surf. Sci. 377 (1997) 343.
2. B. J. Greenblatt, M. T. Zhanni, and D. M. Neumark, Chem. Phys. Lett. 258 (1996) 523.
3. M. T. Zanni, B. J. Greenblatt, A. V. Davis, and D. M. Neumark, J. Chem. Phys. 111 (1999) 2991.
4. B. J. Greenblatt, M. T. Zhanni, and D. M. Neumark, Science 276 (1997) 1675.
5. N. Pontius, M. Neeb, W. Eberhardt, G. Lüttgens, and P. S. Bechthold, Phys. Rev. B 67 (2003) 035425 .
6. N. Pontius, P. S. Bechthold, M. Neeb, and W. Eberhardt, J. Electr. Spectr. Rel. Phenom. 114 (2001) 163.
7. N. Pontius, P. S. Bechthold, M. Neeb, and W. Eberhardt, Phys. Rev. Lett. 84 (2000) 1132.
8. G. Lüttgens, N. Pontius, P. S. Bechthold, M. Neeb, and W. Eberhardt, Phys. Rev. Lett. 88 (2002) 076102.
9. J. R. R. Verlet, A. E. Bragg, A. Kammrath, O. Cheshnovsky, and D. M. Neumark, J. Chem. Phys. 121 (2004) 10015.
10. A. E. Bragg, J. R. R. Verlet, A. Kammrath, O. Cheshnovsky, and D. M. Neumark, J. Chem. Phys. 122 (2005) 54314.
11. G. Ganteför, S. Kraus, and W. Eberhardt, Electr. Spectr. Relat. Phenom. 88 (1997) 35.
12. M. Niemietz, P. Gerhardt, G. Ganteför, and Y. D. Kim, Chem. Phys. Lett. 380 (2003) 99.
13. P. Gerhardt, M. Niemietz, Y. D. Kim, and G. Ganteför, Chem. Phys. Lett. 382 (2003) 454.
14. H. Handschuh, Chia-Yen Cha, P. S. Bechthold, G. Ganteför, and W. Eberhardt, J. Chem. Phys. 102 (1995) 6406.
15. W. A. de Heer, Rev. Mod. Phys. 65 (1993) 611.
16. S. Burkart, N. Blessing, B. Klipp, J. Müller, G. Ganteför, and G. Seifert, Chem. Phys. Lett. 301 (1999) 546.
17. S. Minemoto, J. Müller, G. Ganteför, H. J. Münzer, J. Boneberg, and P. Leiderer, Phys. Rev. Lett. 84 (2000) 3554.
18. G. Ganteför, D. M. Cox, and A. Kaldor, J. Chem. Phys. 93 (1990) 8395.
19. Y. D. Kim and G. Ganteför, Chem. Phys. Lett. 383 (2004) 80.
20. F. Federmann, K. Hoffmann, N. Quaas, and J. P. Toennies, Eur. Phys. J. D 9 (1999) 11.
21. P. Radcliffe, A. Przystawik, T. Diederich, T. Döppner, J. Tiggesbäumker, and K.-H. Meiwes-Broer, Phys. Rev. Lett. 92 (2004) 173403.
22. B. E. Salesbury, W. T. Wallace, and R. L. Whetten, Chem. Phys. 262 (2000) 131.
23. D. Stolcic, M. Fischer, G. Ganteför, and Y. D. Kim, J. Am. Chem. Soc. 125 (2003) 2848.
24. L. D. Socacio-Siebert, J. Hagen, J. Le Roux, D. Popolan, M. Vaida, S. Vajda, T. M. Bernhardt, and L. Wöste, Phys. Chem. Chem. Phys. 7 (2005) 2706.

Femtochemistry VII
Fundamental Ultrafast Processes in Chemistry, Physics, and Biology
A.W. Castleman, Jr. and M.L. Kimble

Time-Resolved Single Molecule Spectroscopy of Conjugated Polymer Nanoparticles

Young Jong Lee, Rodrigo E. Palacios, and Paul F. Barbara*

Center for Nano- and Molecular Science and Technology and Department of Chemistry and Biochemistry, University of Texas at Austin, Austin, Texas 78712, USA

1. Introduction

The dynamics of excitons in conjugated polymer materials play a central role in the mechanism of operation of the many important photonic devices based on these materials, including organic light-emitting diodes, photovoltaic solar cells, and sensors [1-8]. The investigation of dynamics of excitons by ensemble time-resolved spectroscopic techniques are limited by the extreme heterogeneity of conjugated polymers. Recently, single molecule spectroscopy (SMS) has proven to be a unique tool for direct measurement of the inhomogeneous spectral distributions of conjugated polymers and for unraveling complex photodynamics involving triplet states, photo-oxidation, and various charge transfer processes that would be extremely difficult to measure by ensemble methods [9-25].

This proceeding summarizes recent breakthroughs in using time-resolved, single molecule spectroscopy to investigate excitons in conjugated polymers. It includes two main topics. First, we use the recently developed technique of microsecond time-scale *excitation intensity modulation SMS* to study the inhomogeneity in the triplet decay dynamics of the conjugated polymers MEH-PPV at room temperature [16,26]. Second, we report for the first time the use of a femtosecond time-resolved SMS technique to study single exciton dynamics in conjugated polymers. The femtosecond technique monitors the SMS fluorescence as a function of delay between two femtosecond excitation pulses. It achieves both femtosecond time resolution and single molecule sensitivity. It involves the observation of the fluorescence "dip" of single MEH-PPV molecules due to the interaction of singlet excitons through such processes as singlet-singlet (S-S) annihilation, energy transfer, and electron transfer in the single multichromophoric MEH-PPV polymer chains

2. Experimental Setup

Experimental samples were comprised of a glass microscope coverslip that was coated with a 100 nm spin-coated polymer film comprised of an inert polymer matrix of polymethyl methacrylate (PMMA, Aldrich) that contained a very low concentration (~10^{-11} M) of MEH-PPV (poly[2-methoxy,5-(2'-ethylhexyloxy)-*p*-phenylene-vinylene], M_w ~ 10^6, Uniax) conjugated polymer molecules. A gold layer (thickness ~ 200 nm) was evaporated on top of the host polymer film to exclude oxygen, which has been shown to efficiently quench triplet states in polymer thin films.

The experimental apparatus for time-resolved SMS is a home-built sample scanning confocal microscope (see Fig. 1) [26]. The fluorescence signal is directed either to an avalanche photodiode (APD, Perkin-Elmer) or to an LN_2-cooled CCD attached on a monochromator (Acton, SP-150). For the microsecond-scale excitation intensity modulation technique, the transient signal $F(t)$ from the APD is recorded with a multichannel scaler (MCS) that was synchronized to the sequence of excitation pulses. The sequence of excitation pulses $I_E(t)$ was derived from the output of an argon ion laser (488 nm) that was intensity-modulated by an acoustic optical modulator (IntraAction, AOM-40). The input electronic waveform to the modulator was produced by a programmable function generator (Wavetek, Model 29) that was synchronized to the MCS (Fast ComTec, MCA-3).

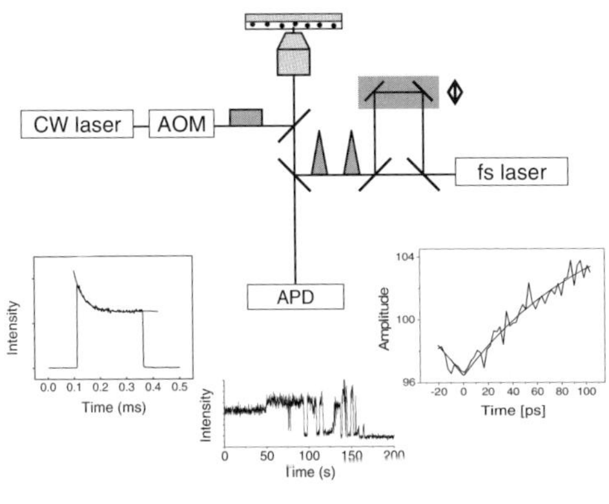

Figure 1. Schematic setups for time-resolved single molecule spectroscopy. For the microsecond-scale excitation intensity modulation technique, the fluorescence intensity measured by an avalanche photodiode (APD) is synchronized with the square excitation pulse generated by an acousto-optic modulator (AOM). The femtosecond pump–pump technique measures the fluorescence intensity change as a function of time delay between the two identical pulses.

The femtosecond laser system consists of a regenerative amplifier (Coherent, RegA 9000) and an optical parametric amplifier (Coherent, OPA 9400), which produced 200 fs pulses centered at 488 nm at 250 kHz repetition rate. For the pump-pump spectroscopy, the laser beam was divided into two beams. The arrival time of one of the pulses was controlled using a fast scanning delay stage (Clark-MXR, ODL-150), whose scanning range is 150 ps.

3. Results and Discussion

3.1. Triplet dynamics on the microsecond timescale

Excitation intensity modulation SMS was previously used to study triplet dynamics on the microsecond timescale in single MEH-PPV conjugated polymer molecules [16,26]. In this approach, the fluorescence intensity of a single polymer molecule is recorded while the molecule is irradiated with repetitive cycles of a sequence of square excitation pulses with variable intensity and duration. The time transient is synchronously averaged over many cycles to yield time-ensemble intensity $F(t')$ as a function of time t' within the cycle. In the present study, we simultaneously measured a time transient and a spectrum for a single molecule to correlate the photodynamics and the spectrum.

Figure 2 shows a set of typical spectra and $F(t')$ of single MEH-PPV molecules in a PMMA matrix at 300 K. The histogram of peak wavelengths in Fig. 2B shows a well-resolved bimodal distribution consistent with previous reports [11,27-30]. For more than 90% of single molecules, the measured $F(t')$ exponentially decays to a steady-state value on a microsecond timescale. Typically, as the excitation intensity increases, the decay time becomes shorter ,and the contrast ratio (the ratio of the initial intensity to the steady-state intensity) becomes larger. Values of the decay time and contrast ratio vary greatly molecule by molecule, as clearly seen in Figs. 2C– 2F.

The $F(t')$ behaviors presented here are consistent with predictions from the recently developed kinetic population state model [16] for localized multichromophoric conjugated polymers. In this model, $F(t')$ is a consequence of triplet-triplet (T-T) annihilation and singlet quenching by triplets in single conjugated polymers. During the excitation-off period, triplet excitons decay through reverse intersystem crossing ($T_1 \rightarrow S_0$), and the initial emission intensity $F(t')$ immediately after the excitation pulse is on reflects the emission of an essentially triplet free molecule. As time progresses, multiple excitations ($S_0 \rightarrow S_1$) occur, causing a build-up of triplet excitons by forward intersystem crossing ($S_1 \rightarrow T_1$). The fluorescence intensity decreases as the triplet population increases due primarily to singlet quenching by triplets ($S_1 + T_1 \rightarrow S_0 + T_n$). The triplet population eventually reaches a steady-state population due to T-T annihilation ($T_1 + T_1 \rightarrow S_1 + S_0$), which corresponds to the plateau in the $F(t')$ data during the excitation on period. The T-T annihilation of MEH-PPV plays a similar dynamic role to the previously reported singlet-triplet annihilation mechanisms for other multichromophoric systems [31-34].

Assuming very rapid T-T annihilation, the conjugated polymer is predicted to spend most of the time in either (0,0) or (0,1) states, where (N_S,N_T) is the population state that has N_S singlets and N_T triplets in a single molecule. The population kinetic scheme can be approximated by the following two-state model,

$$(0,0) \underset{k_B}{\overset{k_F}{\rightleftharpoons}} (0,1) \tag{1}$$

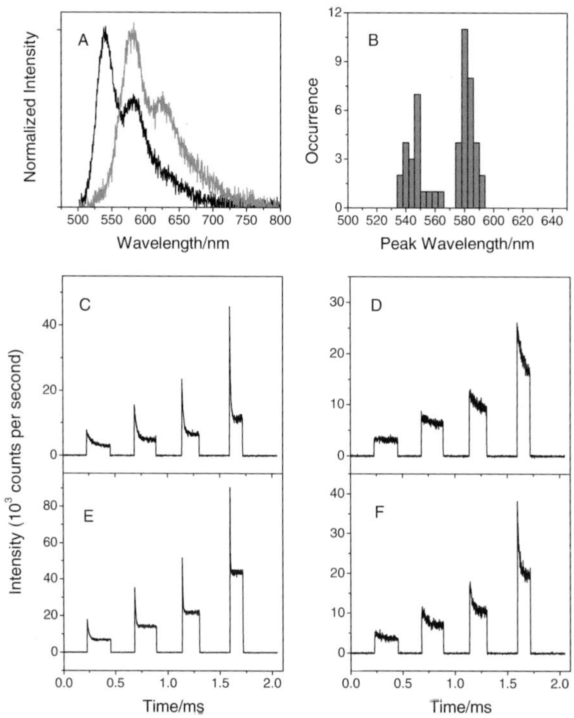

Figure 2. (A) Typical single molecule emission spectra of MEH-PPV (λ_{exc} = 488 nm) measured at 300 K. (B) Peak wavelength distribution of single molecule emission spectra. (C–F) Typical examples of time-averaged fluorescence trajectory comprised of many cycles of an excitation pulse train acquired from single molecules. The intensities of the four sequential excitation pulses are 70, 160, 250, and 500 W cm^{-2}, from left to right, respectively.

Table 1. Summary of the measured (best-fit) absorption cross sections (σ) and rate constants for reverse intersystem crossing (k_{isc}') and quenching of singlets by triplets (k_{QST}).

	Blue	Red
k_{isc}' [10^3 s^{-1}]	6.5 ± 3.0 [c]	6.4 ± 3.1
k_{QST} [10^{13} l mol^{-1} s^{-1}] [a]	1.0 ± 0.9	2.1 ± 1.3
σ [10^{-15} cm^2] [b]	1.1 ± 0.7	7.4 ± 5.8

[a] The molecular volume of 10^3 nm^3 and the τ_{fl} of 300 ps[53] are used.
[b] The quantum yield of intersystem crossing is assumed to be 1.25% from Ref.35.
[c] Standard deviation of the measurements.

with forward and backward first-order rate constants, k_F and k_B, that are given by,

$$k_F = k_{exc} k_{isc} \tau_{fl} \tag{2}$$

$$k_B = k_{isc}' + k_{exc} k_{isc} \tau_{fl}' \tag{3}$$

where k_{exc} is the excitation rate from the ground state to a singlet excited state, k'_{isc} is the reverse intersystem crossing rate, k_{isc} is the intersystem crossing rate, τ_{fl} and τ'_{fl} are the fluorescence lifetime of a singlet exciton without and with a triplet state, respectively. The τ'_{fl} involves an additional relaxation process, which is singlet quenching by triplets (k_{QST}), and thus τ'_{fl} is expected to be shorter than τ_{fl}. We apply this two-state model in the analysis of 49 MEH-PPV molecules at 300 K, summarizing the best-fit kinetic parameters in Table 1.

This excitation intensity modulation SMS technique can measure the triplet lifetime ($1/k'_{isc}$) of an isolated conjugated polymer, which is not accessible by typical ensemble measurements for the bulk film. The present "unintentionally doped" hole polarons (typical density ~ 10^{16} cm^{-3}) diffuse in the bulk film, quenching triplet exciton. The quenching rate is fast enough at room temperature to shorten an apparently measured triplet lifetime. Even at cryogenic temperatures, where the hole polaron diffusion is suppressed, the absorption spectrum of hole polarons overlaps with that of triplet excitons, complicating determination of the triplet exciton lifetime. Unlike the bulk films, an isolated single polymer chain (volume ~ 10^{-18} cm^3) allows an almost hole-free environment since the probability of hole presence in a single molecule is roughly estimated to be ~1% from the hole polaron density and the molecular volume (~10^{16} cm^3). In addition, the measured k'_{isc} value by the excitation intensity modulation SMS is obtained by extrapolation of the exponential decay rate ($k = k_F + k_B$) of $F(t')$ in Equations (2) and (3), being insensitive to other fitting parameters. The measured value for k'_{isc} of (6.5 ± 3.0) ×10^3 s^{-1} is consistent with previously reported values for MEH-PPV [35-42], including 1.1×10^4 s^{-1} in a benzene solution at room temperature [35], 10^4 s^{-1} for in a film at room temperature [36]; and for PPV, 9.1×10^3 s^{-1} for in a frozen benzene solution and 1.4×10^4 s^{-1} in a film at 20 K [41].

It is interesting to compare the photodynamics of the bimodal emission components of MEH-PPV ("blue" and "red") by obtaining fluorescence spectra simultaneously with $F(t')$. As seen in Table 1, the two species do not exhibit a significant difference of the fit parameters as compared to the molecule-to-molecule heterogeneity. However, "red" molecules tend to have a larger absorption cross section than "blue" ones, which leads to faster decay rates of $F(t')$. A larger absorption cross section corresponds to a longer chain length, which leads to higher probability for a specific morphology (chain-chain contact or aggregation) responsible for the "red" emission [27,29,30,43-45].

3.2. Singlet Dynamics on the Femtosecond Timescale

In a multichromophoric molecule like MEH-PPV, the presence of multiple excitons or charge carriers is known to decrease the fluorescence quantum yield by opening new deactivation pathways, such as singlet-singlet (S-S) annihilation ($S_1 + S_1 \rightarrow S_n + S_0$) [31-34], and singlet quenching by triplets ($S_1 + T_1 \rightarrow S_0 + T_n$) [31-34] and by holes ($S_1 + h^+ \rightarrow S_0 + h^+$) [17,46-50]. These deactivation pathways shorten the lifetime of singlet excitons, and the ultrafast

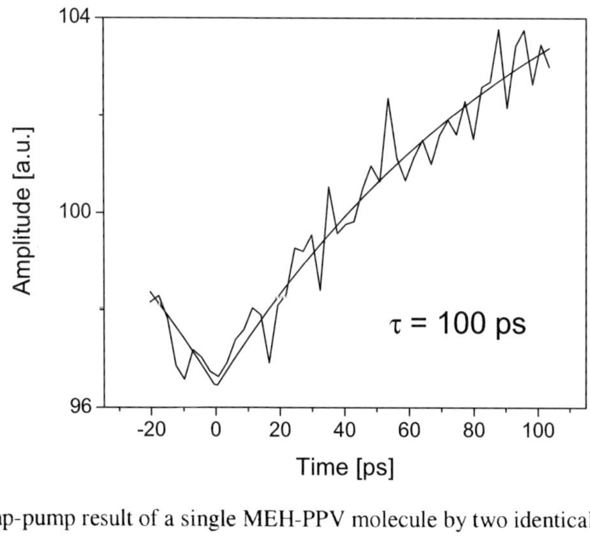

Figure 3. Pump-pump result of a single MEH-PPV molecule by two identical femtosecond pulses. The time resolution is estimated to be 500 fs as a result of dispersion of a 200 fs pulse by the objective lens [52].

photodynamics are often not accessible by conventional time-resolved spectroscopic techniques, such as time-correlated single photon counting, whose typical time resolution is 50 ps.

Recently, a group reported a result of a SMS study on a dye molecule (single chromophore) using femtosecond pump-probe technique, achieving both femtosecond time resolution and single molecule sensitivity [52]. We have utilized a similar experimental scheme to observe the ultrafast singlet exciton dynamics in single conjugated polymer nanoparticles (multichromophore). In a multichromophoric conjugated polymer molecule, when the excitation rate of a single molecule becomes comparable to the singlet exciton lifetime, S-S annihilation possibly reduces fluorescence intensity by opening a deactivation pathway of singlet excitons. The time resolution of our system is estimated to be ~500 fs as a result of dispersion of a 200 fs pulse through the objective lens [52]. When both pump pulses arrive at the same time, the rate of S-S annihilation will be the maximum, and measured fluorescence intensity will be the minimum. As the time difference of the two pulses increases, the measured intensity increases exponentially, converging to the sum of those given by individual pulses, and the time constant corresponds to the singlet lifetime. Figure 3 shows femtosecond pump-pump data of a single MEH-PPV molecule. Despite the limited scanning range of 150 ps, the data exhibits a clear feature of fluorescence intensity recovery, which is well fit with a single exponential of a 100 ps decay constant. This value is in the same order of magnitude as lifetimes of singlet exciton of MEH-PPV measured by ensemble measurements [53].

3.3. Photodamaging of MEH-PPV by ultrashort laser pulses

We have observed that a serious complication in these measurements is photodamage (probably multiphoton ionization) of the single molecules due to the high intensity femtosecond pulses. Indeed, photodamage of single MEH-PPV molecules with femtosecond excitations is observed to be much more severe than the corresponding experiments with CW laser excitation. A significant and irreversible intensity decrease and spectral shift is observed even at a very low intensity and irradiation time of femtosecond pulses. Less than 10% of investigated molecules survive long enough to show the detectable fluorescence "dip" signal. This is probably because the threshold of photodamaging is very close to that of S-S annihilation for ultrashort pulses. In other words, the excess energy of S_n state generated by S-S annihilation ($S_1 + S_1 \rightarrow S_n + S_0$) might efficiently result in photodamaging probably via photo-oxidation.

To further explore the mechanism by which MEH-PPV molecules are photodamaged, we have examined the emission spectra of single MEH-PPV molecules before and after brief irradiation of femtosecond pulses. Since multiphoton processes are strongly dependent on the excitation intensity, irradiation with femtosecond pulses should dramatically increase the quantum

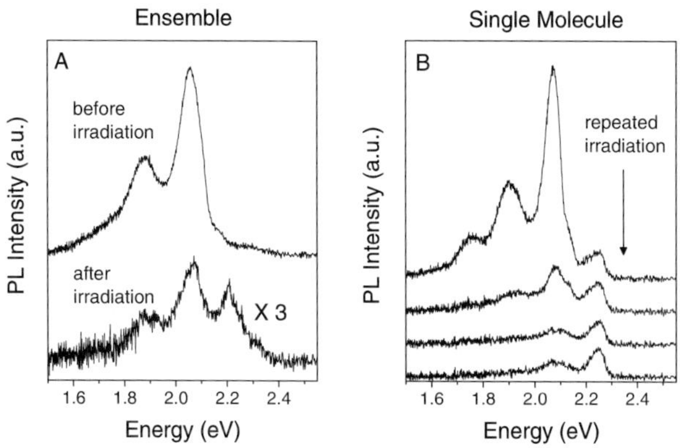

Figure 4. (A) Comparison of the ensemble spectra of 50 single MEH-PPV molecules dispersed in a Au-sealed 110 nm thick PMMA matrix at 20 K measured before and after femtosecond pulse irradiation. The femtosecond pulse was irradiated by scanning the beam over the whole area with an average intensity of 130 W cm^{-2} for an average irradiation time of 0.5 s for a molecule. The CW excitation intensity for spectrum measurement was 160 W cm^{-2}. (B) Irreversible spectral progression of a single MEH-PPV molecule measured between repeated irradiations of femtosecond pulses on the molecule with an intensity of ~10 W cm^{-2} for 3 s.

yield of photo-oxidation. Before irradiation, the emission spectra of our sample are observed to peak at low energy as expected from our previous studies, see Figure 4. In contrast, after irradiation, the ensemble spectra of Figure 4(a) exhibit a large blue shift and a concomitant drop in emission quantum yield. Thus, femtosecond irradiation produced the high energy emitting spectra. This photodamaging is clearly seen in a series of spectra of a single MEH-PPV molecule in Figure 4(b), where each spectrum was measured between repeated femtosecond pulse irradiation. This series of spectra clearly show progression of a photodamaging process appearing in emission spectra of single MEH-PPV molecules.

4. Conclusion

Recently developed time-resolved SMS techniques have proven to be powerful for studying the dynamics of singlet and triplet excitons in conjugated polymer nanoparticles. First, by using the excitation intensity modulation SMS, single MEH-PPV molecules at room temperature have been extensively investigated. The results of this study reveal large inhomogeneity in singlet-triplet interaction and the triplet decay dynamics on the microsecond time scale. In addition, the lifetimes of triplet excitons in single MEH-PPV molecules at room temperature have been also successfully measured without complication due to triplet quenching by hole polarons in a bulk film. Second, a new femtosecond pump-pump SMS study of conjugated polymers has shown preliminary results of the singlet exciton decay dynamics of a single MEH-PPV molecule. This new time-resolved technique will significantly contribute to an understanding of the ultrafast singlet dynamics in conjugated polymers, including S-S annihilation and energy and charge transfer processes. Finally, the comparison of SMS emission spectra before and after femtosecond pulse excitation has shown that femtosecond pulses cause more severe damage to conjugated polymers than CW laser excitation, probably due to higher chances of multiphoton ionization (or photo-oxidation) via the S-S annihilation of the conjugated polymer MEH-PPV.

References

1. A. Kohle, J. S. Wilson, and R. H. Friend, Adv. Mater. 14 (2002) 701.
2. M. A. Baldo, D. F. O'Brien, M. E. Thompson, and S. R. Forrest, Phys. Rev. B 60 (1999) 14422.
3. Y. Cao, I. D. Parker, G. Yu, C. Zhang, and A. J. Heeger, Nature 397 (1999) 414.
4. J. S. Wilson, A. S. Dhoot, A. Seeley, M. S. Khan, A. Kohler, and R. H. Friend, Nature 413 (2001) 828.
5. M. Wohlgenannt, X. M. Jiang, Z. V. Vardeny, and R. A. J. Janssen, Phys. Rev. Lett. 88 (2002)
6. Z. Shuai, A. Ye, D. Beljonne, R. J. Silbey, and J. L. Bredas, Synth. Met. 121 (2001) 1637.
7. P. K. H. Ho, J. S. Kim, J. H. Burroughes, H. Becker, S. F. Y. Li, T. M. Brown, F. Cacialli, and R. H. Friend, Nature 404 (2000) 481.
8. D. Beljonne, A. J. Ye, Z. Shuai, and J. L. Bredas, Adv. Funct. Mater. 14 (2004) 684.
9. F. Kulzer and M. Orrit, Ann. Rev. Phys. Chem. 55 (2004) 585.
10. D. A. Vanden Bout, W.-T. Yip, D. Hu, D.-K. Fu, T. M. Swager, and P. F. Barbara, Science (Washington, D. C.) 277 (1997) 1074.
11. J. Yu, D. Hu, and P. F. Barbara, Science (Washington, D. C.) 289 (2000) 1327.
12. D. H. Hu, J. Yu, K. Wong, B. Bagchi, P. J. Rossky, and P. F. Barbara, Nature 405 (2000) 1030.
13. D. H. Hu, J. Yu, and P. F. Barbara, J. Am. Chem. Soc. 121 (1999) 6936.
14. P. F. Barbara, A. J. Gesquiere, S.-J. Park, and Y. J. Lee, Acc. Chem. Res. 38 (2005) 602.
15. J. Yu, R. K. Lammi, A. J. Gesquiere, and P. F. Barbara, J. Phys. Chem. B 109 (2005) 10025.
16. A. J. Gesquiere, Y. J. Lee, J. Yu, and P. F. Barbara, J. Phys. Chem. B 109 (2005) 12366.
17. A. J. Gesquiere, S.-J. Park, and P. F. Barbara, J. Am. Chem. Soc. 127 (2005) 9556.
18. Z. Yu and P. F. Barbara, J. Phys. Chem. B 108 (2004) 11321.
19. C. Ronne, J. Tragardh, D. Hessman, and V. Sundstrom, Chem. Phys. Lett. 388 (2004) 40.
20. F. Schindler, J. M. Lupton, J. Feldmann, and U. Scherf, Proc. Natl. Acad. Sci. U. S. A. 101 (2004) 14695.
21. W. T. Yip, D. H. Hu, J. Yu, D. A. Vanden Bout, and P. F. Barbara, J. Phys. Chem. A 102 (1998) 7564.
22. J. Hofkens, M. Cotlet, T. Vosch, P. Tinnefeld, K. D. Weston, C. Ego, A. Grimsdale, K. Mullen, D. Beljonne, J. L. Bredas, S. Jordens, G. Schweitzer, M. Sauer, and F. De Schryver, Proc. Natl. Acad. Sci. U. S. A. 100 (2003) 13146.

23. S. S. Sartori, S. De Feyter, J. Hofkens, M. Van der Auweraer, F. De Schryver, K. Brunner, and J. W. Hofstraat, Macromolecules 36 (2003) 500.
24. J. D. White, J. H. Hsu, S. C. Yang, W. S. Fann, G. Y. Pern, and S. A. Chen, J. Chem. Phys. 114 (2001) 3848.
25. T. Huser, M. Yan, and L. J. Rothberg, Proc. Natl. Acad. Sci. U. S. A. 97 (2000) 11187.
26. Y. J. Lee, D. Y. Kim, J. K. Grey, and P. F. Barbara, ChemPhysChem 6 (2005) 2404.
27. S. Wachsmann-Hogiu, L. A. Peteanu, L. A. Liu, D. J. Yaron, and J. Wildeman, J. Phys. Chem. B 107 (2003) 5133.
28. P. Wang, C. M. Cuppoletti, and L. J. Rothberg, Synth. Met. 137 (2003) 1461.
29. P. K. H. Ho, J.-S. Kim, N. Tessler, and R. H. Friend, J. Chem. Phys. 115 (2001) 2709.
30. D. Y. Kim, J. K. Grey, and P. F. Barbara, Synth. Met. (2005) submitted.
31. J. Hofkens, M. Maus, T. Gensch, T. Vosch, M. Cotlet, F. Kohn, A. Herrmann, K. Mullen, and F. De Schryver, J. Am. Chem. Soc. 122 (2000) 9278.
32. P. Tinnefeld, J. Hofkens, D. P. Herten, S. Masuo, T. Vosch, M. Cotlet, S. Habuchi, K. Mullen, F. C. De Schryver, and M. Sauer, ChemPhysChem 5 (2004) 1786.
33. C. G. Hubner, G. Zumofen, A. Renn, A. Herrmann, K. Mullen, and T. Basche, Phys. Rev. Lett. 91 (2003) 093903.
34. F. Schindler, J. M. Lupton, J. Feldmann, and U. Scherf, Adv. Mater. (Weinheim, Germany) 16 (2004) 653.
35. H. D. Burrows, J. S. de Melo, C. Serpa, L. G. Arnaut, M. D. Miguel, A. P. Monkman, I. Hamblett, and S. Navaratnam, Chem. Phys. 285 (2002) 3.
36. S. Sinha and A. P. Monkman, Appl. Phys. Lett. 82 (2003) 4651.
37. A. S. Dhoot, D. S. Ginger, D. Beljonne, Z. Shuai, and N. C. Greenham, Chem. Phys. Lett. 360 (2002) 195.
38. L. P. Candeias, G. Padmanaban, and S. Ramakrishnan, Chem. Phys. Lett. 349 (2001) 394.
39. J. S. Wilson, N. Chawdhury, M. R. A. Al-Mandhary, M. Younus, M. S. Khan, P. R. Raithby, A. Kohler, and R. H. Friend, J. Am. Chem. Soc. 123 (2001) 9412.
40. D. S. Ginger and N. C. Greenham, Phys. Rev. B 59 (1999) 10622.
41. J. Partee, E. L. Frankevich, B. Uhlhorn, J. Shinar, Y. Ding, and T. J. Barton, Phys. Rev. Lett. 82 (1999) 3673.
42. V. Dyakonov, G. Rosler, M. Schwoerer, and E. L. Frankevich, Synth. Met. 84 (1997) 483.
43. T.-Q. Nguyen, I. B. Martini, J. Liu, and B. J. Schwartz, J. Phys. Chem. B 104 (2000) 237.
44. C. J. Collison, V. Treemaneekarn, W. J. Oldham, J. H. Hsu, and L. J. Rothberg, Synth. Met. 119 (2001) 515.
45. P. Wood, I. D. W. Samuel, G. R. Webster, and P. L. Burn, Synth. Met. 119 (2001) 571.
46. A. J. Gesquiere, S.-J. Park, and P. F. Barbara, J. Phys. Chem. B 108 (2004) 10301.
47. S.-J. Park, A. J. Gesquiere, J. Yu, and P. F. Barbara, J. Am. Chem. Soc. 126 (2004) 4116.
48. J. D. McNeill, D. B. O'Connor, D. M. Adams, P. F. Barbara, and S. B. Kammer, J. Phys. Chem. B 105 (2001) 76.
49. J. D. McNeill, D. Y. Kim, Z. Yu, D. B. O'Connor, and P. F. Barbara, J. Phys. Chem. B 108 (2004) 11368.
50. Y. J. Lee, S.-J. Park, A. J. Gesquiere, and P. F. Barbara, Appl. Phys. Lett. 87 (2005) 051906.
51. E. van Dijk, J. Hernando, J. J. Garcia-Lopez, M. Crego-Calama, D. N. Reinhoudt, L. Kuipers, M. F. Garcia-Parajo, and N. F. van Hulst, Phys. Rev. Lett. 94 (2005) 078302.
52. J. B. Guild, C. Xu, and W. W. Webb, Appl. Opt. 36 (1997) 397.

53. L. Smilowitz, A. Hays, A. J. Heeger, G. Wang, and J. E. Bowers, J. Chem. Phys. 98 (1993) 6504.

Femtochemistry VII
Fundamental Ultrafast Processes in Chemistry, Physics, and Biology
A.W. Castleman, Jr. and M.L. Kimble

Influence of different anchor-bridge groups on electron transfer from the molecular chromophore perylene to the wide gap semiconductor TiO_2

R. Ernstorfer, L. Gundlach, S. Felber, R. Eichberger, F. Willig

Hahn-Meitner-Institute, Department: Dynamics of Interfacial Reactions
Glienickerstrasse 100, 14109 Berlin, Germany, willig@hmi.de

1. Introduction

Transport of electrons through molecular units is an important topic in the field of molecular electronics [1] and future device modeling. Most of the research [2] has focused on saturated aliphatic chains or on ring structures with saturated C-C bonds, and only few theoretical papers and experiments have dealt also with conjugated molecular bridges. All the theoretical work has addressed molecular bridges in ultra-high-vacuum (UHV), whereas all the previous measurements have been performed in a solvent environment. We report here on transient absorption experiments in UHV that probed heterogeneous electron transfer from the molecular chromophore perylene via different molecular anchor -bridge groups into nm-structured anatase TiO_2 layers.

2. Experimental

Transient absorption measurements were carried out in UHV with laser pulses of a few nJ resulting in cross correlation functions shorter than 30 fs (FWHM). Pump (central wavelength 440 nm) and probe (570 nm) pulses were generated with novel NOPAs [3] designed for repetition rates above 100 kHz and extremely low input power. The pump pulse wavelength was obtained by frequency doubling the NOPA output operated in the NIR at 880 nm.

Bulky side groups attached to the perylene chromophore (2,5-bi-tertiary-butyl-perylene-9-yl, DTB-Pe) prevented the formation of dimers by neighboring chromphores. Perylene was linked covalently to a carboxylic (-COOH) and alternatively to a phosphonic (-P(O)(OH) $_2$) group acting as anchor unit that binds to the TiO_2 surface. In addition, two different bridge groups, i.e. -C=C- or -C-C-, were inserted between the chromophore and the carboxylic anchor group. The surface area enhancement due to nm-structuring of the anatase TiO_2 film ((101) dominant surface plane) allowed for a good signal to noise ratio with 3 μ thick films. The relative energetic level alignment at the interface was determined from ground state UPS data and two-photon-photoemission data (not shown here).

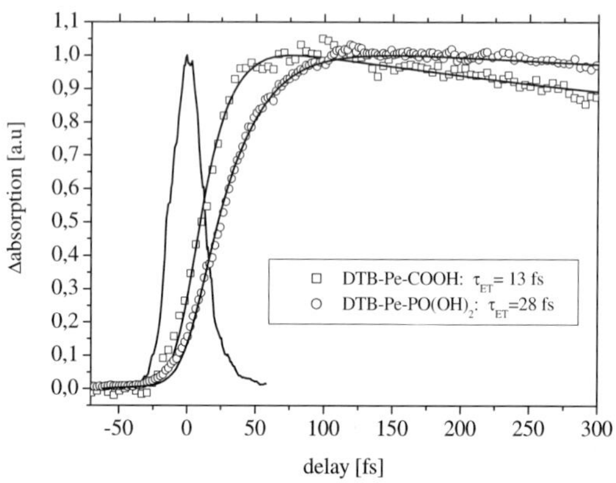

Fig.1. Rise of the absorption of the ionized chromophore, i.e. electron transfer product state, for the two different anchor groups indicated in the inset along with the time constants for the rise.

3. Results and discussion

Electron injection from the first excited perylene singlet state across the different anchor-bridge groups into TiO_2 was measured as the rise of the cation absorption band centered at 570 nm. The transient absorption measurements revealed a characteristic difference in the rise time and also in the initial decay behavior for the two different anchor groups when no additional bridge unit was attached (Fig. 1). The respective fit gave a time constant of 13 fs for the carboxylic and only 28 fs for the phosphonic anchor group. The faster rise for the carboxylic anchor group was accompanied also by a faster initial decay, i.e. disappearance of the cation state via recombination with an electron in TiO_2 to form the neutral ground state of perylene. Although rise and early decay times differ considerably for the two different anchor groups, electron injection and recombination follow a similar pattern for both species where a faster injection time is linked to a faster recombination time. Thus, the effective electronic tunneling barrier for electron transfer via the phosphonic anchor group is larger than for the carboxylic anchor group. The electron affinities of both the isolated acids are repulsive with values around 1eV. A quantum chemical calculation for the two anchor groups suggested a similar reaction distance from the edge (C-atom) of the perylene chromophore to the nearest surface Ti atoms.

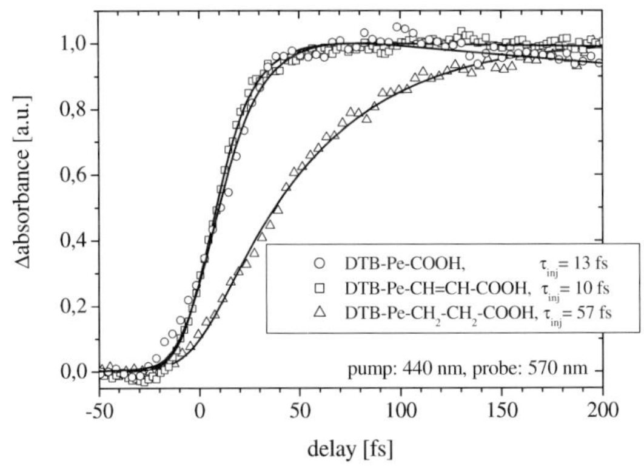

Fig. 2. Rise of the absorption of the product state of heterogeneous electron transfer, i.e. the cation of perylene, for DTB-perylene attached to TiO_2 via different bridge-anchor units, -CH=CH-COOH, -CH-CH-COOH, and –COOH only.

Different electron transfer dynamics (Fig. 2) were found when the two different bridge groups were inserted between the chromophore and the carboxylic anchor group. The electron transfer time was 10 fs with the -C=C- bridge unit and only 57 fs with the -C-C- bridge unit. The spatial separation of the perylene ring structure from the TiO_2 surface, however, was comparable for the conjugated and the saturated bridge unit. The fast injection time can be attributed to the extension of the perylene donor orbital onto the conjugated bridge unit as was suggested by a semi-empirical calculation for the free molecule (Fig. 3). The perylene S_0-S_1 transition has strong HOMO/LUMO character [4] so that the LUMO wavefunction can be considered a sufficient approximation to the excited state wavefunction. The molecular donor orbital of the excited singlet state of perylene was found about 0.7 eV above the lower edge of the conduction band of TiO_2, sufficiently high to fulfill the "wide band limit" [5] for electron transfer. In the latter case, all the individual Franck-Condon factors are available in parallel for the reaction, and the measured rate constants can be directly related to the electronic coupling strength. For the carboxylic anchor group alone, the measured rise time corresponded to an electronic coupling strength in the range of 40 meV, with the same value also applying to the conjugated -C=C- bridge unit, whereas the electronic coupling was only 7 meV via the saturated -C-C- bridge unit.

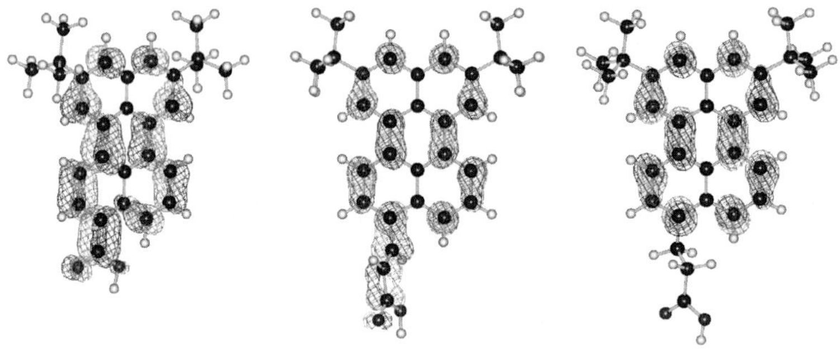

Fig. 3. LUMOs of DTB-Perylene with three different bridge-anchor groups: -COOH (left), -CH=CH-COOH (center), and -CH-CH-COOH (right).

4. Summary

The strength of electronic coupling between a molecular chromophore and a semiconductor can be varied with different molecular bridge units inserted between the chromophore and anchor group. The measured electron transfer times showed conducting or insulating behavior of the respective bridge unit, -C=C- and -C-C-. Heterogeneous electron transfer times were measured with different bridge-anchor groups in ultra-high-vacuum where the wide band limit is realized for the perylene/ TiO_2 system. These experimental data should allow for a more direct comparison with theoretical calculations since they are neither complicated by the presence of a solvent environment nor by specific Franck-Condon factors.

Acknowledgement: We are grateful for support by the German Science Foundation (SFB450).

References

1. Molecular Electronics, J. Jortner and M.A. Ratner (Eds), Blackwell Science, Oxford 1997.
2. R.E. Tanner, Y. Liang, and E.I. Altman, Surface Science 506 (2002) 251 and references in this paper.
3. L. Gundlach, R. Ernstorfer, E. Riedle, R. Eichberger, and F. Willig, Appl. Phys. B, 80 (2005) 727.
4. T. M. Halasinski, J. L. Weisman, R. Ruiterkamp, T. J. Lee, F. Salama, and M. Head-Gordon, J. Phys Chem., A 107 (2003) 3660.
5. S. Ramakrishna, F. Willig, V. May, and A. Knorr, J. Phys. Chem., B 107 (2003) 607; S. Ramakrishna, F. Willig, and V. May, Phys. Rev., B 62 (2000) R16330.

Femtochemistry VII
Fundamental Ultrafast Processes in Chemistry, Physics, and Biology
A.W. Castleman, Jr. and M.L. Kimble

Electron transfer from molecular chromophores to semiconductors probed with two-photon-photoemission

Lars Gundlach, Ralph Ernstorfer, Silke Felber, Rainer Eichberger, and Frank Willig

Hahn-Meitner-Institut Berlin, Dept. SE4, Glienickerstr. 100, D-14109 Berlin, Germany

1. Introduction

Heterogeneous electron transfer from the excited state of a molecular chromophore to empty electronic states of a solid is an important process in a variety of fields ranging from photo-chemistry [1] to molecular electronics [2]. In the last years, numerous experimental and theoretical studies have addressed the dynamics of such processes. Light-induced heterogeneous electron transfer has been extensively investigated mainly with transient absorption in polycrystalline or nano-crystalline samples with anchored chromophores [3]. Transient two-photon photoemission (TR-2PPE) is capable of detecting the signals of sub-monolayer chromophore coverage on single crystal surfaces and gives information about the electron transfer dynamics, the position of the electronic levels at the surface, and the adsorption geometry of the molecules on the surface. Comparing the different 2PPE transients and the corresponding transient absorption measurements for identical systems revealed the origin of unexpected short time constants in the transient absorption measurements.

2. Experimental

The femtosecond laser system was based on a commercial amplified 150 kHz Ti:sapphire oscillator operating at 800 nm which pumped two NOPAs. The NOPA principle, commonly used for 1 kHz systems, was adapted to the high repetition rate by our group in cooperation with the group of E. Riedle. Tunable sub-20 fs pulses in the visible and 15 fs pulses from second harmonic generation (SHG) of a NIR output pulse were generated with the first NOPA. The visible output of a second NOPA was frequency doubled in a 75 μm thick BBO crystal to obtain a spectrally broad UV pulse centered at 280 nm. Due to the limited output power of the laser system, both NOPAs were operated with the same pump pulse [4].

The measurements shown here were carried out with 440 nm pump light, addressing the S_0-S_1 transition of perylene, and 280 nm probe light for the photoionization step. The cross correlation of both pulses was 37 fs.
TiO_2(110) surfaces were cleaned in four consecutive cycles of Ar^+ sputtering (600 eV, 5 μA) and subsequent annealing at 600°C in UHV. The perylene compound was adsorbed from a toluene solution in a specially designed UHV chamber allowing for switching between UHV and inert gas conditions in a well defined manner [7].
The measurement chamber (base pressure $2*10^{-10}$ mbar) was equipped with a time of flight (TOF) spectrometer with an angular acceptance of ±3.5°. Measurements were carried out in normal emission with an angle of incidence for the light pulses of 45° with respect to the sample surface. All measurements were carried out at room temperature with the probe pulse p-polarized.

3. Results and Discussion

The knowledge of the orientation of the injecting molecule with respect to the surface is essential for studying the electron transfer process. The S_0-S_1 transition dipole moment of perylene is oriented along the long molecular axis [5].

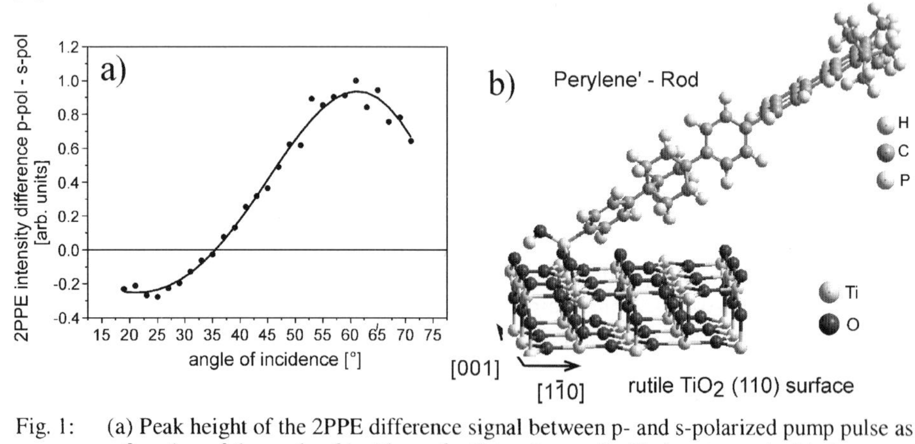

Fig. 1: (a) Peak height of the 2PPE difference signal between p- and s-polarized pump pulse as a function of the angle of incidence for Pe'-rod on rutile (dot), measured with the plane of incidence parallel to [001]. Plot of the fit wit Fresnel's equations (solid). (b) Balls and sticks model of the adsorption geometry of Pe'-rod on TiO_2 rutile as deduced from angular resolved measurements.

Thus, one can measure the orientation of the perylene moiety with respect to the TiO_2 surface by varying the polarization and the angle of incidence of the pump pulse, provided the molecules form an ordered layer on the surface.

Fig. 1(a) shows the peak height of the 2PPE difference signal at t_0 between p- and s-polarized pump pulses as a function of the angle of incidence (dots); the probe pulse was kept p-polarized. The peak heights were deduced from Lorentzian fits of the difference spectra. Fig. 1(a) shows a measurement with the plane of incidence parallel to [001], i.e. parallel to the oxygen rows on the surface. The measurement was fitted by means of Fresnel's equations, neglecting any anisotropy in the emission angel of the electrons.

For the perylene-rod molecule anchored via a phosphonic acid group on TiO_2, the measurement revealed that the long molecular axis is tilted against [110] with an angle of around 58° in the direction perpendicular to the oxygen rows (Fig. 1(b)). This configuration is compatible with a bi- as well as a monodentade binding geometry with one or two hydrogen bonds, respectively, and agrees well with calculations of Nilsing *et al.* [6].

As mentioned above, light-induced heterogeneous electron transfer has been extensively investigated mainly with transient absorption in nano-crystalline samples. For comparison, the same Pe'-rod chromophore was measured adsorbed on a sponge-like TiO_2 anatase colloidal film as well as adsorbed on the (110) surface of rutile single crystals.

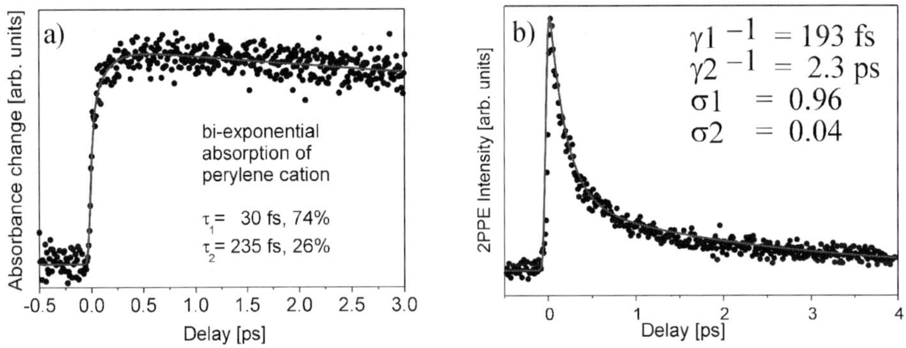

Fig. 2: (a) Transient absorption signal of the Pe'-rod cation formation when adsorbed on an anatase colloidal film (dots) and the fit with a rate model (solid).
(b) Time trace measured via TR-2PPE in an energy window of 400 meV kinetic energy at the excited state position of Pe'-rod (dots) and fit with the rate model described in Ref. [7] (solid).

Fig. 2(a) shows the transient absorption measurement of Pe'-rod adsorbed onto an anatase TiO_2 colloidal film (dots) and a fit with a rate model representing two independent species (line). Fit parameters were: injection times $\tau_1 = 30$ fs and $\tau_2 = 235$ fs, amplitudes $A_1 = 74$ % and $A_2 = 26$ %, and decay times $\tau_{d1} = \tau_{d2} = 15$ ps.

A short injection time constant is dominating the transient signal. The second time constant of 235 fs is reasonable compared with other measurements of our group [3].

To elucidate the origin of the short time constant in the nano-structured TiO_2 film, TR-2PPE measurements were performed. The TR-2PPE measurement of Pe'-rod on a rutile single crystal and a fit with a rate model with two rates is shown in Fig. 2(b). The time constants are τ_1 = 193 fs, τ_2 = 2.3 ps, with weight factors: σ_1 = 0.96, σ_2 = 0.04. The second time constant represents the long-lived background signal ascribed to electrons injected from the chromophore into the conduction band of TiO_2. The short time constant measured via TR-2PPE agrees well with the long time constant of the transient absorption measurement, whereas no time constant of the order of some ten fs is observed.

The short time constant is not present in TR-2PPE measurements on single crystal surfaces. Therefore, it is ascribed to conformational inhomogeneities in the colloidal film, namely to molecules bound in the cavities of the colloidal film such that the chromophore is much closer to an adjacent wall than to the floor where it is fixed with the anchor group. This effect gains importance when the spacer group gets longer. First, for a bigger molecule it is more likely to touch a wall. The molecule discussed here is around 25 Angstrom in its long axis and thus only a factor 4 smaller than the mean diameter of the cavities. Secondly, for molecules with shorter spacers the regular time constant for charge transfer in the direction of the bridge becomes comparable or shorter than the time constant for the shortcut to an adjacent wall. Thus, electron transfer in the direction of the anchor group is the fastest process.

It has been shown that TR-2PPE is a suitable tool to investigate heterogeneous ET from photoexcited chromophores to metal and semiconductor substrates. The sensitivity is sufficient to investigate sub-monolayer coverages on single crystal surfaces. This is important when large molecules are to be investigated, in particular where transient absorption spectroscopy yields erroneous injection times. Furthermore, the binding geometry of the molecules with respect to the surface has been investigated here on single crystal surfaces.

References

1. J. W. Gadzuk, Phys. Rev. Lett.76 (1996) 4234.
2. M. Ratner and J. Jortner, Molecular Electronics, Blackwells, Oxford (1997).
3. C. Zimmermann, F. Willig, S. Ramakrishna, B. Burfeindt, B. Pettinger, R. Eichberger, and W. Storck, Phys. Chem. B 105 (2001) 9245.
4. L. Gundlach, R. Ernstorfer, E. Riedle, R. Eichberger, and F. Willig, Appl. Phys. B 80 (2005) 27.
5. T. M. Halasinski, J. L. Weisman, R. Ruiterkamp, T. J. Lee, F. Salama, and M. Head-Gordon, J. Phys. Chem. A 107 (2003) 3660.

6. M. Nilsing, S. Lunell, P. Persson, and L. Ojamäe, Surf. Sci. 582 (2005) 49.
7. L. Gundlach, Ph.D. thesis, Freien Universität Berlin, (2005), online: http://www.diss.fu-berlin.de/2005/269/.

Femtochemistry VII
Fundamental Ultrafast Processes in Chemistry, Physics, and Biology
A.W. Castleman, Jr. and M.L. Kimble

Interfacial Electron Transfer or Concentration Quenching?

Yu-Xiang Weng. Xin-Hai Chen, Lu-Chao Du

Institute of Physics, Chinese Academy of Sciences, Beijing 100080, China. Email: yxweng@aphy.iphy.ac.cn

Abstract

Excited-state lifetime of bacteria chlorophyll *a*(BChl *a*) assembled onto Au, Ag, and SiO_2 nanoparticles have been studied by fs time-resovled spectroscopy. It has been found that the excited-state lifetime of BChl *a* assembled onto nanoparticales was substantially shortened with respect to that of the free BChl *a* in solution. Within the experimental error, the BChl/Au, BChl/Ag, and BChl/SiO_2 have the same lifetime. We propose that the dominant process for the observed change in the lifetime of BChl *a* when assembled onto nanoparticles arises from the concentration quenching rather than the interfacial electron transfer.

1. Introduction

Organized molecular assemblies are of much interest because of the unique properties not seen in a single molecule. Photosynthesis is among the most typical examples of sophisticated molecular assemblies in nature. The central issue in photosynthesis is the process of light-induced charge separation whereby light energy is converted into chemical energy by green plants. The principal pigment in this transduction of energy is chlorophyll. The essential roles of chlorophyll are to capture solar energy, transfer the excitation energy to special locations, i.e., the reaction centers, and bring about the charge separation for the subsequent electron-transfer processes [1-5]. Of particular interest has been the study of donor-acceptor systems containing chlorophyll that can mimic the photoinduced electron-transfer process of natural photosynthesis.

The ability of chlorophyll to donate electrons from its singlet and triplet states to different quinones has been clearly demonstrated by various authors using fluorescence, laser photolysis, and electron spin resonance techniques [6]. Recently, gold nanoparticles have been drawing a lot of attention because of their unique optical and electronic properties [6-8]. Electrochemical studies have demonstrated the electron-storing properties of gold nanoparticles and their ability to act as an electric relay on a nanotemplate structure. Several examples of excited-state quenching by metal nanoparticles have also been reported [9-12]. Murray and co-workers have reported quenching of fluorescein and dansyl fluorophores attached to monolayerprotected gold clusters [13, 14]. The fluorescence of lissamine dye molecules attached to gold nanoparticles is strongly quenched [12] and so is the fluorescence of conjugated polymers in the presence of gold nanoparticles [15]. Fluorescence quenching in fluorophores bound to gold nanoparticles, as a result of electron transfer from excited fluorophore to gold nanoparticles, has been recently reported by Kamat *et al.* [16, 17].

In the present work we designed the BChl *a* and Au nanopartical system to mimic the natural photosynthesis (Fig.1). Effort has been made to probe the excited state interaction between BChl *a* and gold nanoparticles. The objective of the study is to see if photoinduced electron transfer or energy transfer occurred between BChl *a* and gold nanoparticles. For comparison, we also assembled BChl *a* on Ag and the photoinactived SiO_2 nanoparticles. The excited state of BChl *a*/Ag and BChl *a*/SiO_2 have also been studied.

2. Experimental Section

Sample preparation. Au colloidal nanoparticles were prepared by reduction of Au(III) in $AuCl_4^-$ which was accomplished by adding 0.5% sodium citrate in H_2O, then produced an immediate pale-purple-colored solution after heating the mixture. Ag colloidal nanoparticles were prepared by the reduction of silver ions in aqueous solution with freshly prepared $NaBH_4$ solution and PVA in ice-cooled condition which produced a pale-yellow-colored solution. SiO_2 nanoparticles were purchased from Alfa Aesar. BChl *a* was purchased from J&K Chemical. BChl *a* in CH_3OH was mixed with Au (size: ca.10 nm, c=3.7 mg/L), Ag (size: less than 60 nm, c=5 mg/L), or SiO_2 (size: 6 nm,c=2 mg/L) nanoparticle solution for over 24h for self-assembling with a concentration for BChl *a* of 1.9×10^{-5}M.

Measurement. The femtosecond time-resolved difference absorbance spectrometer has been described elsewhere [18]. Briefly it employs a 1 kHz repetition regenerative amplifier Ti:sapphire (Hurricane, Spectra Physics) as the primary beam source. The output power of the laser is 750 mJ, with a pulse duration of 150 fs at 800 nm. The probe white light continuum was generated by focusing 800 nm beam with a power of 6 μJ/pulse onto a 2.4 mm thick sapphire plate. A dual beam configuration (probe and reference) was adopted for the detection. The probe and reference beams were collected by a pair of objective lenses coupled to a two-branched optical fiber. The delay between pump and probe beams was realized by a computer controlled translation stage. The polarization of 800 nm pump beam was set to the magic angle (54.7^o) in respect to the probe beam.

3. RESULTS AND DISCUSSION

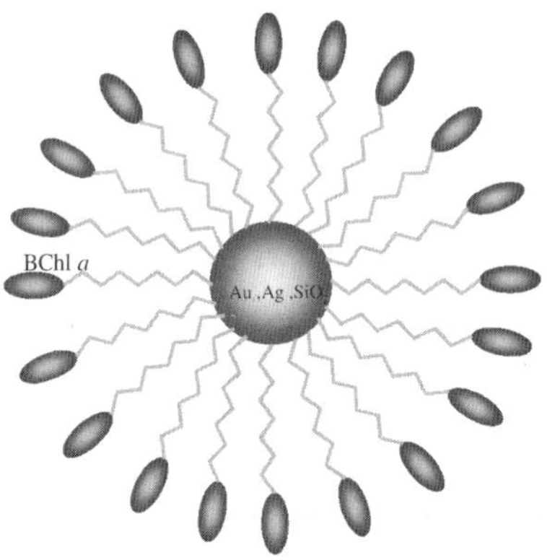

Fig. 1. Schematic diagram for aessembling of BChl *a* onto the nanoparticle. The employed nanoparticles were Au, Ag, and SiO2 in aqueous solution.

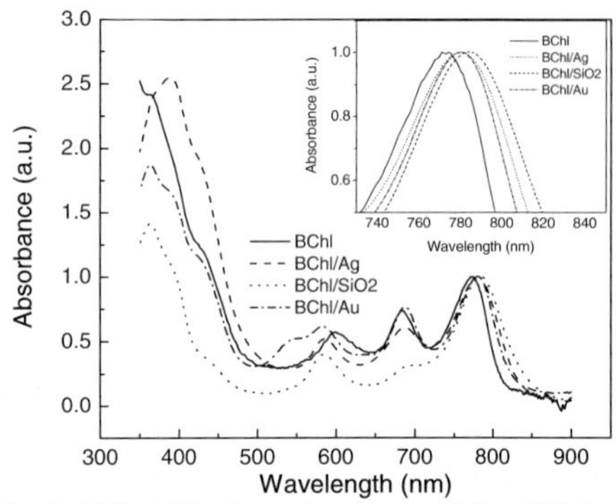

Fig. 2. UV-visible absorption spectra of free BChl *a* and BChl/Au,Ag,SiO_2 normalized at 780nm with that of Bchl *a*.

Fig. 2. shows the UV-visible absorption spectra of BChl *a* assembled onto the corresponding nanoparticles. The graphic inset shows the Q_y region of free BChl *a*, BChl *a*/Au, BChl *a*/Ag, and BChl *a*/SiO_2. It reveals that, when assembled onto Au, Ag, and SiO_2 nanoparticles, the absorption maximum of the Q_y band at 780nm exhibits red shift of 6 nm for Au, 7 nm for Ag, and 10 nm for SiO_2. From Fig.1 we can see that after assembled onto nanoparticles, BChl *a* formed an aggregate-like system on the surface of the nanoparticle. The distance between BChl can be greatly reduced than that of free BChl *a* in solution, where the red shift of the Q_y absorption band reveals the typical symbol of *J*-aggregates [19]. Therefore we conclude that the BChl *a* molecules form *J*-aggregations on Au, Ag, and SiO_2 nanoparticles.

Fig. 3 presents the Q_y band bleaching recovery kinetics recorded at 780nm of free BChl *a* and the corresponding BChl *a*/nanoparticles acquired by femtosecond time-resolved absorbance-difference spectrometer after laser excitation at 800 nm with a pump power of 0.1μ J/pulse. The bleaching recovery kinetic curves can be fitted by a two-component exponential decay with time constants of τ_1 =14.6ps and τ_2= 534 ps for free BChl *a*, τ_1 =4.7ps and

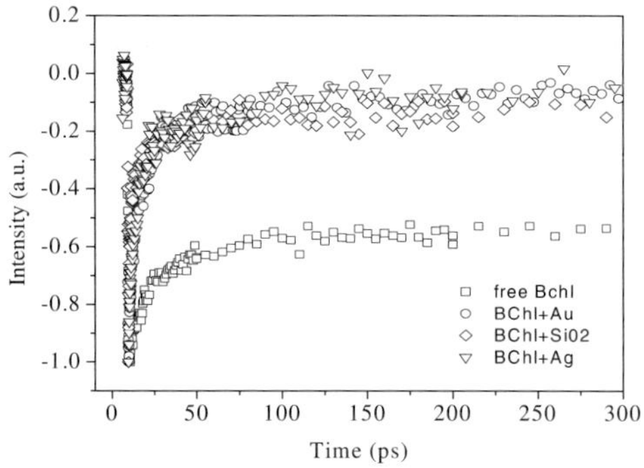

Fig. 3 B780 bleaching recovery kinetic curves of free BChl *a* and the corresponding BChl *a*/nanoparticles acquired by femtosecond time-resolved absorbance-difference spectrometer

τ_2=55.8ps for BChl *a*/Au, τ_1 = 4.3ps and τ_2= 47.8ps for BChl a/Ag, and τ_1 =3.1ps and τ_2= 60.2ps for BChl *a*/SiO_2, respectively. The B780 bleaching recovery kinetic curves reveal that the excited-state lifetime of BChl *a* assembled onto Au, Ag, and SiO_2 nanoparticles are substantially shortened compared to that of the free BChl *a* in solution. From approximately the same decay time constants of BChl *a*/Au, BChl *a*/Ag, and BChl *a*/SiO_2, the conclusion can be drawn that it is the same process that causes the excited-state quenching of BChl *a* when assembled onto nanoparticles.

In general, there are several possible mechanisms which can lead to the quenching of the BChl *a* excited-state, including electron and energy transfer. Gold nanoparticles have the properties of electron-storing and the ability to act as an electric relay on a nanotemplate structure. Further, photochemical and spectroelectrochemical experiments have shown that gold nanoparticles capped with organic molecules exhibit unusual redox activity by readily accepting electrons from a suitable donor. Barazzouk and co-workers have reported electron transfer between Chlorophyll *a* and gold nanoparticles [6]. A number of authors have reported energy transfer to gold nanoparticles [20-23]. However in the present case, the SiO_2 nanoparticle is photoinactive, so the electron and energy transfer between BChl *a* and nanoparticles is not the process that causes the decrease in the BChl *a* excited-state lifetime.

From the illustration in Fig.1 and the red shift of the Q_y band of BChl *a* when assembled onto nanoparticles, it can be inferred that BChl *a* may form J-aggregation on the nanoparticles. With regard to the fact that Linnanto and co-workers have reported exciton interaction caused exciton lifetime shortening of

BChl *a* with J-aggregation [19], we proposed that the dominant process for the observed decrease in the lifetime of BChl *a* when assembled onto nanoparticle is arising from the interaction among the BChl *a* molecules within the aggregation on the nanoparticles, that is, the concentration quenching.

References

1. L. P. Vernon and G. R. Seely (Eds.) The Chlorophylls, Academic Press, New York, 1966.
2. H. Scheer (Ed.) Chlorophylls, CRC Press, Boca Raton, FL, 1995.
3. A. B. J. Parusel and S. Grimme, J. Phys. Chem. 104 (2000) 5395.
4. (a) J. K. Hurley and G. Tollin, Sol. Energy 28 (1982) 187; (b) G. Tollin, J. Phys. Chem. 80 (1976) 2274.
5. M. A. M. J. Van Zandvoort, D. Wrobel, P. Lettinga, G. Van Ginkel, and Y. K. Levine, Photochem. Photobiol. 62 (1995) 299.
6. S. Barazzouk, P. V. Kamat, and S. Hotchandani, J. Phys. Chem. B 109 (2005) 716.
7. H. Imahori, H. Norieda, Y. Nishimura, et al., J. Phys. Chem. B 104 (2000) 1253.
8. T. Hasobe, H. Imahor, P. V. Kamat, et al., J. Am. Chem. Soc. 127 (2005) 1216.
9. P. Avouris and B. N. J. Persson, J. Phys. Chem. 88 (1984) 837.
10. R. Rossetti and L. E. Brus, J. Phys. Chem. 76 (1982) 1146.
11. K. H. Drexhage, M. Flock, H. Kuhn, F. R. Shafer, and W. Spurlig, Ber. Bunsen-Ges. Phys. Chem. 73 (1969) 1179.
12. E. Dulkeith, A. C. Morteani, T. Niedereichcholz, T. A. Klar, J. Feldmann, S. A. Levi, F. C. J. M. van Veggel, D. N. Reinhoudt, M. Mo¨ller, and D. I. Gittins, Phys. Rev. Lett. 89(2002) 203002.
13. A. C. Templeton, D. E. Cliffel, and R. W. Murray, J. Am. Chem. Soc. 121 (1999) 7081.
14. A. Aguila and R. W. Murray, Langmuir 16 (2000) 5949.
15. C. Fan, S. Wang, J. W. Hong, G. C. Bazan, K. W. Plaxco, and A. J. Heeger, Proc. Nat. Acad. Sci. 100 (2003) 6297.
16. B. I. Ipe, K. George Thomas, S. Barazzouk, S. Hotchandani, and P. V. Kamat, J. Phys. Chem. B 106 (2002) 18.
17. P. V. Kamat, S. Barazzouk, and S. Hotchandani, Angew. Chem., Int. Ed. 41 (2002) 2764.
18. L. Zhang, J. Yang, L. Wang, G.-Z. Yang, and Y.-X. Weng, J. Phys. Chem. B 107 (2003) 13688.
19. J. Linnanto, J. A. I. Oksanen, and J. E. I. Korppi-Tommola, Phy. Chem. Chem. Phy. 4 (2002) 3061.
20. H. Imahori, M. Arimura, T. Hanada, Y. Nishimura, I. Yamazaki, Y. Sakata, and S. Fukuzumi, J. Am. Chem. Soc. 123 (2001) 335.
21. H. Imahori, Y. Kashiwagi, Y. Endo, T. Hanada, Y. Nishimura, I. Yamazaki, Y. Araki, O. Ito, and S. Fukuzumi, Langmuir 20 (2004) 73.
22. H. Imahori, Y. Kashiwagi, T. Hanada, Y. Endo, Y. Nishimura, I. Yamazaki, and S. Fukuzumi, J. Mater. Chem. 13 (2003) 2890.
23. S. Fukuzumi, Y. Endo, Y. Kashiwagi, Y. Araki, O. Ito, and H. Imahori, J. Phys. Chem. B 107 (2003) 11979.

6. Protein Dynamics and Photobiology

Femtochemistry VII
Fundamental Ultrafast Processes in Chemistry, Physics, and Biology
A.W. Castleman, Jr. and M.L. Kimble

Femtosecond Spectroscopy for the Study of Initial Reactions in Protein folding

Wolfgang J. Schreier[a], Tobias E. Schrader[a], Florian O. Koller[a], Arne Sieg[a], Christian Renner[b], Luis Moroder[b], Josef Wachtveitl[c], Wolfgang Zinth[a]

[a]*Lehrstuhl für BioMolekulare Optik, Ludwig-Maximilians-Universität München, Oettingenstraße 67, 80538 München, Germany*
[b]*Max-Planck-Institut für Biochemie, AG Bioorganische Chemie, MPI für Biochemie, Am Klopferspitz 18a, D-82152 Martinsried bei München, Germany*
[c] *Institut für Physikalische und Theoretische Chemie der Universität Frankfurt, Marie-Curie-Str. 11, 60439 Frankfurt am Main, Germany*

1. Introduction

In living organisms the genetic information is contained in the DNA and RNA. Here the sequence of the individual base pairs encodes the construction of proteins and the regulation of cellular reaction networks. Proteins perform important functions of life such as transcription of information, gene expression, chemical synthesis, energy conversion, and storage and cellular motion [1]. In spite of their multiple tasks, proteins are built according to a unique and simple principle: they consist of a linear chain of amino acids. Nature uses a set of 20 α-amino acids, differing only in their specific side-groups. These side-groups have various chemical properties: hydrophobic to hydrophilic, low to high pK_a, inducing local flexibility or local rigidity. The sequence of the amino acids in the protein determines the structural arrangement of the different groups and thus leads to the overall three-dimensional structure of its native form and to the specific function of the protein. During biosynthesis of a protein, the information genetically encoded in the m-RNA is translated codon by codon to the defined sequence of individual amino acids. Upon the release of the polypeptide chain from the ribosomes its rearrangement starts. Here rotations around individual bonds of the peptide backbone lead to the transition into the final structure. These processes are driven by the interaction between different amino acids and between amino acids and the surrounding solvent. Foldases such as chaperons, disulfide isomerases, and proline cis-trans isomerases catalyze the folding processes without affecting the final overall structure.

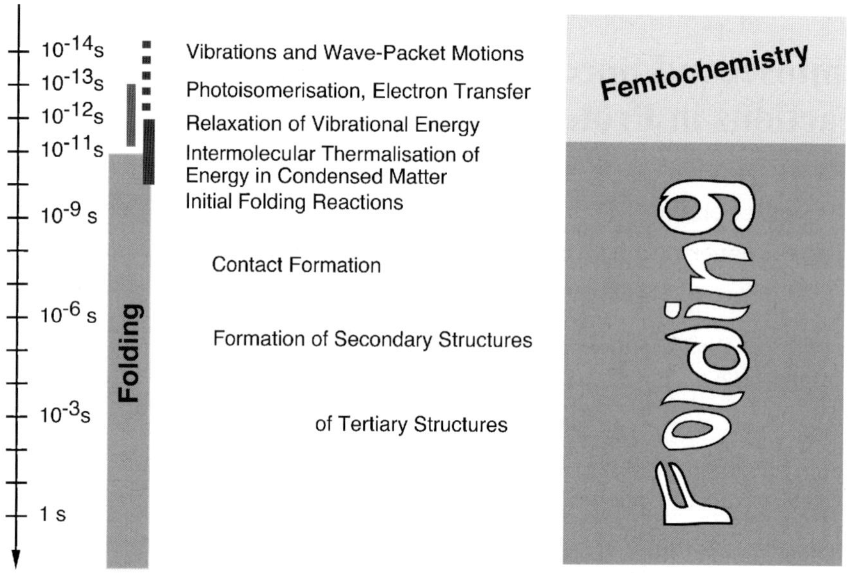

Fig. 1: Time scales of femtochemistry and protein folding.

The folding process occurs on the timescale of 1s. For small proteins it may be completed on the microsecond timescale, whereas several seconds may be required to assemble complex and large proteins into their final form. At present a detailed understanding of folding is missing, even for simple and small proteins. Protein folding is one of the most important - still unresolved – processes in biology. For this reason many scientists intend to gain insight into this important reaction. The understanding of protein folding is closely related to the intra- and intermolecular interactions of the amino acid-chain. Here dynamics occur which cover more than 12 orders of magnitude in time (see Fig. 1). In the time-range typical for femtochemistry (10^{-13} seconds), transitions between different conformations (e.g. isomerizational processes) take place which are driven by large energy quanta. Local relaxation of energy (vibrational energy redistribution) and local heat transfer occur on the timescale of a few picoseconds. Contact formation of different parts of an amino acid chain may be as fast as a few nanoseconds. The formation (and transient destruction) of secondary structure elements should occur on the timescale of nanoseconds to microseconds. The build-up of tertiary structures may span the range of microseconds to many seconds.

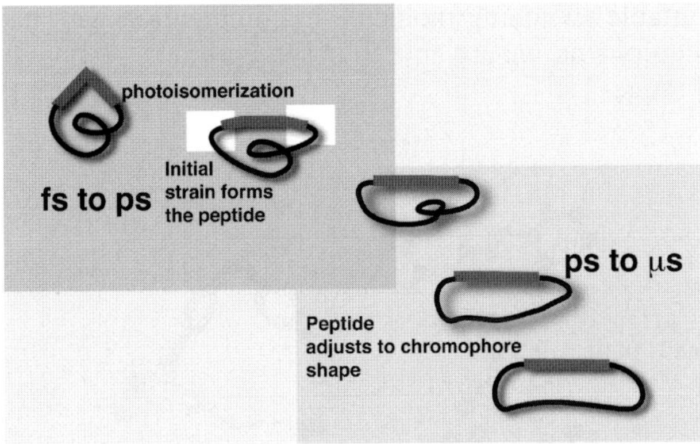

Fig. 2: Ultrafast driven and slower, allosteric reactions in light triggered peptides.

A wide variety of techniques have been used to study the different steps of protein folding [2]. Most experiments have been done in a quasi-steady-state-way by modifying the protein and it's surrounding, e.g. by temperature changes or addition and removal of denaturing agents. However for the detailed understanding of the reaction dynamics, time resolved information is required. Here one has to employ techniques to (i) induce folding or unfolding processes at well defined times and to (ii) allow to determine the structural changes as a function of time. Widely used are techniques on long time scales. For example stopped-flow techniques have supplied important information down to the millisecond range. The time range down to 10^{-6} seconds can be covered by NMR relaxation techniques. Laser flash photolysis and temperature-jump techniques have been employed down to the nanosecond range [3-5]. When even faster structural dynamics have to be addressed, special care has to be taken to use a well-defined trigger process for a synchronized start of the folding dynamics. These procedures have to be supplemented by structure sensitive techniques with ultrafast temporal resolution.

For the study of the fastest folding dynamics we developed a new experimental approach [6-8]. We combine the peptide/protein with a molecular switch which can be operated by light (see Fig. 2). Absorption of light triggers a fast photochemical reaction (e.g. a photoisomerization), which induces constraints on the peptide portion. These constraints then start a series of structural dynamics. The strong conformational restriction exerted by the trigger molecule on the peptide may lead to fast, strongly driven structural changes in different parts of the peptide chain. When these initial constraints are released, the peptide moiety may not yet have reached the final structural arrangement. In allosteric, thermally driven processes the peptide then approaches the final most favorable conformation.

The structural dynamics can be followed by visible spectroscopy (e.g. in the absorption band of the optical switch) and by infrared spectroscopy. The latter technique gives direct access to the dynamics of the peptide moiety [9,

10]. Using a suitable switching molecule in combination with modern infrared spectroscopic equipment, we are able to observe ultrafast dynamics down to the subpicosecond range.

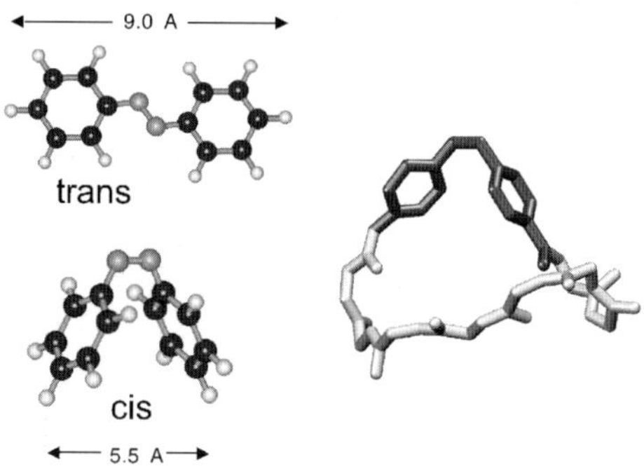

Fig. 3: Photoisomerization of azobenzene causes considerable structural changes.

2. The Azobenzene Peptides

The chromopeptides studied in this paper contain azobenzene as the optical trigger. The schematic view of the trigger molecule azobenzene in Fig. 3 shows that the two aromatic rings of azobenzene where the peptide-chain is attached, are strongly displaced upon isomerization; the photoisomerization changes the distance between the amino and carboxyl –group used for grafting the peptide from 9 Å to 6 Å. For azobenzene the *trans* isomer is stable; at room temperature the *cis* isomer thermally reconverts to the *trans* form on time scales that strongly depend on the solvent and type of azobenzene derivative. For the cyclic azobenzene-peptides used in the present study, $t_{1/2}$ ranging from 8 to 250 h were determined. For the experiments both isomers can be prepared at high concentrations by illumination with light at specific wavelengths. The photoisomerization of azobenzene is ultrafast. The *cis* to *trans* reaction is basically completed within 100 fs [11-13]. The *trans* to *cis* reaction is somewhat slower; in the azobenzene-peptides studied so far, this process is completed within a few picoseconds. Several types of azobenzene derivatives are used to construct the azobenzene-peptides. In this report we show data on two samples:
(i) (4-amino)phenylazobenzoic acid (APB), which forms a trigger molecule where the structural changes induced by photoisomerization are directly coupled with the peptide chain.
(ii) (4-aminomethyl)phenylazobenzoic acid (AMPB) that contains an additional flexible methylene spacer that untightens the constraints between the rigid azobenzene moiety and the peptide chain.

For the peptide part we used the active site fragment of *thioredoxin reductase* (H-Ala-Cys-Ala-Thr-Cys-Asp-Gly-Phe-OH) connected to the trigger molecules forming the cyclic compound. NMR analysis showed that isomerization of the trigger molecule induces distinct structural differences of the peptide part [14, 15]. In the *cis* form of the chromophore, the peptide exhibits a loop-like conformation with variable arrangements. In the *trans* form, the structure is more constrained and the peptide is in a rigidly extended conformation. Additional constraints are introduced by the formation of a disulfide bridge between the two cysteines leading to the bicyclic molecule bcAMPB. In the monocyclic molecule cAPB the cysteine thiol functions are protected to prevent formation of the disulfide bridge.

In previous studies we obtained important information on the primary reaction dynamics of the APB and AMPB peptides for the *cis* to *trans* reaction [8, 16-18]. Experiments in the visible spectral range have shown that the cyclic peptide cAPB exhibits absorption transients in the 50 ps time domain, which are connected with the relaxation of strain between the chromophore and the peptide. Very weak slower absorption transients indicate that long lasting reaction dynamics may be present.

Information on the *cis* to *trans* reaction of the bicyclic bcAMPB sample has been performed by visible and infrared spectroscopy. The visible experiments indicate some weak and long lasting absorption transients extending into the nanosecond range. Surprisingly the transient infrared experiments have shown that the dominant change in the structure sensitive amide I bands (1600 – 1700 cm^{-1}) occurs during the first 20 ps. At that moment the transient absorption change strongly resembles the steady state difference spectrum. Weaker absorption changes occur later on, on the picosecond to nanosecond and even longer time scales. These results were interpreted as strongly driven reactions that lead to a *trans*-like arrangement within 20 ps, followed by a weakly driven not even diffusive motion on much longer time scales. Apparently the bcAMPB molecule has very little degrees of freedom when the azobenzene has acquired the *trans* configuration. Therefore the isomerization motions induce rapidly a conformation, which based on IR data should strongly resemble the final *trans* form. In order to learn more about alternative reaction mechanisms where the molecule is transferred from a well-defined *trans* structure into a *cis* conformation consisting of a subset of different structures, we present in this paper data for the *trans* to *cis* reaction in cAPB recorded by ultrafast IR-spectroscopy.

3. Material and methods

The azobenzene-peptides were prepared as described in [6, 14]. The samples were dissolved in DMSO at concentrations of about 4-10 mM and were circulated through homemade flow cells (pathlength 0.1 mm), which ensure exchange of the illuminated sample volume between consecutive excitation laser pulses. As the peptide molecules adopt the *trans*-azobenzene conformation at room temperature at a timescale of some days, dark adapted molecules were used for the initial *trans* isomeric state yielding a concentration of nearly 100 % *trans* isomers. The experimental setup is described in detail in [19]. We used the pump and probe technique with pulses from a regenerative chirped pulse amplifier operated at 1 kHz. Second harmonic generation leads to the 400 nm excitation pulses. IR probing was in the range of the amide I band with a spectral width of ~100 cm^{-1}. In order to reduce excessive nonlinearities induced by the intense excitation pulses their duration was increased by an additional quartz substrate in front of the sample to 700 fs. The spectrum of the transmitted IR-pulses was recorded by a multichannel detection system.

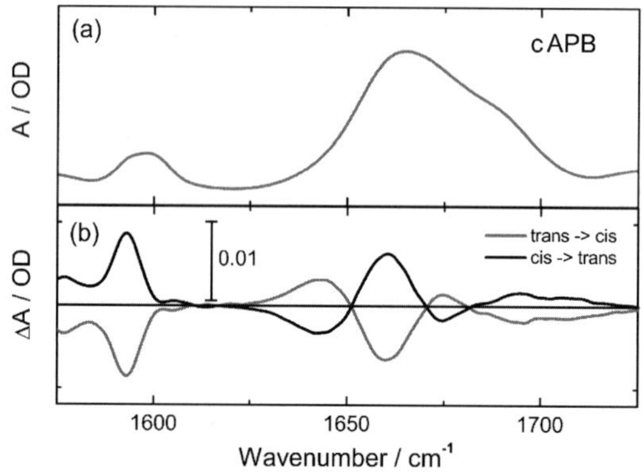

Fig. 4: IR-absorption of cAPB in trans conformation (a) and absorption difference spectra (b) induced by cw-illumination.

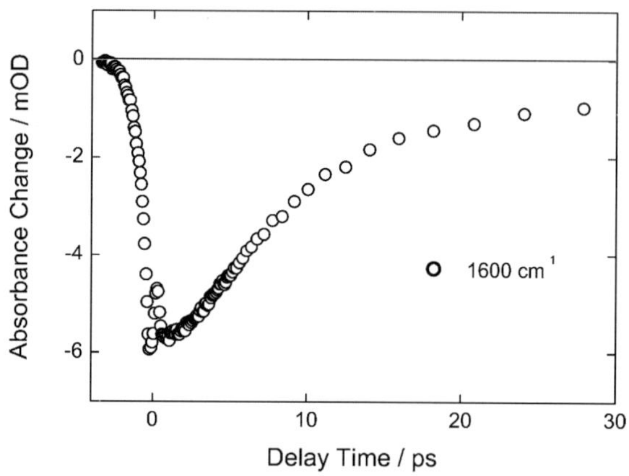

Fig. 5: Transient absorption changes of cAPB at 1600 cm^{-1}.

4. Results and Discussions

The stationary absorption spectrum of cAPB in the mid-infrared range is shown in Fig. 4a. In the range of the amide I band, *trans* cAPB has a strong absorption band peaking around 1667 cm^{-1}. A weaker band around 1600 cm^{-1} originates from the molecular switch APB. Illumination of the *trans* cAPB leads to photoisomerization and the formation of cAPB in its *cis* form. This transition is connected with changes in the infrared absorption spectrum (see gray curve in Fig. 4b). Around the peak of the amide I-band, the absorption is reduced, while there is an absorption increase around 1645 cm^{-1} and 1675 cm^{-1}. These absorption changes point to a rearrangement of the peptide structure. Absorption also changes in the low frequency wing of the APB-band at 1590 cm^{-1}. When the *cis* sample is reconverted (by illumination at $\lambda > 400$ nm) to the *trans* form, the initial absorption spectrum is reestablished (see symmetry of the difference absorption spectra for both switching directions, Fig. 4b).

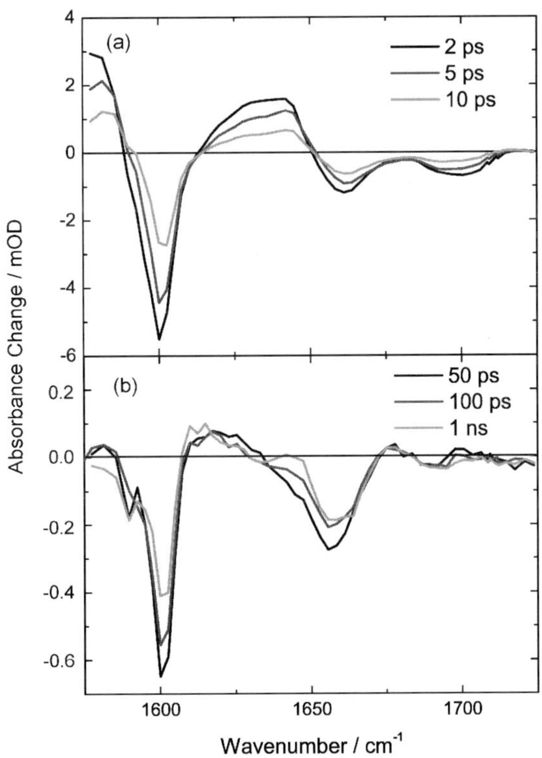

Fig. 6: Transient absorption spectra for cAPB recorded at different delay times.

Results on ultrafast visible/UV-pump, infrared probe experiments are shown in Fig. 5. Here we plotted the absorption change observed at 1600 cm^{-1}

during the first 30 ps. Prior to time zero there is a smooth exponential formation of a bleach as expected from a perturbed free induction decay [20]. After the end of the excitation process we find a strong bleach which partially decays within the 10 ps time-range. The initial process occurs on the same time scale as found with visible probing experiments. More detailed information is obtained when the absorption difference spectrum is plotted as a function of delay time (see Fig. 6). Immediately after excitation of the cAPB molecule there is a strong bleach of the 1600 cm^{-1} band of APB and an absorption increase at lower frequencies. Bleaching is also found in the amide I range at 1660 cm^{-1} and at 1700 cm^{-1}. Between 1610 cm^{-1} and 1650 cm^{-1} we observe an absorption increase. During the first 10 ps all these spectral features behave in a similar way. At later delay times (see Fig. 6b) the absorption shows a band-specific behavior. The absorption decrease at 1665 and 1690 cm^{-1} sharpens up and stays constant. There is continuous decrease of the 1600 cm^{-1} bleach during the same time. The difference spectrum reached after 1 ns is compared in Fig. 7 with the cw-difference spectrum. Similarities are observed but also pronounced differences; at 1 ns the bleach of cAPB around 1600 cm^{-1} is still at higher frequencies (1600 cm^{-1} instead of 1590 cm^{-1}) indicating that cAPB has not yet reached its final, relaxed form. The absorption increase found in the cw-spectrum around 1640 cm^{-1} is still missing. It may be concluded that also the peptide part has not yet reached its final arrangement within 1 ns. The missing 1640 cm^{-1} absorption may indicate that a fraction of amide groups in highly polar surroundings or with hydrogen bonds is not yet fully established.

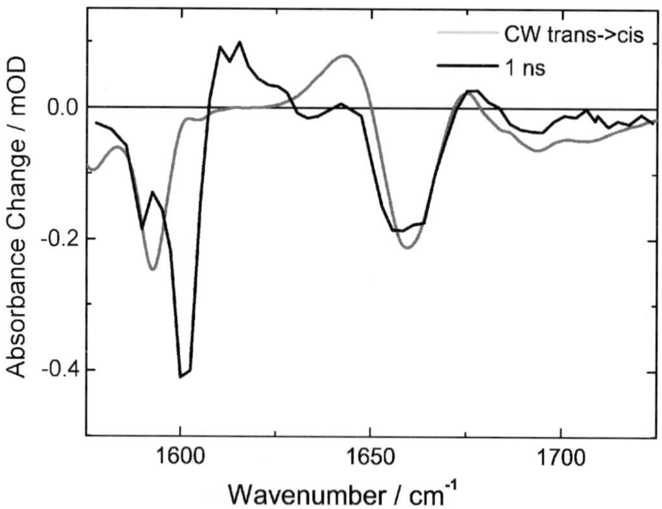

Fig. 7: Comparison of the transient difference spectrum at a delay time of 1 ns with the cw-difference spectrum. Even after 1 ns there remain considerable differences.

5. Conclusions

Transient infrared spectroscopy on light-induced structural changes of the monocyclic azobenzene peptide cAPB has shown a rich combination of ultrafast absorption changes extending from the few picosecond to the nanosecond time range. Light absorption in the molecule in the *trans* form prepares the chromophore azobenzene in the Franck-Condon region of its excited electronic state. The strong absorption changes found in the amide I range at very early times indicate that immediately after light absorption there are structural modifications in the peptide part. This can only be explained if in the *trans* to *cis* isomerization of azobenzene ultrafast structural dynamics occur within the time resolution of the present experiment (700 fs). This observation is in line with previous publications of isomerization dynamics of azobenzene [11-13]. After the relaxation of the initial strain we find reaction dynamics extending until the 1 ns-range. Here very fast structural rearrangements of the peptide part occur. The strong differences between the absorption spectra reached after 1 ns and the cw difference spectrum indicate that important structural rearrangements are not yet finished here and have to occur on much longer time scales. The above-mentioned view of the fast reaction dynamics (see Fig. 2) is supported by the presented experimental data. cAPB shows fast initial reaction dynamics on the time-scale of a few picoseconds together with pronounced, much slower structural changes. While in the previously studied bcAMPB molecule most of the structural dynamics are ultrafast and occur within 20 ps, structural dynamics on longer time scales are more important in cAPB. Since bcAMPB was studied on its way to a conformationally highly restricted state and cAPB on its way to a less constrained structure, we may conclude that the speed of conformational dynamics of these peptides is strongly affected by the size of accessible conformational space, which differs in the presence and absence of the small flexible methylene spacer.

References

1. L. Stryer, Biochemistry, IV ed., W. H. Freeman and Company, New York, 1995.
2. J. Buchner and T. Kiefhaber, Protein Folding Handbook, Wiley-VCH, Weinheim, 2005.
3. S. Williams, T. P. Causgrove, R. Gilmanshin, K. S. Fang, R. H. Callender, W. H. Woodruff, and R. B. Dyer, Biochemistry 35 (1996) 691.
4. P. A. Thompson, V. Munoz, G. S. Jas, E. R. Henry, W. A. Eaton, and J. Hofrichter, J. Phys. Chem. B 104 (2000) 378.
5. O. Bieri, J. Wirz, B. Hellrung, M. Schutkowski, M. Drewello, and T. Kiefhaber, Proc. Natl. Acad. Sci. U. S. A. 96 (1999) 9597.
6. R. Behrendt, C. Renner, M. Schenk, F. Q. Wang, J. Wachtveitl, D. Oesterhelt, and L. Moroder, Angew. Chem.-Int. Edit. 38 (1999) 2771.
7. R. Behrendt, M. Schenk, H.-J. Musiol, and L. Moroder, J. Pept. Sci. 5 (1999) 519.
8. S. Spoerlein, H. Carstens, H. Satzger, C. Renner, R. Behrendt, L. Moroder, P. Tavan, W. Zinth, and J. Wachtveitl, Proc. Natl. Acad. Sci. U. S. A. 99 (2002) 7998.

9. A. Barth and C. Zscherp, Q. Rev. Biophys. 35 (2002) 369.
10. S. Krimm and J. Bandekar, Adv. Protein Chem. 38 (1986) 181.
11. T. Nagele, R. Hoche, W. Zinth, and J. Wachtveitl, Chem. Phys. Lett. 272 (1997) 489.
12. H. Satzger, C. Root, and M. Braun, J. Phys. Chem. A 108 (2004) 6265.
13. B. Schmidt, C. Sobotta, S. Malkmus, S. Laimgruber, M. Braun, W. Zinth, and P. Gilch, J. Phys. Chem. A 108 (2004) 4399.
14. C. Renner, R. Behrendt, S. Sporlein, J. Wachtveitl, and L. Moroder, Biopolymers 54 (2000) 489.
15. C. Renner, J. Cramer, R. Behrendt, and L. Moroder, Biopolymers 54 (2000) 501.
16. J. Bredenbeck, J. Helbing, A. Sieg, T. Schrader, W. Zinth, C. Renner, R. Behrendt, L. Moroder, J. Wachtveitl, and P. Hamm, Proc. Natl. Acad. Sci. U. S. A. 100 (2003) 6452.
17. H. Satzger, C. Root, C. Renner, R. Behrendt, L. Moroder, J. Wachtveitl, and W. Zinth, Chem. Phys. Lett. 396 (2004) 191.
18. J. Wachtveitl, S. Sporlein, H. Satzger, B. Fonrobert, C. Renner, R. Behrendt,D.Oesterhelt, L. Moroder, and W. Zinth, Biophys. J. 86 (2004) 2350.
19. T. Schrader, A. Sieg, F. Koller, W. Schreier, Q. An, W. Zinth, and P. Gilch, Chem. Phys. Lett. 392 (2004) 358.
20. P. Hamm, Chem. Phys. Lett. 200 (1995) 415.

Femtochemistry VII
Fundamental Ultrafast Processes in Chemistry, Physics, and Biology
A.W. Castleman, Jr. and M.L. Kimble

The photochemistry of novel flavin-binding photoreceptors

Magdalena Gauden[a], Wouter Laan[b], Jason M. Key[a], Ivo H.M. van Stokkum[a], Rienk van Grondelle[a], Klaas J. Hellingwerf[b], Peter Hegemann[c] and John T.M. Kennis[a]

[a] *Department of Biophysics, Faculty of Sciences, Vrije Universiteit, Amsterdam, The Netherlands*
[b] *Swammerdam Institute for Life Science, University of Amsterdam, Amsterdam, The Netherlands*
[c] *Experimental Biophysics, Institute of Biology, Humboldt Universitaet zu Berlin, Berlin, Germany.*

1. Introduction

Blue-light sensing using FAD (BLUF) domains are a distinct family of flavin-binding photoreceptors that show no significant relationship to other sensor proteins in sequence or structure. The BLUF domain was first discovered as the N-terminal part of the flavoprotein AppA from *Rhodobacter (Rb.) sphaeroides* [1]. AppA is a two-component protein that can sense blue light at its N-terminal domain, and oxygen tension at its cysteine-rich C-terminal domain, and was shown to control photosynthesis gene expression in response to high-intensity blue light irradiation and variation of the oxygen tension [1,2]. Single BLUF domains which have been characterized include Slr1694 from the cyanobacterium *Synechocystis PCC6803* [3], Tll0078 from *Thermosynechococcus elongatus* [4], and BlrB from *Rhodobacter sphaeroides* [5].

BLUF domains bind a single molecule of oxidized FAD, resulting in absorption bands near 450 and 360 nm. Upon illumination, a long-lived signaling state is formed which shows a ~ 10-15 nm red-shifted absorption spectrum. This red-shifted state has a lifetime of several seconds in the cyanobacterial BLUF domains to 10^3 s in AppA and is thought to correspond to the signaling state of the photoreceptor [1,3,6,7]. Previous FTIR results on

Slr1694 domains were attributed to hydrogen bond rearrangements between the isoalloxazine ring of FAD and the protein, so that the carbonyls of isoalloxazine ring become more strongly hydrogen-bonded with residues in the flavin binding pocket [3]. Based on the AppA X-ray structure and these FTIR data, Anderson *et al.* have proposed a photoactivation model that involves a hydrogen-bond rearrangement accompanying a ~180° rotation of a conserved glutamine. In this view, photon absorption would lead to the breaking of hydrogen bonds from the glutamine amino moiety to the N5 of flavin and to a conserved tyrosine, and formation of a hydrogen bond to the C4=O of the flavin, as shown in Figure 1 [8].

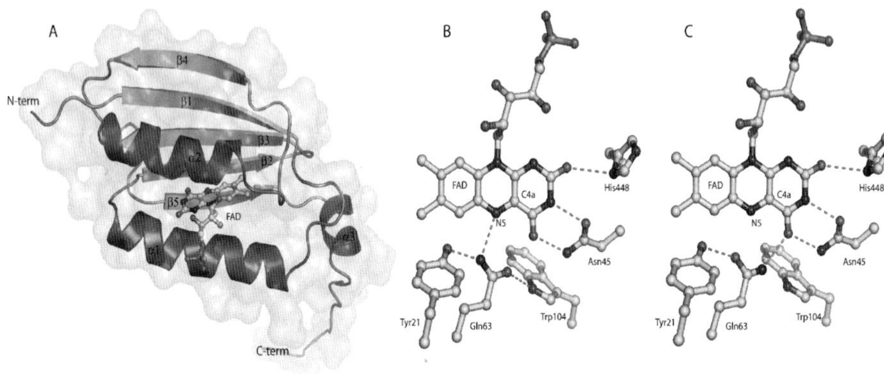

Figure 1: (A) X-ray structure of the *Rhodobacter sphaeroides* AppA BLUF domain, after ref. (8). Panel (B) and (C) represent a close-up of the crystal structure in the vicinity of the FAD chromophore in two proposed orientations of the conserved glutamine.

Here, we have used femtosecond transient absorption spectroscopy to investigate the photochemical events that lead to the formation of the red-shifted signaling state in the AppA and *Synechocystis* Slr1694 BLUF domains.

2. Materials and methods

Femtosecond transient absorption spectroscopy was carried out with a 1 kHz Ti:sapphire based regenerative amplification system described earlier [9]. 400 nm light was obtained by using a frequency doubled output of the Ti:sapphire laser, attenuated to 300 nJ -1.0 1μJ and used as a pump. A white light continuum that was generated by focusing the rest of the amplified 800 nm light on a 1 mm CaF_2 crystal was used as a probe. The polarization of the pump and probe beams was set to the magic angle. After passing through the sample, the probe was dispersed by a polychromator and projected on a diode array detector. The recorded femtosecond transient absorption spectra were globally analyzed in terms of kinetic model with sequentially interconverting evolution-associated difference spectra (EADS), i.e. 1→2→3→... The arrows indicate successive mono-exponential decays of increasing time constant, which can be regarded as the lifetime of each species. The first EADS corresponds to the time zero difference spectrum. This procedure enables us to visualize clearly the evolution of the (excited) states of the system. In general, the EADS are not necessarily associated with pure excited molecular species and often reflect mixtures of molecular species. The instrument response function was fit to a Gaussian of 120 fs FWHM. The global analysis procedures applied in the manuscript have been extensively reviewed in ref. [10].

3. Results

Transient absorption spectroscopy of AppA

Femtosecond time-resolved absorption spectroscopy was performed on the AppA BLUF domain to assess the primary intermediates that lead to formation of the long-lived signaling state. The sample was excited at 400 nm, and the time-resolved absorbance changes were measured over a wavelength range from 430 to 700 nm. Figure 1B shows a kinetic trace recorded at 548 nm, which measures the stimulated emission from the FAD chromophore. The stimulated emission signal rises on a timescale of hundreds of femtoseconds to a picosecond, after which we observe a multiexponential decay on timescales from 100 ps to nanoseconds, closely corresponding to those obtained with a synchroscan streak camera [11]. The multiexponentiality of excited-state decay likely follows from a significant conformational freedom in the dark state, as recently revealed for the AppA BLUF domain by NMR spectroscopy [12].

The time-resolved data were globally analyzed in terms of a kinetic scheme with sequentially interconverting species. The global analysis procedure indicated that six components were required for an adequate description of the time resolved data, with lifetimes of 250 fs, 1.2 ps, 90 ps, 590 ps, 2.7 ns, and a nondecaying component. The EADS that result from the sequential analysis are presented in Figure 2A, while the solid lines in Figure 2B and C denote the fits to the kinetic traces. The EADS with a lifetime of 250 fs (dotted curve) can be assigned to the mixture of initially created singlet excited states of S_1 and S_2 of the FAD chromophore. This EADS evolves in 250 fs to the next EADS, which has a lifetime of 1.2 ps (dash-dotted line). The stimulated emission band has increased, which is due to internal conversion from the S_2 to the S_1 state. The evolution to the next EADS, which is formed in 1.2 ps and possesses a lifetime of 90 ps, corresponds to an increase and blue-shift of the stimulated emission signal near 540 nm. We attribute the spectral characteristics of this process to vibrational cooling of the FAD S_1 state (11). The 90 ps component (solid line) shows a ground state bleach at 450-475 nm, an excited state absorption at 520 nm, a pronounced stimulated emission around 540 nm, and excited-state absorption at wavelengths longer than 600 nm. This component is assigned to the relaxed FAD S_1 singlet excited state and exhibits spectral characteristics similar to that of the FMN singlet excited state in the LOV2 domain of phototropin [13].

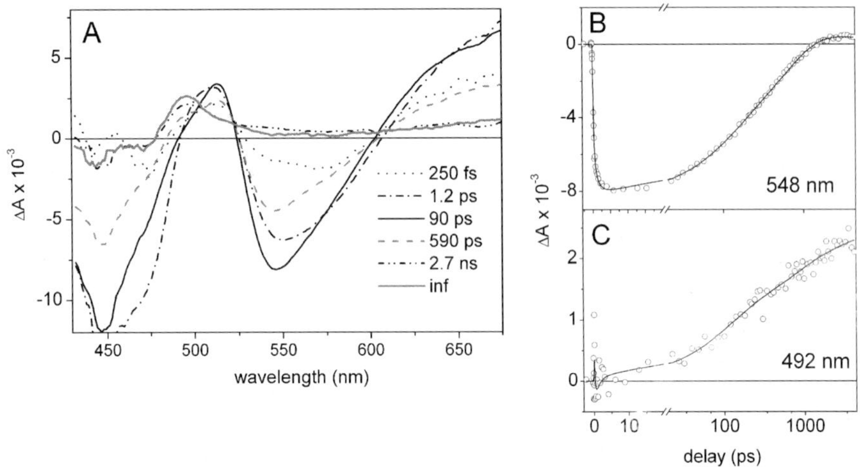

Figure 2 (A) Evolution-associated difference spectra (EADS) and their corresponding lifetimes resulting from global analysis of ultrafast transient absorption experiments on AppA upon excitation at 400 nm. (B) kinetic trace of AppA detected at 548 nm (circles) upon excitation at 400 nm, along with the result of the global analysis (solid line); (C) same as (A) with detection at 492 nm.

The next EADS, which has a lifetime of 590 ps (grey dashed line) shows an overall smaller amplitude compared to 90 ps component, signaling decay of the FAD singlet excited state. In the fourth EADS, which is formed in 590 ps and decays in 2.7 ns (dash-dot-dot line) the signals have largely disappeared. Interestingly, a narrow absorption feature near 497 nm has appeared. This feature resembles that of the difference spectrum of the long-lived AppA signaling state (*1*). In addition, a weak, broad absorption band is observed between ~550 and 700 nm. The evolution that follows from the fifth to the final, nondecaying EADS (solid grey line) in 2.7 ns only shows a slight increase of the 497 nm absorption band.

These experiments thus show that the long-lived signaling state in the AppA BLUF domain is formed on the sub-ns timescale. This is illustrated clearly in the kinetic trace at 492 nm (Figure 2C), where the photoproduct absorbs and the difference spectrum of the FAD singlet excited state shows an isosbestic point. The signal starts at zero, and rises in the course of hundreds of picoseconds to a few nanoseconds. Close inspection of the 590 ps EADS (Fig.

2A, dashed grey line) reveals a shoulder at 495 nm which is absent in the 90 ps EADS (solid line). This indicates that formation of the product state takes place in parallel with FAD excited-state decay, i.e. with time constants of 90 ps and 590 ps. An additional species with a broad absorption in the red (~650) is formed as well and can be assigned to the triplet state of the protein-bound FAD chromophore [14].

Transient absorption spectroscopy of Slr1694

To further investigate the photochemistry of BLUF domains, femtosecond time-resolved absorption spectroscopy was performed on the *Synechocystis* Slr1694 BLUF domain. Global analysis indicated that five components were required for the description of the time resolved data, with lifetimes of 0.9 ps, 7 ps, 22 ps, 123 ps, and a component that did not decay on the time scale of our experiment. The resulting EADS are presented in Figure 3A, while selected kinetic traces at 483, 550, 610, and 710 nm are shown in Figure 3C-F as the open circles.

The first EADS (dotted curve) in Figure 2A can be assigned to the initially created "hot" S_1 singlet excited state, the FAD chromophore, and is similar to that observed in AppA (Fig. 2A). This EADS evolves in 0.9 ps to the next EADS (solid line), which has a lifetime of 7 ps. As in AppA, this evolution involves a small blue-shift of the stimulated emission band and is assigned to a vibrational cooling process of the excited state of FAD. The system evolves in 7 ps to the next EADS (dashed line), which has a lifetime of 22 ps. This component shows a decrease of stimulated emission band, an increase of absorption around 510 nm, and the appearance of a new absorption around 580 nm. This component is assigned to a remaining fraction of FAD singlet excited state but with a contribution of another molecular species involved in the BLUF photoreaction. As will become evident from deuteration experiments, this species corresponds to a neutral flavin radical (*vide infra*). The evolution in 22 ps to the next EADS (dash-dotted line), which has a lifetime of 123 ps, corresponds to a further loss of the stimulated emission band and partial loss of the ground state bleach. The excited state absorption bands at 510 nm and 610 nm are still present, and additionally a shoulder around 487 nm is formed. As with the previous EADS, it corresponds to a mixture of FAD singlet excited state and the radical intermediate species. In the fifth, non decaying EADS (grey line), the ground state bleach largely has disappeared and a narrow absorption feature near 483 nm has appeared, which resembles the absorption band of the long-lived signaling state [3].

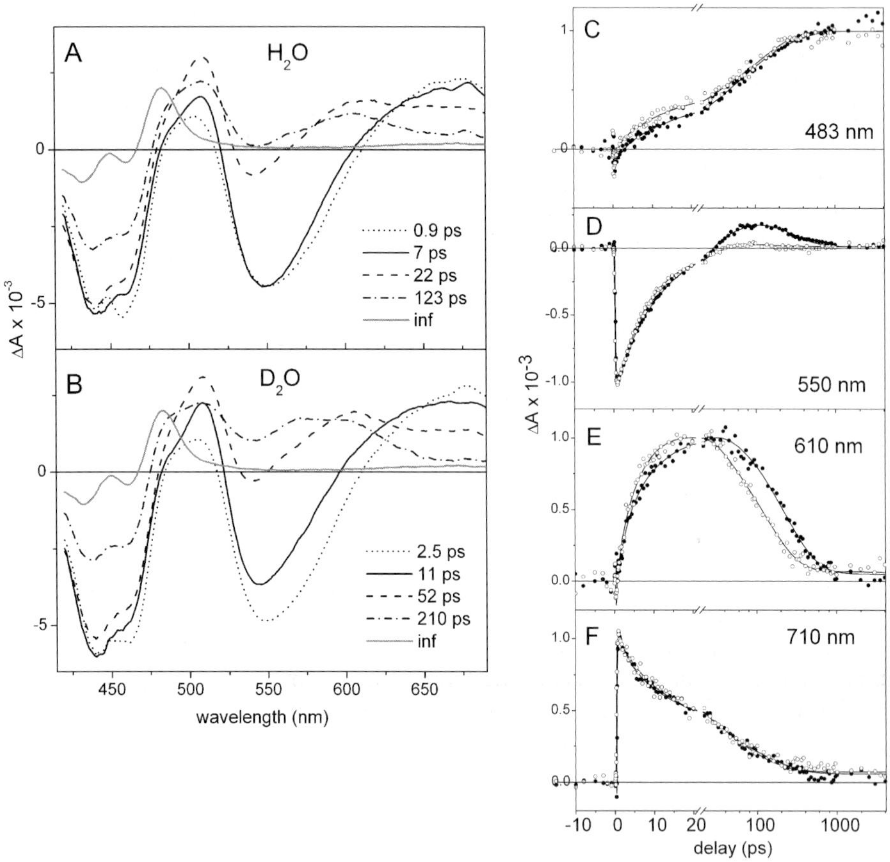

Figure 3. (A) Evolution-Associated Difference Spectra (EADS) and their corresponding lifetimes resulting from global analysis of ultrafast transient absorption experiments on Slr1694 in H_2O buffer upon excitation at 400 nm. B. Same as (A) for Slr1694 in D_2O buffer; Kinetic traces of Slr1694 in H_2O (open circles) and D_2O (closed circles) upon excitation at 400 nm, with detection at (C) 483 nm; (D) 550 nm; (E) 610 nm and (F) 710 nm. The solid lines denote the result of the global analysis. Note that the time axis is linear from -10 to 20 ps, and logarithmic thereafter.

The signaling state intermediate rises in 123 ps which is consistent with the results on AppA, where the long-lived intermediate is formed on the sub-ns timescale. Similar to AppA, the excited state of FAD in Slr1694 decays multiexponentially but with lifetimes that are much shorter than in AppA, as judged by decay of the stimulated emission band near 550 nm and the kinetic trace taken at 550 nm (Fig 2D, open circles). In contrast to the situation in

AppA, the appearance of additional reaction intermediates preceding the signaling state is observed in Slr1694.
To investigate whether the reaction intermediates are kinetically coupled via proton transfer, ultrafast experiments were performed with deuterated Slr1694.

The EADS that resulted from the sequential analysis are similar to those obtained with Slr1694 in H_2O and presented in Fig 3B. Kinetic traces are shown in Figures 3C-F as the solid circles. The lifetimes of 2.5 ps, 11 ps, 52 ps, and 210 ps were consistently longer than in Slr1694 in H_2O, indicating a pronounced kinetic isotope effect in the photoreaction of Slr1694. The 210 ps EADS (dash-dotted line) differs from 123 ps EADS of Slr1694 in H_2O, indicating that the relative concentration of coexisting species is different. In fact, the contribution by the FAD excited state seems to be negligible in this EADS. Interestingly, the shape of this EADS in the spectral region from 500-700 nm strongly resembles that of a neutral flavin semiquinone radical $FADH^{\bullet}$ [15].

We conclude there is an isotope effect on the formation of intermediate radical species and signaling state (22 ps versus 52 ps and 123 ps versus 210 ps in hydrated and deuterated samples, respectively), which indicates that proton transfer is involved in the process. In the kinetic traces at 710 nm, where FAD* has a pronounced absorption but where most other flavin species like semiquinone radicals have little or no absorption, the decay phases of Slr1694 in H_2O and D_2O can be overlaid, indicating that no or little kinetic isotope effect applies to the FAD* lifetime. A similar observation was reported for AppA [16]. At 610 nm, FAD* has an isosbestic point, giving a clean view on the kinetic behavior of the intermediate neutral radical species. The trace shows slower rise and slower decay in D_2O. In particular, in Slr1694 in D_2O, the decay of the radical intermediate takes approximately twice as long as in H_2O and occurs in ~200 ps. The kinetic isotope effect is observed also at 483 nm where the rise in D_2O is slower than in H_2O. Thus, the decay of the intermediate is kinetically coupled to the rise of the long-lived product, which demonstrates that the intermediate is on the reaction coordinate towards product formation.

4. Discussion

Ultrafast transient absorption experiments on the AppA BLUF domain demonstrate that, concomitant with the decay of excited FAD, a red-shifted product state is formed that strongly resembles the long-lived signaling state of AppA. After formation of the signaling state on ultrafast timescales of 90 ps and 590 ps, no spectroscopically distinguishable changes can be detected with UV-visible spectroscopy over 12 decades of time [11]. In the absence of reaction intermediates, it is difficult to propose a photochemical reaction mechanism for AppA. However, given the ultrafast timescales, the changes that lead to the red-shifted product must be very local and confined to the immediate environment around the chromophore.

More insights on the photochemistry of BLUF domains follow from the ultrafast experiments on hydrated and deuterated Slr1694 BLUF domains. A clear signature of a FAD neutral radical is observed, and the effect of deuteration on its lifetime and on the rise of the long-lived product demonstrates that it is a reaction intermediate on the pathway to signaling state formation. Thus, photoactivation of BLUF domains proceeds via a radical pair mechanism. The observation that the FAD excited-state lifetime is insensitive to deuteration suggests that electron transfer is the rate-limiting event in neutral radical formation and is followed by proton transfer. Based on the femtosecond time-resolved data, a qualitative photocycle scheme may be constructed for the photoactivation of BLUF domains, shown in Figure 4.

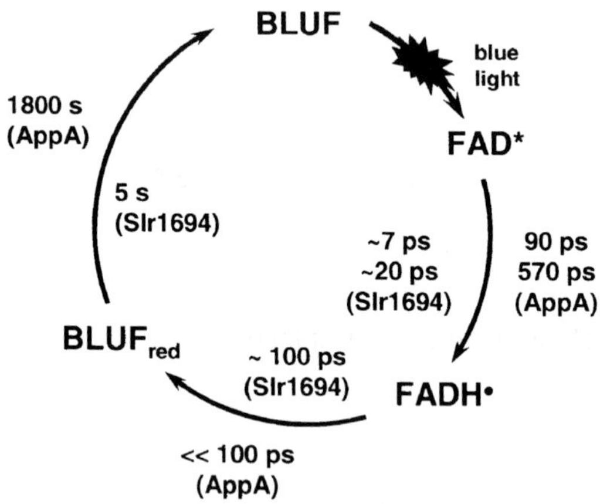

Figure 4: The photocycle scheme of BLUF domains. $BLUF_{red}$ denotes the long-lived signaling state. See text for details.

In Slr1694, excited state FAD (FAD*) decays multiexponentially through electron transfer from a nearby aromatic residue to form the anionic semiquinone FAD radical, with approximate time constants of 7 and 20 ps. This, in turn, is rapidly protonated to form the neutral semiquinone radical $FADH^{\bullet}$. Very likely the conserved tyrosine acts as the electron and proton donor, resulting in a radical pair $FADH^{\bullet}$ - $Tyr^{\bullet}$. It is well established that protonation of $FADH^{\bullet}$ takes place at the N5 atom of the flavin, which inevitably leads to a breaking of the hydrogen bond between the conserved glutamine and N5 (shown in Figure 1B), which in turn may trigger a reorientation of the conserved glutamine to result in a newly formed hydrogen bond with O4 (Figure 1C). The signaling state of Slr1694 (indicated as $BLUF_{red}$) is formed in

~100 ps from the neutral semiquinone radical FADH$^{\bullet}$ via radical-pair recombination with the conserved tyrosine. A kinetic isotope effect of ~2 is associated with this reaction, which is reasonable because the signaling state corresponds to an oxidized FAD, and thus a hydrogen needs to be removed from FADH$^{\bullet}$.

Given the similarities in structure and sequence between all BLUF domains studied so far, a similar picture very likely holds for AppA. The reason that no reaction intermediates are observed in AppA probably stems from a rate-limiting initial electron transfer rate. This is illustrated in Figure 4, where formation of the FADH$^{\bullet}$ neutral radical occurs with the observed time constants of 90 and 590 ps, which subsequently decays into the long-lived product on much faster (<<100 ps) timescale. In this way, the FADH$^{\bullet}$ intermediate does not transiently accumulate during the reaction and therefore escapes detection.

Acknowledgements

M.G. and W.L were supported by the Netherlands Organization for Scientific Research through the Chemical Sciences Council (NWO-CW). J.M.K. was supported by the Life Sciences Council of the Netherlands Organization for Scientific Research (NWO-ALW) via the 'Molecule to Cell' programme. J.T.M.K. was supported by NWO-ALW via a VIDI fellowship. We thank Tina Schiereis for excellent technical assistance.

References

1. S. Masuda and C. E. Bauer, Cell 110 (2002) 613.
2. M. Gomelsky and G. Klug, G. Trends Biochem. Sci. 27 (2002) 497.
3. S. Masuda, K. Hasegawa, A. Ishii, and T. Ono, Biochemistry 43 (2004) 5304.
4. A. Kita, K. Okajima, Y. Morimoto, M. Ikeuchi, and K. Miki, J. Mol. Biol. 349 (2005) 1.
5. A. Jung, T. Domratcheva, M. Tarutina, Q. Wu, W. H. Ko, R. L. Shoeman, M. Gomelsky, K. H. Gardner, and L. Schlichting, Proc. Natl. Acad. Sci. USA 102 (2005) 12350-.
6. W. Laan, M. A. van der Horst, I. H. van Stokkum, and K. J. Hellingwerf, Photochem. Photobiol. 78 (2003) 290.
7. B. J. Kraft, S. Masuda, J. Kikuchi, V. Dragnea, G. Tollin, J. M. Zaleski, and C. E. Bauer, Biochemistry 42 (2003) 6726.
8. S. Anderson, V. Dragnea, S. Masuda, J. Ybe, K. Moffat, and C. Bauer, Biochemistry 44 (2005) 7998.
9. C. C. Gradinaru, J. T. M. Kennis, E. Papagiannakis, I. H. M. van Stokkum, R. J. Cogdell, G. R. Fleming, R. A. Niederman, and R. van Grondelle, Proc. Natl. Acad. Sci. USA 98 (2001) 2364.
10. I. H. M. van Stokkum, D. S. Larsen, and R. van Grondelle, Biochim. Biophys. Acta (Bioenerg.) 1657 (2004) 82.

11. M. Gauden, S. Yeremenko, W. Laan, I. H. M. van Stokkum, J. A. Ihalainen, R. van Grondelle, K. J. Hellingwerf, and J. T. M. Kennis, Biochemistry 44 (2005) 3653.
12. J. S. Grinstead, S. T. D. Hsu, W. Laan, A. M. J. J. Bonvin, K. J. Hellingwerf, R. Boelens, and R. Kaptein, Chem. Bio. Chem., in press.
13. J. T. M. Kennis, S. Crosson, M. Gauden, I. H. M. van Stokkum, K. Moffat, and R. van Grondelle, Biochemistry 42 (2003) 3385.
14. M. Sakai and H. Takahashi, J. Mol. Struct. 379 (1996) 9.
15. F. Muller, M. Brustlein, P. Hemmerich, V. Massey, and W. H. Walker, Eur. J. Biochem. 25 (1972) 573.
16. W. Laan, M. Gauden, S. Yeremenko, R. van Grondelle, J. T. M. Kennis, and K. Hellingwerf, J. Biochemistry, in press.

Femtochemistry VII
Fundamental Ultrafast Processes in Chemistry, Physics, and Biology
A.W. Castleman, Jr. and M.L. Kimble

Electronic 2D Spectroscopy of Light Harvesting

Tobias Brixner[a,b], Jens Stenger[a], Harsha M. Vaswani[a], Minhaeng Cho[c], Robert E. Blankenship[d], and Graham R. Fleming[a]

[a]*Department of Chemistry, University of California, Berkeley, and Physical Biosciences Division, Lawrence Berkeley National Laboratory, Berkeley, California 94720, USA*
[b]*Present address: Physikalisches Institut, Universität Würzburg, Am Hubland, 97074 Würzburg, Germany*
[c]*Department of Chemistry and Center for Multidimensional Spectroscopy, Division of Chemistry and Molecular Engineering, Korea University, Seoul 136-701, Korea*
[d]*Department of Chemistry and Biochemistry, Arizona State University, Tempe, Arizona 85287, USA*

1. Introduction

Optical spectroscopy is increasingly employed for the investigation of complex systems, and femtosecond techniques have been very successful in clearing up the time-dependent dynamics of numerous processes in physics, chemistry, and biology. In the pump-probe scheme, the evolution of quantum level occupations can be monitored on an ultrashort time scale. However, it is often difficult to obtain mechanistic insight because the fundamental reasons for population transfer processes remain hidden. This is true especially in large biosystems such as natural light-harvesting antennas for which it is hard to disentangle the individual site energies and couplings between the involved chromophores.

Two-dimensional (2D) spectroscopy is a newly developed technique that measures the complete third-order optical response function of quantum-mechanical systems. This method was introduced for electronic transitions and discussed in detail by Jonas *et al.* [1,2] and in parallel employed for vibrational transitions by the groups of Hochstrasser [3], Tokmakoff [4], Hamm [5], and others. Theoretical discussions [6,7] have shown for some time the potential of this type of experiment. Most common is the use of a phase-matched photon-echo geometry with heterodyne detection of the signal field. This makes it possible to obtain the full amplitude and phase information and thus to recover the response function. Couplings are visible directly in the 2D traces [8], yielding much richer information than conventional techniques alone. One of the difficulties is the required phase stability and interferometric pulse-delay accuracy of the optical setup. Here the development of diffractive-optic-based detection schemes [9-12] has been very important for the successful implementation of 2D Fourier spectroscopy in the visible.

2. The Technique

Our setup for phase-stabilized and frequency-tunable 2D spectroscopy in the visible spectral range has been described in detail elsewhere [10,11]. The basic principle of this four-wave mixing experiment is shown in Fig. 1. Three femtosecond excitation laser pulses irradiate the sample from different directions with wavevectors $\mathbf{k}_1$, $\mathbf{k}_2$, and $\mathbf{k}_3$. The emerging signal in the phase-matched direction $\mathbf{k}_s$ is overlapped collinearly with a fourth laser pulse ("local oscillator") at a fixed temporal delay, and interferences are detected with a spectrometer. Fourier analysis of these fringes [2,11] delivers the signal amplitude and phase as a function of the detection frequency ω_t (corresponding to the signal time t). This procedure is carried out for any given population time T while scanning the coherence time τ. The desired 2D traces are then obtained by Fourier-transforming along the τ axis, to deliver (ω_τ, ω_t)-plots for each parameter T. The real parts of these spectra can be interpreted roughly as the electric field amplitude emitted at frequency ω_t, after excitation at frequency ω_τ, and waiting period T.

As indicated in Fig. 1, the system evolves in electronic coherence states (i.e., as off-diagonal elements in the quantum density matrix ρ) after the first and third pulse interactions and in an electronic population state after the second pulse (at least for two-level systems). In general, there are a number of possible excitation pathways that all contribute to the total signal, and the response-function formalism developed by Mukamel [13] can be used for calculations. However, if electronic couplings are present in the system (as is the case in light-harvesting pigment-protein complexes), the situation gets significantly more complicated but nonetheless treatable [14]. The power of 2D spectroscopy is that the couplings and energy-transfer signatures are visible directly as cross peaks in the experimental spectra.

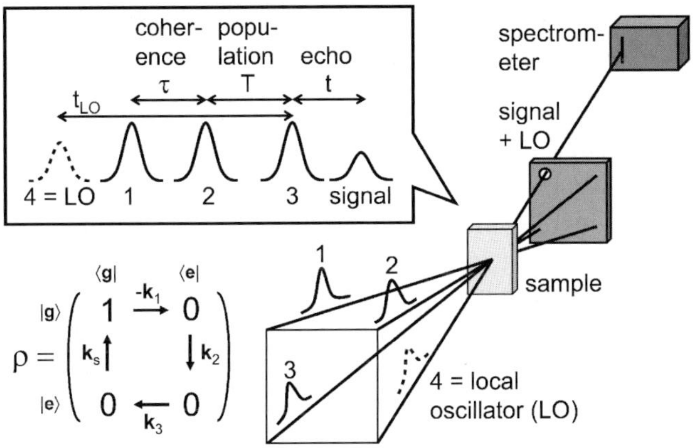

Fig. 1: Two-dimensional electronic spectroscopy. Heterodyne-detected three-pulse photon echo spectroscopy is carried out using a diffractive-optic-based setup. Adjustable pulse delays τ and T are provided by movable glass wedges with 10 as timing precision. As an example, one of the possible Liouville-space pathways (rephasing, R_2) is shown in reduced quantum density-matrix notation ρ, with electric field interactions labeled by the corresponding wavevectors. Signals are detected in folded-box phase-matching geometry $\mathbf{k}_s = -\mathbf{k}_1 + \mathbf{k}_2 + \mathbf{k}_3$. Spectral interference with a local oscillator (LO) yields the complete signal in amplitude and phase as a function of frequency.

3. Results

The 2D spectroscopy in our lab was first applied to two limiting exemplary cases: a monomeric dye molecule without intermolecular electronic couplings [11] and molecular J-aggregates with very strong (excitonic) couplings. It was possible to determine correlations between different electronically excited states, to distinguish between different types of line-broadening mechanisms, and to retrieve frequency-dependent exciton relaxation rates. The next step was to apply 2D spectroscopy on a sample from the intermediate coupling regime, in which energetic variations of individual chromophore site energies are on the same scale as interchromophore coupling parameters.

For this purpose, we investigated the Fenna-Matthews-Olson (FMO) light-harvesting complex found in green sulphur bacteria [8,14]. It serves both as a direct light harvester and as a "funnel" for directing the energy collected in the so-called chlorosome antennas to the reaction center. As in all photosynthetic systems, the flow of energy is driven by electronic couplings between molecules. In the case here, seven bacteriochlorophyll *a* molecules are contained in one protein shell, and three of these subunits make up one FMO complex. For a correct description, one has to take into account both the individual chlorophyll site energies as well as their mutual couplings. This interaction leads to characteristic excited-state wavefunctions ("excitons") that may be delocalized spatially over several chlorophyll molecules. Fluctuation of the optical transition frequency at the location of one particular chlorophyll (e.g., caused by protein vibrational motion) hence affects in a correlated manner all excitons to which this specific chlorophyll contributes, thus facilitating excitonic energy transfer.

The experimental result for 2D spectroscopy of FMO at population time $T = 0$ is shown in Fig. 2. Three main peaks are observed along the diagonal (dashed line), corresponding to the features of the linear absorption spectrum shown at the bottom. As can be seen, the laser spectrum covers all investigated transition frequencies. Just as in the extreme cases of vanishing or very strong electronic coupling mentioned above, the shapes of the diagonal peaks here can be analyzed to deliver information on inhomogeneous broadening. The elongation along the diagonal points at a correlation between excitation and emission frequency and thus indicates preserved memory.

Most important, however, is the presence of cross peaks at off-diagonal positions that indicate coupling. Strongest coupling is observed for exciton levels 1-5, 2-3, and 2-5 (black crosses). This is in nice agreement with a simple estimate based on the known structure of FMO and the corresponding dipole-moment prefactors of the different response-function contributions which predict the strongest couplings exactly for the observed positions. The asymmetry with respect to the diagonal and the presence of negative regions in the 2D trace (white crosses) are due to the effects of excited-state absorption (two-exciton contributions). This has to be taken into account in a more detailed analysis.

For larger population times $T > 0$, energy transport can be observed (results not shown here). The transport rates are determined by the electronic couplings and by the transition-frequency fluctuations (spectral density). Reproducing the experimental data with self-consistent Frenkel-exciton calculations and modified Förster/Redfield theory [8,14], we arrive at a simple picture for how the energy flows through the FMO complex as a function of space and time. Large transfer rates are seen between those exciton states that are delocalized at least partly over the same chlorophyll molecules; stronger wavefunction overlap means faster transfer.

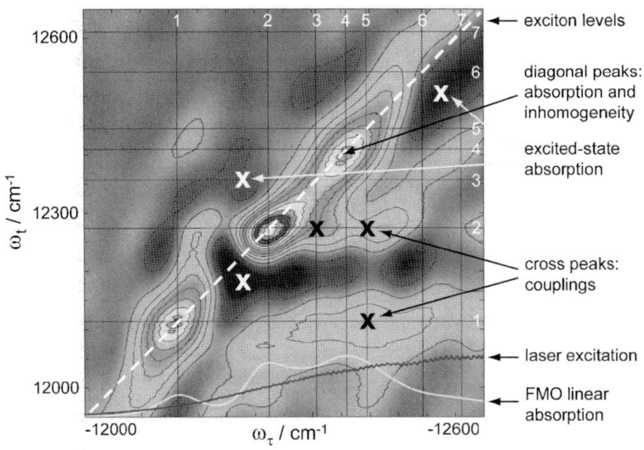

Fig. 2: Experimental 2D spectrum of FMO for population time $T = 0$ and temperature 77 K. Diagonal peaks indicate linear absorption positions, and cross peaks indicate electronic coupling between different exciton states.

4. Conclusion

The new technique of optical 2D spectroscopy for electronic transitions remedies the deficiency of conventional types of spectroscopy in which only the evolution of populations over time can be determined directly but not the mechanisms that underlie these changes. Specifically, one can determine couplings, relaxation pathways and rates, and the spatial properties and overlap relationships between exciton states. This makes it possible to follow energy flow on a molecular length scale with femtosecond time resolution, yielding microscopic insight into reaction mechanisms. In the case of FMO, it was found that energy is not simply transferred stepwise down the energy ladder; rather, the delocalization of wavefunctions over several bacteriochlorophyll molecules is exploited to move energy in a smaller number of steps but larger energy portions than previously supposed.

We expect this methodology to be useful for similar investigations in a number of photosynthetically active systems, antenna complexes, and reaction centers. But on more general terms, electronic couplings and transport processes can be studied in any photoactive assembly, macromolecule, or other nanoscale system.

References

1. S. M. Gallagher, A. W. Albrecht, J. D. Hybl, B. L. Landin, B. Rajaram, and D.M. Jonas, J. Opt. Soc. Am. B 15 (1998) 2338.
2. D. M. Jonas, Annu. Rev. Phys. Chem. 54 (2003) 425.
3. P. Hamm, M. H. Lim, and R. M. Hochstrasser, J. Phys. Chem. B 102 (1998) 6123.
4. M. Khalil, N. Demirdoven, and A. Tokmakoff, J. Phys. Chem. A 107 (2003) 5258.
5. V. Cervetto, J. Helbing, J. Bredenbeck, and P. Hamm, J. Chem. Phys. 121 (2004) 5935.
6. S. Mukamel, Annu. Rev. Phys. Chem., 51 (2000) 691.
7. M. Cho, Phys. Chem. Comm. 5 (2002) 40.
8. T. Brixner, J. Stenger, H. M. Vaswani, M. Cho, R. E. Blankenship, and G. R. Fleming, Nature 434 (2005) 625.
9. G. D. Goodno, G. Dadusc, and R. J. D. Miller, J. Opt. Soc. Am. B 15 (1998) 1791.
10. T. Brixner, I. V. Stiopkin, and G. R. Fleming, Opt. Lett. 29 (2004) 884.
11. T. Brixner, T. Mancal, I. V. Stiopkin, and G. R. Fleming, J. Chem. Phys. 121 (2004) 4221.
12. M. L. Cowan, J. P. Ogilvie, and R. J. D. Miller, Chem. Phys. Lett. 386 (2004) 184.
13. S. Mukamel, Principles of Nonlinear Optical Spectroscopy, Oxford, New York, 1995.
14. M. Cho, H. M. Vaswani, T. Brixner, J. Stenger, and G. R. Fleming, J. Phys. Chem. B 109 (2005) 10542.

Femtochemistry VII
Fundamental Ultrafast Processes in Chemistry, Physics, and Biology
A.W. Castleman, Jr. and M.L. Kimble

Intermediates in the ultrafast repair of DNA by DNA photolyase

Zhanjia Hou, Goutham Kodali, Madhavan Narayanan, Kongsheng Yang, and Robert J. Stanley

Department of Chemistry, Temple University, Philadelphia, PA 19122

1. Introduction

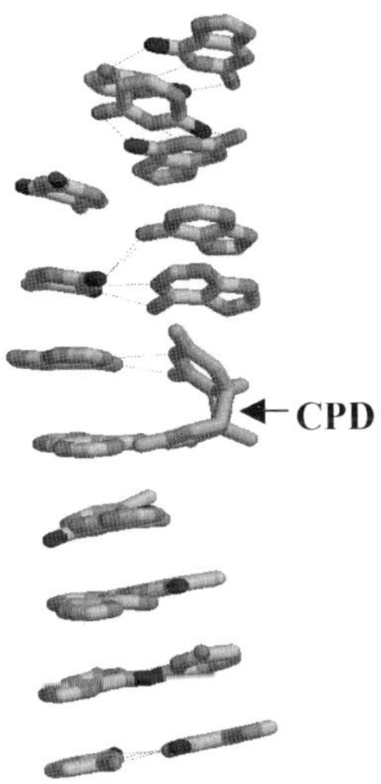

Figure 1: x-ray crystal structure of a dodecamer DNA duplex containing a centrally located CPD

It is widely known that ultraviolet (UV) light causes DNA damage. The most common form of UV lesion found in DNA is the *cis-syn* pyrimidine dimer (CPD) in which adjacent pyrimidines on a DNA strand undergo a 2+2 photocycloaddition to generate a cyclobutane ring joining the two rings across the 5- and 6-carbon atoms of the pyrimidine ring. The structure of a thymidine-thymidine CPD lesion has been solved most recently by x-ray crystallography for a dodecamer duplex [1]. (See Figure 1). The 5' thymidine of the CPD lesion has a longer hydrogen bond distance to its complementary adenine (2.4 Å) than the 3' thymidine (1.9 Å). Base stacking is also significantly more disrupted at the 5' end of the CPD, and this may be the motif that causes mutation or regulation problems that lead to disease or death.

DNA photolyase is a widely distributed CPD repair protein [2]. It is unique in that it is the only repair protein of its kind that directly reverses the CPD, without excision or backbone modification of the DNA itself.

In addition, it is a light-driven protein, utilizing a flavin adenine dinucleotide (FAD) cofactor. The flavin cofactor is required for substrate binding and for repair. The reaction is thought to occur through photoinduced electron transfer from the reduced anionic $^{1}FADH^{-}$ to the CPD, leaving the flavin cofactor in its one electron reduced form, the neutral semiquinone, $FADH^{\bullet}$. Formation of the CPD radical anion ($CPD^{\bullet-}$) leads to rapid repair at physiological temperatures in about 2 nanoseconds. In spite of significant effort by several ultrafast spectroscopy groups over more than 15 years [3-8], direct evidence of this electron transfer mechanism has been lacking. In fact, no direct evidence of any radical intermediates has been obtained.

The protein also contains an additional chromophore, which can be either methynyltetrahydrofolate (MTHF) or deazaflavin, depending on the organism. In *E. coli* photolyase (PL), the object of this study, the cofactor is MTHF. It has been shown that this chromophore has a higher extinction than the $FADH^{-}$ cofactor and acts as a modest light-harvesting antenna, with singlet-singlet energy transfer from $^{1}MTHF$ to $FADH^{-}$ taking place in less than 100 ps [9]. *In vitro* preparations of *E. coli* photolyase typically contain a sub-stoichiometric amount of MTHF, and crystal structures of the protein show that MTHF is loosely bound on the periphery of the protein [10]. However, the protein does not require the antenna pigment for repair. Selective photodecomposition of the MTHF molecule can be accomplished using 384 nm light [11].

Substrate binding by photolyase is facilitated by a cavity in the protein having the approximate dimensions of the CPD. This cavity was first reported in 1995 by Park *et al.* [10], and it was hypothesized that base flipping of the CPD was required to bring the substrate close enough to the FAD cofactor for rapid electron transfer to compete with excited state decay processes. This has been corroborated by biochemical [12] and biophysical approaches [13] and most recently by a protein:product crystal structure [14]. Several groups have measured binding constants for the substrate as a function of substrate length and have reported that the dimer itself is a poor substrate. Binding efficiency increases dramatically when the substrate length exceeds 4 nucleotides [15].

The importance of these studies were confirmed in a comparative study of repair rates *vs.* substrate length by our group [8]. This finding was significant because previous groups had used a two nucleotide substrate d(TpT) to determine the electron transfer rate constant and, indeed, to observe and characterize the putative $FADH^{\bullet}$ and $CPD^{\bullet-}$ (and other) intermediates. Most of these experiments were performed with > 30 ps time resolution using a white light continuum probe, but no definitive evidence of these intermediates was obtained. In this contribution we have used a pentameric substrate with subpicosecond time resolution to provide direct evidence that ultrafast electron transfer is operative in DNA photolyase. However, the identity of other intermediates still remains elusive.

2. Experimental

A more complete description of this work will be published elsewhere. Briefly, *E. coli* photolyase was purified as described in Gindt *et al.* [16] from MS09 cells obtained as a gift of Aziz Sancar (University of North Carolina, Chapel Hill). The MTHF chromophore was photodecomposed, as described by Heelis *et al.* [11], as ascertained by the loss of absorption at 384 nm where folate absorbs (but see below). Pentameric thymidine substrate, $(T\text{<>}T)_5$, was prepared as described in [8]. No attempt was made to isolate various isomers (e.g. T<>TTTT, TTT<>TT, etc.). In some experiments, a T_{12} oligo containing one or more CPDs was used.

Buffered solutions of 200-300 μM protein with approximately 10-fold more substrate were photoreduced in a sealed fused silica cuvette (2 x 10 mm) using red light from a xenon arc lamp filtered through a monochromator and a water filter. The extent of photoreduction was followed by UV/Vis spectroscopy (disappearance of the 384 nm folate band). A small magnetic stirring bar was included in the cuvette before reduction.

The laser system is as described in [8] with the exception that the probe consisted of a white light continuum generated in a rotating CaF_2 flat. The continuum and *ca.* 390 nm pump beam (~300 nJ/pulse) were focused into the cuvette, which was placed in a temperature-controlled holder (5°C) and stirred by an external rotating magnet.

The pump beam was filtered both spatially and optically, and the probe beam was focused into an optical fiber and fed into the slit of a ¼ meter spectrograph. The dispersed continuum was detected by a CCD detector. A chopper was used to modulate the pump beam so that ΔI spectra could be constructed on a shot-by-shot basis. Data was obtained at a repetition rate of 20-100 Hz (10-50 Hz for the chopped pump beam). These spectra were averaged in a computer for further processing. The computer also controlled an optomechanical delay line to achieve *ca.* 3.7 ns scan length.

A cuvette containing a toluene-naphthalene mixture was used to measure the time resolution of the apparatus from its optical Kerr response to the pump and continuum pulses. The instrument response function was approximately 300 fs, and the spectral chirp on the pulse was approximately ± 0.75 ps. The data analysis is described below.

3. Results and Discussion

Transient absorption data of the photolyase:product mixture and the photolyase:substrate complex with a small amount of contaminating folate are shown in Figure 2 (top and bottom respectively). The data are the result of averaging several data sets taken from -20→3,700 ps and 400-700 nm. The protein:substrate data were obtained first, followed by photorepair of any

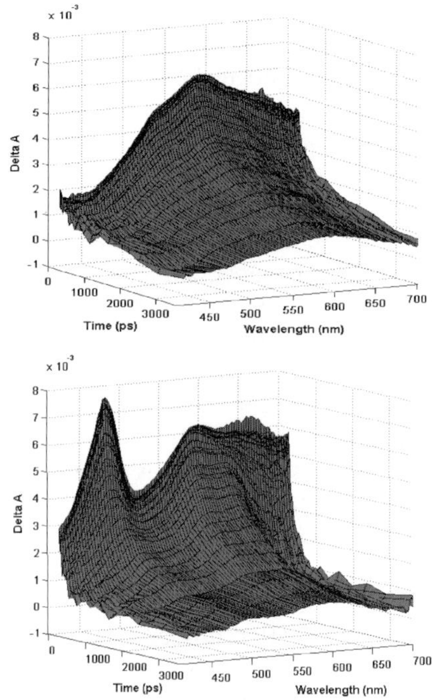

Figure 1: Transient absorption spectra of the photolyase: product mixture (top) and photolyase:substrate (T_5) complex (bottom).

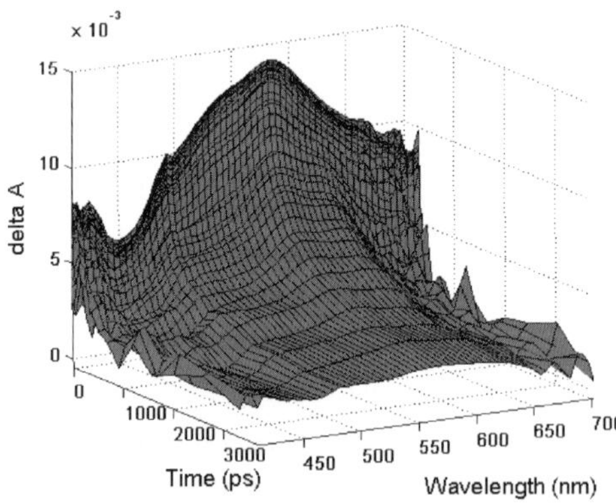

Figure 2: MTHF-depleted protein:substrate (T_{12}) transient absorption spectra

remaining substrate using water-filtered white light. The protein:product data set was obtained immediately following this treatment.

The protein:product transient absorption spectrum (2, top) shows characteristics of excited state reduced flavin, with a maximum at about 550 nm. This feature has been observed in other reduced flavoproteins and in reduced flavins in simple solvents [17] and has been assigned to an excited state absorption. No evidence of semiquinone formation is seen, indicating that photolyase has a very slow electron transfer rate compared to its excited state decay in the absence of substrate. A deeper analysis is given below.

The protein:substrate data are clearly different, with the most noticeable difference being a large feature at approximately 475 nm in the protein:substrate data. Comparison of this data with those obtained by Kim *et al.* [9] suggests that this feature is due to a small amount of intact folate chromophore. This artifact was due to the use of red light, not white light, for the photoreduction of semiquinone photolyase. Its disappearance in the protein:product spectrum is due to the use of white light to repair any remaining substrate, which had the effect of photodecomposing the MTHF.

Data with the folate completely decomposed are given in Figure 3. The rapidly decaying absorption at *ca.* 410 nm and 700 nm seen in this data set is clearly due to the presence of substrate. A set of peaks are clearly observed around 600 nm, indicative of the formation of semiquinone FADH$^{\bullet}$ following photoexcitation and electron transfer. The transient absorption in this region disappears within about 3.5 ns, suggesting that the electron is back-transferred from the product radical anion ($T^{\bullet -}$) to FADH$^{\bullet}$, closing the catalytic cycle.

These protein:substrate data were analyzed using a unidirectional sequential model. This model (Scheme 1) assumes that excitation of the ground state complex (**1**) generates excited state (**2**). This state can decay by nonradiative channels with constant k_{nr}, or electron transfer can occur with rate constant k_{eT} to produce the semiquinone $PL^{\bullet}_{sq}$ and the CPD radical anion, T<>T$^{\bullet -}$ (**3**). Repair follows electron transfer with rate constant k_{repair} (**4**), followed by charge recombination (k_{rec}) to close the catalytic cycle. While this is not the only possible scheme, it is a minimal scheme that produces a reasonable fit to the data.

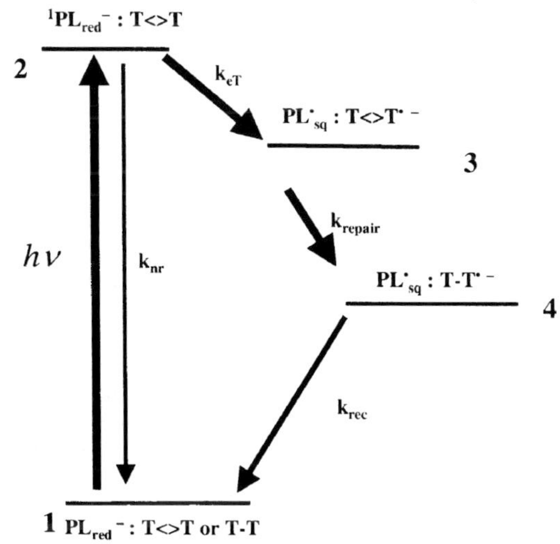

Scheme 1

The transient absorption spectrum is calculated assuming *n* intermediates, each with its own absorption (extinction) spectrum, weighted by its time-dependent concentration profile,

$$\Delta A(\lambda,t) = \sum_{i=1}^{n} c_i(t)\varepsilon_i(\lambda)$$

The nonlinear part of the fit begins with a calculation of the concentration profiles:

$$C(t) = C_0 e^{Kt}$$

given an initial set of concentrations, C_0, and a rate constant matrix, K, which includes the rate constants from the scheme that would be used to solve the set of coupled differential equations resulting from an analysis of Scheme 1. Once the time-dependent concentrations have been calculated, the intermediate spectra, E(λ), are obtained by scaling the pseudo-inverse of C(t) with the experimental data, $\Delta A(\lambda,t)$. A fit to the data is achieved by minimizing the quantity

$$\{\Delta A(\lambda,t) - C(E - \varepsilon_0)\}^2$$

where ε_0 is the ground state absorption spectrum, entering here as a ground state bleach.

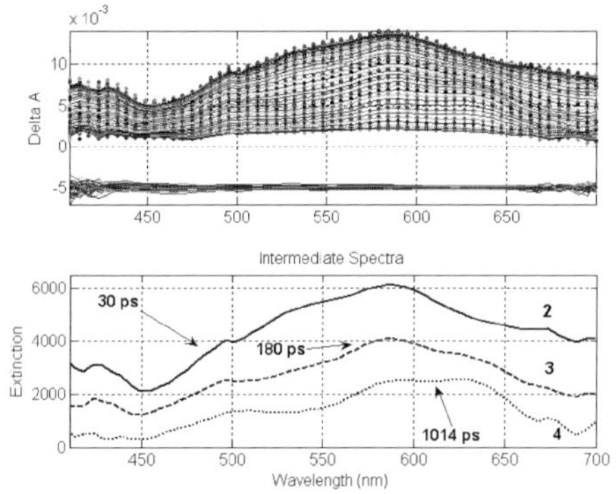

Figure 4: Fit to 4 state unidirectional sequential model

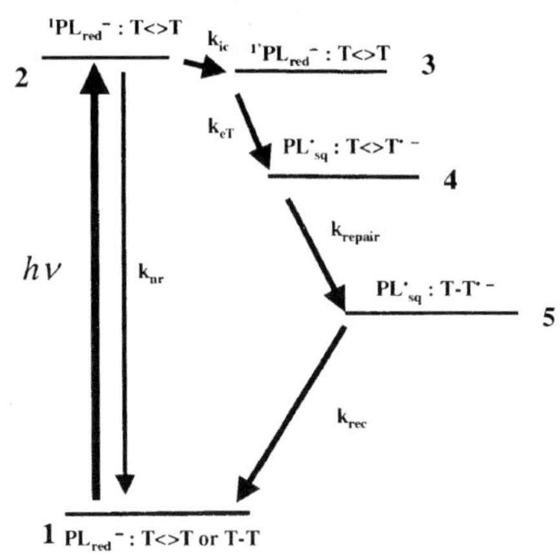

Scheme 2

The fit to the data and the intermediate spectra are shown in Figure 4, top and bottom, respectively. The residuals of the fit are offset from the fit and data by -5 mOD. The spectra of the intermediate states **2-4** are denoted with lifetimes from the fit. The excited state spectrum of the photolyase protein (**2**) decays in about 30 ps to an intermediate that has features corresponding to the flavosemiquinone form, $PL^{\bullet}_{sq}$, including a peak at *ca.* 640 nm where the semiquinone absorbs. This state may be a combination of relaxation of $^1FADH^-$ and electron transfer. State **3** decays in about 180 seconds to form the completely charge separated state, as suggested by the clear $FADH^{\bullet}$ spectral signature in the 600 nm region. There appears to be evidence for a ground state bleach in the 400-500 nm range, though the details of this assignment are still in question. Finally, state **4** decays in about 1,000 ps, demonstrating that the semiquinone returns back to the ground state reduced form of the protein.

This minimal scheme is clearly inadequate, as it cannot separate the excited state relaxation of the $^1FADH^-$ from the electron transfer step. Excited state relaxation on the picosecond time scale has been observed for other

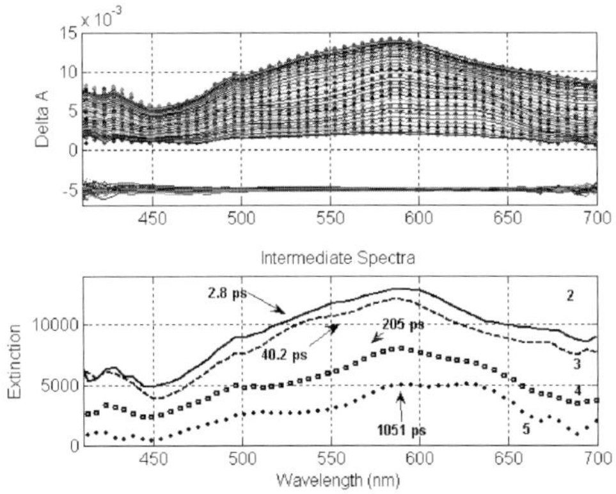

Figure 5: 5 state sequential model and fit

reduced flavoproteins and for flavins in simple solvents [17-19]. A second scheme seems warranted, involving a 5th state (Scheme 2). The fit to this model is shown in Figure 5.

The lifetimes for the intermediate spectra are given in the figure, with the exception of the internal conversion of **2→1**, which has a lifetime of about 92 ps. This would lead to a low quantum yield of repair if it were not for the ultrafast relaxation of **2→3**, which has a 2 ps lifetime, decaying into relaxed $^{1'}FADH^-$ (**3**). This is similar to relaxation times observed for other flavoproteins (~ 4 ps), as noted above. This relaxation step is followed by electron transfer in about 40 ps to form the flavosemiquinone (peaks at 580 nm and 640 nm) and other as yet unidentified intermediates quantitatively (**4**). This initial charge separated state decays into state **5** in about 200 ps, with some loss of absorption at 550 nm. The species responsible for this absorbance change has yet to be elucidated. State **5** decays back to ground state $FADH^-$ with a lifetime of about 1,050 ps. This analysis is consistent with all previous data from transient absorption studies and corroborates our 30 ps electron transfer lifetime estimate for *A. nidulans* photolyase made by blue/UV pump-probe experiments [8].

While this analysis is consistent with what is known, there are still many unresolved questions. The CPD radical anion should exhibit absorption in the blue[20], but no obvious feature is evident in the transient absorption data or in the intermediate spectra derived from the sequential model. Indeed, this reaction scheme is only one of many plausible mechanisms. However, the signal to noise level of transient absorption data will need to be substantially better before a more detailed analysis is possible.

In comparison to the protein:substrate analysis, the analysis of the protein:product data is well fitted using a three state model: **2→3→1** (Figure 6), where excitation into state **2** is lost through relaxation through an intermediate state **3** with a lifetime of about 67 ps, followed by relaxation to the ground state in about 1,500 ps. The average of these two constants gives about 800 ps, similar to what has been measured by fluorescence upconversion for the lifetime of the excited protein without substrate [21].

4. Conclusions

Transient absorption spectra of photolyase with and without substrate have been obtained in the visible region of the spectrum with subpicosecond time resolution. The spectrum of the flavosemiquinone is observed clearly for the first time, establishing without a doubt that photolyase functions via photoinduced electron transfer, leading to reductive cleavage of the CPD lesion. The decay of this state by charge recombination to the initial reduced flavin was also observed. Substrate intermediates have not yet been identified.

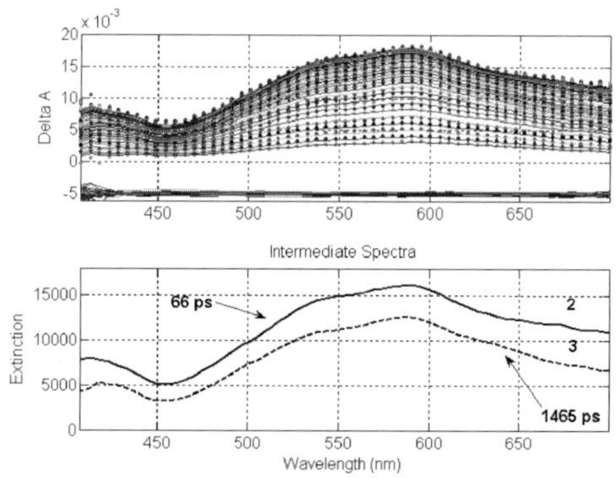

Figure 3: protein:product 3 state fit and intermediate spectra.

References

1. H. Park, K. Zhang, Y. Ren, S. Nadji, N. Sinha, J.-S. Taylor, and C. Kang, Proc. Nat. Acad. Sci. USA 99 (2002) 15965.
2. A. Sancar, Chem. Rev. 103 (2003) 2203.
3. T. Langenbacher, X. Zhao, G. Bieser, P. F. Heelis, A. Sancar, and M. E. Michel-Beyerle, J. Am. Chem. Soc. 119 (1997) 10532.
4. S. T. Kim, M. Volk, G. Rousseau, P. F. Heelis, A. Sancar, and M. E. Michel-Beyerle, J. Am. Chem. Soc. 116 (1994) 3115.
5. S. T. Kim, P. F. Heelis, and A. Sancar, Biochem. 31 (1992) 11244.
6. T. Okamura, A. Sancar, P. F. Heelis, T. P. Begley, Y. Hirata, and N. Mataga, J. Am. Chem. Soc. 113 (1991) 3143.
7. P. F. Heelis, T. Okamura, and A. Sancar, Biochem. 29 (1990) 5694.
8. A. W. MacFarlane, IV and R. J. Stanley, Biochem. 42 (2003) 8558.
9. S. T. Kim, P. F. Heelis, T. Okamura, Y. Hirata, N. Mataga, and A. Sancar, Biochem. 30 (1991) 11262.
10. H.-W. Park, S. T. Kim, A. Sancar, and J. Deisenhofer, Science 268 (1995) 1866.
11. P. F. Heelis, G. Payne, and A. Sancar, Biochem. 26 (1987) 4634.
12. B. J. V. Berg and G. B. Sancar, J. Biol. Chem. 273 (1998) 20276.
13. K. S. Christine, A. W. MacFarlane, IV, K. Yang, and R. J. Stanley, J. Biolog. Chem. 277 (2002) 38339.
14. A. Mees, T. Klar, P. Gnau, U. Hennecke, A. P. M. Eker, T. Carell, and L.-O. Essen, Science 306 (2004) 1789.
15. M. S. Jorns, G. B. Sancar, and A. Sancar, Biochem. 24 (1985) 1856.
16. Y. M. Gindt, E. Vollenbroek, K. Westphal, H. Sackett, A. Sancar, and G. T. Babcock, Biochem. 38 (1999) 3857.
17. M. Enescu, L. Lindqvist, and B. Soep, Photochem. Photobiol. 68 (1998) 150.
18. R. J. Stanley and H. Jang, J. Phys. Chem. A. 103 (1999) 8976.
19. A. W. MacFarlane, IV and R. J. Stanley, Biochem. 40 (2001) 15203.
20. S. R. Yeh and D. E. Falvey, J. Am. Chem. Soc. 113 (1991) 8557.
21. H. Wang, C. Saxena, D. Quan, A. Sancar, and D. Zhong, J. Phys. Chem. B 109 (2005) 1329.

Femtochemistry VII
Fundamental Ultrafast Processes in Chemistry, Physics, and Biology
A.W. Castleman, Jr. and M.L. Kimble

Ultrafast protein dynamics

Dongping Zhong

Departments of Physics, Chemistry, and Biochemisitry, Programs of Biophysics, Chemcial Physics, and Biochemistry, 191 West Woodruff Avenue, The Ohio State University, Columbus, OH 43210, USA

1. Introduction

Understanding biological function is determined by our ability to map out structures and dynamics of macromolecules. The past decades have witnessed great progress in the determination of structures, but the study of dynamics is much less advanced. One of the major challenges is to characterize such dynamics at the atomic scale and thus elucidate the relevant molecular mechanisms. With femtosecond resolution, it is now possible to observe atomic motions in real time. This temporal resolution is crucial for a *complete* description of the elementary dynamics which is an essential determinant of the function at longer times, as observed in vision and photosynthesis. At this ultrafast time scale, phenomena are numerous; twisting motion in photoreceptors takes less than one picosecond to ensure efficient energy funneling; electron and proton transfer in enzymatic reactions usually occurs on the picosecond time scale through quantum tunneling, energy transport in many proteins also occurs in tens of picoseconds, and protein configurations themselves fluctuate often in subnanosecond for a variety of biological functions such as molecular recognition.

In this contribution, we report our two focused areas. One is centered on the fundamental understanding of a longstanding biological question---how water controls protein structure, dynamics, and function. Here the major effort concentrates on the elucidation of hydration dynamics around proteins at the local molecular level. Understanding of protein-water interactions has far-reaching implications relating to protein plasticity and recognition, protein folding and aggregation, and enzyme catalysis. The other area focuses on a series of photo-driven biological processes with the aim of understanding how nature efficiently converts solar energy to perform critical biological functions.

The systems considered encompass photoenzymes for repairing damaged DNA and photoreceptors for synchronizing biological timing. We integrate *femtosecond spectroscopy* and *molecular biology techniques* to reveal these dynamics. The temporal resolution covers the range from femtoseconds to nanoseconds to follow the functional evolution and thus elucidate the complex dynamics with unprecedented detail. Molecular biology methods, especially site-directed mutagenesis, are used to manipulate protein structures and identify key functional residues. These studies with femtosecond temporal resolution and single-residue spatial resolution will elucidate the function of macromolecular structures at the most fundamental level.

2. Protein Hydration Dynamics

Water is a natural solvent and almost all proteins are inactive in the absence of water. In the hydration layer of a protein, biological water molecules (protein-associated) are very heterogeneous and have different spatial and temporal characteristics. The dynamically ordered water controls functionally important protein fluctuations. Thus, elucidation of the structure and dynamics of biological water is fundamental to biology. A range of experimental techniques have been used over the past decades to tackle this important problem, reporting certain ordered-water structures but having limited temporal resolution to detect mobile water at protein surfaces. The femtosecond time resolution is ideal for studying hydration dynamics on the time scale of the actual water motions, both for "free" water in the vicinity of proteins in less than one picosecond and for bound water in the hydration layer on the picosecond time scale.

Various femtosecond-resolved probing strategies have been used to study protein solvation. However, chemical modification of an intrinsic amino acid residue (cysteine or lysine) in proteins is often restricted, and incorporation of a synthetic one is sophisticated. Direct use of the intrinsic amino acid tryptophan (Trp or W) as a local optical probe has been established very recently and applied to study protein hydration [1-3]. Ultrafast hydration dynamics in tens of picoseconds at protein surfaces has been revealed. In what follows, we report our recent studies of protein solvation (hydration) in several systems in correlation with their structures and functions.

2.1. *Folding and Aggregation: Melittin*

The studies of melittin have been published elsewhere [3, 4] and a summary is given here. Melittin, an amphipathic peptide from honeybee venom, consists of twenty-six amino acid residues and adopts different conformations from a random coil, to an α-helix, and to a self-assembled tetramer under certain aqueous environments. We have carried out systematic studies of the hydration dynamics in these conformations using single intrinsic tryptophan (W19) as a molecular probe. With femtosecond resolution, we observed the solvation dynamics occurring in 0.62 and 14.7 ps in a random-coiled primary structure. The former represents bulk-like water motion, and the latter reflects surface-type hydration dynamics of proteins. At a membrane-water interface, melittin folds into a secondary α-helical structure, and the interfacial water motion was found to take as long as 114 ps, indicating a well-ordered water structure along the membrane surface. In high-salt aqueous solution, the dielectric screening and ionic solvation promote the hydrophobic core collapse in melittin aggregation and facilitate the tetramer formation. This self-assembled tertiary structure is also stabilized by the strong hydrophilic interactions of charged C-terminal residues and associated ions with water molecules in the two assembled regions. The hydration dynamics were observed to occur in 87 ps, significantly slower than typical water relaxation at protein surfaces but similar to water motion at membrane interfaces. Thus, the observed time scale of ~100 ps probably implies appropriate water mobility for mediating formation of high-order structures of melittin in an α-helix and a self-assembled tetramer. These results elucidate the critical role of hydration dynamics in peptide conformational transitions and protein structural stability and integrity.

We have also studied solvent-assisted supramolecular folding by examining supramolecular solvation and local solvent structure [5]. The supramolecule (a crown ether derivative, Fig.1) in gas phase or nonpolar solvent is believed to be in a stretched form, but in polar solvent (acetonitrile), we found that it is in a folded structure. We observed a very large Stokes shift, slow

N,N'-Bis(2-(3-indolyl)ethyl)-4,13-diaza-18-crown-6

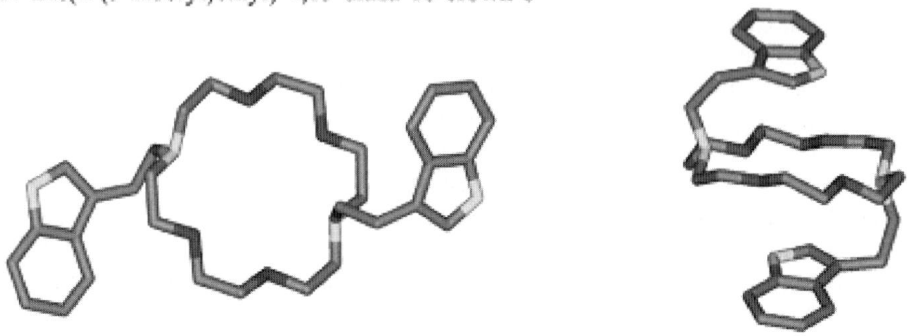

Fig. 1. (Left) Schematic representation of the molecular structure in tubular model in gas phase or nonpolar solvent. (Right) Folded molecular structure induced by polar solvent molecules (acetonitrile).

microsolvation dynamics, and fast rotational relaxation, revealing a highly ordered local solvent structure sandwiched between the two rings with a folded configuration. This folded complex structure results in enhancement of the local polarization and thus the large solvent stabilization energy (large Stokes shift in emission). We also studied the solvent mediation in recognition of cations by this supramolecule [5].

2.2. Conformational Transitions and Site-Selected Recognition: Human Serum Albumin

The structure and hydration of human serum albumin (HSA) are important for the transport of fatty acids and for binding a great variety of metabolites, drugs, and organic compounds in the circulatory system. It is a single polypeptide chain consisting of 585 amino acids. Under physiological conditions (pH~7), HSA adopts a heart-shaped 3D structure with three homologous domains I-III; each domain contains two subdomains A and B which consist of 4 and 6 α-helices, respectively. The X-ray structure shows that two halves of the albumin molecule form a 10-Å wide deep crevice which is filled with water molecules. The only single tryptophan residue W214 in the protein is located in the binding site IIA at the bottom of the crevice.

With change in pH, HSA undergoes reversible conformational isomerization. At neutral pH, the conformation of HSA is in its common physiological state, referred as a norm form (**N**); an abrupt transition occurs at pH value of less than 4.3, changing the **N** form to the so-called fast migrating form (**F**); and when the pH is less than 2.7, a further transition takes place from the **F** form to the extended form (**E**). On the other side of the scale, when the pH value becomes above 8, the **N** conformation changes to the basic form (**B**), and at pH above 10 the structure transforms to another aged form (**A**). These transitions are summarized below [6].

$$\begin{array}{lccccccccc} \textit{Conformation} & E & \longleftrightarrow & F & \longleftrightarrow & N & \longleftrightarrow & B & \longleftrightarrow & A \\ \textit{pH transition} & & 2.7 & & 4.3 & & 8 & & 10 & \end{array}$$

The protein has been extensively studied in every aspect due to its physiological importance, its potential as a drug delivery vehicle, and because different acidic and basic isomers may have certain biological function. Most studies have focused on binding interactions with various ligands and revealed that HSA is an assembly of squirmy and resilient components which frequently change in conformation through opening and closing of major crevices. With this "breathing" motion, HSA assimilates and releases a variety of substances during transportation in the circulation. Many complex structures have also been resolved at high x-ray resolution. The flexibility to adopt different conformations with independent segmental movements has also been shown through certain independent and sequential folding/unfolding of individual subdomains. The tryptophan W214 and extrinsic labeled dye molecules have been used as optical probes to study the conformational dynamics through changes in fluorescence lifetime, resonance energy transfer, and time-resolved anisotropy. The significant change in the lifetime of W214 for a series involving mutation in the binding site IIA clearly showed the local flexibility.

We have carried out systematic studies of solvation dynamics and local rigidity in these conformations using a single intrinsic tryptophan (W214) residue as a local molecular probe (Fig.2). With femtosecond resolution, we observed a robust bimodal distribution of time scales for solvation dynamics of all conformational isomers (Fig.3). The initial solvation occurs in several picoseconds, representing the local librational/rotational motions, followed by the dynamics, in the tens to hundreds of picoseconds, which result from the more bonded water in tryptophan crevice. Under the physiological condition of neutral pH, we measured ~100 ps for the decay of the solvation correlation function and observed large wobbling motion of W214 at the binding site which is deeply buried in a crevice, revealing the softness of the binding pocket and the large plasticity of the native structure. At the acidic pH, the albumin molecule transforms to an extended conformation with a large charge distribution at the surface, and a similar temporal behavior was observed. However, at the basic pH, the protein opens up the crevice and tightens its globular structure, and we observed significantly faster dynamics, 25-45 ps. These changes in the solvation dynamics are correlated with the conformational transitions and related to their structural integrity (Fig.4), suggesting that the "lock-and-key or induced-fit" concept of molecular recognition is incomplete without knowledge of dynamical solvation and plasticity of the protein [7].

2.3. Native and Molten Globule States: Apomyoglobin

Sperm whale myoglobin is a 17.7 kDa globular protein consisting of eight helical segments (named A-H). For all studies, we will use the apo-form to avoid resonance energy transfer from the excited tryptophan residue to the prosthetic heme group. At pH=6, apomyoglobin is in the native state, but at pH=4, it becomes a molten globule structure. Among 153 amino acid residues, there are two tryptophans, W7 and W14, in the A helix. To achieve site-specific detection, we first mutated tryptophan by tyrosine one at a time, and two mutant proteins W7Y and W14Y have been overexpressed and purified from *E. coli* in a large quantity (the plasmid pMb122, containing sperm whale myoglobin gene, was generously provided by Prof. Sligar's group at the University of Illinois at Urbana-Champaign). All mutants are carefully checked by various methods before and after the experiments to make sure that the protein did not change during measurements. By placing tryptophan at desirable positions along the protein with site-directed mutation and using a tryptophan scan, we map out global water motions around protein surfaces in the native and molten globule

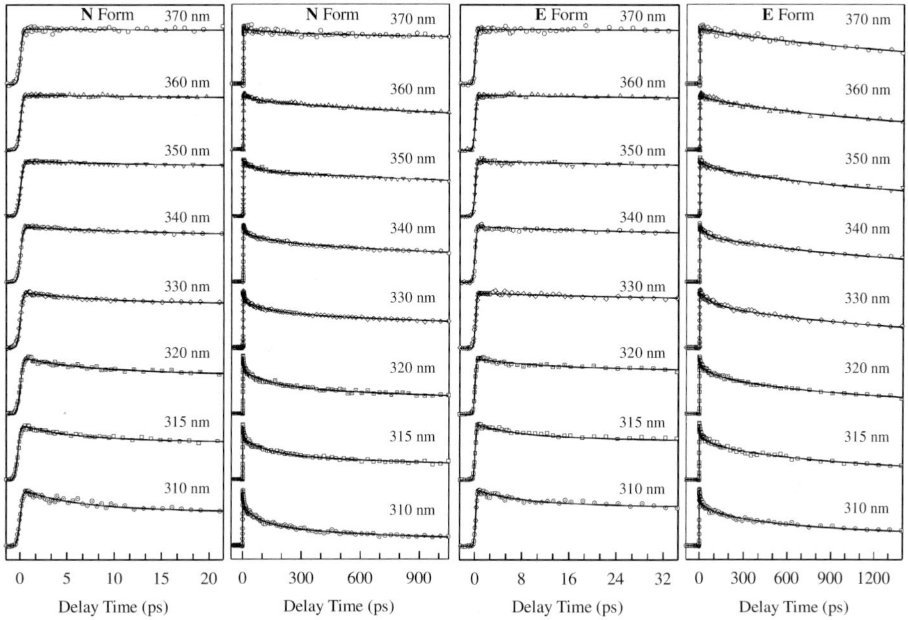

Fig. 2. Normalized, femtosecond-resolved fluorescence transients of the **N** and **E** isomers in the short and long time range, with a series of gated fluorescence emissions.

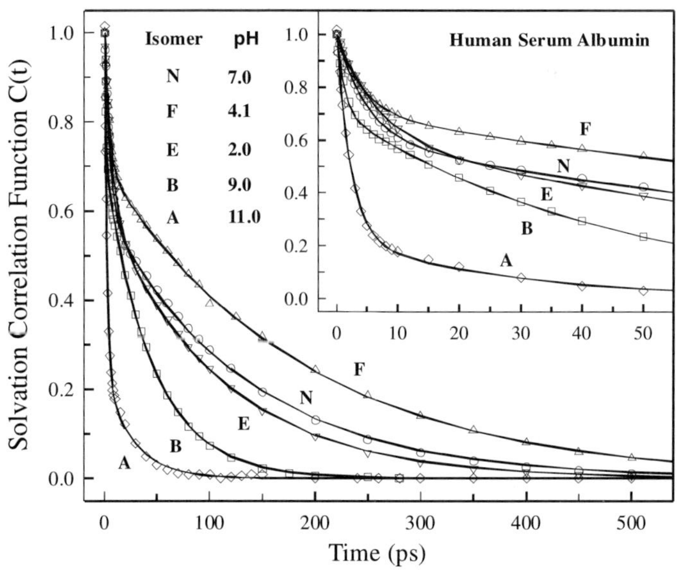

Fig. 3. Solvation correlation function probed by tryptophan (W214) in the five conformational isomers studied. The inert shows the correlation function in the short-time range.

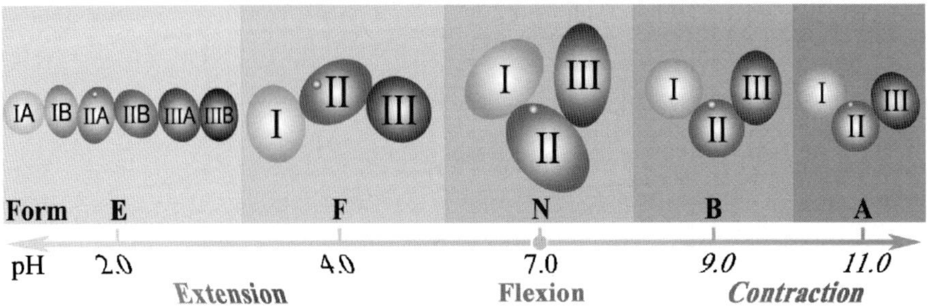

Fig. 4. Schematic representation of conformational transitions from the contracted configuration in basic pH, to the flexible globular structure at neutral pH, and to the extended form in acidic pH. The location of W214 is indicated by the white dot in domain IIA, and the observed solvation dynamics correlate well with these structural transitions.

states. We have already screened more than twenty triple mutants (two intrinsic tryptophan residues were mutated to W7YW14F), each with one surface-tryptophan residue. We have also examined water motions at several sites in the two states and clearly observed their different dynamic behaviors. Figure 5 shows the results of W12 (H12W/W7Y/W14F mutant) in the two states. The steady-state fluorescence emission peaks at 328 nm for the native (pH=6) and at 336 nm for the molten globule state (pH=4). Clearly, the transients from the molten globule state decay faster than those of the native one, indicating that water is less mobile (ordered) in the native state. The derived solvation correlation functions are presented in Figure 6. Both correlation functions show a picosecond solvation dynamics with a biphasic distribution: for the native

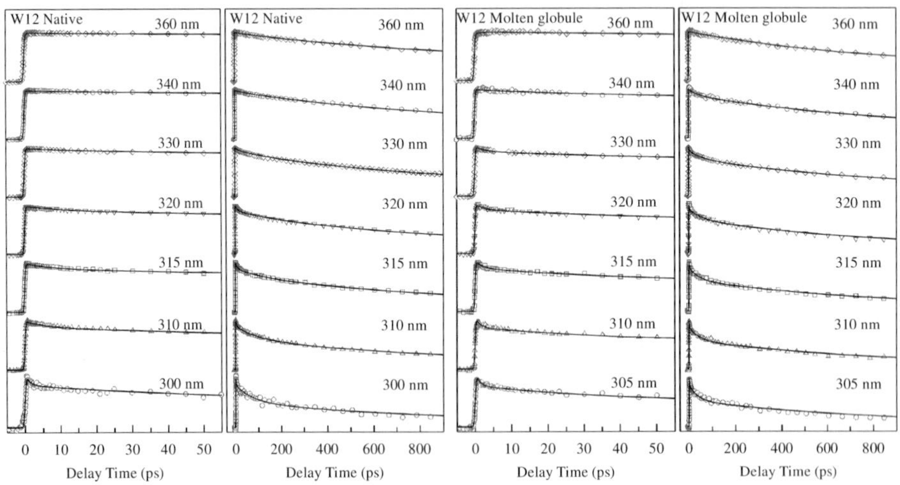

Fig. 5. Normalized, femtosecond-resolved fluorescence transients of W12 (H12W/W7Y/W14F mutant) in the native and molten globule states in the short and long time range with a series of gated fluorescence emissions.

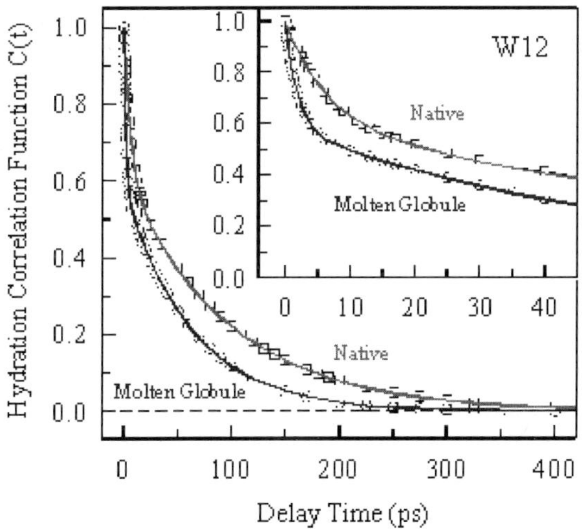

Fig. 6. Hydration correlation functions of the native and molten globule states probed by tryptophan W12 of A-helix in apomyoglobin. The inset shows the correlation function in the short-time range.

state, 6.2 ps (38%) and 98 ps (62%), and for the molten globule state, 1.9 ps (42%) and 63 ps (58%). These relaxation dynamics are dominantly from the ordered-water motions.

The observed different water dynamics show that, in the molten globule state, water has not been completely squeezed out of the local protein structure, and the structure is still partially solvated. The protein probably experiences a correlated desolvation at the late folding stage. In parallel with these experimental studies, molecular dynamics simulations are being carried out to obtain a mechanistic understanding of protein-water interactions [8].

3. Dynamics of Photoenzyme: DNA Repairs by Photolyase

We report here our recent studies of photolyase, a photoenzyme utilizing photon energy for a series of electron-transfer reactions to repair ultraviolet radiation damaged DNA in many organisms [9]. In what follows, we summarize our direct observation of active-site solvation and dynamic control of DNA repair efficiency; the detailed studies can be found in refs 10-14.

We have characterized the solvation dynamics in the active site of photolyase using the intrinsic cofactor $FADH^-$ and examined the dynamic role in the catalytic reactions. By gating a series of fluorescence wavelengths of the weak $FADH^{-*}$ emission from the blue to the red side, we obtained the fluorescence dynamics (Fig. 7) with systematic decays from 475 to 620 nm, showing significant solvation dynamics at the active site from several picoseconds until nanoseconds. This observation of continuous solvation processes, consistent with the highly polar active site from the x-ray structure, reveals a dynamic active site upon function initiation. The observed continuous motion of the active site directly controls the catalytic reactions of DNA repair through modulating the redox potentials of reaction species [13].

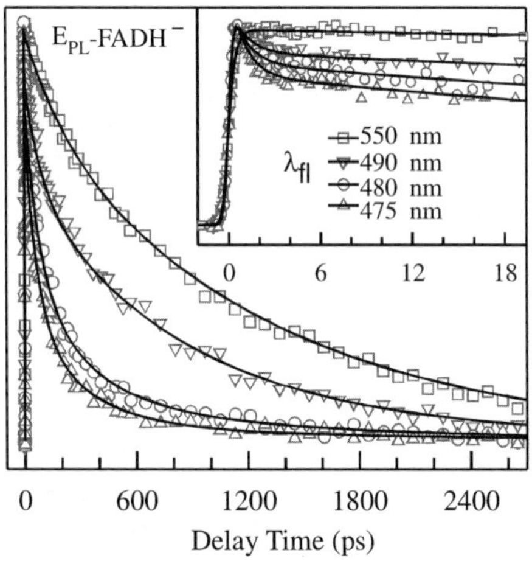

Fig. 7. Four typical gated fluorescence transients, with systematic decay from 475 to 550 nm, reflecting significant solvation at the active site. Inset shows the transients in the short time range.

Fig. 8 summarizes the results from our direct mapping of DNA repair by photolyase [13, 14]. The photocycle of the repair is through a radical mechanism and completed on subnanosecond time scale with no net change in the redox state of the flavin cofactor. Upon function initiation by a blue-light photon, the active site starts a continuous dynamical motion, which strongly couples with catalytic electron transfer reactions. To achieve high repair efficiency, the quantum yields of both charge-separation and ring-splitting reactions must be optimized to minimize the nonproductive pathway of charge recombination before ring cleavage. Unlike other sensitizer systems which have a low repair quantum yield, the active-site solvation in photolyase appears to be critical to strategically timing the forward charge separation (170 ps) and recombination, by dynamically tuning the redox potentials of reaction species and stabilizing the charge-separated radical intermediates, leaving enough time to cleave the cyclobutane ring (560 ps). These synergetic motions in the active site of the damaged DNA-enzyme complex, optimized by evolution, control a perfect timing for biological function and reveal its structural integrity and dynamical locality, a functional correlation between structure and dynamics, to ensure maximum repair efficiency (0.87) on ultrafast time scale of 560 ps.

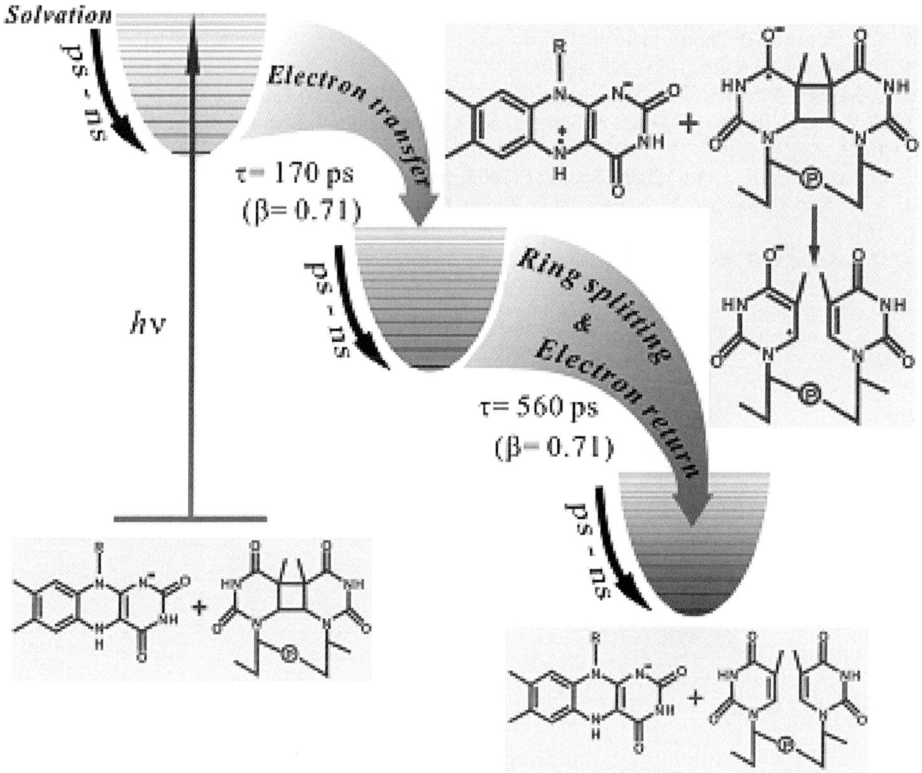

Fig. 8. Evolution of catalytic reactions of DNA repair by photolyase along the reaction coordinate. Active-site solvation strongly modulates the charge-separation, ring-splitting, and electron-return processes, resulting in charge separation in 170 ps and a stretched-single-exponential decay dynamics (β=0.71). The charge recombination must be slower than the complete ring splitting (560 ps) to eliminate possible ring reclosure and achieve a maximum repair quantum yield (0.87).

This work was supported by the National Science Foundation (CHE-0517334), the Petroleum Research Fund (PRF-42734-G4) and the Packard fellowship. Thanks to Ya-Ting Kao and Luyuan Zhang for preparation of some figures.

References

1. S. K. Pal and A. H. Zewail, Chem. Rev. 104 (2004) 2099.
2. W. Lu, J. Kim, W. Qiu, and D. Zhong, Chem. Phys. Lett. 388 (2004) 120.
3. W. Qiu, *et al.*, J. Phys. Chem. B 109 (2005) 16901.
4. W. Qiu, L. Zhang, L. Wang, and D. Zhong, in this book.
5. W. Lu, *et al.*, Chem. Phys. Lett. 394 (2004) 415.
6. V. M. Rosenoer, *et al.* (Eds), Albumin Structure, Function and Uses, Oxford, 1977.

7. W. Qiu, *et al.*, J. Phys. Chem. B, submitted.
8. Y.-T. Kao and D. Zhong, to be published.
9. A. Sancar, Chem. Rev. 103 (2003) 2203.
10. C. Saxena, A. Sancar, and D. Zhong, J. Phys. Chem. B 108 (2004) 18026.
11. H. Wang, C. Saxena, D. Quan, A. Sancar, and D. Zhong, J. Phys. Chem. B 109 (2005) 1329.
12. C. Saxena, *et al.*, J. Am. Chem. Soc. 127 (2005) 7984.
13. Y.-T. Kao, C. Saxena, A. Sancar, and D. Zhong, Proc. Natl. Acad. Sci. USA 102 (2005) 16128.
14. C. Saxena, Y.-T. Kao, A. Sancar, and D. Zhong, in this book.

Femtochemistry VII
Fundamental Ultrafast Processes in Chemistry, Physics, and Biology
A.W. Castleman, Jr. and M.L. Kimble

A Mechanism of Photo-Labile (De)-Protection in Chemical and Biological Systems Studied by Femtosecond Vibrational Spectroscopy

Stefan Laimgruber[a], Wolfgang Schreier[a], Peter Gilch[a]

[a]*Lehrstuhl für BioMolekulare Optik, Department für Physik, LMU München, Oettingenstr. 67, D-80538 München, Germany*

1. Introduction

Aromatic nitrocompounds are known to undergo photochemical transformation since the seminal work of Ciamician and Silber in 1901 [1]. In their paper "Chemische Lichtwirkungen" (Chemical Effects of Light) they describe the light induced reaction of *o*-nitrobenzaldehyde (NBA) yielding *o*-nitrosobenzoic acid or the ester of that acid. This work initiated modern photochemistry [2] and of course the photochemistry of aromatic nitrocompounds. It is now well documented that aromatics with diverse substituents in the *ortho* position to the nitro group feature photochemical activity [3]. This activity makes these compounds very attractive for protecting group chemistry [3]. Protecting groups render certain functional groups inactive. Usually the re-activation of these groups requires a chemical agent like an acid or hydrogen plus a catalyst. With an aromatic nitro compound as a protecting group this re-activation just requires to shed light on the sample. An example is the (de-)protection of a hydroxy function of the sugar residue of a DNA base by an *o*-nitrobenzyl type residue (Figure 1) [4].

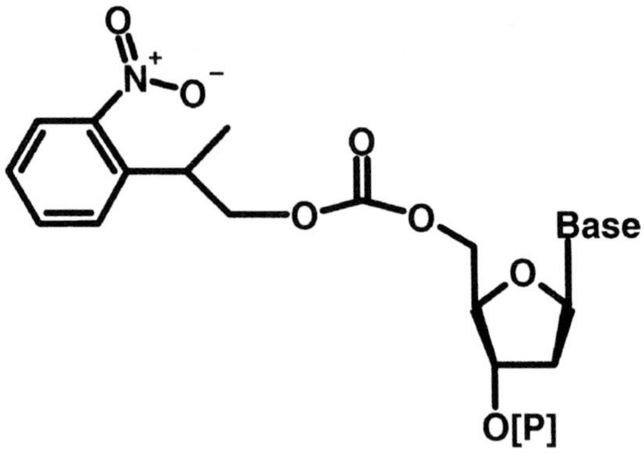

Figure 1 Example of a DNA nucleoside protected at its 5´ terminus by 2-(2-nitrophenyl)-propoxycarbonyl. Shedding light on this compound generates a free 5´ hydroxy terminus which can bind the next nucleotide. This allows the on spot synthesis of DNA single strands with a desired sequence. Example taken from ref. (4).

This example leads us to an important application of such protection groups. It is the basis for the in-situ synthesis of specific DNA sequences in high density DNA chips which are used for efficient genome analysis [4]. For this synthesis, nucleosides are immobilized by covalently linking them to a glass substrate via their 3' terminus. A photo-labile protecting group renders the 5' terminus in-reactive. Areas the size $25 \times 25\ \mu m^2$ or smaller are addressed photo-lithographically. Thereby, the 5'-ends of the oligonucleotides are de-protected; they now carry a free OH-group. The chip is then flushed with a solution of the desired nucleoside (A,C,G, or T). The nucleoside is activated at its 3' end so that it will attach to the 5' end of the oligonucleotide and is protected at its 5' end which allows to further extend the oligonucleotide.

The properties of the photo-labile protection group of course strongly affect the production process. For instance, the extinction coefficient of the group and the quantum yield settle the light doses necessary in the de-protection step. Side reaction can generate aggressive by-products that might attack the olignucleotides and so on. Thus, details on the photo-kinetics and the mechanism of these reactions are highly desirable. So it is quite surprising that femtosecond data on the photo-chemistry of aromatic nitro compounds are scarce (one example is the study of Schwörer and Wirz on nitrotoluene [5]). We have recently started to fill in this blank by looking in detail at the prototype of this reaction – the photo transformation of NBA [6]. In this study we lay particular emphasis on elucidating structures of intermediates relying on femtosecond vibrational spectroscopy.

2. Experimental Results

Measurements described in the following were performed on solutions of NBA in ethanol. As NBA in alcohols are slowly transformed to the respective hemi-acetale [7], freshly prepared sample solutions were employed in all experiments. The photoreaction was initiated with ~ 260 nm femtosecond laser pulses obtained from an amplified titanium sapphire laser systems by frequency tripling. Spectroscopic changes induced by absorption of these pump pulses were monitored in pump probe fashion by three spectroscopy techniques.

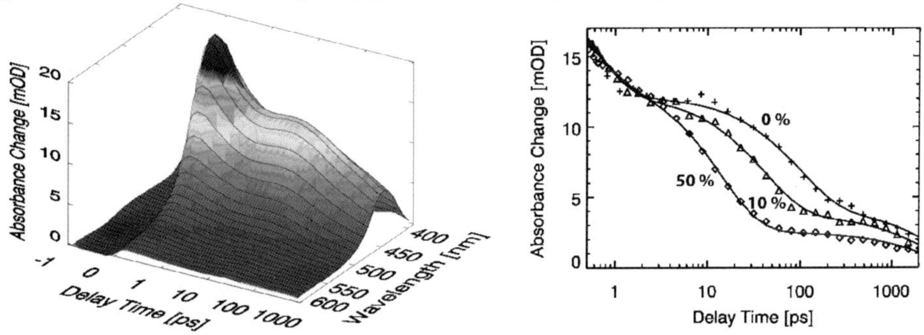

Figure 2 Femtosecond UV/Vis experiment on the photoreaction of NBA in ethanol. Left: Pump induced absorption changes versus delay time and detection wavelength. Right: Impact of the water content of the solvent on the difference absorption time trace recorded at 475 nm.

Multi-channel UV/Vis transient absorption spectroscopy gives an overview of the photo-kinetics. An induced absorption band centered around 500 nm rises within the time resolution of the experiment (~ 200 fs) and evolves on multiple time scales (Figure 2). In a global analysis at least three exponential terms are required to model the data, the best fit time constants being 400 fs, 90 ps, and 3 ns. The 400 fs constant characterizes the decay of the photo-excited singlet state; the two other constants indicate that at least two further intermediates are involved (referred to as **I** and **II**). Inspired by picosecond studies of Yip and Sharma [8], the dependence of the lifetime of intermediate **I** on the water content of solvent was studied. Indeed, the presence of water accelerates the decay of this intermediate. In a 1:1 water ethanol solution, the decay time is as short as 13 ps. The lifetime of intermediate **II** is hardly affected. Based on the water dependence Yip and Sharma [8] suggested that intermediate **I** is a ketene that results when a hydrogen atom is transferred from the aldehyde to the nitro substituent (cf. Figure 6). Femtosecond IR spectroscopy will deliver strong evidence for a ketene intermediate.

Ketenes feature very distinctive IR resonance around 2100 cm^{-1}, stemming from their anti-symmetric stretching vibration [9]. The ketene form of NBA trapped in an argon matrix was identified via this resonance [10]. In our UV pump IR probe experiment, an absorption band emerges ~ 400 fs after photo-excitation in the 2100 cm^{-1} region (Figure 3).

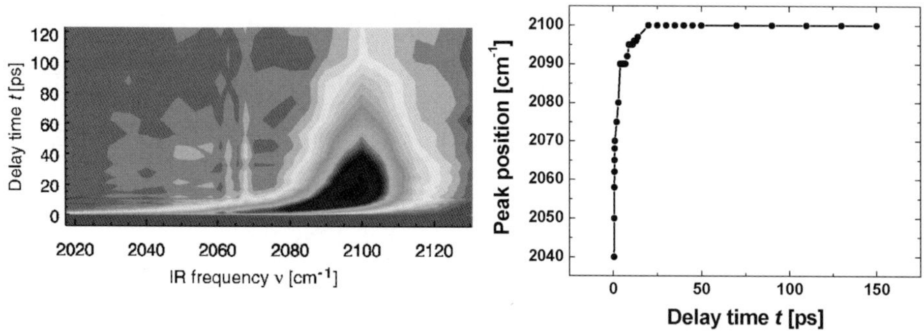

Figure 3 IR absorption changes in the 2100 cm^{-1} region induced by 260 nm excitation of NBA in ethanol. In the contour representation on the left dark shading stands for a strong induced absorption. The experiment reports on the formation and decay of the ketene intermediate. The resonance frequency of the ketene vibration changes with time – seen as a tail in the graph on the left. In the graph on the right this is highlighted by plotting the peak of the ketene band versus delay time.

The center of this band is first located around 2040 cm^{-1} and then up-shifts to an equilibrium value of 2100 cm^{-1} in around 20 ps. The band decays with a characteristic time of ~ 60 ps. This can be considered as an experimental proof that within 400 fs after photo-excitation NBA forms a ketene which then (in ethanol) transforms into intermediate **II** with a time constant of 60 – 90 ps. The origin of the pronounced frequency shift will be addressed below.

Time resolved IR spectroscopy also holds information on the structure of intermediate **II**. In the spectral area of carbonyl stretching vibrations around 1700 cm^{-1} one first observes a bleach induced by the photoexcitation of NBA (Figure 4). The bleached band is due to the carbonyl vibration of the aldehyde. Concomitant with the bleach recovery on a time scale of 10 – 100 ps, a band located at ~ 1730 cm^{-1} emerges. The rise of a band at this location points to a carbonyl function with a rather high resonance frequency in intermediate **II**. This suggests that intermediate **II** carries an ester or carboxy acid function [11].

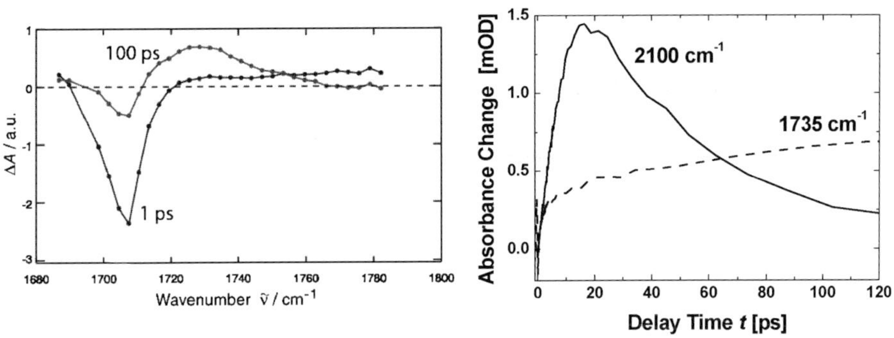

Figure 4 Carbonyl region of the IR difference spectra of NBA in ethanol (excitation at 260 nm). In spectra on the left negative signals stand for a reduction of the absorption (bleaching) and positive signals for induced absorptions of intermediates. Spectra before (1 ps) and after (100 ps) the reaction of the ketene are depicted. The bleach spectrum after 1 ps peaking around 1705 cm^{-1} coincides with the C=O stretching mode of the aldehyde group in NBA. The initial bleach is due to the change of the eigen frequency of this mode in the electronically excited state. With the decay of the excited state this bleach partially recovers (note that the quantum yield of the photo-reaction is smaller than one). Following that recovery an absorption band located 1730 cm^{-1} rises in a multi-exponential fashion on the time scale of 10 – 100 ps. The kinetics of the rise (1735 cm^{-1}) is compared with that of the decay of the ketene band (2100 cm^{-1}) in the graph on the right. The 1735 cm^{-1} trace has been displaced vertically for illustrative purposes. The initial rise of the ketene band is caused by vibrational cooling. During this cooling period the rate of the formation of intermediate **II** is accelerated.

Further structural details are revealed by Raman spectroscopy. This proves to be particularly sensitive to the bonding conditions of the phenyl ring. Depending on the reaction the ketene undergoes, the phenyl ring of intermediate **II** can either be aromatic or adopt an *ortho*-quinoniod configuration (Figure 6). The latter as other *ortho*-quinones should feature a pronounced Raman resonance around 1400 – 1500 cm^{-1} [12]. In recording the time resolved Raman spectra, we relied on the technique of femtosecond stimulated Raman scattering [13,14]. In this technique a femtosecond white light pulse overlaps in space and time with an intense laser pulse (Raman pump). Due to stimulated Raman scattering, the white light spectrum is modulated. The spectral characteristics of these modulations map the Raman spectrum of the sample. A third pulse (centered at 260 nm in our case) induces the photo-process, and photo-induced changes can be monitored in the usual pump probe fashion. The technique has several advantages over conventional time resolved Raman spectroscopy as an efficient fluorescence background suppression, higher signal levels, and better spectral and temporal resolutions. Still, as in any type of vibrational spectroscopy the problem remains that solvent resonances dominate the spectrum, and one is mainly interested in the resonances of the solute. The problem can be partially overcome by resonance enhancement as done here by using a 390 nm Raman pump pulse. Stimulated Raman spectra of intermediates **I** and **II** both contain pronounced resonances around 1500 cm^{-1}, indicative of an *ortho*-quinoniod structure (Figure 5).

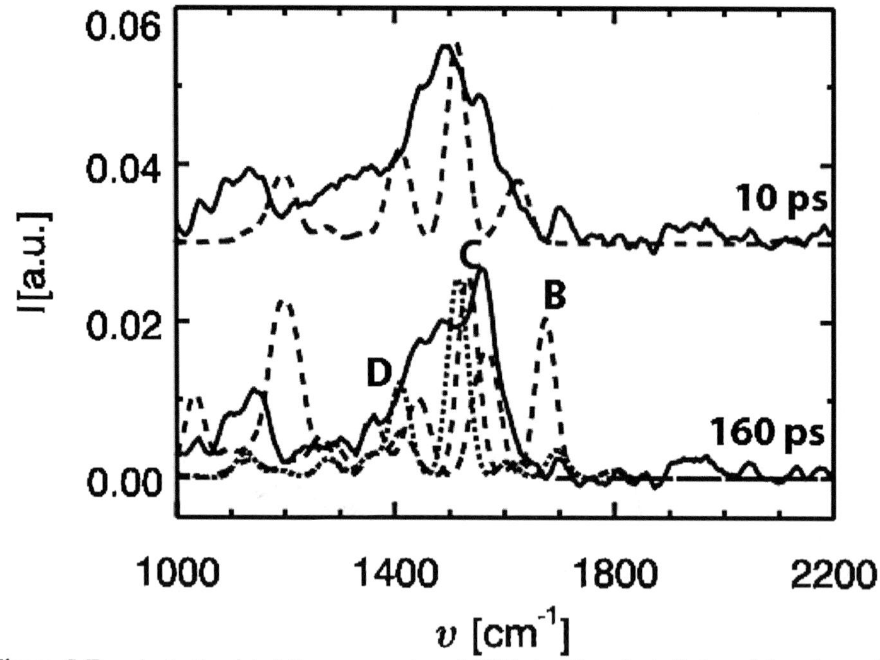

Figure 5 Transient stimulated Raman spectra of NBA in ethanol at distinct delay times after 260 nm photoexcitation. Transient spectra were calculated from experimentally obtained difference spectra by correcting for the bleach contribution. The 10 ps spectrum (intermediate **I**, solid line) features pronounced Raman resonance around 1500 cm^{-1} due to its *ortho*-quinoniod structure. The assignment is supported by a DFT calculation (dashed line). Intermediate **II** (160 ps, black line) also features such a resonance giving some confidence that this motif is retained when **I** reacts to form **II**. **B**, **C**, and **D** refer to spectra (dashed lines) calculated for the structures given in Figure 6.

Note that intermediate **I** (the ketene) *must* adopt an *ortho*-quinoniod structure. This can be taken as an indication that intermediate **II** also adopts such a structure. Density functional calculations support this notion. The spectrum of the ketene intermediate is nicely reproduced by these calculations. For the intermediate **II**, calculated spectra of a molecule adopting an *ortho*-quinoniod structure match better with the experimental spectrum than aromatic molecules.

3. Devising a Reaction Mechanism

The present spectroscopic findings give unequivocal evidence that, within 400 fs after photo-excitation, NBA transforms into a ketene (Figure 6). The transformation requires the hydrogen atom to shift from the aldehyde substituent to the nitro group. The x-ray structure of NBA shows that the hydrogen atom of the aldehyde substituent points to the nitro group that should promote a fast hydrogen transfer [15,16]. The hydrogen transfer does not seem to proceed via a radical mechanism [17], as we have found no evidence for a bi-radical intermediate. This points to an electrocyclic mechanism involving singlet states.

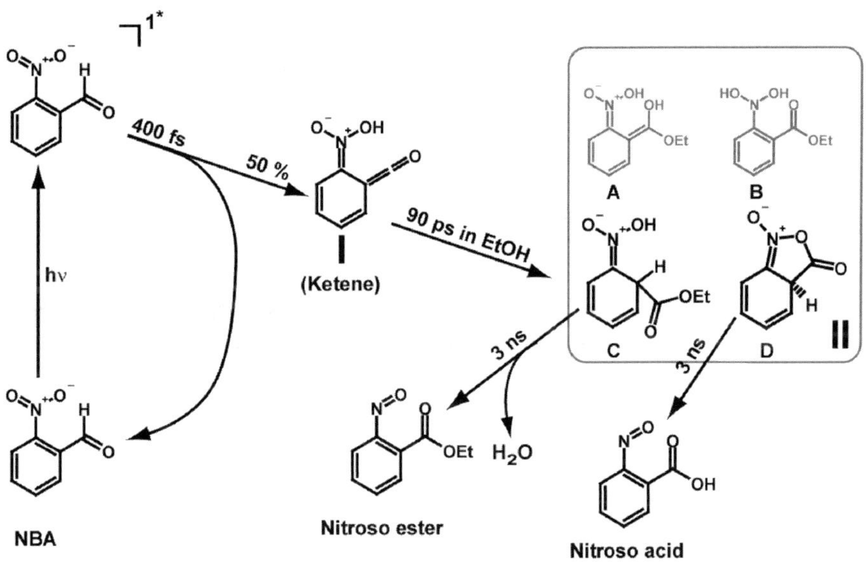

Figure 6 Scheme graphically summarizing kinetic and structural data retrieved from femtosecond experiments. Intermediate **I** has been identified as a ketene. Several potential structures of intermediate **II** can be ruled out. Structure **A** lacks a carbonyl group and is not in line with the IR results. The phenyl ring in structure **B** is in conflict with the Raman measurements. Structures **C** and **D** resulting from an inter- or intramolecular reaction, respectively, are both in line with the spectroscopic data. Based on the product distribution we currently favor structure **D**.

In line with results of Yip and Sharma [8], no spectroscopic indications for triplet states have been seen. The 400 fs reaction not only generates the ketene intermediate but also partially recovers the starting material. The quantum yield of the photo-redox reaction of NBA is around 50 % more or less independent of the nature of the solvent [17]. Since the kinetics ensuing the formation of the ketene are strongly solvent dependent (see ref. [8] and Figure 2), it has to be the formation of the ketene intermediate which settles the overall quantum yield of the reaction. A likely scenario for this process is that the motion of the hydrogen atom brings the potential energy surfaces of the ground and excited state in energetic vicinity. There, internal conversion to the ground state takes places which goes along with a branching between the ketene and the starting material.

Immediately after internal conversion, the ketene will be vibrationally highly excited. Due to anharmonic effects, vibrational excitations result in frequency shifts of IR and Raman resonances (see e.g. ref. [18]). The motion of the ketene band in Figure 3 towards higher frequency, thus, reports vibrational relaxation and cooling which here takes around 10 – 20 ps. The natures of the vibrational excitation and relaxation still needs further investigation. For instance, it is presently not clear whether the shift of the ketene is mainly due to the vibrational excitation of this mode or due to other modes anharmonically coupled. An interesting effect seems to go along with this vibrational excitation. Based on the UV/Vis and IR data in the 2100 cm^{-1} region, a reaction time of the ketene of 60 – 90 ps is deduced. When monitoring the rise of ketene product in the 1700 cm^{-1} region, multi-exponential kinetics are observed with components significantly shorter than the 60 – 90 ps. A rationale for these kinetics could be that the reaction of the ketene involves a barrier. At early times, the ketene has a high internal temperature, and surmounting the barrier is easier and the reaction fast. As the internal temperature goes down, the reaction time increases.

So which reaction do the ketenes undergo? In general ketenes are known to be subject to nucleophilic additions. The nucleophil |NuH can in principle add to the C=C or the C=O double bond. It is commonly believed that the C=O addition is favored. In the present case, the reaction of the ketene forms a molecule with a C=O double hinting to a C=C addition (or less likely a C=O addition followed by a rapid tautomerization.). The indications for an ortho-quinoniod structure of intermediate **II** shows that a 1,2 addition occurs, i.e. the proton of |NuH adds to the carbon atom adjacent to the central atom of the ketene function.

A question surprisingly difficult to answer is whether the 1,2 addition to the C=C double bond is an intra- or intermolecular process. In an intramolecular reaction, the OH group of the ketene form takes the role of the nucleophil. For the intermolecular process, the solvent (ethanol here) would act as |NuH. Based on our present spectroscopic indicator, we cannot decide upon this issue. Two arguments can be put forth for the intermolecular process. First, the strong solvent dependence of the reaction time can be easily rationalized when assuming that the ketene reacts with the solvent. Second, already Ciamician and Silber observed that in ethanol NBA photo-reacts to form mainly the ester and not the free acid [1]. The ester obviously seems to be the product of an intermolecular reaction. Both arguments can be questioned. The solvent could also act catalytically by stabilizing for instance an ionic transition state via hydrogen bonding. The formation of the ester could result from a photo-reaction of the hemi-acetale. As mentioned above, NBA reacts with the ethanol solvent to form a hemi-acetale [7]. The time scale of this reaction is comparable with time it takes to photo-transform NBA in a steady state experiment. Since these hemi-acetales are known to photo-transform to nitroso-esters [7], it could well be the hemi-acetale that is responsible for the ester formation and not NBA itself. A simple titration experiment underscores this view. We illuminated a solution of NBA in 2-butanol. In 2-butanol the hemi-acetale formation is strongly suppressed. By acid-base titration, we determined the portion of acid formed in relation to the NBA amount. We found a lower boundary of 75 % for this portion. This gives some support for the intramolecular pathway. Further time resolved work is under way to clarify this issue. For either pathway, the final process is a re-arrangement which re-establishes the aromaticity of the phenyl ring and forms the nitroso group. In neat ethanol this step takes about 3 ns.

In summary, by means of time resolved vibrational spectroscopy we have mapped out the kinetics of crucial intermediates in the photochemistry of NBA. One intermediate is a ketene formed by an ultrafast (~ 400 fs) hydrogen transfer process. The transfer is most likely an electrocyclic reaction and proceeds with less than unity quantum yield. Since the hydrogen transfer is regarded as the canonical reaction step in the photochemistry of all aromatic nitro compounds, our results will help understand photo-labile protecting groups based on these compounds and in the long run help optimize them.

References

1. G. Ciamician and P. Silber, Chem. Ber. 35 (1901) 2040.
2. H. D. Roth, Pure Appl. Chem. 73 (2001) 395.
3. A. P. Pelliccioli and J. Wirz, Photochem. Photobiol. Sci. 1 (2002) 441.
4. X. L. Gao, E. Gulari, and X. C. Zhou, Biopolymers 73 (2004) 579.
5. M. Schworer and J. Wirz, Helv. Chim. Act.a 84 (2001) 1441.
6. S. Laimgruber, W. J. Schreier, T. Schrader, F. Koller, W. Zinth, and P. Gilch, Angew. Chem., Int. Ed. in press (2005).
7. A. A. Pashayan, A. L. Prokhoda, and S. A. Sarkisyan, High Energy Chem. 11 (1977) 41.
8. R. W. Yip and D. K. Sharma, Res. Chem. Intermed. 11 (1989) 109.
9. J. L. Duncan, A. M. Ferguson, J. Harper, K. H. Tonge, and F. Hegelund, J. Mol. Spectrosc. 122 (1987) 72.
10. S. Kuberski and J. Gebicki, J. Mol. Struct. 275 (1992) 105.
11. B. Schrader, Infrared and Raman Spectroscopy, VCH, Weinheim, New York, Basel, Cambridge, Tokyo, 1995.
12. R. T. Packard and R. L. McCreery, J. Phys. Chem. 92 (1988) 6345.
13. M. Yoshizawa and M. Kurosawa, Phys. Rev. A 6001 (2000) art. no.

14. D. W. McCamant, P. Kukura, S. Yoon, and R. A. Mathies, Rev. Sci. Instrum. 75 (2004) 4971.
15. P. Coppens, Acta Crystallographica 17 (1964) 573.
16. P. Coppens and G. M. J. Schmidt, Acta Crystallographica 17 (1964) 222.
17. M. V. George and J. C. Scaiano, J. Phys. Chem. 84 (1980) 492.
18. T. Schrader, A. Sieg, F. Koller, W. Schreier, Q. An, W. Zinth, and P. Gilch, Chem. Phys. Lett. 392 (2004) 358.

Femtochemistry VII
Fundamental Ultrafast Processes in Chemistry, Physics, and Biology
A.W. Castleman, Jr. and M.L. Kimble

Photochemistry and Dynamics of Base Pairs

Elena Samoylova, V.R. Smith, H.-H. Ritze, W. Radloff, I. V. Hertel, Thomas Schultz

Max Born Institute Berlin, Max Born Str. 2A, 12489 Berlin

1. Introduction

The genetic information of all living organisms on earth is encoded in DNA, and the DNA bases form the characters carrying this information. To ensure survival in harsh conditions, DNA and its building blocks must be quite robust against degradation or chemical modification. We want to understand the stability of DNA against photochemical damage arising from irradiation with ultraviolet light. Investigating small building blocks of DNA, we hope to obtain a detailed understanding of photochemical and photophysical processes which may allow a 'bottom-up' description of the photochemical properties of DNA. Here we present our recent advances investigating base pairs and micro-hydrated clusters composed of the bases adenine (A) and thymine (T), as well as the model base pair aminopyridine dimer. For our investigations, we used time-resolved excitation-ionization spectroscopy with 1 kHz, ~100 fs pulses in the ultraviolet for photoexcitation and 400 or 800 nm pulses for multiphoton-ionization.

2. Model Base Pair

Aminopyridine dimer served as a model compound for a Watson-Crick DNA base pair (Fig. 1). *Ab initio* calculations predicted that a rapid electron-proton transfer (net H-transfer) in the excited state provides an efficient relaxation pathway to the ground state. We identified this pathway experimentally [1], but the dimer lifetime of 55-85 ps indicated either a considerable barrier on the H-transfer coordinate or an unexpected weak coupling for the electronic transition from the $\pi\pi^*$ charge-transfer state to the electronic ground state. We now characterized the isotope effect for this reaction in the dimer of N,N-dideutero-aminopyridine (Fig. 1, right hand side).

The replacement of all amino-protons prolonged the excited state lifetime by a factor of $\tau_D/\tau_H \approx 3$. This considerable isotope effect indicates a barrier on the H-transfer coordinate, which may explain the picosecond lifetime of the excited state.

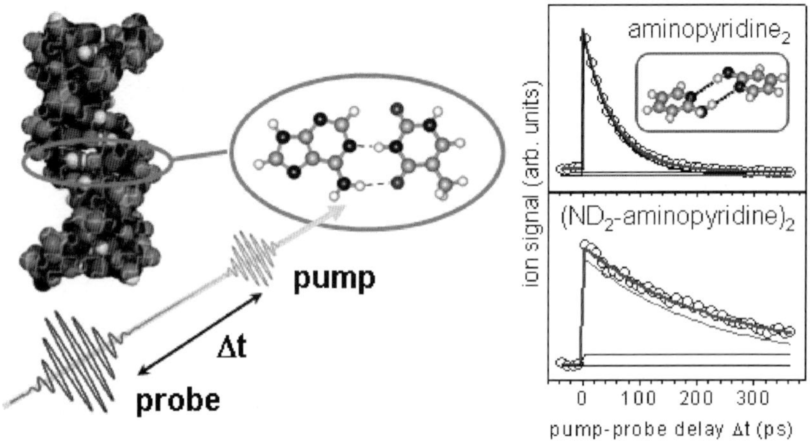

Fig. 1: Investigation of aminopyridine dimer as a model for a Watson-Crick DNA base pair (e.g. adenine-thymine in inset, center) by pump-probe spectroscopy with 293 nm excitation and 800 nm multiphoton ionization. The short lifetime $\tau = 55$ ps of the dimer (top right) was assigned to an electron-proton transfer in the excited state [2]. The deuterated N,N-dideutero-aminopyridine showed a considerable isotope effect (bottom right), prolonging the lifetime to $\tau \approx 150$ ps.

3. DNA Bases and Clusters

For the Watson-Crick AT base pair, a similar intermolecular H-transfer reaction was predicted by theory [2]. In experiments, however, we identified intra-monomer relaxation mechanisms involving a locally excited $\pi\sigma^*$ state for the AA, AT, and TT base pairs [3]. In the clusters, the $\pi\sigma^*$ state is stabilized relative to the photoexcited $\pi\pi^*$ and $n\pi^*$ states which dominate the relaxation in the monomers. Therefore, signal contributions from the $n\pi^*$ state are strongly suppressed in the base pairs, as readily apparent for the adenine dimer in Fig. 2. The discrepancy between the theoretically predicted intermolecular mechanism and our experimentally observed intra-monomer relaxation may be due to the structure of the AT base pair; experiment and theory [4] indicate that the Watson-Crick structure is less stable than other H-bound structures in the molecular beam. An experimental confirmation of the existence of an H-transfer mechanism in real DNA base-pairs is therefore currently unavailable.

Upon hydration, the hydrogen-bonding sites of DNA bases are occupied by water molecules, and DNA base pairs adopt a π-stacked structure. *Ab initio* calculations predicted that such π -stacked structures can also be observed for microhydrated clusters with two or more water molecules [5]. We investigated the excited state lifetime of the microhydrated adenine dimer and identified a long-lived excited state (x* in Fig. 2, bottom right). Due to extensive fragmentation, this long-lived state was observed in all $A_2(H_2O)_n$ clusters with n = 0-5 but vanished when we reduced the width of the cluster distribution to exclude clusters with n > 2 water molecules (see traces for A_2 and $A_1(H_2O)_3$ in Fig. 2). We therefore assign the x* state to adenine dimers with n ≥ 2, in agreement with the theoretical prediction of π-stacked structures. We tentatively assign this state to an excimer with delocalized excitation spread over the π-system of two chromophores. Similar observations of a long-lived state were made for single-stranded poly(A) and double-stranded poly(a)-poly(T) in water [6], indicating that this state might play a role in larger systems which are closer to biology.

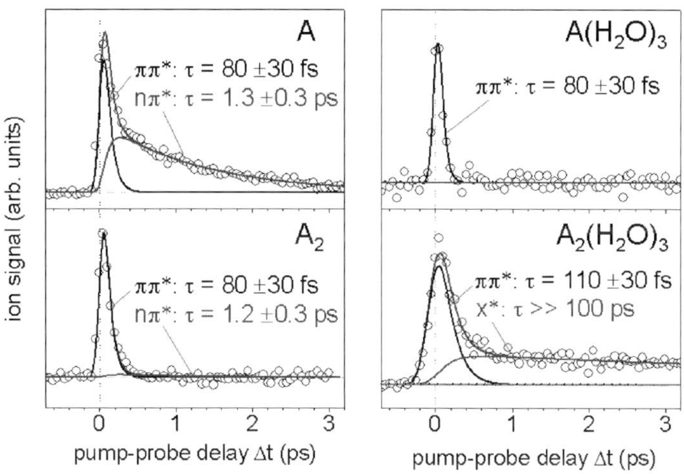

Fig. 2: Time-resolved ion signals for clusters of adenine and water with 266 nm excitation, 800 nm ionization. The spectra are from separate measurements, optimized for minimal fragmentation in the ion channel of interest. The monomer shows fast ($\tau_{\pi\pi^*} \approx$ 80 fs) internal conversion of the photoexcited ππ* state to the nπ* state and subsequent internal conversion to the ground state ($\tau_{n\pi^*} \approx$ 1.3 ps). In A_2 and $A(H_2O)$, contributions from the nπ* state are suppressed due to a relaxation pathway via the πσ* state. $A_2(H_2O)_3$ shows a novel, long-lived excited state x*, probably due to excimer states in the π-stacked base pair.

Electronic spectra from femtosecond electron-ion coincidence spectroscopy (FEICO) allow the direct observation of electronic and nuclear structure along the reaction pathway [7]. Only a single ionization event per laser shot can be observed to allow the assignment of an electron to a cluster species. We therefore work with small ionization rates of ~10^{-1} events per laser shot (eps) and usually count <10^{-3} eps (<<1 count per second) in the cluster channel of interest. Fig. 3 shows a first result: the FEICO spectrum for 266 nm one-color, two-photon ionization of toluene, adenine, and adenine-water clusters. Our measured vertical ionization potentials of 8.92 eV for toluene and 8.44 eV for adenine (calibrated against the known ionization potentials of NO) identify the ionization process as $\pi\pi^* \rightarrow \pi^1$ transition and correspond to known one-photon ionization potentials of 8.86 and 8.48 eV. A pronounced red shift of the ionization potentials (~50 meV per water molecule) reflects a stabilization of the ionic charge by polar water molecules.

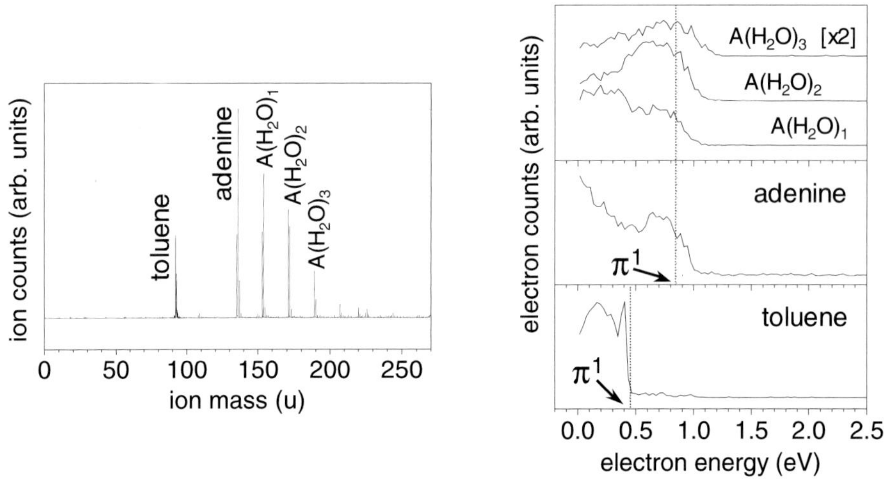

Fig. 3: FEICO spectra of toluene, adenine and adenine-water clusters. Left: mass spectrum and the assignment of ions and ionic clusters. Right: photoelectron spectra of selected ions and clusters. Ionization into the π^1 ionic ground state is observed for adenine (IP = 8.48 eV) and toluene (IP = 8.86 eV). Clustering with water lowered the ionization potential of adenine by 50-150 meV.

Further FEICO experiments are currently performed to experimentally characterize the electronic states involved in the relaxation processes of the DNA base pairs and model base pairs mentioned above. The combination of time-resolved mass and FEICO spectroscopy of isolated clusters promises detailed insights into ultrafast processes and offers a powerful tool to verify theoretical predictions.

References

1. T. Schultz, E. Samoylova, W. Radloff, I. V. Hertel, A. L. Sobolewski, and W. Domcke, Science 306 (2004) 1765.
2. Oral communications from S. Perun, A. L. Sobolewski, and W. Docke at this conference.
3. E. Samoylova, H. Lippert, S. Ullrich, I. V. Hertel, W. Radloff, and T. Schultz, J. Am. Chem. Soc. 127 (2005) 1782.
4. (a) C. Plützer, I. Hünig, K. Kleinermanns, E. Nir, and M. S. deVries, Chemphyschem 4 (2003) 838; (b) M. Kabelac and P. Hobza, J. Phys. Chem. B 105 (2001) 5804.

5. M. Kabelac and P. Hobza, Chem. Eur. J. 10 (2001) 2067.
6. C. E. Crespo-Hernández, B. Cohen, and B. Kohler, Nature 436 (2005) 25.
7. V. Stert, W. Radloff, C. P. Schulz, and I. V. Hertel, Eur. Phys. J. D 5 (1999) 97.

Femtochemistry VII
Fundamental Ultrafast Processes in Chemistry, Physics, and Biology
A.W. Castleman, Jr. and M.L. Kimble

2D Electronic Spectroscopy of the B800-B820 LH3 Light-Harvesting Complex

Donatas Zigmantas,[a,b] Elizabeth L. Read,[a,b] Tomáš Mančal,[a,b] Tobias Brixner,[a,b,d] Alastair T. Gardiner,[c] Graham R. Fleming[a,b]

[a]*Department of Chemistry, University of California, Berkeley, CA 94720-1460, USA*
[b]*Physical Biosciences Division, Lawrence Berkeley National Laboratory, Berkeley, CA 94720, USA*
[c]*Division of Biochemistry and Molecular Biology, Institute of Biomedical and Life Science, University of Glasgow, Glasgow G128QQ, UK*
[d]*Current address: Physikalisches Institut EP1, Universität Würzburg, 97074, Würzburg, Germany*

Abstract

Two-dimensional optical spectroscopy is a potent tool for studying correlations between different vibrational and electronic transitions as well as other non-linear properties of molecules and materials. We utilized recently developed phase-stabilized two-dimensional (2D) electronic Fourier-transform spectroscopy to probe electronic couplings between different chromophores and explore the distribution of instantaneous molecular environments in photosynthetic light-harvesting complexes. Here we report a study on the B800-B820 (LH3) complex, a spectroscopic variant of the B800-B850 LH2 peripheral light-harvesting complex from purple bacteria *Rhodopseudomonas (Rbs) acidophila* strain 7050.

1. Introduction

Light harvesting is a crucial part of photosynthesis, ensuring efficient adaptation of photosynthetic organisms to environments with low light conditions. Herein we consider the elegant ring shaped light-harvesting antennae of purple bacteria, which exhibit nearly unity efficiency of energy transfer to the photochemical reaction center. Experimental and theoretical studies of bacterial light-harvesting complexes [1,2] have been greatly facilitated by determination in recent years of the crystal structures [3-5]. The peripheral light harvesting complex (LH2) of *Rbs acidophila* has two concentric rings of 9 and 18 bacteriochlorophyll (BChl) molecules labeled B800 and B850, respectively, corresponding to their Q_Y transition band centers. Some purple bacteria, when grown under stressed conditions corresponding to low light illumination and/or low temperatures, synthesize a spectroscopic variant of LH2 known as LH3 with Q_Y absorption maxima at 800 and 820 nm. The observed spectral shift of the lower-energy Q_Y band is attributed to differences in the primary structure of the apoproteins [6]. B800 to B820 energy transfer in LH3 occurs in ~.75 ps at room temperature and is comparable to the interband transfer rate in LH2 [7]. In this study LH3 was investigated by means of 2D electronic spectroscopy, which allows full characterization of the third-order nonlinear response by a single experiment.

2. Results and Discussion

The LH3 linear absorption spectrum (Fig. 1) shows the B800 and B820 bands located at 804.5 nm and 823.5 nm.

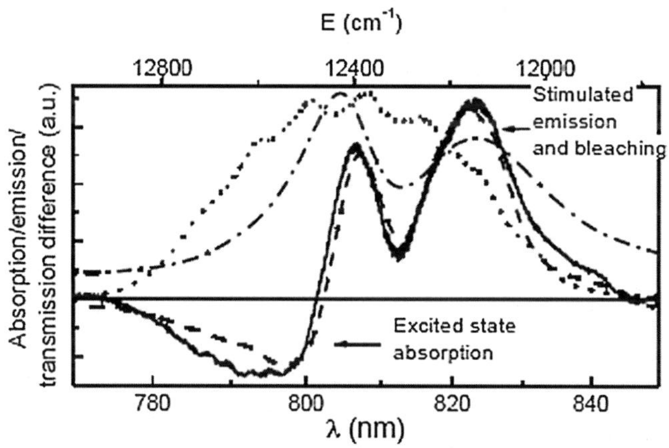

Figure 1. LH3 absorption at 77 K (dash dotted line), transient electric field (solid line), phased 2D spectrum projection (dashed line), laser spectrum (dotted line).

The laser spectrum used in the measurements (dotted line in Fig. 1) covers most of the BChl Q_Y transitions, except for the red tail above 840 nm. The separately measured, frequency-resolved transient electric field amplitude (analogous to a transient absorption measurement) features two bleaching/stimulated emission bands at 807 nm and 823 nm. The projection of the 2D spectrum phased to the transient electric field spectrum is also shown.

2D Fourier transform electronic spectroscopy (2D FOTRES) [8,9] has proved to be a very informative spectroscopic technique for investigating multichromophoric biological complexes [10]. Briefly, the three pulse echo signals generated by focusing a sequence of three variably-delayed laser pulses (pulselength ~40 fs, spectral bandwidth ~30 nm) on the sample was heterodyned with a fourth, local oscillator (LO) pulse and detected by spectral interferometry, giving the dependence of the signal electric field on the detection frequency, ω_t. By scanning the coherence (first) delay between pulses for a fixed population (second) delay and performing Fourier analysis, the signal dependence on the coherence frequency ω_τ is obtained, resulting in 2D spectra for each population time T (Fig. 2). The 2D spectra corresponding to the real part of the signal electric field were phased by comparing their projections onto the detection frequency axis with the frequency resolved transient electric field amplitudes (Fig. 1).

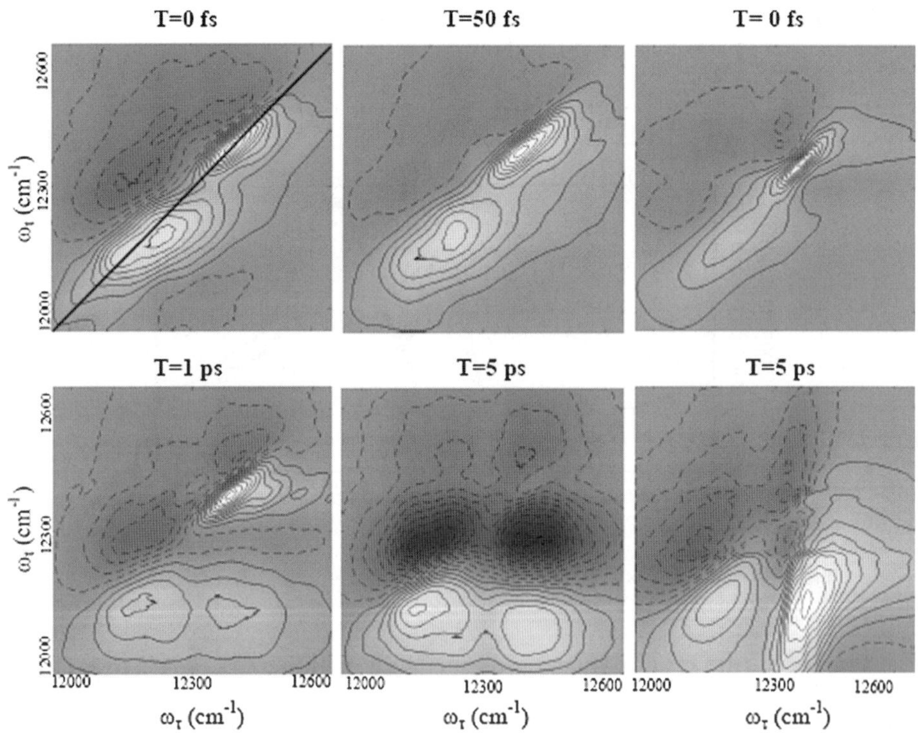

Figure 2. 2D LH3 experimental spectra (left and central columns) and theoretical spectra (right column) at population times, T, as indicated above the spectra. The top right diagonal peak corresponds to the B800 band and lower left to the B820 band.

In this way, full amplitude and phase information for the third-order polarization induced in the sample is determined. Several features of the system are apparent in the T = 0 2D spectrum (top left image in Fig. 2). Because of the static disorder, the inhomogeneous (diagonal) linewidth is larger than the homogeneous (antidiagonal) linewidth for both the B800 and B820 diagonal peaks. Furthermore, the homogenous linewidth of the B820 band is broader than that of the B800 band due to the much stronger interactions between BChl molecules in B820 as compared to B800, which cause very fast relaxation between B820 excitonic levels. Also, the B820 peak is shifted to lower frequencies off the diagonal, indicating a larger reorganization energy as compared to the B800 ring. Dynamics occurring on two time scales are observed. Evolution of coherent dynamics happens in B820 within 50 fs, as seen in 2D spectra with T = 0 and 50 fs in Fig. 2, wherein a double-peak feature of the B820 band disappears as the B800 peak remains unchanged. Energy transfer from B800 to B820 occurs with a time constant of ~1 ps and is marked by the appearance of a crosspeak at lower detection frequency and of corresponding negative excited state absorption features, as seen in the 2D spectra with T = 1 and 5 ps. The peaks grow rounder with increasing population time as electronic transitions lose memory of their initial transition frequencies, caused by the intraband energy transfer and protein environment fluctuations.

Theoretical simulation of the experiment is underway employing a Frenkel exciton Hamiltonian including energetic disorder to model the BChls and surrounding protein. 2D signals are calculated using the third-order nonlinear response function formalism in the impulsive limit with a Brownian oscillator transition frequency correlation function. Intra- and interband relaxation rates are calculated by modified Redfield and generalized Förster methods, respectively. The calculations match the LH3 linear absorption spectrum well and qualitatively reproduce general features of the 2D data (Fig. 2, right panels). A comprehensive, quantitative analysis of system parameters including electronic couplings, static disorder, reorganization energies, and coherence correlation times requires further calculation and will be reported in the future.

References

1. V. Sundstrom, T. Pullerits, and R. van Grondelle, J. Phys. Chem. B 103 (1999) 2327.
2. G. D. Scholes and G. R. Fleming, J. Phys. Chem. B 104 (2000) 1854.
3. G. Mcdermott, S. M. Prince, A. A. Freer, A. M. Hawthornthwaitelawless, M. Z. Papiz, R. J. Cogdell, and N. W. Isaacs, Nature 374 (1995) 517.
4. J. Koepke, X. C. Hu, C. Muenke, K. Schulten, and H. Michel, Structure 4 (1996) 581.
5. A. W. Roszak, T. D. Howard, J. Southall, A. T. Gardiner, C. J. Law, N. W. Isaacs, and R. J. Cogdell, Science 302 (2003) 1969.
6. K. McLuskey, S. M. Prince, R. J. Cogdell, and N. W. Isaacs, Biochemistry 40 (2001) 8783.
7. Y. Z. Ma, R. J. Cogdell, and T. Gillbro, J. Phys. Chem. B 102 (1998) 881.

8. T. Brixner, T. Mancal, I. V. Stiopkin, and G. R. Fleming, J. Chem. Phys. 121 (2004) 4221.
9. T. Brixner, I.V. Stiopkin, and G. R. Fleming, Opt. Lett. 29 (2004) 884.
10. T. Brixner, J. Stenger, H. M. Vaswani, M. Cho, R. E. Blankenship, and G. R. Fleming, Nature 434 (2005) 625.

Femtochemistry VII
Fundamental Ultrafast Processes in Chemistry, Physics, and Biology
A.W. Castleman, Jr. and M.L. Kimble

Spectrally Resolved Femtosecond Nonlinear Spectroscopy for the Study of Vibrational Dynamics of Carotenoid Light Harvesting Molecules

Lap Van Dao[a], My Thi Tra Do[a], Peter Hannaford[a], Junji Akahane[b] and Yasushi Koyama[b]

[a]*Centre for Atom Optics and Ultrafast Spectroscopy, ARC CoE for Coherent X-Ray Science, Swinburne University of Technology, Melbourne, Australia 3122*
[b]*Faculty of Science and Technology, Kwansei Gakuin University, Sanda 669-1337, Japan*

1. Introduction

In photosynthesis, carotenoid (Car) molecules play a number of roles: (1) light harvesting via singlet-state energy transfer to Bacteriochlorophyll, (2) photon protection via the quenching of the chlorophyll singlet state, and (3) scavenging of singlet oxygen. In the antenna complexes of photosynthesis, Car molecules absorb the photon energy and transfer it with high efficiency (up to 100% [1]) to a photochemical reaction center where a light-induced electron-transfer reaction takes place. Because of the C_{2h} symmetry of the molecular structures the ground state of Car molecules is the $1A_g^-$ state, and the lowest optically allowed singlet state is the $1B_u^+$. Two or three other optically forbidden states, e.g. $3A_g^-$, $1B_u^-$, $2A_g^-$, exist below the $1B_u^+$ state [2,3] depending on the length of the molecular chain.

Many time resolved spectroscopy techniques such as four-wave mixing (FWM) or photon echo peak shift give very high time resolution but little or no spectral information. The addition of frequency resolution can be used to obtain the complete time evolution of the system. Different pulse sequences are used to form wave packets in the ground and excited states [4] and to control the dynamics of the molecule in a certain state. We have found that spectrally resolved measurements allow us to study the dynamics of complex systems following different pulse sequences and wavelengths.

In this paper we report the study of ultrafast dynamics in lycopene, a carotenoid molecule with n = 11 conjugated C=C bonds in the linear conjugated backbone. In particular, we study the vibrational relaxation dynamics of photon excited molecules in the $1B_u^+$ state with different excess energy. The dynamics are reflected in the observed nonlinear spectrum generated in the phase-matching direction and controlled by the wavelength and the sequence of the applied laser pulses.

2. Results and discussion

The ultrafast laser pulses are generated by two independently tunable optical parametric amplifiers (OPA) (100 fs, 1-30 μJ) pumped by a 1 mJ amplified pulse (80 fs, 800 nm, 1 kHz). The output of one OPA is split into two beams, which act as the pump pulses ($\mathbf{k}_1$ and $\mathbf{k}_2$), and the output of the second OPA acts as the probe pulse ($\mathbf{k}_3$). Three beams with time delays t_{12} (between pulses $\mathbf{k}_1$ and $\mathbf{k}_2$) and t_{23} (between pulses $\mathbf{k}_2$ and $\mathbf{k}_3$) are aligned in a triangular configuration and focussed by a 20 cm focal length lens into a 300 μm diameter spot where the sample is placed [5]. The signals generated by the sample in the phase-matching directions $\mathbf{k}_4 = -\mathbf{k}_1 + \mathbf{k}_2 + \mathbf{k}_3$ and $\mathbf{k}_6 = -\mathbf{k}_3 + \mathbf{k}_1 + \mathbf{k}_1$ are detected simultaneously by spectrometers equipped with CCD arrays having a spectral resolution < 1 nm. The spectra are measured over a range of different fixed times t_{12} or times t_{23} by scanning the other delay time t_{23} or t_{12}, respectively.

The absorption spectrum of a lycopene sample in cyclohexane [1,2] shows strong vibronic structure around 400-530 nm. Three clear peaks at 503, 472 and 445 nm correspond to transitions between the ground vibrational level (v = 0) of the $1Ag^-$ state and the ground and excited vibrational levels v' = 0, 1, 2 of the first optically excited state $1Bu^+$. The energy gap ~ 1300 cm^{-1} between the vibronic levels results from a combination of two symmetric vibrational modes of C=C stretch (v_1 ~ 1500-1600 cm^{-1}) and C-C stretch (v_2 ~1150 cm^{-1}). It has been shown [1-3] that the C=C stretching vibration is an important active mode in Car molecules. The extremely weak shoulder in the red wing of the absorption band confirms that the one-photon transition from the ground state $1Ag^+$ to other low energy states is optically forbidden.

Laser pulse wavelengths of 503 nm (corresponding to the transition from $1Ag^-$ to $1Bu^+$ for v = 0; v' = 0) and 472 nm (v = 0; v' = 1) have been used in this study. Figure 1 shows the signal spectrum versus delay time t_{12} in the $\mathbf{k}_4$ direction for a fixed delay t_{23} = 0 fs (when the two pulses overlap) and t_{23} = 120 fs (when the two pulses are well separated in time). For the fixed delay t_{23} = 120 fs (Fig. 1a,c), a nearly symmetric time evolution of the intensity profile about the center wavelength (470 and 503 nm) is observed. The decay time (~ 35 - 40 fs) of the signals for both negative and positive t_{12} (Fig. 1a and 1b) is the same as the duration of the applied pulses. In this case, the observed signal is due mostly to the diffraction of the probe pulse from the grating induced by the two pump pulses, and the time evolution of the signal reflects the creation of a transient grating in the ground state. For the fixed delay t_{23} = 0 fs, the time evolution of the signal is an asymmetric function with a longer decay time for negative t_{12} (40 fs at 470 nm and 70 fs at 503 nm). The signal spectrum is broadened for $t_{12} < 0$ when the laser wavelength is 503 nm. For times $t_{12} < 0$, the pulse sequence is $\mathbf{k_1}$, $\mathbf{k_2}$, $\mathbf{k_3}$, and the observation of a photon echo is expected. The photon echo signal should be more sensitive to the redistribution of the excited carriers because the generation of the signal involves rephasing processes. Since the two transitions (470 and 503 nm) involve the same level in the ground state, the difference of the time and spectral evolution of the signal reflects the dynamics of the excited state. The decay times of 40 fs and 70 fs are the effective lifetimes of the vibrational levels v' = 1 and v' = 0 of $1Bu^+$. Since the pulse duration is 80 fs, the actual vibrational relaxation time of the level v' = 1 is less than 40 fs. The very short vibrational relaxation time of the v' = 1 level causes a small variation of the signal spectrum when the delay t_{12} is fixed at t_{12} = 0 fs and the t_{23} is scanned (Fig. 2a). The long decay time of the signal (2.5 ps, not shown here) versus t_{23} is the recovery time of the ground electronic state. With the wavelength of the laser pulse at 503 nm, a delayed signal is observed at long wavelength (Fig. 2b) due to vibrational cooling in the excited state.

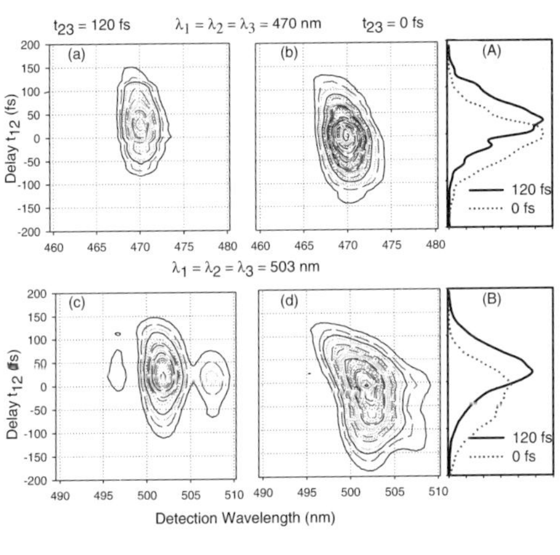

Fig. 1 Spectra of the nonlinear signal in the $\mathbf{k_4} = \mathbf{k_3} + \mathbf{k_2} - \mathbf{k_1}$ direction versus coherence time t_{12} with fixed t_{23} = 120 fs (a,c) and t_{23} = 0 fs (b,d). The wavelength of the laser pulses is $\lambda_1 = \lambda_2 = \lambda_3$ = 470 nm (a,b) and $\lambda_1 = \lambda_2 = \lambda_3$ = 503 nm (c,d). Intensity profile at 470 nm (A) and 503 nm (B) with a spectral window of 1 nm.

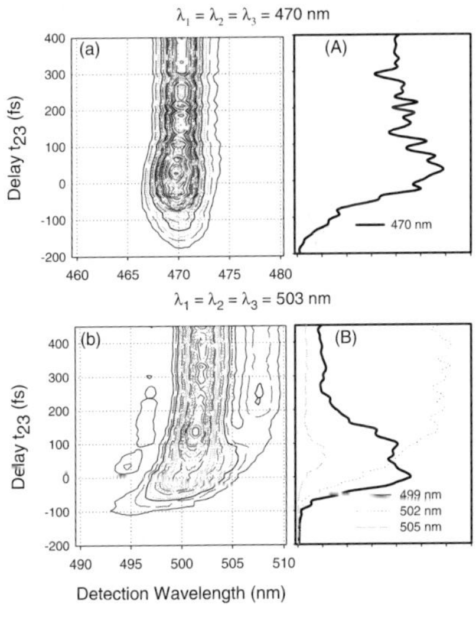

Fig 2. Spectra of the nonlinear signal in the $\mathbf{k_4} = \mathbf{k_3} + \mathbf{k_2} - \mathbf{k_1}$ direction versus population time t_{23} with fixed t_{12} = 0 fs. The wavelength of the laser pulses is $\lambda_1 = \lambda_2 = \lambda_3$ = 470 nm (a) and $\lambda_1 = \lambda_2 = \lambda_3$ = 503 nm (b). Intensity profile at different detection wavelengths with a spectral window of 1 nm.

Figures 3 and 4 show the time evolution of the observed signal in the $\mathbf{k_4}$ and $\mathbf{k_6}$ directions versus coherence time (t_{12} for $\mathbf{k_4}$ and t_{23} for $\mathbf{k_6}$) with fixed other time. The wavelengths of the pump and probe pulses are 472 nm and 503 nm, respectively. Two vibrational levels v' = 1 and v' = 2 in the $1B_u^+$ state and two vibrational levels v = 0 and v = 1 in the $1A_g^-$ state are involved in the generation of the signal in the $\mathbf{k_6}$ direction at a wavelength around $\lambda_6 \approx 2\lambda_1 - \lambda_3$ (~ 440 nm). For the fixed time t_{12} = - 120 fs (Fig. 4a), the observed signal versus t_{23} in the $\mathbf{k_6}$ direction reflects the dynamics of molecules in the v' = 1 and v = 1 levels. The broadening of the spectrum around t_{23} = 0 indicates a long phase memory and strong coupling of the vibronic levels in the ground state. For $t_{23} < 0$, two levels v = 0 and v = 1 contribute to the observed signal in the $\mathbf{k_4}$ direction, and the splitting in the observed spectrum corresponds to the Stokes shift between the potential energy surface of the ground and excited states. The strong broadening of the signal confirms again the long phase memory of the ground state.

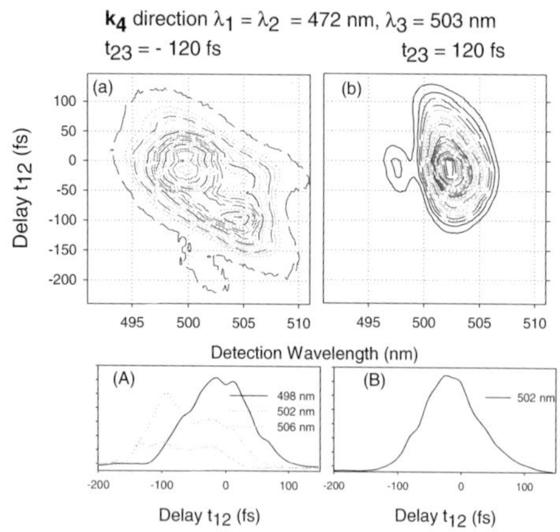

Fig.3 Spectra of the nonlinear signal in the $\mathbf{k_4} = \mathbf{k_3} + \mathbf{k_2} - \mathbf{k_1}$ direction versus coherence time t_{12} with fixed a) t_{23} = -120 fs, and b) t_{23} = 120 fs. The wavelength of the laser pulses is $\lambda_1 = \lambda_2$ = 470 nm and λ_3 = 503 nm. Intensity profile at different detection wavelengths with a spectral window of 1 nm (A,B).

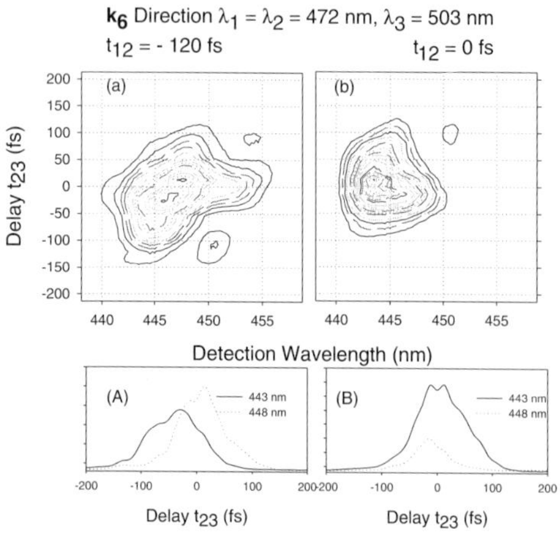

Fig 4. Spectra of the nonlinear signal in the $\mathbf{k_6} = \mathbf{k_1} + \mathbf{k_2} - \mathbf{k_3}$ direction versus coherence time t_{23} with fixed a) t_{12} = -120 fs, and b) t_{12} = 0 fs. The wavelength of the laser pulses is $\lambda_1 = \lambda_2$ = 470 nm and λ_3 = 503 nm. Intensity profile at different detection wavelengths with a spectral window of 1 nm (A,B).

In conclusion, using different colors and pulse sequences for the pump and probe pulses allows the separation of the vibrational dynamics in the ground and excited states. The time (and spectral) evolution of the nonlinear signal can be used to study the vibrational relaxation and coupling of vibronic levels.

References

1. A. N. Macpherson and T. Gillbro, J. Phys. Chem. A 103 (1998) 5049.
2. K. Nishimura, F.S. Rondonuwu, R. Fujii, J. Akahane, Y. Koyama, and T. Kobayashi, Chem. Phys. Lett. 392 (2004) 68 and references therein.
3. R. Fujii, T. Inaba, Y. Watanabe, Y. Koyama, and J. P. Zhang, Chem. Phys. Lett. 369 (2003) 165.
4. I. Pastirk, V. V. Lozovoy, B. I. Grimberg, E. J. Brown, and M. Dantus, J. Phys. Chem. A 103 (1999) 10226.
5. L. V. Dao, C. N. Lincoln, R. M. Lowe, and P. Hannaford, J. Chem. Phys. 120 (2004) 8434.

Femtochemistry VII
Fundamental Ultrafast Processes in Chemistry, Physics, and Biology
A.W. Castleman, Jr. and M.L. Kimble

Ultrafast Studies of the Electronic Structure and Dynamics of B_{12} Cofactors

Andrew B. Stickrath, D. Ahmasi Harris, Roseanne J. Sension

FOCUS Center, Department of Chemistry and Department of Physics, University of Michigan, Ann Arbor, MI 48109-1055, USA

1. Introduction

Adenosylcobalamin (AdoCbl, Figure 1) dependent enzymes catalyze a variety of chemically unusual carbon skeletal rearrangement reactions [1]. The catalysis proceeds via the generation of organic free radicals formed through homolysis of the active Co-C bond to produce cob(II)amin (Cbl(II)) and an adenosyl radical. When bound to protein, the homolysis rate is observed to increase 10^{12}-fold compared with free AdoCbl in solution [2]. However, the influence of the protein and the cofactor environment in general on priming the Co-C bond for homolysis is not well understood, despite experimental and theoretical studies directed at this problem. Glutamate mutase, which catalyzes the carbon-skeletal rearrangement of L-glutamate to L-threo-3-methylaspartate, is one of the best-characterized AdoCbl dependent enzymes. Crystal structures with both the coenzyme and substrate bound in the active site have been solved for glutamate mutase [3,4]. This system provides a useful paradigm for protein-cofactor interactions and for the enzymatic generation and control of free radicals.

Time-resolved UV-Visiblespectroscopy provides a tool to study the excited electronic state structure of cobalamins in regions away from the ground state equilibrium geometry and to compare the influence of environment on this electronic structure. Recent theoretical calculations of Cbl(II) and Cbl(III) species help to elucidate the influence of the cobalt oxidation state and the axial ligands on the electronic structure of cobalamins [5,6]. In this study time-resolved spectroscopy was used to investigate the electronic states and photoinduced homolysis of a series of alkylcobalamins including AdCbl, MeCbl, EtCbl, and PrCbl with adenosyl, methyl, ethyl, and n-propyl axial ligands respectively, in water and ethylene glycol, in addition to AdoCbl bound to glutamate mutase.

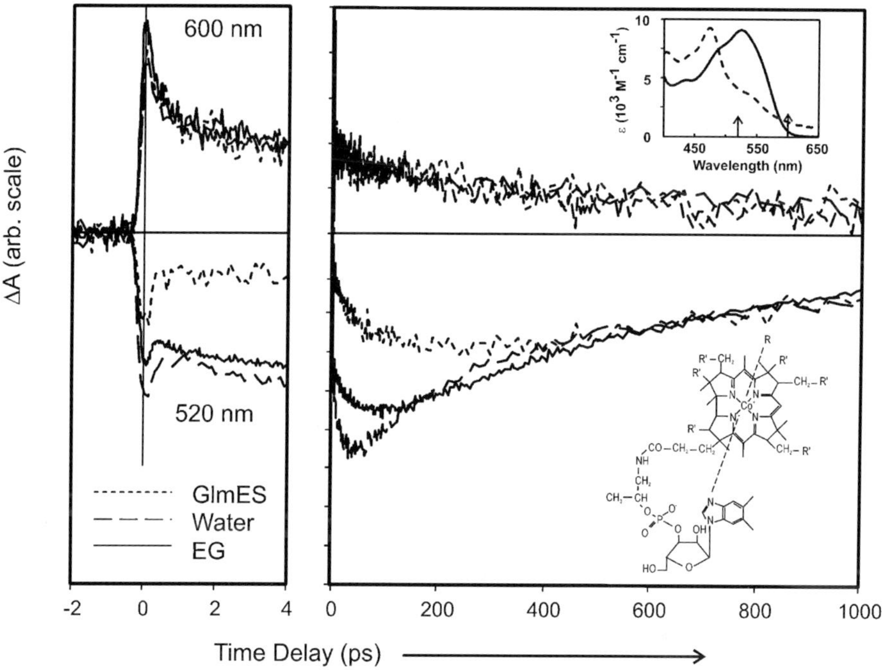

Figure 1. Transient absorption signals measured at 520 nm and 600 nm following excitation of AdoCbl at 400 nm in water, ethylene glycol, and bound to GlmES. For reference steady state spectra of AdoCbl (solid line) and Cbl(II) (dashed line) in water are shown in the inset. A sketch of the cobalamin structure is also shown.

2. Experimental Methods

Transient absorption measurements were performed using a kHz amplified Titanium:Sapphire laser system. The resulting laser beam was centered at approximately 800 nm, providing 400 μJ, 50 fs pulses at 1 kHz repetition rate. Tunable probe pulses were generated by sending the 800 nm laser pulse into a home-built non-collinear optical parametric amplifier (NOPA). Samples of AdoCbl, MeCbl, EtCbl, and PrCbl in water and ethylene glycol were prepared as described previously [7]. Samples of protein bound AdoCbl used an engineered 'single-subunit' glutamate mutase, designated GlmES, which was expressed and purified as a recombinant protein from E. coli. An excess of GlmES protein in 50 mM potassium phosphate buffer, pH

7.0, was reconstituted with AdoCbl immediately prior to the transient absorption scans. [8] All samples were circulated using a small volume flow cell and a peristaltic pump. A 1.5 m delay stage with 1μm step size was used to control the relative timing of the pump and probe pulses. This stage permits measurements of spectral dynamics spanning six decades in time from 10 fs to 10 ns.

3. Results

Time-resolved absorption dynamics for AdoCbl in water, ethylene glycol, and bound to GlmES are plotted in Figure 1 for probe wavelengths of 600 nm and 520 nm. These data illustrate the influence of environment on the photolysis of AdoCbl. When the probe is centered at 520 nm, near the peak of the AdoCbl ground state absorption band, clear differences in the signals are observed. These differences are discussed in greater detail elsewhere [8]. In contrast, the spectral changes probed at 600 nm are essentially independent of the environment over the entire temporal range. In fact, the changes probed at 600 nm are also independent of the nature of the alkyl ligand. Data for several alkylcobalamins in water and ethylene glycol are compared in Figure 2.

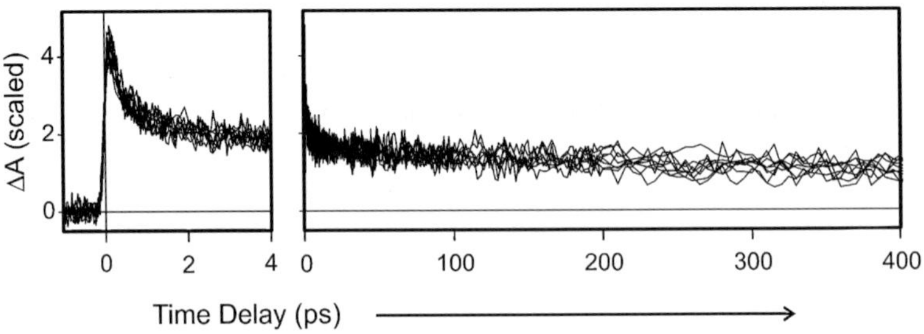

Figure 2. Transient absorption traces obtained with a 600 nm probe for AdoCbl, MeCbl, EtCbl, and PrCbl in water and in ethylene glycol and AdoCbl bound to GlmES.

All nine traces exhibit a fast decay (ca. 0.2-0.6 ps) followed by a longer, small amplitude decay (3-5 ps). On even longer time scales (>500 ps) the decay varies with alkyl radical and environment and reflects ground state recovery via geminate recombination of the radicals or internal conversion [7-9].

4. Discussion and Conclusions

For time delays greater than a couple hundred picoseconds, the transient difference spectra of all species except MeCbl are dominated by the bleach of the Cbl(III) ground state and the formation of the Cbl(II), alkyl radical pair. Recent DFT calculations reported by Stich et al. identify several long wavelength transitions of Cbl(II) [6]. These transitions are primarily characterized as Co 3d→ 3d or Co 3d → cor π* transitions, although some transitions of cor $\pi \rightarrow$ Co 3d/cor π^* character are also identified. The assignment of the red absorption to ligand field transitions is consistent with the time-dependent absorption changes. The insensitivity of the transient absorption signal in the red tail of the spectrum, as epitomized by the observation at 600 nm, to the nature of the cobalamin compound initially excited and to the environment of the corrin ring suggests that excitation of the alkylcobalamins results in a rapid internal conversion from the initial $\pi\pi^*$ excited state to a π3d state. This internal conversion process occurs on a subpicosecond timescale (0.2 to 0.6 ps). The 600 nm absorption is insensitive to further change because absorption from the π3d excited states are dominated by Co 3d$\rightarrow$ 3d or Co 3d $\rightarrow$ cor π^* transitions similar to those of the Cbl(II) radical. The low-lying $\pi 3d_{z^2}$ excited state of alkylcobalamins facilitates bond cleavage accounting for the high quantum yield for formation of the radical pair in most cobalamins.

The $\pi 3d_{z^2}$ state of cobalt porphyrins and nonalkylCbl(III) compounds is generally characterized as a ligand-to-metal (LMCT) state (corrin $\pi \rightarrow$ Co) in contrast to the apparent metal-to-ligand (MLCT) state observed following excitation of AdoCbl bound to GlmES [8] and MeCbl in all environments tested [9,10]. On the other hand, a significant difference between the molecular orbitals calculated for MeCbl and those calculated for H_2OCbl and CNCbl is a substantial mixing of the Me $2p_z$ and Co $3d_{z^2}$ orbitals [5]. This mixing increases the electron density on the alkyl ligand and may account for the MLCT character of the intermediate excited state in MeCbl and in AdoCbl bound to GlmES. An interesting hypothesis is that the protein environment acts to modify the mixing of the C $2p_z$ and Co $3d_{z^2}$ for AdoCbl bound to GlmES. This observation may help to elucidate the mechanism by which the protein influences the reactivity of the Co-C bond.

5. Acknowledgement

This research was supported by grants from the NSF (CHE 0078972) and NIH (DK53842).

References

1. E. N. G. Marsh and C. L. Drennan, Curr. Opin. Chem. Biol. 5 (2001) 499.
2. E. N. G. Marsh and D. P. Ballou, Biochemistry 37 (1998) 11864.
3. R. Reitzer, K. Gruber, G. Jogl, U. G. Wagner, H. Bothe, W. Buckel, and C. Kratky, Structure 7 (1999) 891.
4. K. Gruber, R. Reitzer, and C. Kratky, Angew. Chemie Int. Ed. 40 (2001) 3377.
5. T. A. Stich, A. J. Brooks, N. R. Buan, and T. C. Brunold, J. Am. Chem. Soc. 125 (2003) 5897.
6. T. A. Stich, N. R. Buan, and T. C. Brunold, J. Am. Chem. Soc. 126 (2004) 9735.
7. A. G. Cole, L. M. Yoder, J. J. Shiang, N. A. Anderson, L. A. Walker, II, M. M. Banaszak Holl, and R. J. Sension, J. Am. Chem. Soc. 124 (2002) 434.

8. R. J. Sension, D. A. Harris, A. Stickrath, A. G. Cole, C. C. Fox, E. N. G. Marsh, J. Phys. Chem. B 109 (2005) 18146.
9. R. J. Sension, D. A. Harris, A. G. Cole, J. Phys. Chem. B 109 (2005) in press.
10. J. J. Shiang, L. A. Walker, II, N. A. Anderson, A. G. Cole, and R. J. Sension, J. Phys. Chem. B 103 (1999) 10532.

Femtochemistry VII
Fundamental Ultrafast Processes in Chemistry, Physics, and Biology
A.W. Castleman, Jr. and M.L. Kimble

A simple artificial light harvesting dyad as a mimic of nonphotochemical quenching in green plants

Rudi Berera[a], Ivo H.M van Stokkum[a], Christian Herrero[b], Mikas Vengris[a], Gerdenis Kodis[b], Rodrigo E. Palacios[b], Herbert van Amerongen[c], Rienk van Grondelle[a], Devens Gust[b], Thomas A. Moore[b], Ana L. Moore[b], John T.M Kennis[a]

[a]*Department of Biophysics, Division of Physics and Astronomy, Faculty of Sciences, Vrije Universiteit, 1081 HV, The Netherlands*
[b]*Department of Chemistry and Biochemistry and the Centre for the early events in Photosynthesis, Arizona State University Tempe Arizona USA*
[c]*Laboratory for Biophysics, Wageningen University, Wageningen 6703 HA, The Netherlands*

1. Introduction

Activation of the thermal dissipation of excess energy in plants PSII (nonphotochemical quenching) is linked to the conversion of a 9 double-bond carotenoid (violaxanthin) into an 11 double-bond carotenoid (zeaxanthin) [1]. It has been suggested that the increase in the conjugation length turns the carotenoid from a non-quencher into a quencher of chlorophyll excited states, but there is no unequivocal evidence for this. We present results from transient absorption spectroscopy on a model system made up of a phthalocyanine molecule (Pc) covalently linked to carotenoids with conjugation lengths of 9, 10, or 11. We show that a carotenoid can quench the Q_y energy of Pc and the conjugation length is critical to the process. By using target analysis of the time-resolved data, we show that the quenching mechanism is energy transfer.

2. Materials and Methods

Femtosecond transient absorption spectroscopy was carried out with a setup described in [2]. The data were globally analyzed [3] using a model consisting of sequentially interconverting evolution-associated difference spectra (EADS), 1→2→... where the arrows indicate successive mono-exponential decays of increasing lifetimes. Because EADS may reflect mixtures of molecular species, target analysis was also applied in order the estimate the species-associated difference spectra (SADS), i.e.the spectra of pure molecular states.

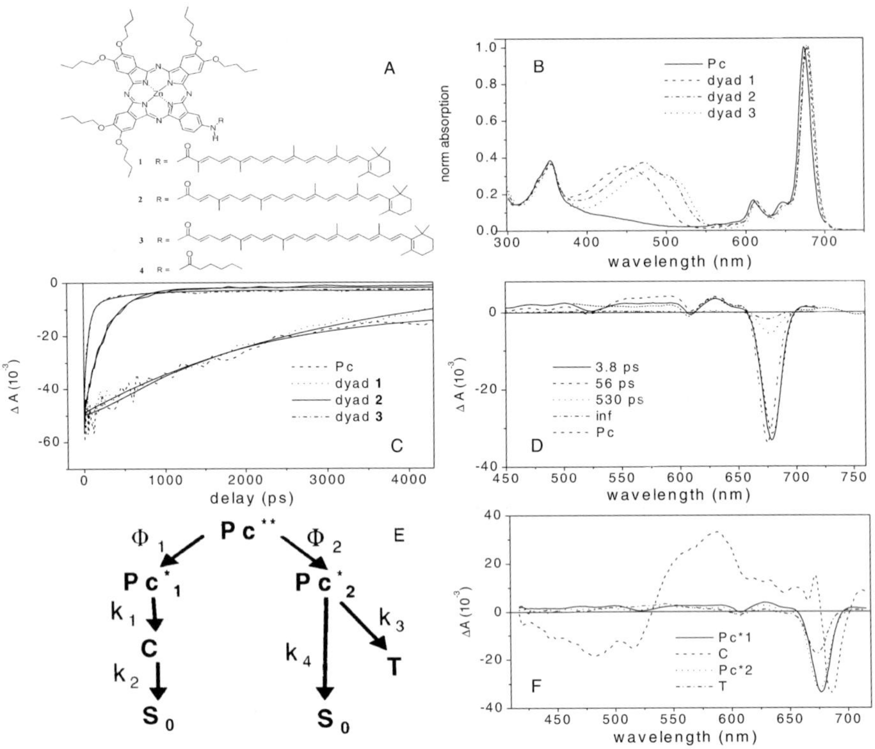

Fig 1. (A) Molecular structures of the dyads. A zinc-phthalocyanine is covalently linked to a carotenoid with 9 (**1**), 10 (**2**), and 11 double bonds (**3**). Model phthalocyanine **4** bears a hexanoyl group instead of a polyene. (B) Absorption spectrum in THF for **1** (dashed line), **2** (dash-dotted line), **3** (dotted line) and model Pc **4** (solid line). (C) Kinetic traces at 680 nm for **4** (dashed line), **1** (dotted line), **2** (solid line) and **3** (dash-dotted line) in THF, along with the global analysis fit (thin solid line). (D) Evolution-associated difference spectra (EADS) from a global analysis of data for **3**. The difference spectra corresponding to the vibrationally relaxed, fully solvated Qy state for model Pc **4** is also shown. (E) Kinetic model used in target analysis of data on **3**. (F). SADS from the target analysis for **3** in THF.

3. Results

Figure 1 shows the structure of dyads **1**, **2**, **3**, and model Pc **4** (A) and their absorption spectra in THF (B). Figure 1(C) shows kinetic traces, upon Qy excitation, at 680 nm, for **1**, **2**, **3**, and **4**. The switch from non-quenching to quenching upon an increase in the conjugation length from 9 (**1**) to 10 double bonds (**2**) is clear. In **3** (11db), the carotenoid is an even stronger quencher. Figure 1(D) presents the EADS for **3** in THF. The first EADS, with a lifetime of 100 fs, is not shown. The second EADS (solid line) displays Q_x and Q_y bleach at 610 and 680 nm, as well as excited state absorption features. At 525 nm there is a negative-going feature that we assign to a band shift/ground state bleach of the carotenoid ground state absorption upon Pc excitation. The second EADS evolves into the third (dashed line) in 3.8 ps. In this EADS, there is an increase of the positive signal above 540 nm and a small decrease of the Q_y bleach. Below 540 nm, there is a decrease and blue shift of the signal. With a 56 ps time constant, it evolves into the fourth EADS (dotted); the Q_y bleach and the excited state absorption have dramatically decreased and most of Pc is now in its ground state. The remaining signal decays in 530 ps to the non-decaying component (dash-dotted line), characterized by a small population of triplet species. The evolution from the second to the third EADS suggests that the increase of negative-going amplitude below 540 nm may correspond to depopulation of the carotenoid ground state, while the increase of the signal above 540 nm must originate from the absorption of a species different from the Pc Q_y state (fine dashed line). For 2, similar phenomena occur (not shown) but are much less pronounced.

Target Analysis: identification of the quenching species

To identify the quenching state, we analyzed the data with a target analysis. To account for the biexponential decay of the Q_y state in 3 the model requires two independent Pc excited state populations. The model (Figure 1(E)), consists of five compartments: Pc^{**} is the unrelaxed Pc excited state which decays in 100 fs into two independent Pc excited state populations, Pc^*_1 and Pc^*_2. A fraction Φ_1 relaxes to Pc^*_1, and a fraction Φ_2 to Pc^*_2. Pc^*_1 evolves to C, the carotenoid state (rate constant k_1). C decays entirely to the ground state (rate constant k_2). Pc^*_2 decays with rate constant k_3 to a long lived mix of Pc and carotenoid triplet T, and to the ground state with a rate constant k_4. Figure 1(F) shows the results for **3** in THF. The Pc^{**} spectrum is not shown; 82 % of Pc^{**} decays to Pc^*_1 (solid line), while 18 % decays to Pc^*_2 (dotted line). Pc^*_1 decays to C (dashed line) with a rate constant of $(56\ ps)^{-1}$.C decays to the ground state with rate constant $(3.8\ ps)^{-1}$. Pc^*_2 decays to T (dash-dotted line) with a rate constant of (840 ps)$^{-1}$ and to the ground state with a rate constant of (1.3 ns)$^{-1}$.The Species Associated Difference Spectrum (SADS) of Pc$^{*}_{1}$ shows Q_x and Q_y bleach as well as excited state absorption features. The SADS of C has the typical shape of the carotenoid S_1 state [4]. The strong bleach in the 680-690 nm region is an artifact.

4. Discussion

Our results show that a carotenoid can efficiently quench the Q_y energy of Pc. Target analysis provides evidence for the pivotal role of the carotenoid excited state in the quenching by showing that the spectrum of the quenching species resembles the carotenoid S_1 spectrum [4]. However we believe that energy transfer involving the carotenoid S_1 state alone cannot be solely responsible for the quenching since the process is solvent polarity dependent (not shown). To gain further insight into the process, we performed a transient absorption measurement on a model carotenoid with 10 double bonds similar to that of **2** (not shown). Solvent polarity-dependent shape changes that cannot be ascribed to the S1 state were detected. Similar changes were reported for several substituted carotenoids [5, 6] and assigned to an intramolecular charge transfer state (ICT). Thus, we think the quenching state is a mixture of S_1 and ICT states. Our results show that carotenoids can quench tetrapyrrole singlet excited states by means of energy transfer to optically forbidden carotenoid states; expanding the conjugated system of the carotenoid by one double bond turns the carotenoid from a non-quencher into a strong quencher. In LHCII, a similar phenomenon may take place: the addition of two double bonds to violaxanthin may turn it into a strong quencher, zeaxanthin.

References

1. P. Muller, X. P. Li, and K. K.Niyogi, Plant Physiol. 125 (2001) 1558
2. C. C. Gradinaru, J. T. M. Kennis, E. Papagiannakis, I. H. M. van Stokkum, R. J. Cogdell, G. R. Fleming, R. A. Niederman, and R. van Grondelle, Proc. Nat. Acad. Sci. 98 (2001) 2364.
3. I. H. M. van Stokkum, D. S. Larsen, and R. van Grondelle, Ciochimica Et Biophysica Acta 1657 (2004) 82.
4. T. Polivka and V. Sundstrom, Chem. Rev. 104 (2004) 2021.
5. H. A. Frank, J. A. Bautista, J. Josue, Z. Pendon, R. G. Hiller, F. P. Sharples, D. Gosztola, and M. R. Wasielewski, J. Phys. Chem. B 104 (2000) 4569.
6. D. Zigmantas, R. G. Hiller, F. P. Sharples, H. A. Frank, V. Sundstrom, and T. Polivka, Phys. Chem. Chem. Phys. 6 (2004) 3009.

Femtochemistry VII
Fundamental Ultrafast Processes in Chemistry, Physics, and Biology
A.W. Castleman, Jr. and M.L. Kimble

Ultrafast Dynamics of Fluorescent DNA Intercalators

Alexandre Fürstenberg,[a] Marc D. Julliard,[a] Todor G. Deligeorgiev,[b] Nikolai I. Gadjev,[b] Aleksey A. Vassilev,[b] and Eric Vauthey[a]

[a]*Department of Physical Chemistry, University of Geneva, 30 Quai Ernest-Ansermet, CH – 1211 Geneva 4, Switzerland*
[b]*Faculty of Chemistry, University of Sofia, 1 James Bourchier Avenue, BG – 1126 Sofia, Bulgaria*

1. Introduction

The synthesis of two dyes, YOYO-1 and TOTO-1, which form extremely stable and highly fluorescent complexes with double-stranded DNA has been considered as a breakthrough in the development of DNA probes for molecular biology [1]. They are homodimeric derivatives of oxazole yellow (YO, or YO-PRO-1) and thiazole orange (TO, or TO-PRO-1), which are asymmetric cyanine dyes with a monomethine bridge connecting a benzo-1,3-oxazole and a benzo-1,3-thiazole moiety, respectively, with a quinoline group. The interest for this class of dyes arose mainly from the fact that they display negligible fluorescence when free in solution and very strong fluorescence when bound to DNA (essentially by intercalation [2]), making them useful to detect DNA at sensitivity comparable to that of radioactive labels [3].

The extreme difference in quantum yield between the free and bound forms, which is observed in both monomeric and dimeric dyes, has been ascribed to the fact that the non-radiative relaxation induced by the isomerization around the internuclear bridge between the benzoxazole benzothiazole, and quinoline moieties, respectively, in the free form is hindered upon intercalation in DNA [4]. Such a decay mechanism, which governs the excited-state lifetime of the free dye, is a common feature of cyanine dyes and has been the subject of many investigations [5, 6]. In YOYO-1, this process was predicted to require 1-5 ps in water on the basis of the quantum yields of the free and bound forms and of the radiative lifetime [7]. We report here on femtosecond-resolved fluorescence measurements in aqueous buffer solution (PBS-EDTA) which confirm the occurrence of this isomerization process on a time-scale of a few picoseconds for YO and two newly synthesized monomeric and dimeric YO derivatives (Fig. 1) [8].

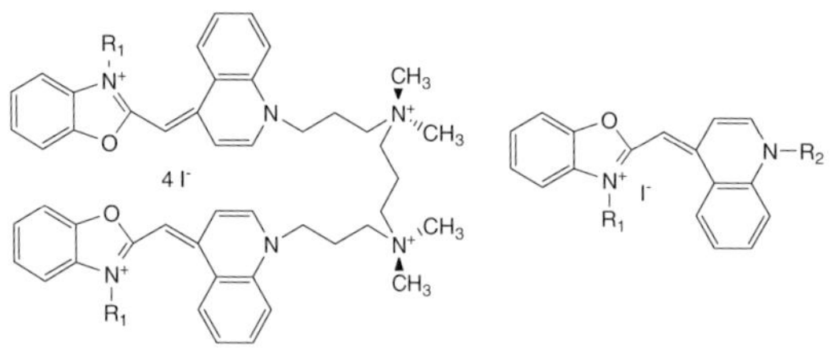

Dye	R_1	R_2
YOYO-1	$-CH_3$	---
YOYOSAC	$SCOCH_3$	---
YO-PRO-1	$-CH_3$	I^- $N^+(CH_3)_3$
YOSACPrPyr	$SCOCH_3$	I^- N^+

Fig. 1. Structure of dimeric (left) and monomeric (centre) YO derivatives. The names of th dyes are given in the table on the right.

2. Experimental Part

Time-resolved fluorescence measurements were performed with all investigated dyes on an up-conversion setup with an instrument response function of ca. 250 fs [9]. Time-resolved fluorescence data were analysed by iterative reconvolution of the instrument response function with trial functions (sum of exponentials) using a nonlinear least-squares fitting procedure (MATLAB, The MathWorks, Inc.). Measurements were carried out at several detection wavelengths from 480 to 630 nm (except anisotropy experiments where the detection wavelength was only 510 nm) and over 4 to 5 time scales accurately covering the span of the fluorescence decay. The data were then normalized and analyzed globally so that the different lifetimes could be extracted.

3. Results and Discussion

For all investigated dyes, the kinetics needed at least 4 exponential functions (Table 1) to be properly reproduced (Fig. 2). The two fastest components can be related to the solvation dynamics, either of the free dye in water (less than 200 fs and ca. 1 ps, as found for many other probes in water [10-13]), or of the dye inside DNA [14]. Their amplitudes are positive in the blue and negative in the red part of the spectrum.

A 3-6 ps component is found only in the free dyes. Its amplitude is positive at all investigated wavelengths, strongly dependent on the DNA

Table 1. Lifetimes τ_i extracted from the fitting procedure on the ultrafast fluorescence dynamics of bound YOYOSAC and free YOYOSAC, YO-PRO-1, and YOSACPrPyr. Accuracy: 10-20%.

System	τ_1	τ_2	τ_3	τ_4
YOYOSAC	<200 fs	1.3 ps	5.9 ps	33 ps
YOYOSAC – DNA	640 fs	1.9 ps	—	64 ps
YOSACPrPyr	<200 fs	1.4 ps	4.5 ps	17 ps
YO-PRO-1	<200 fs	0.8 ps	3.4 ps	—

concentration (Fig. 2, right panel), and ultimately drops to zero when all dye molecules are bound to DNA, while its value is influenced by the viscosity of the solvent (data not shown). This component can thus be attributed to the ultrafast isomerization process governing the lifetime of the free dye. With bound dyes, as shown for YOYOSAC, the decay of the excited state population takes place only on a nanosecond timescale with lifetimes of 1.8 ns (31 %) and 3.9 ns (69 %). The two nanosecond components might reflect the two ways the dye binds to the DNA double strand, namely bisintercalation and groove binding [2].

Upon intercalation in DNA, the large amplitude motion of the fluorescent probes is strongly hindered, and the free rotation cannot take place. Despite this, the fluorescence anisotropy decay was found to have an ultrafast component (5.6 ps, 58 %) in YOYOSAC. This observation is an evidence of the occurrence of an ultrafast intramolecular excitation energy hopping between two dimer subunits intercalated in DNA.

When YOYOSAC is titrated with DNA, its absorption spectrum undergoes dramatic changes and displays two isosbestic points (Fig. 3). The bands centred at 457 and ca. 520 nm (shoulder) are attributed to the free form in which exciton coupling of the two chromophores of the dimer causes a splitting of the energy levels, whereas the band centred at 491 nm corresponds to the bound dye.

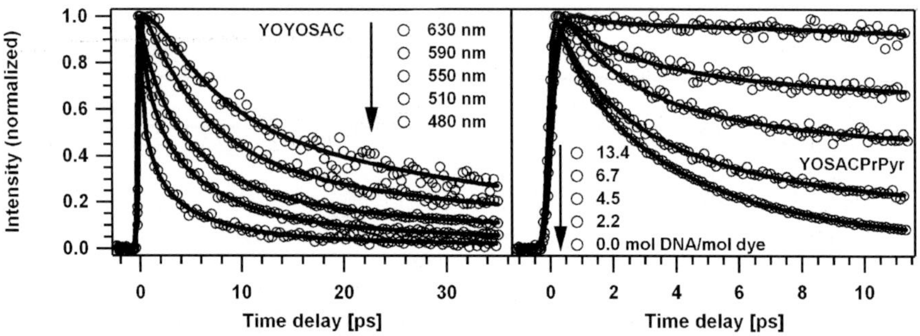

Fig. 2. Left: Fluorescence dynamics of free YOYOSAC at different detection wavelengths. Right: Fluorescence dynamics of YOSACPrPyr at different DNA-dye ratios upon detection at 520 nm. Excitation was performed at 400 nm. One mole of DNA is considered as one mole of DNA bases.

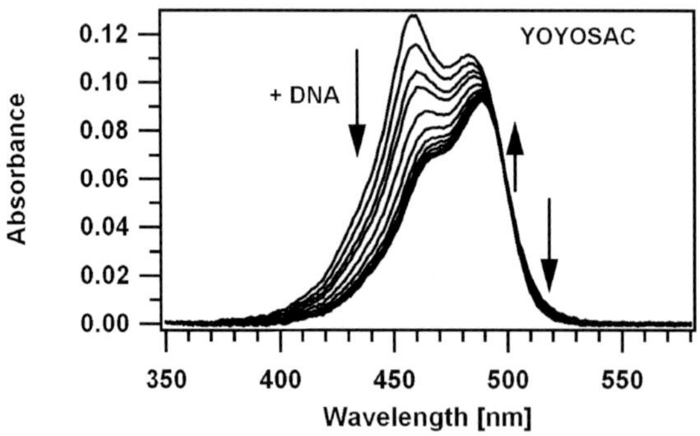

Fig. 3 Absorption spectrum of YOYOSAC titrated with salmon sperm DNA up to a dye:DNA bases ratio of 1:15.

Intramolecular aggregation of the two dimer subunits of free YOYOSAC induces excitonic coupling, which in turn further reduces emission. In the lowest excitonic state, the transition dipole moments of the two chromophores more or less cancel out, resulting in an almost complete loss of oscillator strength for the lower energy transition. Other evidences of this coupling can be found in circular dichroism spectra and in the fact that, in glycerol, the fluorescence intensity is only about half as intense as that of the intercalated dye.

References

1. P. Selvin, Science 257 (1992) 885.
2. A. Larsson, C. Carlsson, M. Jonsson, and B. Albinsson, J. Am. Chem. Soc. 116 (1994) 8459.
3. H. S. Rye, S. Yue, D. E. Wemmer, M. A. Quesada, R. P. Haugland, R. A. Mathies, and A. N. Glazer, Nucleic Acids Res. 20 (1992) 2803.

4. C. Carlsson, A. Larsson, M. Jonsson, B. Albinsson, and B. Norden, J. Phys. Chem. 98 (1994) 10313.
5. V. Sundström and T. Gillbro, Chem. Phys. 61 (1981) 257.
6. B. Sauerwein, S. Murphy, and G. B. Schuster, J. Am. Chem. Soc. 114 (1992) 7920.
7. T. L. Netzel, K. Nafisi, M. Zhao, J. R. Lenhard, and I. Johnson, J. Phys. Chem. 99 (1995) 17936.
8. T. Deligeorgiev, N. Gadjev, A. Vasilev, K.-H. Drexhage, and S. M. Yarmoluk, Dyes Pigments (2006), in press.
9. A. Morandeira, L. Engeli, and E. Vauthey, J. Phys. Chem. A 106 (2002) 4833.
10. A. Fürstenberg and E. Vauthey, Photochem. Photobiol. Sci. 4 (2005) 260.
11. M. L. Horng, J. A. Gardecki, A. Papazyan, and M. Maroncelli, J. Phys. Chem. 99 (1995) 17311.
12. R. Jimenez, G. R. Fleming, P. V. Kumar, and M. Maroncelli, Nature 369 (1994) 471.
13. S. K. Pal, J. Peon, B. Bagchi, and A. H. Zewail, J. Phys. Chem. B 106 (2002) 12376.
14. A. I. Kononov, E. B. Moroshkina, N. V. Tkachenko, and H. Lemmetyinen, J. Phys. Chem. B 105 (2001) 535.

Femtochemistry VII
Fundamental Ultrafast Processes in Chemistry, Physics, and Biology
A.W. Castleman, Jr. and M.L. Kimble

Two-photon excited-state and conformation dynamics of NADH binding with dehydrogenases

Yuexin Liu[a], Florly S. Ariola[b], Hye-Ryong Kim[c], Qianru Yu[a], Ronn Walvick[a], Ahmed Heikal[a,b,c]

[a]Department of Bioengineering, [b]Huck Institues of the Life Sciences, [c]Biomaterials and Bionanotechnology Summer Institute,Pennsylvania State University, 231 Hallowell Building, University Park, PA 16802, USA

Abstract

Reduced β-nicotinamide adenine dinucleotide (NADH) is the principal electron donor in glycolytic and oxidative energy metabolism in living cells. Recently, there has been major interest in using native NADH fluorescence to monitor energy metabolism in living cells. Recent studies indicate distinctive differences between the excited state dynamics of free NADH in solution and its intrinsic counterpart in living cells or brain slices. To reconcile these differences, we investigated the two-photon fluorescence lifetime and rotational anisotropy of NADH, under equilibrium binding conditions with malate dehydrogenase (MDH) and lactate dehydrogenase (LDH), in solution. These results indicate an enhancement of the fluorescence lifetime of NADH by a factor of ~2 upon MDH enzyme binding (association constant 3±1 μM), which is consistent with literature values using other biochemical techniques. Furthermore, equilibrated NADH-MDH reveals a complex rotational anisotropy (or associated anisotropy) at enzyme concentrations around the K_d value. Analogous measurements on NADH binding with LDH reveal similar trends, albeit different kinetic parameters. The results are compared with lipoamide dehydrogenase where flavin adenine dinucleotide (another electron carrier in the oxidative phosphorylation pathway) is covalently bound to the enzyme backbone.

1. Introduction

Reduced nicotinamide adenine dinucleotide (NADH) is an essential element in the electron transport chain (ETC) and, therefore, in the cellular synthesis of adenine triphosphate (ATP) via the oxidative phosphorylation pathway for energy metabolism [1,2]. Intrinsic NADH has been recently used for monitoring the mitochondrial respiration in brain slices of a young rat under different physiological conditions [3]. The two-photon (2P) autofluorescence of the brain slices was assigned to intrinsic NADH based on the functional response under hypoxia and cyanide. Most importantly, the averages of 2P-autofluorescence polarization anisotropy have indicated an associated anisotropy behavior, which was used to quantify the free and enzyme-bound fractions of NADH population in intact, live tissues for the first time [3]. Furthermore, the fitting parameters of the observed associated anisotropy exhibit distinctive differences between hypoxic and normoxic brain slices. A variety of enzymes are involved in cellular energy metabolism, where NADH functions as a cofactor. For instance, malate dehydrogenase (MDH) has two NADH binding sites (one on each subunit) [8]. MDH catalyze the reduction of malate to oxaloacetate in the final reaction of the Krebs cycle, with NADH acting as a cofactor in the reaction [4]. Mitochondrial MDH can bind with Complex I of the electron transport chain (ETC) such that NADH is channeled from the Krebs cycle directly to the ETC, which could enhance the rate of these processes. Lactate dehydrogenase (LDH), which consists of four subunits with NAD^{+}/NADH binding site each [9,10], catalyzes the reversible redox reaction of pyruvate to lactate in the presence of NADH. Both LDH and MDH presumably have similar reaction mechanisms with their respective substrates and cofactor [6]. Glycolysis, which yields pyruvate as a product, represents an alternative pathway for energy metabolism, especially for cells with dysfunctional mitochondria such as cancer cells. Due to the ability of LDH to overproduce lactate aerobically, elevated LDH levels in the blood often indicate some form of cancer due to an alteration of the glycolytic pathway. Recently, it has been suggested that cancer cells undergo impaired mitochondrial metabolism in which enzyme activities, like those studied here, are skewed from normal (*e.g.*, LDH activity is increased whereas MDH activity is decreased) [13].

The underlying catalytic mechanism of these enzymes includes the cofactor binding at first, followed by substrate binding in an ordered reaction mechanism that leads to its ternary complex [6]. The rate-limiting step of this mechanism is the unbinding of the cofactor from the enzyme [6]. Therefore, in-depth understanding of the enzyme-cofactor binding kinetics becomes essential in studying cellular energy metabolism. In this report, we use time-resolved two-photon (2P) fluorescence and polarization anisotropy of NADH as a function of LDH and MDH binding in solution, using femtosecond IR excitation laser pulses with ~15 ps temporal resolution for detection. The results are compared with flavin adenine dinucleotide (FAD), which is another cofactor for enzymes involved in oxidation-reduction reactions, and lipoamide dehydrogenase (LipDH). In the mitochondrial matrix, electron transfer flavoprotein and LipDH are known to contribute significantly to cellular flavoprotein fluorescence and the redox state of its FAD cofactor in direct equilibrium with the mitochondrial NAD^+/NADH pool [5].

2. Materials and Methods

NADH and FAD were purchased from Sigma-Aldrich and used without further purification; both were resuspended in 100 mM Tris buffer (pH 8.5). Both MDH and L-lactate dehydrogenate (LDH), from bovine heart, were purchased as crystalline suspensions in ammonium sulfate from Sigma-Aldrich and dialyzed (three times) against chilled 100 mM Tris buffer (pH 8.5) at 4°C. After dialysis, the MDH and LDH concentrations and extinction coefficients were determined using a Beckman Coulter DU800 Spectrophotometer. Details of the experimental setup, which includes the femtosecond laser system (Mira-900F, Coherent), microscope (IX81, 1.2NA water immersion objective, Olympus), and time-correlated single-photon counting module (SPC-830, Becker & Hickl) with ~15 ps temporal resolution, will be published elsewhere.

3. Results and Discussion

Callender and colleagues [9,10] applied laser-induced temperature-jump spectroscopic techniques to examine the kinetics of NADH binding to LDH over microsecond timescales. It was reported that the NADH binding site on LDH (from pig heart) has a dissociation constant of 6.8 μM. However, very little is understood about the dynamic nature of enzyme binding on the fast timescale (ps – ns). Statistical analysis of the molecular binding dynamics would provide insights into the structure-function relationship in these enzymes. In this paper, we systematically investigate the excited-state dynamics and rotational anisotropy of NADH in solution, titrated with MDH and LDH enzymes. The 2P-fluorescence (detected around 450 nm) of free NADH (pH 7.6, excited at 740 nm) decays as a biexponential (τ_1=380 ps, a_1=0.89, τ_2=785 ps, and a_2=0.11) in solution with an estimated average lifetime of ~420 ps. These fast time constants were assigned to different NADH conformations (folded and stretched) with efficient non-radiative pathways (i.e., low fluorescence quantum yield) that include isomerization. Upon MDH titration, the average fluorescence lifetime increases with an estimated association constant of 3±1 μM, which is consistent with literature values using other techniques [7]. The 2P-fluorescence of a fully bound $(NADH)_2 \cdot$MDH complex decays as a triexponential (τ_1=324 ps, a_1=0.18, τ_2=792 ps, and a_2=0.71, τ_3=1.72 ns, and a_3=0.12) with an estimated average lifetime of ~840 ps. Multiple exponential decays of $(NADH)_4 \cdot$LDH, on the other hand, were observed with a slower average lifetime (~1.1 ns) than the $(NADH)_2 \cdot MDH$ system (~840 ps). The enhanced fluorescence lifetime of NADH in the enzyme environment indicates a reduction of the isomerization (and other non-radiative) channels. Currently, we are modeling the reaction kinetics of NADH binding with MDH and LDH using these results. Insights into the excited-state and conformation dynamics of NADH-enzyme complexes are critical for future quantitative analysis of cellular autofluorescence as a reporter for energy metabolism.

Time-resolved 2P-fluorescence polarization anisotropy of free NADH decays as a single exponential of 380 ps with an additional minor component of 785 ps (11%), supporting the possibility of multiple conformations (i.e., folded and stretched). In contrast, the 2P-fluorescence anisotropy decay of partially-bound NADH (i.e., at [MDH]~K_d) is best described using an associated anisotropy model, due to the existence of two fluorescent species, namely free and enzyme-bound NADH [3,12]. The negative amplitude (i.e., rise) and the corresponding buildup time constant in the associated anisotropy (Figure 1) decrease as the concentration ratio increases slightly above the K_d-value. To examine whether the associated anisotropy is unique to NADH-MDH binding, we also investigated the NADH-LDH system, under similar titration conditions, and observed analogous behavior.

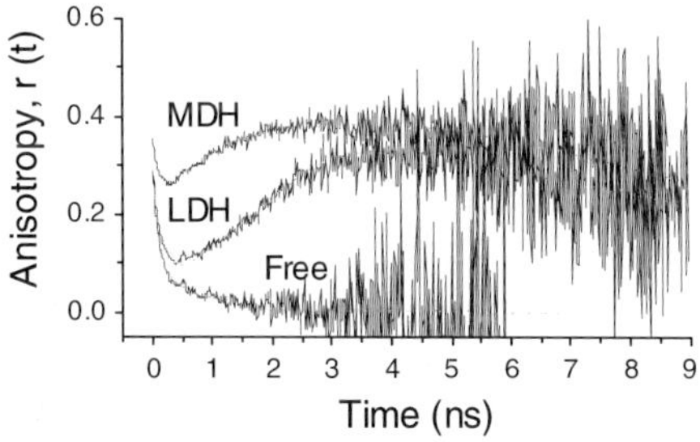

Figure 1: Time-resolved 2P-fluorescence polarization anisotropy of partially bound NADH with enzymes, under equilibrium, reveals associated anisotropy. The anisotropy of free NADH decays as a biexponential (φ_1=99 ps, r_{01}=0.20, φ_2=899 ps, and r_{02}=0.08). However, upon partial binding [i.e., [NADH]/[enzyme]~1.0:0.13], associated anisotropy of NADH was observed for MDH and LDH, albeit the different fitting parameters.

In contrast, the 2P-fluorescence polarization anisotropy of a fully bound $(NADH)_2 \cdot MDH$ system (i.e., [MDH]>>K_d) decays as a single exponential with an estimated rotational time of ~25 ns and, therefore, hydrodynamic volume of ~115 nm^3 (Figure 2). A solution ratio of [NADH]:[MDH] ≥1.0:0.5 indicates a fully bound NADH-MDH complex by exhibiting a single exponential decay, supporting previous conclusions that there are two binding sites in MDH [6] where NADH can be tightly constrained without segmental mobility. Comparative studies on the NADH-LDH binding reaction indicate similar associated anisotropy, albeit different fitting parameters and binding kinetics. The uncertainty of these rotational time measurements is limited by the relatively short 2P-excited-state lifetime of NADH-enzyme complexes.

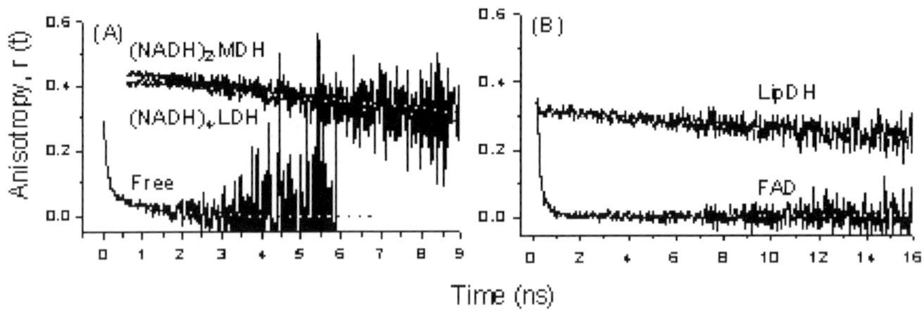

Figure 2: Time-resolved 2P-fluorescence polarization anisotropy of (A) $(NADH)_2$·.MDH and $(NADH)_4$·LDH, and (B) LipDH decays as single exponential with rotational times consistent with their respective molecular weights. Anisotropy decays of free NADH and FAD (pH 7.4) are also shown for significant comparison.

The 2P-fluorescence anisotropy of FAD and LipDH was also measured for comparative purposes and is best described as a single exponential with rotational time ~180 ps (free FAD) and ~52 ns for LipDH. Based on this rotational time, the estimated hydrodynamic volume of LipDH is ~236 nm^3, assuming a spherical shape for the enzyme. The corresponding initial anisotropy of these biomolecular systems (Figure 2) is lower than the theoretical maximum (0.57), possibly due to non-radiative processes that take place on a much faster timescale than our temporal resolution (~15 ps). These measurements are the crucial first step towards quantitative analysis of the concentration and conformation of these biomolecules as a probe of the respiratory state of cells or tissues under different physiological conditions.

In summary, the two-photon excited state dynamics are used to investigate the enzyme binding effects on the fluorescence properties and structural conformations of NADH and FAD, key cofactors in energy metabolism. Under IR femtosecond laser excitation, MDH and LDH are used to bridge the observed differences in excited-state dynamics of cellular autofluorescence and free NADH and FAD in solution. Near the K_d value, the time-resolved 2P-fluorescence anisotropy of $(NADH)_2$.MDH and $(NADH)_4$.LDH best fit the "associated anisotropy" model with picoseconds and nanoseconds decay components. In contrast, the 2P-fluorescence anisotropy of covalently bound FAD as an integral part of the lipoamide dehydrogenase structure decays as a single exponential with a rotational time consistent with its molecular weight. Currently, we are carrying out similar measurements at the single cell level while modeling the observed associated anisotropy to examine the possibility of slight segmental mobility of NADH around the binding sites of these enzymes under equilibrated binding reactions. These studies will allow for quantitative analysis of metabolic activities in living cells using NADH and FAD as intrinsic reporters, under different physiological conditions.

This work was supported by the National Science Foundation (MRSEC/PSU, COT/Lehigh, and MRI/PSU) and the Huck Institutes of the Life Sciences (PSU). HK was supported by BBSI (NIH and NSF undergraduate program; Bioengineering, PSU). The authors also express their gratitude to Coherent Lasers, Inc. for their loan of a pulse picker (MIRA9200, Coherent).

References

1. B. Alberts, D. Bray, A. Johnson, J. Lewis, R. Martin, K. Roberts, and P. Walter, Essential Cell Biology, Garland Publishing Inc., New York, 2003.
2. T. D. Pollard and C. E. William, Cell Biology, Saunders, Philadelphia, 2002.
3. H. D. Vishwasrao, A. A. Heikal, K. A. Kasischke, and W. W. Webb, J. Biol. Chem. 208 (2005) 25119
4. I. Scheffler, Mitochondria, Wiley-Liss, New York, 1999.
5. S. Huang, A. A. Heikal, and W. W. Webb, Biophys. J. 82 (2002) 2811.
6. A. Lodola, J. D. Shore, D. M. Parker, and J. J. Holbrook, Biochem. J. 175 (1978) 987.
7. J. D. Shore, S. A. Evans, J. J. Holbrook, and D. M. Parker, J. Biol.Chem. 254 (1989) 9059.
8. N. E. Labrou, E. Eliopoulos, and Y. D. Clonis, Biochem. J. 315: (1996) 687.
9. H. Deng, N. Zhadin, and R. Callender, Biochemistry 40 (2001) 3767.
10. M. Gulotta, H. Deng, H. Deng, R. B. Dyer, and R. H. Callender, Biochemistry 41(2002) 3353.
11. J. R. Lakowicz, Principles of Fluorescence Spectroscopy, 2nd edition, Kluwer Academic, NY, 1999.
12. L. Brand, J. R. Knutson, L. Davenport, J. M. Beechem, R. E. Dale, D. G. Walbridge, and A. A. Kowalczyk, in: Spectroscopy and The Dynamics of Molecular Biological Systems, P. M. Bayley and R. E.Dale (Eds.) Academic Press Inc, London, 1985.
13. A. Uppal and P. K. Gupta, Biotechnology and Applied Biochemistry 37 (2003) 45.

Femtochemistry VII
Fundamental Ultrafast Processes in Chemistry, Physics, and Biology
A.W. Castleman, Jr. and M.L. Kimble

Transient IR and 2D-IR spectroscopy of thiopeptide isomerization

Valentina Cervetto, Rolf Pfister, Jens Bredenbeck, Peter Hamm, Jan Helbing

Universität Zürich, Physikalisch-Chemisches Institut, Winterthurstrasse 190, CH-8057 Zürich, Switzerland

1. Introduction

In thiopeptides, the oxygen atom of one or more backbone carbonyl groups is replaced by sulphur. This one atom substitution renders the peptide unit photoswitchable without significantly perturbing the original peptide structure [1]. The thiopeptide unit can be selectively photoisomerized from the trans to the metastable cis/conformation because its $\pi-\pi*$ and n-$\pi*$ transition energies are significantly lower than for oxopeptides [2-3]. Here we use the isomerization of a thiopeptide bond to induce a conformational transition in a small model peptide, Boc-Ala-Gly(=S)-Ala-Aib-OMe (Fig. 1a), where two amide I oscillators are separated by a thioswitch. The molecule was designed to explore the potential of using transient 2D-IR spectroscopy for resolving conformational changes in peptides.

2. Time Resolved UV-IR spectroscopy

Fig. 1b shows that the thiopeptide Boc-Ala-Gly(=S)-Ala-Aib-OMe is reversibly photoisomerizable from the trans to the cis configuration upon S_2 ($\pi-\pi$*) excitation. Irradiation at 248 nm leads to a red shift of the UV absorption spectrum, indicating formation of the cis isomer. The original spectrum completely recovers due to thermal relaxation back to trans (Fig. 1 b). We have monitored the isomerization after 260 nm excitation by transient IR spectroscopy in the spectral range of 1600 cm^{-1} - 1760 cm^{-1} (Fig. 1d). The infrared absorption spectrum in methanol exhibits two resolved bands of the amide I modes at 1665 cm^{-1} and 1683 cm^{-1}. CO stretch bands of the urethane and ester protection groups are observed at 1726 cm^{-1} and 1746 cm^{-1} respectively (Fig. 1c).

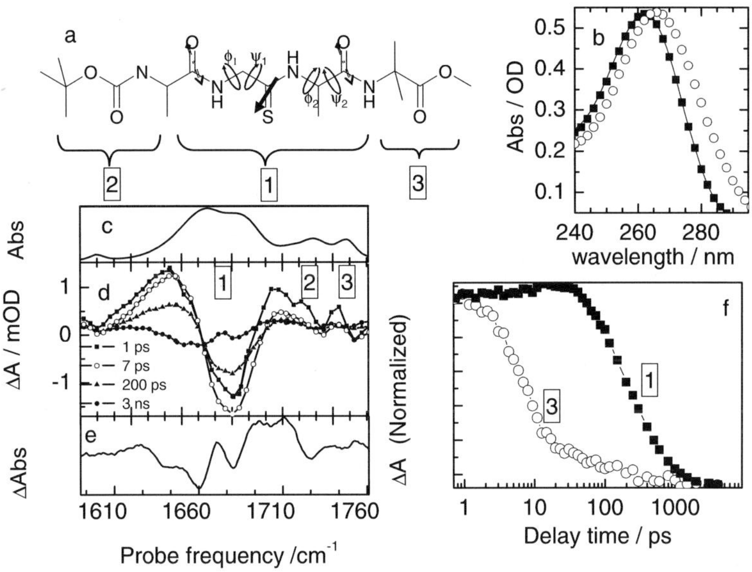

Fig.1 a) Thiopeptide sequence showing the amide I and π–π* transition dipole moments in red and in purple. b) UV absorption before irradiation (black line), after 248 nm irradiation (red line) and after thermal relaxation (solid squares). c) FTIR absorption. d) Magic angle transient IR absorption spectra at different delay times for 260 nm excitation. e) FTIR difference absorption spectrum. f) Decay of the transient absorption signal for the amide I band (1) and the ester protection group (3).

At very short delays, the transient spectrum can be well reproduced by a red shifting and broadening of the amide I bands and a (smaller) red shift of the ester and urethane vibrations. The signal due to the protection groups subsequently diminishes strongly with a 7 ps time constant, while only minor spectral shifts are observed in the amide I region on this timescale (Fig. 1f). The decay of the amide I signal takes place with a $\approx$ 250 ps time constant. Isomerization is complete after 3 ns, when the transient spectrum resembles the FTIR difference spectrum (Fig. 1e). The spectrum of long delay can be well reproduced by slightly blue-shifted amide I absorption bands.

To understand the observed spectral changes, we recall the isomerization mechanism established for *trans*-methylthioacetamide (NMTAA) [3], which can be regarded as a model for thiopeptide switching. In the case of NMTAA, S_2 excitation leads to population trapping on the singlet S_1 or triplet T_1 excited state surface, which are both essentially flat along the torsional coordinate of trans-cis isomerization. Decay of this excited state on a 250 ps timescale leads to the formation of approximately 60% cis and 40% trans molecules in the electronic ground state.

The red-shifted amide I absorption in the thiopeptide can therefore be attributed to molecules in the long-lived electronically excited state. The shift is mainly caused by a Stark effect. Indeed, the calculated static dipole moment of *trans*-NMTAA in the S_0, S_1, and S_2 states is 4.9, 2.7, and 5.3 Debye, respectively [3]. Thus the observed red-shift of the C=O vibrations is compatible with a lowering of the local electric field when the thiopeptide is in S_1 or a triplet state of similar dipole moment. The protection groups, which are at a larger distance from the thioswitch, are only weakly affected by the change in dipole moment and their strong red-shift at early time delays can instead be attributed to a sudden rise in temperature due to the excess photon energy, which is then dissipated to the solvent on a 5-10 ps timescale. The amide I band is equally affected by the heating and cooling of the molecule via anharmonic coupling to low frequency vibrations, which explains the spectral changes observed between 1 and 7 ps. Both the significant amount of dissipated heat and the immediate Stark shift of the amide I band confirm that relaxation from S_2 to the lower lying electronic states S_1 or T_1 must be ultrafast [3, 4]. In these states, the molecules are almost free to rotate along the thiopeptide bond, and isomerization most likely takes place before the molecule finds a conical intersection or intersystem crossing point and relaxes to the electronic ground state.

3. 2D-IR and Transient 2D-IR spectroscopy

Two dimensional infrared spectroscopy can measure the coupling and relative orientation of amide I (C=O stretch) transition dipole moments in peptides. This can be used to determine the dihedral angles between neighboring peptide units and thus the conformation of the molecules [5]. In our implementation, the method consists of two pulses, a spectrally tuneable narrow band IR pump and a broad band IR probe pulse. The center frequency of the IR pump pulse is tuned across the amide I bands to construct 2D-IR spectra as a function of the IR pump and IR probe frequency. Along the diagonal of a 2D spectrum, one observes for each absorption band the bleach and stimulated emission (of the 0-1 transition) as negative signal (blue) and the excited state absorption (1-2 transition) as a positive signal (red). For inhomogeneously broadened absorption lines, the signals are tilted along the diagonal, while they are vertical in the case of purely homogenous broadening. If the amide I modes are coupled, excitation of one gives rise to a spectral change of the other, which is observed as a cross peak in the off diagonal region of the 2D spectrum.

In Figure 2b, we show the 2D-IR spectrum of the thiopeptide recorded with parallel polarization of pump and probe beams. The 2D-IR signal is elongated along the diagonal, which reflects the inhomogeneity of the amide I band. The weighted difference between the 2D-IR spectra recorded with parallel and perpendicular polarization of pump and probe pulses is shown in Fig 2a. Due to the different anisotropy of cross peaks and the diagonal peaks, this difference spectrum contains only the cross peaks [5], which indicate significant coupling between the two amide I modes, although they are separated by the thioswitch.

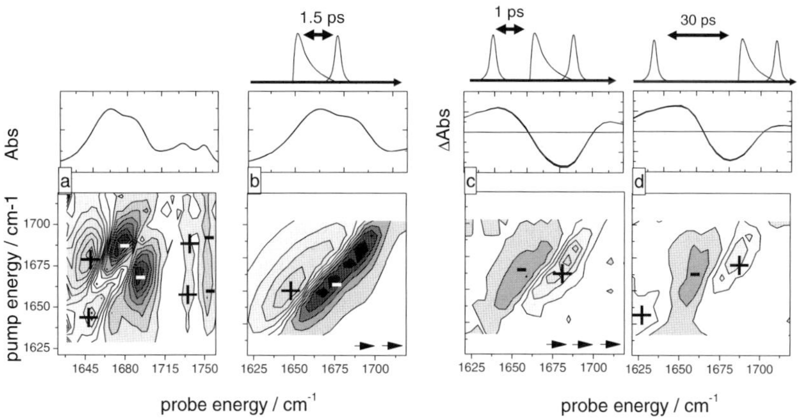

Fig.2 a) Cross peaks in the thiopeptide 2D-IR spectrum (weighted difference between spectra recorded with parallel and perpendicular polarization). b) 2D-IR spectrum for parallel polarization. c) Transient 2D-IR spectrum at a UV-IR delay of 1 ps. d) T2D-IR spectrum at a UV-IR delay of 30 ps. FTIR and UV-IR pump-probe spectra are shown for comparison on top.

A potential application of transient 2D-IR spectroscopy is to resolve *changes* in orientation and coupling between these modes in the course of a conformational transition.

Transient 2D-IR spectroscopy consists of three pulses: a UV pump, a narrow-band IR pump, and a broad-band IR probe pulse. The UV laser pulse triggers the conformational transition. Subsequently 2D-IR spectra of transient species are recorded [6]. The transient 2D-IR spectrum is given by the difference between the 2D-IR spectrum recorded with the UV pulse minus the spectrum recorded without UV pulse. This eliminates contributions of molecules which have not absorbed a UV photon. T2D-IR opens up the possibility to resolve transient structures and to distinguish between homogeneous and inhomogeneous broadening in transient absorption spectra.

We have recorded T2D-IR spectra for Boc-Ala-Gly(=S)-Ala-Aib-OMe for time delays of 1 and 30 ps between UV and IR pump pulses (Fig.2c and d). The spectrum at early delay is very similar to the 2D-IR spectrum without photoexcitation, except for the sign. In this transient 2D-IR spectrum, we do not observe any distinct contribution from the excited state, although it is clearly seen in the corresponding UV-IR pump-probe spectrum (Fig. 2c top). After 30 ps the pump-probe spectrum (Fig. 2c top) has changed only very little in the amide I region compared to the 1 ps delay. However, at the bottom left in the T2D-IR spectrum of Fig. 2d, a new positive band is now observed which is due to the photoexcited molecules. Moreover, the T2D-IR signal is oriented more vertically with respect to the early delay time.

We have discussed the possible reason for a similar observation in detail in a paper on the T2D-IR probing of the conformational transition of a cyclic octapeptide [7]. The weak contribution to the transient 2D-IR spectrum of a strong transient absorption band is an indication that the absorption of the photoproduct has a much larger homogeneous linewidth than the absorption of the molecule before excitation (a large homogeneous width causes the signals due to the (1-2) and (0-1) vibrational transitions to overlap more strongly and partially cancel in the 2D-IR spectrum). In the present case, the stronger homogeneous broadening of photo excited thiopeptides after 1 ps is clearly due to larger amplitude fast fluctuations of the vibrational transition frequencies in the hot molecules. After the molecules have cooled down in an electronically excited state, the amplitudes of fast fluctuation are reduced, and the excited state signal becomes visible in the 30 ps T2D-IR spectrum.

More refined experiments in the future should allow us to track the changes in homogeneous broadening and thus the flexibility of the thiopeptide in the excited state over more time points. Equally promising is the fact that T2D-IR spectra with little contribution from the photoproduct at early delays are essentially ground state 2D spectra of a non-isotropic ensemble of molecules. The orientation of the UV dipole with respect to the amide I modes can thus be used to gain the necessary angle information in order to determine the thiopeptide structure.

5. References

1. J. H. Miwa, A. K. Patel, N. Vivatrat, S. M. Popek, and A. M. Meyer, Organic Letters 3 (2001) 3373.
2. J. Zhao, D. Wildemann, M. Jakob, C. Vargas, and C. Schiene-Fischer, Chem. Commun. (2003) 2810.
3. J. Helbing, H. Bregy, J. Bredenbeck, R. Pfister, P. Hamm, R. Huber, J. Wachtveitl, L. De Vico, and M. Olivucci, J. Am. Chem. Soc. 126 (2004) 8823.
4. H. Satzeger, C. Root, P.Gilch, W. Zinth, D. Wildemann, and G. Fischer, J. Phys. Chem. B 109 (2005) 4775.
5. S. Woutersen and P. Hamm, J. Phys.: Condens. Matter 14 (2002) R1035.
6. J. Bredenbeck, J. Helbing, and P. Hamm, J. Chem.Phys. 121 (2004) 5943.
7. J. Bredenbeck, J. Helbing, C. Renner, R. Behrendt, L. Moroder, J. Wachtveitl, and P. Hamm, J. Phys. Chem. B 107 (2003) 8654.

Femtochemistry VII
Fundamental Ultrafast Processes in Chemistry, Physics, and Biology
A.W. Castleman, Jr. and M.L. Kimble

Direct observation of DNA repair by photolyase

Chaitanya Saxena[a], Ya-Ting Kao[a], Lijuan Wang[a], Aziz Sancar[b] and Dongping Zhong[a]*

[a]*Departments of Physics, Chemistry, and Biochemistry, Programs of Biophysics, Chemical Physics, and Biochemistry, 191 West Woodruff Avenue, The Ohio State University, Columbus, OH 43210, USA*
[b]*Department of Biochemistry and Biophysics, University of North Carolina School of Medicine, Mary Ellen Johns Building, CB 7260, Chapel Hill, NC 27599, USA*

1. Introduction

One of the detrimental effects of UV radiation on the biosphere is the formation of cyclobutane pyrimidine dimers (Pyr<>Pyr) between the two adjacent bases in DNA. Pyr<>Pyr dimers bring DNA polymerases to a standstill and may result in mutation or cell death. Photolyase, a photoenzyme that exists in all three branches of life, harnesses blue or near-UV light energy to cleave the cyclobutane ring of Pyr<>Pyr and thus prevents the harmful effects of UV radiation. Photolyase from *E. coli* is a monomeric protein with two noncovalently bound cofactors: one is a light-harvesting photoantenna, a pterin molecule in the form of methenyltetrahydrofolate (MTHF), and the other one is the catalytic cofactor, a fully reduced deprotonated flavin molecule ($FADH^-$). In the hypothesis proposed for the catalysis, the enzyme binds a Pyr<>Pyr in DNA, independent of light. The antenna chromophore MTHF harvests UV/blue-light photon and transfers the excitation energy (dipole dipole interaction) to $FADH^-$. Excited $FADH^{-*}$ then transfers an electron to the Pyr<>Pyr, which results in the dimer splitting into two pyrimidines and hence repairs the damaged DNA. The repair cycle ends when the excess electron returns from the repaired pyrimidine back to the nascent-formed $FADH^•$ and regenerates the active $FADH^-$ form. These elementary reactions of energy and electron transfer in DNA repair by photolyase are reported in this contribution.

* For correspondence: dongping@mps.ohio-state.edu

2. Dynamics of Antenna Initiation and Cofactor Reduction

The function of the enzyme is triggered by absorption of one photon, and thus the dynamics can be well synchronized with the functional process by femtosecond laser pulse initiation. Femtosecond-resolved fluorescence up-conversion and transient absorption methods were combined to determine the actual time scales of the energy and electron transfer dynamics.

Dynamics of Resonance Energy Transfer from MTHF to FADH⁻.* By gating the fluorescence decay profile of MTHF* under physiological conditions, the excitation energy transfer from MTHF* to $FADH^-$ was found to occur in 292 ps. The orientation factor for MTHF-$FADH^-$ pair was determined to be 0.11. This value, unfavorable for energy transfer, indicates the existing structural constraints probably placed by three functional binding sites. But, the large absorption coefficient of MTHF enhances the DNA repair efficiency.

Intraprotein Electron Transfer of FADH• Photoreduction. The catalytic active cofactor flavin in photolyase *in vivo* is in the fully reduced form $FADH^-$. During purification of the enzyme, flavin was oxidized to its neutral radical form, $FADH^•$. However, excited $FADH^•$ can be reduced to $FADH^-$ by accepting one electron from neighboring aromatic amino acid residues. By monitoring the transient absorption decay of neutral excited $FADH^•$, the photoreduction of $FADH^•$ to $FADH^-$ was revealed to evolve along two electron-transfer pathways: one is along a tryptophan triad (W306→W358→W382) with the initial electron hop in 11 ps; the other route starts with an initial electron separation in 42 ps through the neighboring phenylalanine (F366) followed by either tunneling along an α-15 helix or hopping through the tryptophan triad again. Immediate back electron transfer was observed to occur within ~3 ns. All the measured dynamics are summarized in Figure 1 [1, 2].

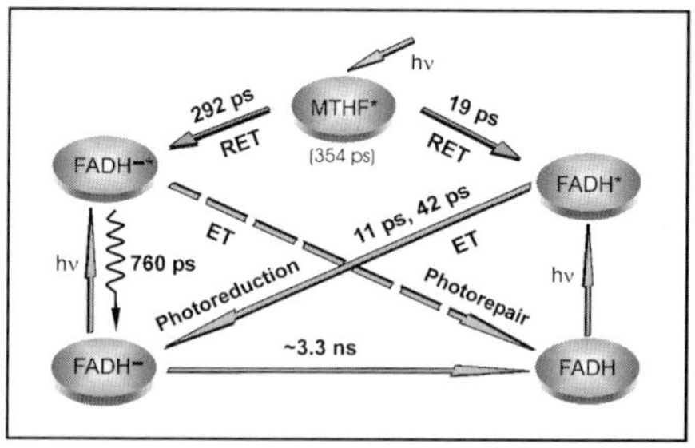

Fig. 1. Time scales determined by measurements of energy and electron transfer dynamics.

The results revealed significant details of the nature of the ultrafast dynamics in DNA photolyase and laid a solid foundation for further studies of mapping out the entire DNA repair process, which is given below.

3. Dynamics of DNA Repair and Elucidation of Photocycle

The model for the catalytic reaction states that the excited flavin cofactor transfers an electron to Pyr<>Pyr to generate a charge-separated radical pair ($FADH^{\bullet}$ + Pyr<>$Pyr^{\bullet -}$). The anionic ring of the dimer is split by a [2 + 2] cycloreversion, and the excess electron returns to the flavin radical to restore the catalytically competent $FADH^-$ and closes the catalytic photocycle.

Active-Site Solvation Dynamics. The photophysics of the cofactor $FADH^-$ in the active site of the enzyme was first characterized in the absence of substrate. MTHF-depleted enzyme was used in studies to avoid the overlapping spectral complexity. It is well established that removal of MTHF cofactor does not hamper the ability of the enzyme to repair the damaged DNA. By gating a series of fluorescence wavelengths of the weak $FADH^{-*}$ emission, wavelength-resolved fluorescence dynamics of $FADH^{-*}$ were obtained. The results showed significant solvation dynamics at the active site after the sudden change of the cofactor dipole upon excitation. By monitoring the blue-side fluorescence emission, a decay component in 2 ps was observed, indicating ultrafast libration/rotation motions of local protein residues and trapped water molecules. Subsequently, a slow solvation process from the time scale of 60 ps until merging with the cofactor lifetime of 1.3 ns in the entire emission was observed, revealing a dynamic active site upon function initiation. Although the substrate recognition may push certain water out of the pocket, the similar solvation dynamics with repaired thymine monomers in the active site were observed. It was concluded that the observed active-site solvation directly controls the catalytic reactions of DNA repair [3, 4].

DNA Repair and Photocycle. In the presence of substrate, the fluorescence emission of $FADH^{-*}$ was quenched as shown in Figure 2. This faster decay transient was attributed to direct electron transfer from the excited $FADH^-$ to the Pyr<>Pry dimer. The transient was best fit by either a stretched single exponential decay with a time constant of 170 ps and $\beta = 0.71$ or a double-exponential decay of 60 and 335 ps with 45% and 55% of the total amplitude, respectively. The dynamics do not follow a single-exponential decay because of its strong coupling with the slow active-site solvation. Thus, the observed 170 ps is forward electron transfer time from $FADH^{-*}$ to Pyr<>Pyr. Ejected electron from $FADH^-$ to the dimer results in Pyr<>Pyr splitting, and, after the repair, the excess electron returns back to $FADH^{\bullet}$ and regenerates the active $FADH^-$. Transient absorption studies by probing at 625 or 510 nm showed a clear formation of $FADH^{\bullet}$, in accord with the decay of excited $FADH^-$. The observed decay of $FADH^{\bullet}$ gives a time scale of 560 ps, representing back electron transfer from the repaired pyrimidine to $FADH^{\bullet}$ [Fig. 2].

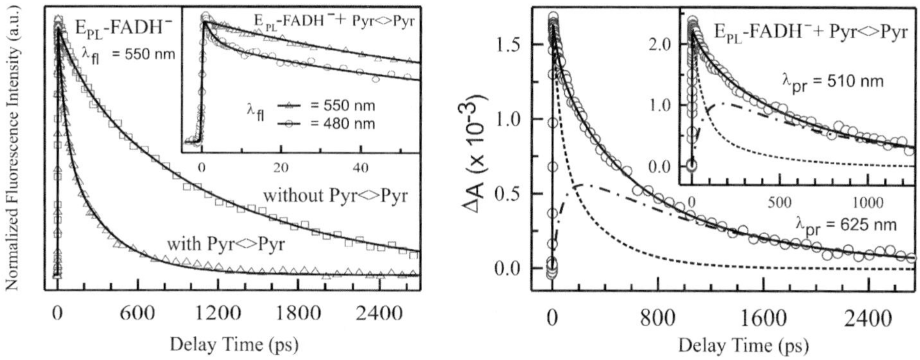

Fig. 2. (Left) Time-resolved fluorescence dynamics of $FADH^{-*}$ of photolyase in the absence and presence of Pyr<>Pyr. A fast decay at 480 nm (Inset) results from initial solvation dynamics. (Right) Absorption transients probed at 625 and 510 nm (Inset). The $FADH^{-*}$ decay is shown by a dashed line and the intermediate $FADH^{\bullet}$ dynamics is represented by a dash-dotted line.

This study reported a direct observation of pyrimidine dimer repair in DNA by photolyase. With femtosecond resolution, the photocycle was followed in real time through mapping out the temporal evolution of entire catalytic reactions. Direct observation of the intermediate of flavin radical ($FADH^{\bullet}$) proved the electron-transfer radical mechanism, which was proposed approximately two decades ago. Active-site solvation was observed to occur on picosecond-to-nanosecond time scales and to play a critical role by the continuous modulation of catalytic reactions. These synergistic motions in the active site of the damaged DNA–enzyme complex, optimized by evolution, revealed a perfect correlation of structural integrity and dynamical locality to ensure maximum repair efficiency on the ultrafast time scales [3].

Acknowledgements

This work was supported by the National Institute of Health Grant GM31082 (to A.S.), the Packard Fellowship (to D.Z.) and the American Heart Association predoctoral fellowship (to C.S.).

References

1. C. Saxena, A. Sancar, and D. Zhong, J. Phys. Chem. B 108 (2004) 18026.
2. H. Wang, C. Saxena, D. Quan, A. Sancar, and D. Zhong, J. Phys. Chem. B 109 (2005) 1329.
3. Y.T. Kao, C. Saxena, L. Wang, A. Sancar, and D. Zhong, Proc. Natl. Acad. Sci. USA 102 (2005) 16128.
4. D. Zhong, in this book.

Femtochemistry VII
Fundamental Ultrafast Processes in Chemistry, Physics, and Biology
A.W. Castleman, Jr. and M.L. Kimble

Ultrafast hydration dynamics in protein conformational transitions

Weihong Qiu, Luyuan Zhang, Lijuan Wang and Dongping Zhong*

Departments of Physics, Chemistry, and Biochemistry, Programs of Biophysics, Chemical Physics, and Biochemistry, 191 West Woodruff Avenue, The Ohio State University, Columumbs, Ohio 43201, USA

1. Introduction

Biological water in the hydration layer plays a critical role in structural stability, flexibility, and integrity of proteins. Water motion in the immediate vicinity of proteins directly affects protein conformational changes such as from the denatured to native state or from the misfolded structure to the eventual protein aggregate. Thus, elucidation of water dynamics with spatial and temporal resolution is essential to understanding the underlying molecular mechanism of protein conformational transitions, leading to a potentially better control over early protein misfolding and aggregation that is responsible for most neurodegenerative diseases.

The intrinsic amino acid tryptophan has recently been established as an optical probe to examine hydration dynamics of proteins with single-residue resolution [1, 2]. In what follows, we will present our recent experimental studies of hydration dynamics of melittin in different conformations using tryptophan as the local optical probe with femtosecond (fs) time resolution.

2. Hydration Dynamics of Melittin Conformers

Melittin, a peptide toxin found in honeybee venom and one of the most studied archetypes of membrane proteins, consists of 26 amino acid residues with W19 being the lone tryptophan. It typically exists as a random coil in water while readily forms an α-helix in alcohol solvents or when interacting with membrane. In aqueous solutions of high ionic strength, high peptide

* For correspondence: dongping@mps.ohio-state.edu.

concentration, or high pH, four melittin peptides aggregate into a tetramer. The conformational versatility of melittin and its biological implications have drawn extensive studies both theoretically and experimentally.

We report here the hydration dynamics around W19 of melittin in different conformational states. Fig. 1 shows the temporal fluorescence decay of aqueous tryptophan and melittin tetramer formed in high-salt solution. For aqueous tryptophan, the observed ultrafast decay components at the blue side of the emission and initial rises at the red end are direct indications of hydration dynamics rather than electronic relaxation between ^1L_b and ^1L_b or vibration cooling [1-3]. In melittin tetramer, compared to aqueous tryptophan, the decay and rise in blue and red emission wavelengths, respectively, became much longer, indicating a significant slowing down of water dynamics around W19 residue. A series of fs-resolved fluorescence up-conversion transients were also collected for melittin in the conformational states of a random coil and a membrane-bound α-helix. All these transients were further analyzed according to the method developed by Zhong and co-workers to extract the corresponding hydration correlation function $C(t)$ [1].

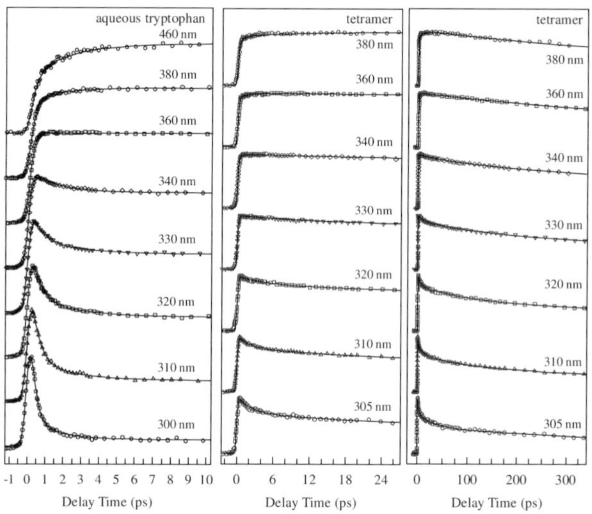

Fig. 1. Normalized fs-resolved fluorescence transients of tryptophan in bulk water (left) and W19 in melittin tetramer in both short (middle) and longer (right) time ranges.

The derived $C(t)$ functions are shown in Fig. 2. In the random-coiled monomer, the hydration dynamics of W19 in melittin can be represented by a double exponential decay: 0.62 ps (68%) and 14.7 ps (32%). For the membrane-bound α-helical melittin, the relaxation process becomes drastically slower: 1.25 ps (35%), 14 ps (27%), and 114 ps (38%). When the melittin peptides self-assemble into a tetramer at high salt concentration, the hydration dynamics also becomes much slower: 3 ps (47%) and 87 ps (53%). Because the hydration dynamics of tryptophan in high salt solution only take about 10-20 ps, the observed time scale of 87 ps must be related to the local tetramer conformation.

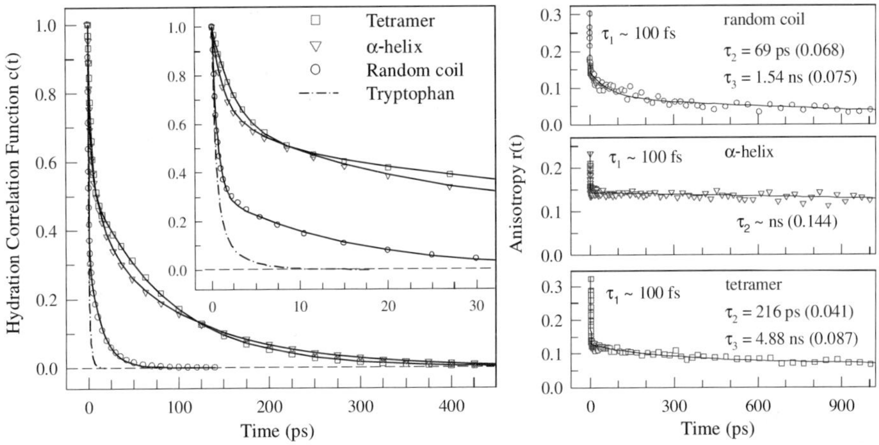

Fig. 2. (left) Hydration dynamics probed by tryptophan in water and by W19 in three different conformations of melittin. The insert shows the correlation functions in short time range. (right) Time-resolved fluorescence anisotropy of W19 in all three conformations of melittin. The ~100-fs decay results from the ultrafast internal conversion between ^1L_a and ^1L_b. Note that all observed rotational relaxations are at least three times longer than the corresponding hydration dynamics.

For all conformational states, we carefully measured the tryptophan anisotropy dynamics to examine the local rigidity, as also shown in Fig. 2. In all conformations, the local wobbling motions of tryptophan are at least three times longer than the corresponding solvation dynamics, suggesting that the local side-chain motion, if any, would take a much longer time than the observed solvation dynamics. *With other observations [1-3], the constructed solvation dynamics here predominantly describe the hydration dynamics of local water molecules around tryptophan.*

The overall hydration dynamics around W19 in random-coiled melittin is longer than that of aqueous tryptophan but similar to that at protein surfaces [2]. Because melittin in water is essentially a denatured protein with all constituting residues projecting into the aqueous environment, the flexible W19 thus probes the hydration dynamics occurring at a floppy random-coiled protein surface. The observed 620-fs component represents the ultrafast relaxation of bulk-like water molecules with inertial/librational motions while the 14.7-ps dynamics arises from the relaxation of local-ordered biological water. Thus, the observed time scale of 10-20 ps reflects the fluctuating nature of the ordered water surrounding the surface of denatured melittin.

Inside the membrane nanochannel of 50 Å in diameter, various evidences [3] indicate that melittin forms an α-helix and penetrates into the membrane until W19 is spatially hindered by the interfacial polar head-groups. The last six hydrophilic residues at the C-terminus protrude into the water layers within the channel. Thus, the observed three distinct time scales of 114, 15, and 1.25 ps represent three different water structures near the membrane interface: well-ordered, quasi-free, and bulk-like. This significant observation reveals a dynamical ordering of local water in the membrane nanochannel. The ~100 ps of the water dynamics close to the channel wall is necessary to maintain the global structural stability of lipid bilayers as well as the protein structural integrity of the α-helical content in the membrane-bound melittin.

At high ionic strength, ions are solvated by water molecules, and such processes reduce the availability of water molecules to interact with melittin, promoting the formation of α-helical content. Unique to its amino acid sequence, a large hydrophobic patch is formed in the α-helical monomer, driving the tetramerization to form a hydrophobic core in the center of the tetramer. Four highly charged C-termini form two crossing joints, and each of these two regions has eight positively charged residues. Besides the salt dielectric screening of the interchain electrostatic repulsion, water molecules solvate these charged residues and associated ions and form a highly ordered structure around two assembled regions, therefore stabilizing the local tetramer structure. The time scale of ~87 ps, longer than most surface hydration dynamics reported so far and similar to ordered-water motion in membrane interfaces [1-3], again confirms the importance of appropriate water mobility for the self-assembly of both lipid bilayers and high-order structures of melittin.

3. Summary

In this contribution, we reported the hydration dynamics of melittin in conformational transitions with femtosecond temporal and single-residue spatial resolution. The observed time scale of ~100 ps represents a type of necessarily dynamical ordering of water molecules to mediate high-order structure formation of melittin in the α-helix and self-assembled tetramer. A similar study of the hydration dynamics of human serum albumin (HSA) in different conformations induced by pH variations has also been carried out recently and presented elsewhere [4, 5].

This work was supported by the National Science Foundation (CHE-0517334).

References

1. W. Lu, J. Kim, W. Qiu and D. Zhong, Chem. Phys. Lett. 388 (2004) 120.
2. S. K. Pal and A. H. Zewail, Chem. Rev. 104 (2004) 2099.
3. W. Qiu, L. Zhang, Y.-T. Kao, W. Lu, T. Li, J. Kim, G. M. Sollenberger, L. Wang and D. Zhong, J. Phys. Chem. B 109 (2005) 16901.
4. W. Qiu, L. Zhang, O. Okobiah, Y. Yang, L. Wang, D. Zhong and A. H. Zewail, J. Phys. Chem. B submitted.
5. D. Zhong, in this book.

Femtochemistry VII
Fundamental Ultrafast Processes in Chemistry, Physics, and Biology
A.W. Castleman, Jr. and M.L. Kimble

Intramolecular electron transfer beyond solvent control: probing sub-100 fs photoinduced electron transfer in triarylmethane lactones

Jerzy Karpiuk,[a] Stefan Lochbrunner[b], Eberhard Riedle[b]

[a] *Institute of Physical Chemistry, Polish Academy of Sciences, Kasprzaka 44/52, 01-224 Warsaw, Poland*
[b] *Lehrstuhl für BioMolekulare Optik, Department für Physik, Ludwig-Maximilians-Universität, Oettingenstr. 67, 80538 München, Germany*

1. Introduction

Ultrafast charge separation (CS) processes in molecular systems ensure effective transformation of the absorbed photon energy into the energy of a charge transfer (CT) state $D^+–A^-$. With 6-fs and sub-3-fs [1,2] electron transfer (ET) times reported for dye/semiconductor systems, the research on heterogeneous ET has reached the low femtosecond limit. Though a number of studies demonstrate that the rate of *intramolecular* [3,4] barrierless ET reactions in solution can be as fast as 10^{13} s^{-1}, the studies on intramolecular ET dynamics faster than diffusional solvation are very scarce. Exploring intramolecular ET in condensed phase in this time domain can provide substantial insight into the initial stages of ET. Ultrafast ET is promoted by high-frequency intramolecular vibrational modes and by ultrafast (inertial) solvation. A better understanding of the interplay between these two driving forces requires elaboration and exploration of D-A systems capable of undergoing ET on a sub-100 fs time scale. Such ultrafast ET processes with theoretically predicted k_{ET} up to 10^{14} s^{-1} have been proposed and discussed by Jortner *et al.* [5], but until now there have been very few experimental reports on intramolecular ET on the time scale of single vibrations [6]. A suitable system for such studies could be one with a CT state lying below the locally excited (LE) state and being available from the latter upon excitation to the LE state without necessary stabilization by solvation dynamics.

2. Triarylmethane lactones (LTAMs) as ultrafast donor-acceptor systems

We have recently found that lactone forms of the well-known triarylmethane dyes (LTAMs) are very good candidates for such studies [7-9]. They are simple D-A systems built around a tetrahedral carbon atom (spiro-like D-A architecture) with the lowest absorption transitions localized on the structural subunits. Malachite green lactone (MGL, Fig.1) and phenolphthalein (PP, Fig. 2) have nearly orthogonal electron donating groups relative to the acceptor moiety. The lengths of all C–C bonds of the spiro carbon (1.509 to 1.525 Å) prove that the delocalization of the electrons from the aromatic rings does not extend to the central C atom. Absorption spectra of MGL and PP in the low energy region are a superposition of the spectra of their structural subunits, indicating that the ground state interactions between the structural components can be neglected. Both LTAMs have no direct optical CT absorption transitions.

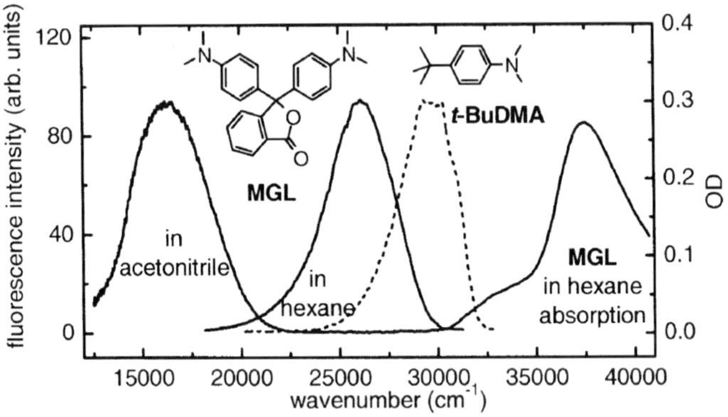

Fig. 1: Absorption and fluorescence spectra of MGL in hexane and acetonitrile compared with the fluorescence spectrum of *t*-butyldimethylaniline (*t*-BuDMA) in hexane (dashed line).

Upon excitation, LTAMs undergo rapid ET to form a highly polar ^{1}CT state with a structure formally consisting of radical ions of the electron donating and accepting moieties. The charges recombine on the ns time scale via radiative and nonradiative channels. The emitting ^{1}CT state in both LTAMs lies below the excited S_1 state of each structural subunit and is strongly stabilized by the interaction with a highly polar environment, as reflected in a huge Stokes shift (e.g., >16000 cm^{-1} for MGL in DMSO). The large dipole moment of the ^{1}CT state (25.0 and 21.0 D for MGL and PP) corresponds to ET distances of 5.2 and 4.4 Å and indicates full electron transfer from the dimethylaniline (DMA) group to the phthalide (Pd) moiety in MGL and from the phenol group to Pd in PP.

The absence of any fluorescence from the locally excited chromophore in both LTAMs (Fig. 1 compares emission of MGL with that of *t*-BuDMA imitating locally excited DMA in MGL) is due to a rapid conversion of the primarily excited LE state into the CT state, i.e. the intramolecular ET. For MGL in acetonitrile, an upper limit for the transfer time was estimated to 130 fs [7]. Effective and extremely fast CS in nonpolar solvents suggests that the solvation is not a precondition to reach the CT state and indicates a significant role of high-frequency intramolecular vibrational modes in the CS process.

3. Intramolecular ion pair formation and fs kinetics of electron transfer

The transient absorption spectrum of PP in acetonitrile (Fig. 2(a)) recorded 1 ps after excitation (see [9] for experimental details) is dominated by a broad band centered at 440 nm which decays exponentially with a decay time of 0.8 ns. The band can neither be ascribed to the $S_n \leftarrow S_1$ absorption of phenol (maximum at 600 nm) nor to the $S_n \leftarrow S_1$ absorption of Pd, but it matches very well the known absorption band of the phenol radical cation. We identify the observed spectral feature with the formation of the phenol radical cation in PP and consider its appearance in the transient absorption spectrum at very short delay times as an evidence for completion of the photoinduced CS in PP in less than 1 ps.

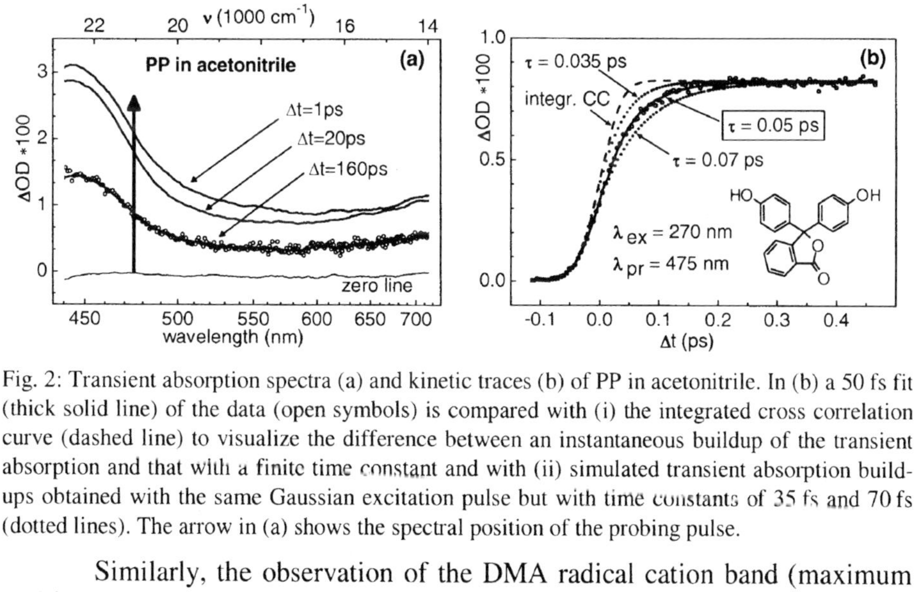

Fig. 2: Transient absorption spectra (a) and kinetic traces (b) of PP in acetonitrile. In (b) a 50 fs fit (thick solid line) of the data (open symbols) is compared with (i) the integrated cross correlation curve (dashed line) to visualize the difference between an instantaneous buildup of the transient absorption and that with a finite time constant and with (ii) simulated transient absorption build-ups obtained with the same Gaussian excitation pulse but with time constants of 35 fs and 70 fs (dotted lines). The arrow in (a) shows the spectral position of the probing pulse.

Similarly, the observation of the DMA radical cation band (maximum at 465 nm) in the transient absorption spectra of MGL indicates ultrafast CS in MGL. Transient absorption spectra for both LTAMs prove the extremely rapid formation of an intramolecular "ion pair". The CS rates have been determined by measuring the formation kinetics (buildup) of the radical cation from the electron donating groups in PP and MGL (Fig. 2(b)). The extremely fast CS after excitation of the S_1 state of PP (50 fs in acetonitrile, 80 fs in ethyl acetate) and only slightly slower CS in MGL excited to the S_2 state (150 fs in acetonitrile, <100 fs in cyclohexane) [9] provide clear evidence for intramolecular ET in LTAMs that occurs from a vibronically non-equilibrated level and is markedly faster than the diffusive and inertial solvation time scales. This indicates negligible control by solvent dynamics and the dominant role of high-frequency intramolecular vibrational degrees of freedom in the CS process which act as the ET promoting modes. In LTAMs, such a source of vibronic coupling is provided by the C–O bond in the lactonic ring, as electron capture by or transfer to the Pd molecule results in excitation of C–O stretching vibrations which leads to a C–O bond instability and a high probability of its cleavage. In addition, the excess electronic energy resulting from electron capture by Pd was found to be preferably deposited in the C–O bond [10].

4. Conclusions

We find that the photoinduced intramolecular charge separation in LTAMs is faster than the inertial solvation dynamics and can proceed within 50 fs. Such a rapid intramolecular electron transfer cannot be explained by a solvent induced lowering of the CT state. We therefore propose that it is due to vibronic coupling to the C–O lactonic bond. The precise mechanism of the coupling will have to be determined by further ultrafast measurements in combination with high level quantum chemical calculations. In total, LTAMs are excellent model systems for studies on ultrafast ET, solvation dynamics, and the initial ET stages in organic donor acceptor systems.

Acknowledgements: We thank Tanja Bizjak for her valuable contributions.

References

1. R. Huber, J. E. Moser, M. Grätzel, and J. Wachtveitl, J. Phys. Chem. B 106 (2002) 6494.
2. J. Schnadt, P. A. Brühwiler, L. Patthey, *et al.*, Nature 418 (2002) 620.
3. F. Pöllinger, H. Heitele, M. E. Michel-Beyerle, C. Anders, M. Fuscher, and H. A. Staab, Chem. Phys. Lett. 198 (1992) 645.
4. G. C. Walker, E. Åkesson, A. E. Johnson, N. E. Levinger, and P. F. Barbara, J. Phys. Chem. 96 (1992) 3728.
5. J. Jortner, M. Bixon, H. Heitele, and M. E. Michel-Beyerle, Chem. Phys. Lett. 197 (1992) 131; M. Bixon and J. Jortner, Adv. Chem. Phys. 106 (1999) 35.
6. E. Baigar, P. Gilch, W. Zinth, M. Stöckl, P. Härter, T. von Feilitzsch, and M. E. Michel-Beyerle, Chem. Phys. Lett. 352 (2002) 176; D. H. Son, P. Kambhampati, T. W. Kee, and P. F. Barbara, J. Phys. Chem. A 106 (2002) 4591; N. Mataga, H. Chosrowjan, S. Taniguchi, Y. Shibata, N. Yoshida, A. Osuka, T. Kikuzawa, and T. Okada, J. Phys. Chem. A 106 (2002) 12191; J. M. Giaimo, A. V. Gusev, and M. R. Wasielewski, J. Am. Chem. Soc. 124 (2002) 8530.
7. J. Karpiuk, Phys. Chem. Chem. Phys. 5 (2003) 1078;
8. J. Karpiuk, J. Phys. Chem. A 108 (2004) 11183.
9. T. Bizjak, J. Karpiuk, S. Lochbrunner, and E. Riedle, J. Phys. Chem. A 108 (2004) 10763;

10. B. G. Zykov, Y. V. Vasil'ev, V. S. Fal'ko, A. N. Lachinov, V. I. Khvostenko, and N. G. Gileva, JETP Lett. 64 (1996) 439.

Femtochemistry VII
Fundamental Ultrafast Processes in Chemistry, Physics, and Biology
A.W. Castleman, Jr. and M.L. Kimble

Fluorescence Dynamics of Coumarin on Montmorillonite Structure

M. Zitnan[a], J. Bdzoch[a], I. Bugar[b], V. Szocs[a], T. Palszegi[c], M. Janek[d], D. Chorvat[b], D. Velic[a, b]

[a]*Komensky University, Mlynska dolina CH-1, 842 15 Bratislava, Slovakia*
[b]*International Laser Center, Ilkovicova 3, 812 19 Bratislava, Slovakia*
[3]*Slovak University of Technology, Radlinskeho 9, 812 37 Bratislava, Slovakia*
[4]*Institute for Technical Chemistry, Water- and Geotechnology, 76021 Karlsruhe, Germany*

1. Introduction

The confinement effect is a driving force for reactivity and function of chemical systems. Cyclodextrin-coumarin represents a confined system in a form of supramolecular host-guest inclusive complex [1-3]. In this contribution, the interaction between coumarin C522, fluorescence probe, and montmorillonite, clay material, was studied by means of fluorescence dynamics. The negative layer charge of montmorillonites can be modified by the fixation of small exchageable cations [4-6]. Lithium was used as a monovalent charge compensating cation on the surface and within the interlayer space to produce the reduced charge montmorillonite (RCM) structure. The dynamic fluorescence Stokes shift of coumarins in the field of ions (1 M salt solution) in liquid or solid matrix was studied [7]; however time-resolved fluorescence of non-cationic organic compound on RCM has not been, to our knowledge, investigated yet.

2. Results and Discussion

The concentration of coumarin C522 (Radiant Dyes Accessories) in solution was approximately 10^{-7} M. Sodium saturated bentonites were received from Sued-Chemie. Purification, fractionation, and lithium intercalation were used [6] to prepare the samples. The concentration of RCM in water was 4 weight %. Increasing the content of exchangeable lithium resulted in the higher intercalation of lithium ions and thus balanced the overall negative charge of montmorillonites. The contents of lithium were 0, 20, 40, 60, 80, and 100 molar %, and therefore the samples were marked as 0, 0.2, 0.4, 0.6, 0.8, and 1.0

mol/dm^3 concentrations. Note that the high content of lithium resulted in a high level of intercalated lithium into the RCM structure and only low concentration of lithium remained on the surface of RCM [6]. Vice versa, the low content provided a low level of intercalation and a high surface concentration. A role of the lithium surface concentration is presumably important, since the interaction of coumarin with intercalated lithium is hindered.

Steady-state fluorescence spectroscopy was performed by using Fluorolog (3-11, Jobin-Yvon/SPEX), and the coumarin-montmorillonite samples were characterized by excitation and emission spectra. The increase of lithium concentration caused the shift of excitation and emission maxima from 419 nm (23 866 cm^{-1}) and 529 nm (18 904 cm^{-1}) in sample without lithium cations to 438 nm (22 831 cm^{-1}) and 513 nm (19 493 cm^{-1}) in 1M lithium sample, respectively. The Stokes shifts changed from 4963 cm^{-1} to 3338 cm^{-1} (110 to 75 nm) as shown in Fig 1.

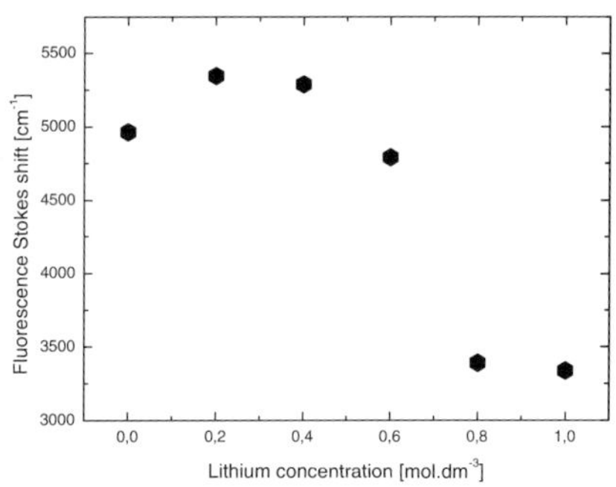

Fig. 1. Stokes shift as a function of lithium content

Time-resolved spectra, measured by using up-conversion system (CDP TiF50, 100 fs, 80 MHz, 0.3 W, 800 nm, and FOG100), provided fluorescence dynamics of coumarin where the excitation wavelength varied from 480 to 560 nm in 10 nm steps. Different models have been used for the microscopic explanation of the static and dynamic Stokes shifts [1-3]. The simpliest approach to analyze the dynamic Stokes shift is the biexponential fit of the experimental curves, providing short and long time components as shown in Eq. 1

$$I(t) = a_1 \exp(-t/\tau_1) + a_2 \exp(-t/\tau_2) \qquad (1)$$

where τ_1 mimics the short relaxation time and τ_2 the overall decay, both shown in Fig. 2 as experimental results.

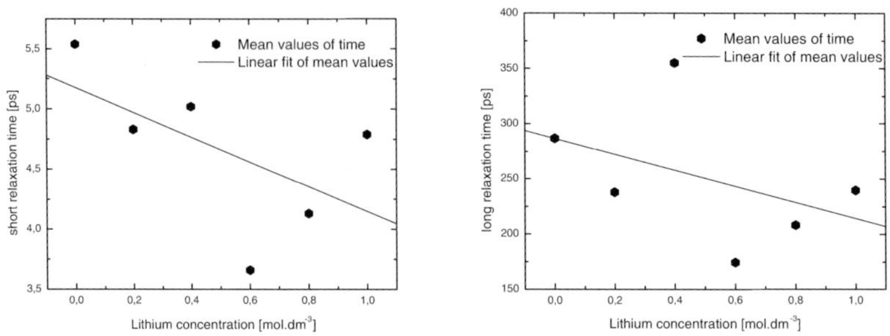

Fig. 2. Left part shows the short component and right part the long component of biexponential analysis, along with the linear trends

The short time component τ_1 changed from 5.5 to 3.7 ps and the long time component τ_2 from 355 to 208 ps. The linear fits of mean times suggested a trend of dependence of coumarin fluorescence dynamics on the lithium concentration. Note that the linear fits were used to only guide an eye without providing a statistical weight. These fits show the decreasing tendency of the short and long times as a function of lithium concentration. The interaction between coumarin and water molecules plays a significant role in the initial phase of the relaxation process. Coumarin has a hydrophobic character with low solubility in water and is presumably adsorbed on the vacant RCM surface. Lithium cations are physisorbed on the surface of RCM. Fig. 3 shows a schematic view of the interaction between RCM and coumarin.

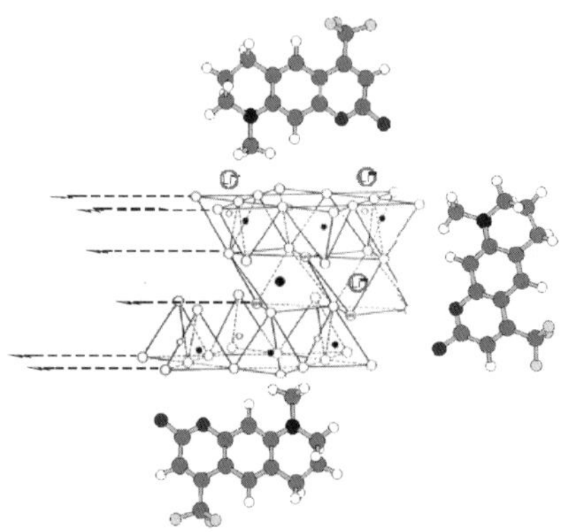

Fig. 3. Schematic interaction between montmorillonite, two sheets of SiO_4 tetrahedra sandwiching a sheet of octahedrally coordinated cations, and coumarin.

Lithium cations have strong hydrophilic character, and the hydration enthalpy is the highest one from the alkaline cations. Non-intercalated lithium cations laying on surface decrease the free surface area and bind water molecules into one solvation shells. The low concentration of lithium on the surface increases the surface vacancies, and the water molecules are left near the surface for interaction with coumarin. Presumably due to this interaction, the fluorescence relaxation of coumarin was faster.

3. Conclusions

The interaction between coumarin C522, fluorescence probe, and montmorillonite, clay material was studied by means of fluorescence dynamics. The short time component τ_1 changed from 5.5 to 3.7 ps and the long time component τ_2 from 355 to 208 ps. The fits showed the decreasing tendency of the short and long times as a function of lithium concentration. The high concentration of lithium on the surface decreased the surface vacancies, and water molecules were bound into lithium ions solvation shells. The low concentration of lithium on the surface increased the surface vacancies, and the water molecules were left presumably free to faster quench the fluorescence of coumarin.

Acknowledgement

The financial supports of VEGA 1/0216/03, APVT-20-029804, and NATO PST.EAP.CLG 980634 grants are appreciated.

References

1. A. Douhal, Chem. Rev. 104 (2004) 1955 .
2. S. Vajda, R. Jimenez, S. Rosenthal, V. Fidler, and G. R. Fleming, J. Chem. Soc. Far. Trans. 91 (1995) 867 .
3. P. Bodis, I. Bugar, T. Palszegi, D. Velic, and D. Chorvat, in: Femtochemistry and Femtobiology, M M. Martin and J. T. Hynes (Eds.) Fluorescence Dynamics of Coumarin C522 in Water and in Cyclodextrim Cavity, Elsevier, Amsterdam, 2004, 237.
4. P. Komadel, J. Madejova, and J. Bujdak, Clays and Clay Minerals 53 (2005) 313.
5. J. Bujdak, N. Iyi, Z. Kaneko, A. Czimerova, and R. Sasai, Phys. Chem. Chem. Phys. 5 (2003) 4680.
6. M. Zemanova, G. Link, S. Takayama, R. Nuesch, and M. Janek, Applied Clay Science, in press.
7. C. F. Chapman and M. Maroncelli, J. Phys. Chem. 95 (1991) 9095.

Femtochemistry VII
Fundamental Ultrafast Processes in Chemistry, Physics, and Biology
A.W. Castleman, Jr. and M.L. Kimble

Fluorescence anisotropy of photoactive yellow protein (PYP) and related systems: Is the low-lying excited state of the PYP quasi-degenerate?

Seiji Taniguchi, Haik Chosrowjan, Noboru Mataga

Institute for Laser Technology, c/o Technical Research Center, Kansai Electric Power Company, 3 – 11 - 20 Nakoji, Amagasaki, Hyogo 661 – 0974, JAPAN
e-mail: stani@ile.osaka-u.ac.jp

1. Introduction

Photoactive Yellow Protein (PYP) is a blue light photoreceptor responsible for the negative phototaxis of the purple sulfur bacterium *Ectothiorhodospira halophila*. Photoexcitation of PYP with a blue light triggers a photocycle that involves several intermediate states. Trans/cis photoisomerization of its chromophore (*p*-coumaric acid) buried in the protein has been identified as the overall primary event triggering its photocycle. However, the complex ultrafast dynamics of PYP's photocycle is not fully understood and efforts are made to separate and characterize the intra- and interchromophore factors affecting the early processes. Ideally, such studies should be curried out in environments from gas to solution and protein phases to elucidate how the protein environment controls and directs the primary photodynamics. Recently, high-resolution gas-phase fluorescence excitation experiments were performed on the neutral form of PYP's *p*CA chromophore [1]. Excitation wavelength dependent photochemistry was observed and explained by the existence of a low-lying second electronic state (n→ π* origin). C. Ko and coworkers [2] theoretically evaluated the excitation wavelength dependent dynamics of neutral trans-*p*CA in vacuum by *ab initio* QM and *ab initio* multiple spawning (AIMS) dynamics calculations. They predicted the existence of a low-lying second electronic state (π→ π* origin), although questioning the original n→ π* assignment [1, 2]. It remains to be seen whether it is also the case for deprotonated *p*CA in solvated or protein environments, and if so, whether the quasi-degenerate excited state plays any significant role in primary photodynamics.

Recently, it was suggested that the excited state of the PYP itself (also denatured PYP and its free chromophore model pCT^-) consists of two close-lying (quasi-degenerate) electronic states [3]. Based on steady state and transient absorption measurements, it was concluded that upon excitation, simultaneous population of those states may occur and the complex photophysics of PYP arise due to competing relaxation pathways from those states. In this work, using femtosecond time-resolved fluorescence up-conversion spectroscopy (instrumental response ~120 fs) [4], we have attempted to further elucidate the quasi-degenerate nature of the PYP's excited state. Namely, we have carried out detailed fluorescence anisotropy measurements on native PYP, 3,4-dihydroxy cinnamic acid (caffeic acid) analogue and several mutant PYP systems (E46Q, T50V, E46Q/T50V, R52Q, P68A, E46Q/R52Q).

2. Results and Discussion

It is worth mention that fluorescence anisotropy (*r(t)*) is a fine tool of testing the electronic origin of the fluorescent state. For single excited states in an isotropic and non-reactive environment, the theoretical value of *r(t)* is 0.4. However, a degenerate (or quasi-degenerate) state, among other factors, may cause a deviation from that value because de-phasing by incoherent coupling (loss of coherence), equalization of the populations of those two levels, etc. could affect *r(t)* [5]. Furthermore, in site-directed mutant proteins, due to a change of the local environment, those close-lying electronic states may further

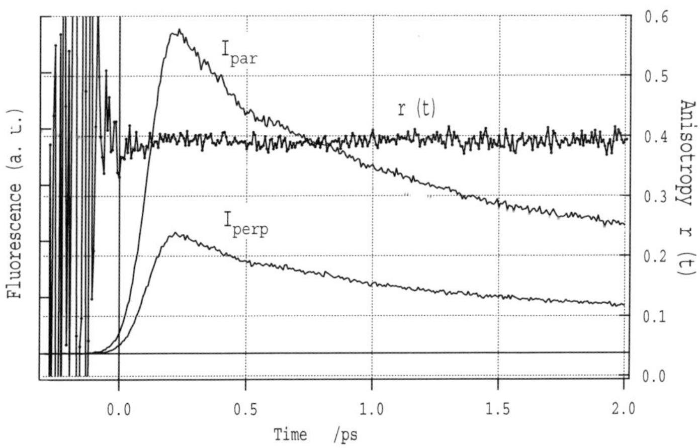

Fig. 1. Fluorescence decay dynamics of native PYP excited at 410 nm and monitored at 500 nm for parallel (I_{par}) and perpendicular (I_{perp}) polarization excitations. The fluorescence anisotropy *r(t)* has been calculated using the formulae $r(t) = (I_{par} - I_{perp})/(I_{par}+2*I_{perp})$

"split", causing additional changes in *r(t)*. In our experiments on native (Fig. 1) and mutant PYPs (data not shown), the fluorescence anisotropy was *always* constant and close to the theoretical value ~ 0.4, although in some mutants the anisotropy value was a little smaller (E46Q/T50V ~0.36, R52Q ~0.37, P68A ~0.38). Furthermore, the fluorescence decay dynamics of the denatured PYP (excluding the contribution of the solvation effect) (Fig. 2) was single exponential (~ 11.5 ps), indicating that in the denatured PYP, only one electronic state is emissive. Based on the above results, the following conclusions may be true: (I) the excited states of the native, mutant, and denatured PYPs are not degenerate (or quasi-degenerate), (II) there is very weak or no interaction between those close lying (quasi-degenerate) electronic states, or (III) electronic quenching from the higher- to the lower- excited state occurs within the time resolution of our experiments. In this case, the excitation to the higher level could be equivalent to the exciting lower level with an appropriate amount of excess vibrational energy. The dynamical simulations carried out by C. Ko and coworkers [2] support the third possibility, i.e. when excited to the highest S_2 level of the system, almost all of the population quenches to S_1 within ~30 fs, and the neutral trans-*p*CA molecule remains completely planar during that quenching process [2]. We think that such a scenario may be working also in the case of deprotonated *p*CA, because in some mutants, the anisotropy values were a little smaller than 0.4, offering the speculation that an ultrarapid decay may have preceded. The small differences in the final *r(t)* values of those mutants can be explained by the small differences in coupling constants and population equalization ratios between those close lying states. A comparative study of various PYP mutants by transient absorption spectroscopy may shed more light on the discussed problem.

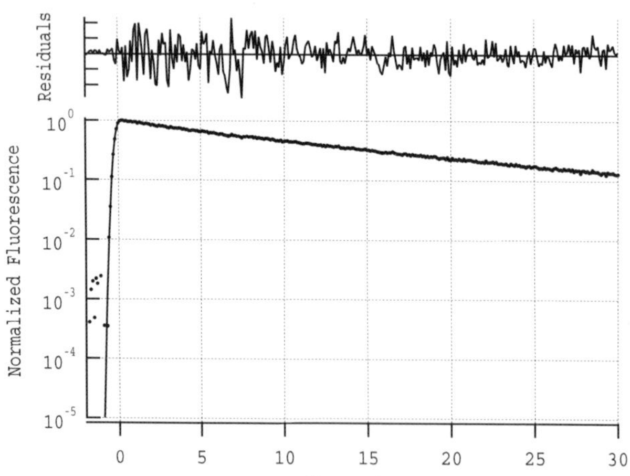

Fig. 2. Fluorescence decay dynamics of denatured PYP (denatured by addition of guanidinium HCl at pH 9.5) excited at 400 nm and monitored at 490 nm.

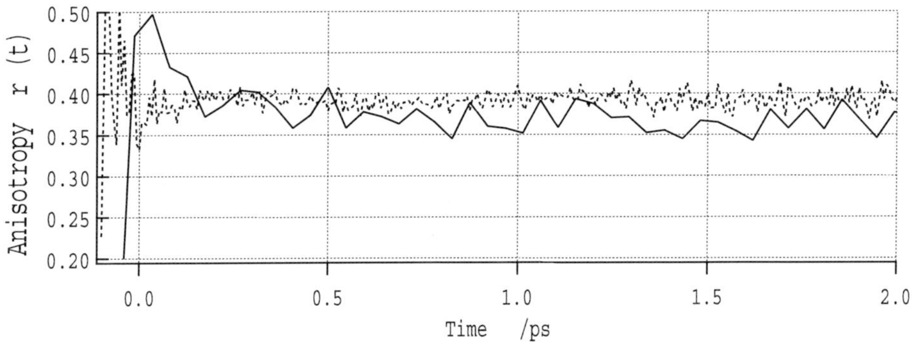

Fig. 3. Comparison between fluorescence anisotropies of native PYP (dotted line) and its 3, 4 - dihydroxy cinnamic acid (caffeic acid) analogue (full line).

On the other hand, in the case of the 3,4-dihydroxy cinnamic acid (caffeic acid) analogue, the initial value of the anisotropy was ~0.5; it relaxed rapidly and remained constant at ca. 0.36. The ultrafast (~50 fs) decay of the fluorescence anisotropy (Fig. 3) in this case cannot readily be explained in terms of rotational diffusion but could be due to the existence of two close lying electronic states as predicted for the native PYP. The ultrafast decay to the 0.4 value may have originated due to the completely dephasing of those close lying states and further to the level of 0.36 after some equalization of the populations of those levels. In an earlier study on the caffeic acid analogue [6], quite different wavelength dependence of fluorescence dynamics, compared with the native PYP, was observed. Furthermore, over-damped oscillations could be recognized in the fluorescence dynamics, but no clear-cut explanation was suggested for this behavior. Involving the existence of a quasi-degenerate excited state for the caffeic acid analogue may explain our earlier observations.

References

1. W. L. Ryan, D. J. Gordon, and D. H. Levy, J. Am. Chem. Soc. 124 (2002) 6194.
2. C. Ko, B. Levine, A. Toniolo, L. Manohar, S. Olsen, H.-J. Werner, and T. J. Martinez, J. Am. Chem. Soc. 125 (2003) 12710.
3. P. Changenet-Barret, A. Espagne, P. Plaza, K. J. Hellingwerf, and M. M. Martin, New J. Chem. 29 (2005) 527.
4. H. Chosrowjan, S. Taniguchi, N. Mataga, M. Unno, S. Yamauchi, N. Hamada, M. Kumauchi, and F. Tokunaga, J. Phys. Chem. B 108 (2004) 2686.
5. G. G. Gurzadyan, T.-H. Tran-Thi, and T. Gustavsson, Proceedings of SPIE 4060 (2000) 96.
6. N. Mataga, H. Chosrowjan, and S. Taniguchi, J. Photochem. Photobiol. C: Photochemistry Reviews 5 (2004) 155.

Femtochemistry VII
Fundamental Ultrafast Processes in Chemistry, Physics, and Biology
A.W. Castleman, Jr. and M.L. Kimble

Light-Driven Enzymatic Reaction in Thermophilic Protochlorophyllide Oxidoreductase

O. Sytina[*], M.L.Groot[*], R. van Grondelle[*], D.J. Heyes[**], C.N. Hunter[**]

[*]*Section Biophysics, Vrije Universiteit, De Boelelaan 1081, 1081 HV Amsterdam, The Netherlands, email address: O.Sytina@few.vu.nl*
[**] Department of Molecular Biology and Biotechnology, University of Sheffield, Firth Court, Western Bank, Sheffield, S10 2TN, UK

1. Introduction

Within the chlorophyll biosynthetic pathway NADPH:protochlorophyllide oxidoreductase (POR; EC 1.3.1.33) catalyzes the light-dependent *trans* addition of hydrogen across the C17-C18 double bond of the *D*-ring of protochlorophyllide (Pchlide) to produce chlorophyllide (Chlide) (*1*). The fact that POR is light activated means the enzyme:substrate complex can be formed in the dark, removing the diffusive components out of the reaction. It was previously shown that the POR-catalyzed reaction can be observed in real time on a picosecond timescale after initiating catalysis with a 50 fs laser pulse. (2) In the present work, POR, from the thermophilic cyanobacterium *Thermosynechococcus elongatus* BP-1, has been analyzed in the same way.

2. Materials and Methods

Pchlide and POR from the cyanobacteria *Thermosynechococcus elongatus* BP-1 were produced and purified as previously described (1,3). The samples contained 0.5mM Pchlide, 0.5mM POR, and 15mM NADPH in activity buffer (50 mM Tris pH 7.5, 300 mM NaCl, 2 % Genapol, 0.1 % β-mercaptoethhanol). The laser setup has been described previously (4). All complexes were excited at 475 nm, at the red edge of the Soret band (S0→S2 transition) of Pchlide where the absorption of the product Chlide is minimal. Detection was in the wavelength region 600 to 700 nm, probing the S0→S1 transition of Pchlide, and products. The sample is filled in a CaF_2 cell of 200μm path length, contained in a Lissajous sample scanner that moved at such speed that a new part of the sample was illuminated at every shot. The same volume was illuminated again after 1-2 minutes. One scan of the pump-delay line, recording data from -15 ps to 6 ns, took ~40 seconds, so in the first scan each sample spot had not been illuminated before. To obtain a sufficient signal to noise ratio, several subsequent scans are averaged over 3 different fresh samples. All experiments were performed at room temperature. The data was analyzed using a global fit algorithm.

3. Results

Figures 1a,b,c show the series of time-dependent data in the form of species-associated-difference spectra (SADS) for thermophilic POR that were taken in consecutive scans. The thermophilic enzyme appears to show dynamics on an ultrafast timescale. There is a clear difference between the first-scan and the last-scan kinetics. A global analysis of the datasets using a sequential model with increasing lifetimes reveals the presence of 4 kinetic components with lifetimes 0.3ps-0.5ps, 9-11ps, 460-520ps, 3-4ns.

The SADS of the thermophilic POR show bleaching and stimulated emission from excited Pchlide, which has an initial peak at 646 nm, and the formation of stimulated emission from an excited Chlide product around 670-680nm in the later scans. Formation of product Chlide in its excited state was already observed in (2) on the mesophilic form of the POR enzyme.

Now in the first few scans, where the product Chlide band is not formed at all, the TPOR SADS show only a progressive, multi-exponential, loss of red signal, due to quenching of stimulated emission from excited Pchlide, in combination with decay of PChlide to the ground state (Fig.1a). This spectral change is accompanied by the decay of broadband excited state absorption above 660nm and below 630nm. Thermophilic enzyme shows the first negative signal around 670-680nm, indicating product formation in later sample scans. To emphasize the difference between the first-scan and the last-scan kinetics, time-traces at 677nm are shown in Fig. 1d. Initial kinetics reveals appearance of Chlide signal on the ns-timescale, whereas Chlide is formed after 1ps in final kinetics. It is important to note that there is negligible direct initial excitation of the product Chlide with laser pulse in later scans, when product is being accumulated in the sample, since the signal around 677nm is flat and positive in the time-zero spectra. Continuous increase in product yield is seen with scan number, i.e. number of excitations. We exclude that the increasing yield effect and its acceleration of formation are caused by changing sample conditions during one delay-line scan and entire experimental period as well; we recorded a reversed-time direction scan, and we can exclude temperature effects also[*].

---

[*] A simple calculation shows that negligible water heating occurs in our experiment, we need 8.10^6 laser shots to heat the water 1 degree, which takes, at 500 Hz, 4 hrs. We can calculate with convection equations that if the sample is at 5°C when it is filled into the cell, it takes only about 77sec to reach room temperature. So the enzyme is at RT during the experiments.

Therefore, we conclude that the scan-dependent kinetics translate into *light-dependent kinetics*.

4. Discussion

Short time scan measurements allowed us to control the amount of light an enzyme has seen, or in other words, how often it has cycled through the excited state, and to monitor the concomitant reaction kinetics. The presented data clearly show that the yield and rate of the Chlide product formation increase with scan number. Since we have eliminated both temperature effects and scan direction as possible causes for this effect, we interpret the changes as due to prolonged illumination of the sample with increasing scan number, i.e. "light adaptation" of the enzyme complex occurs. It can, for example, be due to adjustments of the transition state conformation of the active site during the excited state lifetime, leading to a more favorable configuration of the enzyme to form products, which persists on the seconds time scale. Initially, protein complexes in the solution may have slightly different subconformational states and not all of them are optimal for the reaction. Light-induced redistribution of electrons initiates slow protein domain movements in a long time scale.

It has been previously shown that two ways of the reaction coexists in POR. A catalytic mechanism of POR from Synechocystis proceeds with time constants of 3 ps and 400 ps upon concerted (3 ps) or sequential (400 ps) transfer of a proton and a hydride to the Pchlide molecule. Now for TPOR, we resolved an ultrafast component with broad spectrum and lifetime less than 1 ps with max at 646nm. Broadening and red shift of excited Pchlide band does not allow to follow clearly the evolution of substrate and hides the actual reaction pathway. It is also possible that the presence in the solution or binding of pigment aggregates by protein reduces and slows down its catalytic functionality. In TPOR, without broadening product yield is observed under the same experimental conditions (data not shown).

So, summarizing the main observations in the experiment on thermophilic POR, we observe (1) Ultrafast product formation in thermophilic POR catalysis, (2) The blue shift of Pchlide stimulated emission and its decay, and (3) Faster and more product Chlide formation occurs in samples which have been excited before.

References

1. D. J. Heyes, A. V. Ruban, and C. N. Hunter, Biochem. 42 (2003) 523.
2. D. J. Heyes, C. N. Hunter, I. H. M. van Stokkum, R. van Grondelle, and M. L. Groot, Nature Struct. Bio. 10 (2003) 491.
3. D. J. Heyes and C. N. Hunter, Biochem. 43 (2004) 8265.
4. M. L. Groot, J. Breton, L. van Wilderen, J. P. Dekker, and R. van Grondelle, J. Phys. Chem. 108 (2004) 8001.

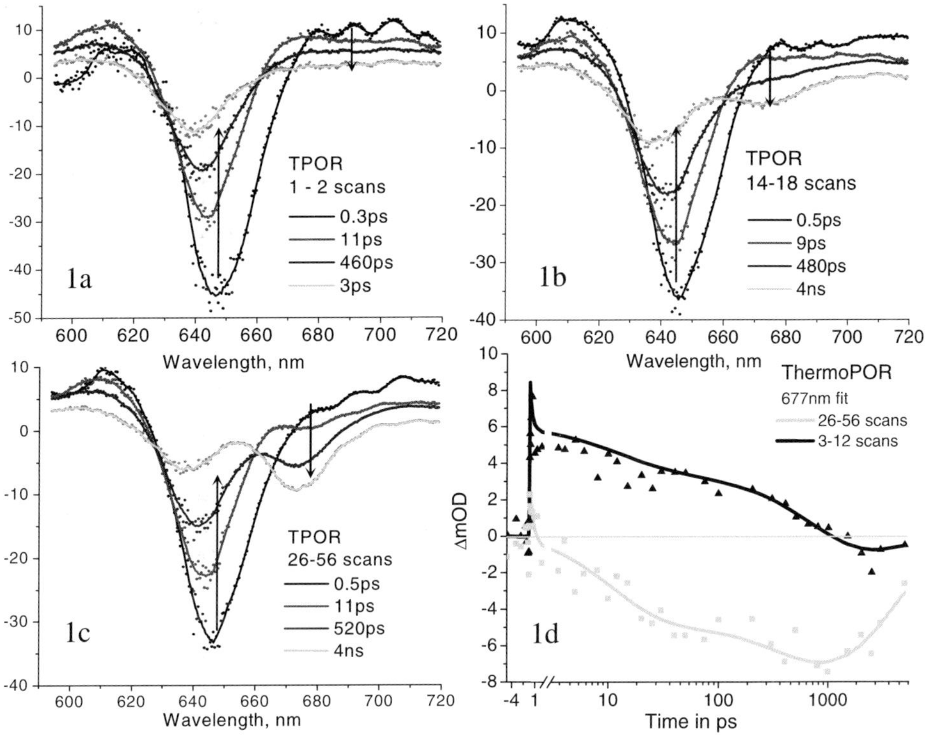

Fig. 1. a,b,c, Species-associated-difference spectra (SADS) of the exponential lifetimes of the thermophilic POR transient absorption changes. Vertical axis corresponds to ΔmOD units. Arrows show a time evolution of the species; d- Experimental data and multi-exponential fit of 677 nm signal in initial and final sample scans.

7. Quantum Control

Femtochemistry VII
Fundamental Ultrafast Processes in Chemistry, Physics, and Biology
A.W. Castleman, Jr. and M.L. Kimble

Converting Concepts and Dreams of Coherent Control into Applications

Marcos Dantus[a], Vadim V. Lozovoy[a], Igor Pastirk[b], Johanna M. Dela Cruz[a]

[a] *Department of Chemistry and Department of Physics and Astronomy, Michigan State University, East Lansing, MI 48824, USA*
[b] *Biophotonic Solutions, Inc., Okemos, MI 48864. Pastirk@biophotonicsolutions.com*

1. Introduction

Since the laser was invented, scientists dreamed of focusing its energy to transform matter in a controlled fashion. Four decades later, the hardware required (femtosecond lasers and pulse shapers) has reached the critical point where many of those early dreams can be converted into robust applications. This brief report summarizes recent progress from our group in the field of laser control of nonlinear optical physicochemical processes and the development of early applications. For a full description of the research, the relevant background and relevant references, the reader is referred to the original articles.

2. Using the known nonlinear optical response of a material to characterize the spectral phase of a laser pulse

This dream application would allow characterization of a laser pulse without using the conventional methods which involve some type of autocorrelation and/or interferometry. Given that the nonlinear optical response of materials is strongly dependent on spectral phase, we realized that this dependence can be used to fully and accurately characterize femtosecond laser pulses at the position of the sample.

The realization of this dream required a number of insights. First, we needed to introduce a known and well-calibrated phase to evaluate the unknown phase distortions in the beam. This concept is similar to the Wheatstone Bridge, a device using a known resistor to measure an unknown resistance [1]. Here, a known spectral phase is introduced by the pulse shaper, and the unknown phase distortions in the laser are measured. Second, because a given response can be caused by a number of different spectral phases, we needed to obtain sufficient information to avoid ambiguities. Third, we found that two scans provided us with great stability and robustness even in situations where the signal to noise ratio was below one. Finally, because the method uses a pulse shaper to introduce the reference phase, it is possible to correct phase distortions, once they have been measured, in order to achieve transform limited pulses. Iteration of the measurement and compensation greatly improves the sensitivity and performance, making this application one of the most accurate methods available for spectral phase characterization [2].

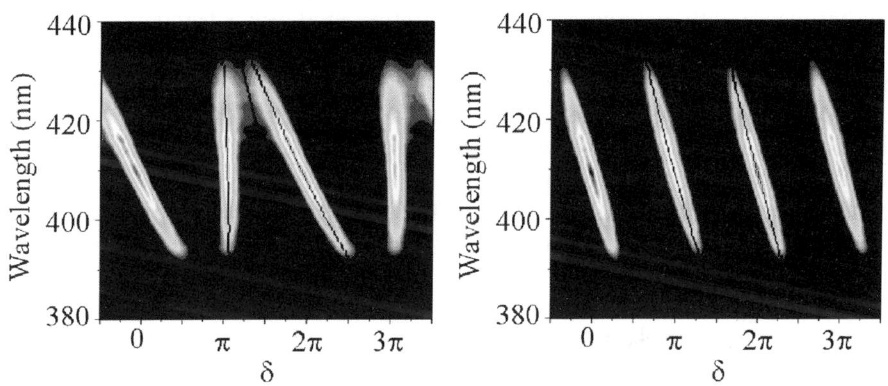

Figure 1. Experimental MIIPS data obtained for femtosecond pulses with spectral phase distortions (left) and the same laser pulses after one compensation iteration (right), showing that the spectral distortions have been compensated and the pulses are now much closer to being transform limited.

In Figure 1, we show multiphoton intrapulse interference phase scan (MIIPS) data for pulses of 15 fs duration. The vertical axis corresponds to the spectrum of the resulting frequency doubled pulse in a thin second harmonic generation crystal. The parameter delta in the horizontal axis scans the reference sinusoidal phase function across the spectrum of the fundamental pulse in the pulse shaper. This sinusoidal function causes a local maximum in the second harmonic spectrum that scans from longer to shorter wavelengths. Distortions in these lines indicate spectral phase distortions. Transform limited pulses correspond to straight parallel lines (see Figure 1 right). Typically, three or four iterations are needed to reduce spectral distortions below 0.1 radian. This application has been shown capable of correcting phase distortions introduced by high numerical aperture objectives even for sub-10 fs pulses [3].

3. Maximizing the nonlinear response at one frequency while minimizing it elsewhere

The dream would be to achieve two-photon excitation spectroscopy with high resolution at specific wavelengths with a single broad-bandwidth ultrashort pulse. With a series of phase modulated pulses, we would be able obtain an entire excitation spectrum without tuning the laser. The main advantage of phase modulation over tuning the laser is that the spectrum of the fundamental does not change; therefore, no additional optimization in alignment of optics and spectral phase compensation would be needed at each point.

The realization of this application required us to find out what would be the optimal phase that could be used to achieve intense two photon excitation at one frequency while suppressing the excitation at all other frequencies. Because we are dealing with nonlinear excitation, a simple slit on the spectrum that cuts out the undesired portions in the spectrum results in a drastic loss of two-photon excitation efficiency. The signal intensity scales quadratically with the width of the slit [4]. Our solution involved three steps. First we had to cause multiphoton intrapulse interference [5,6]. We then realized that binary phase shaping would greatly reduce the number of possible phase functions that need to be considered to solve the problem, without losing at least some of the optimum solutions [7]. The relative phase between two frequencies would either be zero or π. The problem was then reduced to building binary phase functions that had a minimum autocorrelation. This procedure is described in Refs [4,7-8].

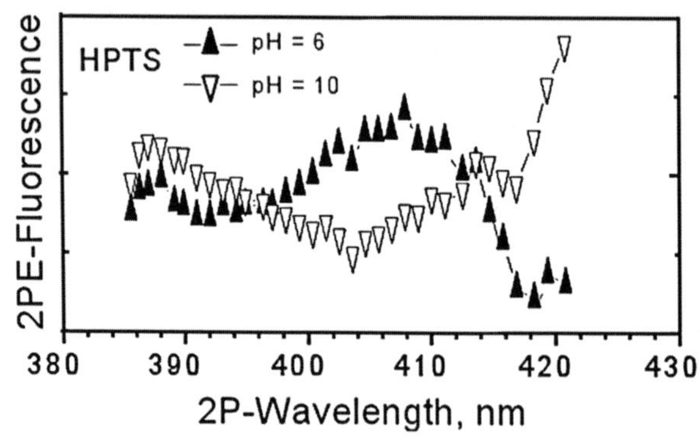

Figure 2. Two-photon excitation spectrum of HPTS, a pH sensitive fluorescent probe in acidic and basic aqueous solutions.

In Figure 2, we demonstrate the application of selective nonlinear optical excitation as it is used to obtain the two-photon excitation spectrum of 8-hydroxypyrene-1,3,6-trisulfonic acid (HPTS) dissolved in acidic and basic solutions. The data were obtained without tuning the laser [8].

4. Achieving selective two-photon excitation through scattering biological tissue

This dream application would allow functional imaging and selective photodynamic therapy. The goal would be to activate the target molecules without causing damage to healthy tissue and to gain selectivity towards diseased cells. This would require the use of phase shaped pulses as indicated above. The challenge here is to make sure that the shaped pulses propagate through scattering biological tissue without losing their nonlinear optical selectivity properties.

The realization of this application involved the preparation of a sample that could be used to test selective excitation through biological tissue [9,10]. For this purpose we used three capillary tubes with an acidic HPTS solution and submerged them in a cell with a basic solution of HPTS at the same concentration. The experimental sample setup can be seen in the left panels of Figures 3A and B (with tissue). When the sample is imaged with transform limited pulses (Figures 3C and D), the walls of the capillary tubes are visible, but there is no distinction of pH between the inside of the capillaries and the solution surrounding them.

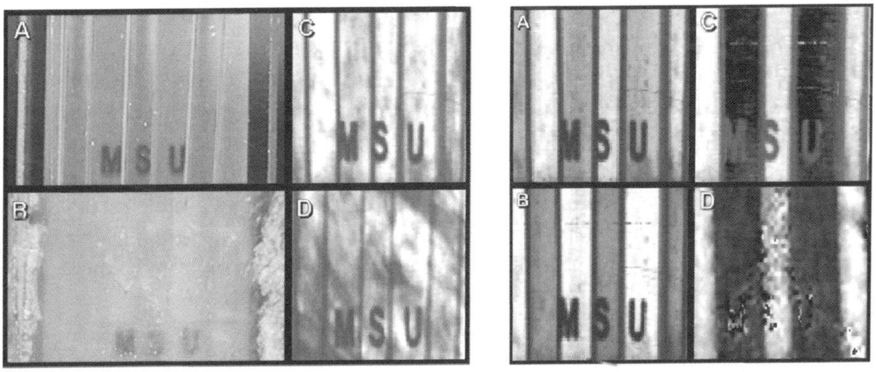

Figure 3. Experimental setup and results showing functional imaging through biological tissue using shaped femtosecond laser pulses (see text). (Reprinted with permission from ref. 8, Copyright 2004, The National Academy of Sciences of the USA)

When we used binary phase shaped pulses to optimize two-photon excitation of HPTS in acidic (or basic) solution, we achieved the desired contrast. We were able to take advantage of changes in the pH-sensitive two-photon cross-section of HPTS. The resulting images obtained with shaped pulses optimized for acidic (basic) HPTS excitation are shown in the right hand panels, Figure 3A (Figure 3B). The functional images shown were obtained by dividing the signals obtained from the differently shaped laser pulses without (Figure 3C) and with (Figure 3D) a 1 mm slice of biological tissue placed between the laser and the sample. The functional images show a clear distinction between the two solutions, bright regions correspond to acicic solution (inside the capillaries) and dark regions to alkaline solution; thus demonstrating that it is possible to deliver selective nonlinear excitation through biological tissue [9,10].

5. Preparing phase-shaped femtosecond pulses that can be used to recognize molecules, including stereoisomers

This dream application would allow us to develop an analytical method that in less than one second can identify molecules from complex mixtures. More interestingly, this development would give us the ability to distinguish stereoisomers that are indistinguishable by conventional electron-impact mass spectrometry.

The realization of this application required us to identify specific phase functions that enhanced the differences between isomeric compounds. To achieve this goal we opted for an exhaustive experimental evaluation of 10-bit binary phase functions. This allowed us to determine the phases that were optimally suited for isomer identification within a few minutes. Once we found the optimum 10-bit phases, we were able to check for reproducibility and to obtain statistical data regarding their use for isomer identification.

This procedure has been used to identify and quantify a number of isomeric pairs. In Figure 4 we show how this procedure results in a number of binary phases that yield high and low ratios for the fragment ions (m/z 90 and m/z 39) of ortho- and para-cresol (see left panel of Fig 4). By picking the highest and lowest ion ratio, we obtain a range of values that can be used to quantify the percentage of each isomeric species in a mixture [11]. Figure 4 (right panel) compiles a series of results that demonstrate the general application of this idea to a number of isomeric pairs.

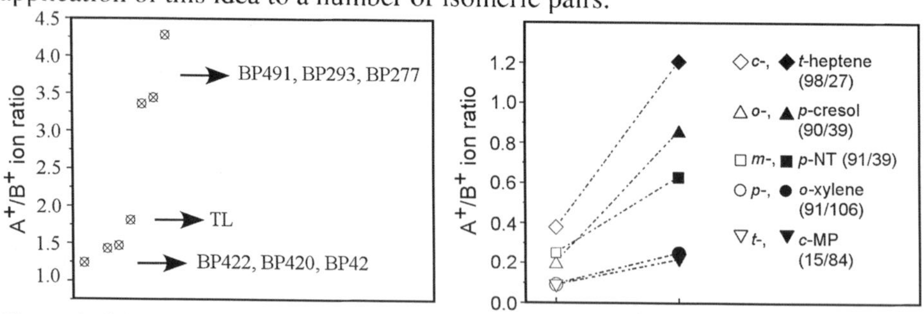

Figure 4. (left) Experimentally obtained ratios for the abundance of m/z 90 divided by m/z 39 resulting from the controlled ionization and fragmentation induced by different binary phase shaped femtosecond pulses. Notice that the data points show an overall factor of four increase in the ratio achieved by the phase shaped pulses. (right) This figure summarizes the results obtained for a number of geometric and stereoisomeric pairs. Each point in the graph represents the average value of 128 measurements.

6. Conclusion

This article summarizes work from our group aimed at taking some of the early concepts and dreams of laser control and turning them into robust applications. This work required a method to ensure that we could deliver a specific phase to the sample. Although this sounds like a simple task, most of the pulse characterization methods cannot be used in situ and require one to send the laser to a device with different dispersion characteristics than those surrounding the experimental sample. The development of MIIPS in our group gave us this advantage [2,3].

The second breakthrough in our work was the introduction of binary phase modulation for the control of nonlinear laser-matter excitation [4]. Although other groups had used binary phase functions, we formalized this practice with a rigorous theoretical foundation where we demonstrated that under certain conditions, binary phase functions provided ideal optimization. This was a critical step. We showed that we could reduce the search space for an optimum laser pulse by hundreds of orders of magnitude without loss of the optimum solution.

The experiments shown here are initial examples of how the technology we have developed can be used to optimize laser induced excitation and how this can be used for purposes of identification. Our group is actively working on methods that will further enhance our ability to find optimum phase functions. We are also pursuing a number of new and exciting applications.

7. Acknowledgements

Work on fundamental pulse shaping and coherent control are funded by the Chemical Sciences, Geosciences and Biosciences Division, Office of Basic Energy Sciences, Office of Science, U.S. Department of Energy. Projects on gas phase chemical reactions and imaging are funded by the NSF grant CHE-0500661.

References

1. M. Dantus, V. V. Lozovoy, and I. Pastirk, OE Magazine 9 (2003) 15.
2. V. V. Lozovoy, I. Pastirk, and M. Dantus, Opt. Lett. 29 (2004) 775.
3. B. Xu, J. M. Gunn, J. M. Dela Cruz, V. V. Lozovoy, and M. Dantus, J.Opt. Soc. Am. B, in press.
4. M. Comstock, V. V. Lozovoy, I. Pastirk, and M. Dantus, Opt. Express 12 (2004) 1061.
5. K. A. Walowicz, I. Pastirk, V. V. Lozovoy, and M. Dantus, J. Phys. Chem 106 (2002) 9369.

6. V. V. Lozovoy, I. Pastirk, K. A. Walowicz,and M. Dantus, J. Chem. Phys. 118 (2003) 3187.
7. V. V. Lozovoy and M. Dantus, Chem. Phys. Chem. 6 (2005) 1971.
8. V. V. Lozovoy, B. Xu, J. C. Shane, and M. Dantus, Phys. Rev. Lett., to be published.
9. J. M. Dela Cruz, I. Pastirk, M. Comstock, and M. Dantus, Opt. Express 12 (2004) 4144.
10. J. M. Dela Cruz, I. Pastirk, M. Comstock, V. V. Lozovoy, and M. Dantus, P. Nat. Acad. Sci. USA 101 (2004) 16996.
11. J. M. Dela Cruz, V. V. Lozovoy, and M. Dantus, J. Phys. Chem. A 109 (2005) 8447.

Femtochemistry VII
Fundamental Ultrafast Processes in Chemistry, Physics, and Biology
A.W. Castleman, Jr. and M.L. Kimble

Design of circularly polarized laser pulses for periodic electron circulation in Mg-porphyrin

Ingo Barth, Jörn Manz

Freie Universität Berlin, Fachbereich Biologie, Chemie, Pharmazie, Institut für Chemie und Biochemie, Takustr. 3, 14195 Berlin, Germany

Abstract

Simulations of the electronic wave packet dynamics driven by circularly polarized laser pulses with durations $\tau < 4$ fs and peak intensities $I_{max} < 6 \cdot 10^{12}$ Wcm^{-2} demonstrate ignition of electron circulation in the oriented ring-shaped molecule, Mg-porphyrin. We design laser pulses for perfectly periodic electron circulation. Starting from non-perfect so-called π and $\pi/2$ laser pulses for reference, the goal is achieved, ultimately, by re-optimized $\pi/2$ laser pulses which yield half population transfer from the ground state $|X\rangle = |1\ {}^1A_{1g}\rangle$ to a selective component $|j\ {}^1E_{u\pm}\rangle$ of excited degenerate states with E symmetry, with energies E_X and E_j, respectively. The resulting period of electron circulation is determined by a Heisenberg type relation in terms of the energy gap. We compare various approachs to these laser pulses. By extension, combinations of sequential re-optimized pulses yield selective electron circulations represented by analogous hybrid states which consist of two components, $|j\ {}^1E_{u\pm}\rangle$ and $|k\ {}^1E_{u\pm}\rangle$, with different periods.

1. Introduction

Femtosecond laser pulses with durations $\tau > 5$ fs have been employed successfully for analysis and control of molecular reaction dynamics [1–3]. In particular, so-called π laser pulses have been designed for complete population transfer between vibrational or vibronic states, see e.g. [4–6]. Likewise, $\pi/2$ laser pulses achieve half population transfers between initial and target nuclear states. Combinations of sequential linearly polarized π and/or $\pi/2$ laser pulses have also been constructed for various purposes, including selective transitions from reactant via intermediate to product states [7, 8] or the preparation of pure enantiomers from an oriented non-rotating racemate [9, 10]. Most relevant for the present paper are extensions to π and $\pi/2$ circularly polarized pulses for control of the nuclear wave packet dynamics, from an oriented but possibly rotating racemate to pure enantiomers [11].

The advent of ultrashort laser pulses with durations from say three fs to hundred attoseconds [12–14], and control of circular polarization [15], has opened analogous possibilities for analysis and control of electron dynamics, in the time domain where the nuclei are essentially frozen. Prominent recent examples include the localization of oscillatory electronic wavepackets [16], attosecond control of electron recollision [17], and control of strong field processes [18]. In particular, Cao, Bardeen, and Wilson have predicted that one may design π laser pulses with durations shorter than ca 5 fs for complete population transfer between electronic states, however without explicit evaluation of the laser driven electron dynamics [19]. (For slightly longer pulses, this process would be disturbed due to the onset of coupled nuclear dynamics, but complete population transfer could still be achieved by means of chirping, see also [20]. By extrapolation, even longer laser pulses, e.g. continuous wave lasers, would be prohibitive.) The corresponding electron dynamics of complete population transfer induced by linearly polarized π pulses has been simulated recently by Krause *et al.* [21]. By analogy, we could simulate complete population transfer from the ground electronic state $|X\rangle = |1\ {}^1A_{1g}\rangle$ to the excited state $|4\ {}^1E_{u+}\rangle$ of Mg-porphyrin by means of a circularly polarized π laser pulse [22]. Accordingly, the chirality of the laser pulse is transferred into a state-selective uni-directional stationary electronic ring current. Last but not least, we could also demonstrate half population transfer from $|X\rangle$ to different excited electronic states, $|2\ {}^1E_{u+}\rangle$ or $|5\ {}^1E_{u+}\rangle$, of Mg-porphyrin by means of circularly polarized $\pi/2$ laser pulses, with re-optimized laser parameters [23]. Here, the chirality of the laser pulse is transferred to periodic electron circulation, different from stationary electronic ring currents.

The purpose of this paper has been to investigate various strategies for the design of re-optimized circularly polarized $\pi/2$ laser pulses in order to induce perfectly periodic electron circulation in Mg-porphyrin. Exemplarily, we construct laser pulses for the corresponding half population transfer from $|X\rangle$ to $|4\ {}^1E_{u+}\rangle$. The present comparison of various methods for the design of re-optimized $\pi/2$ laser pulses, and the specific application, are complementary to our work presented in Ref. [23], see also Ref. [22]. In addition, we shall design series of re-optimized pulses for control of electron circulations, with large variations of the periods of circulation.

The model and techniques are described in section 2, the results are presented and discussed in section 3, and the conclusions are in section 4.

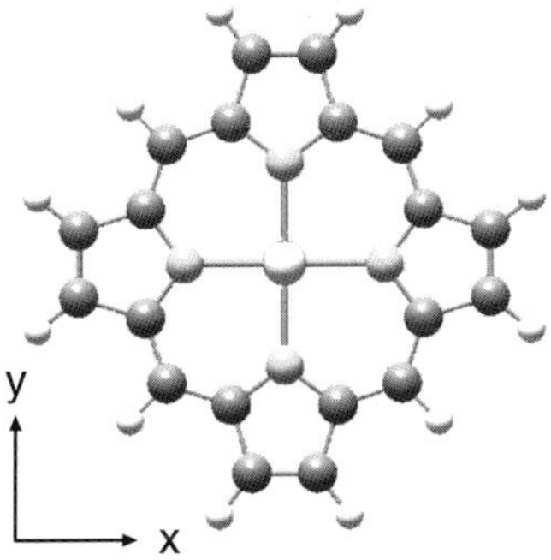

Figure 1. Bird's eye view of Mg-porphyrin.

2. Model and techniques

We consider a model of Mg-porphyrin with fixed orientation, e.g. by means of the methods of Refs. [24, 25], as shown in Figure 1. The nuclei are frozen in their equilibrium positions, implying molecular D_{4h} symmetry. The laser driven electron dynamics of Mg-porphyrin is then described as coherent wave packet

$$|\Psi(t)\rangle = \sum_i C_i(t)|\Psi_i\rangle e^{-iE_i t/\hbar} \tag{1}$$

where E_i denote the electronic eigenenergies of the electronic states $|\Psi_i\rangle$

$$H_{el}|\Psi_i\rangle = E_i|\Psi_i\rangle \tag{2}$$

and H_{el} is the electronic Hamilton operator

$$H_{el} = -\frac{\hbar^2}{2m_e}\sum_{i=1}^{N}\nabla_i^2 + \frac{e^2}{4\pi\varepsilon_0}\left(\sum_{\substack{\alpha,\beta=1\\ \beta>\alpha}}^{N'}\frac{Z_\alpha Z_\beta}{|\underline{R}_\alpha - \underline{R}_\beta|} - \sum_{i=1}^{N}\sum_{\alpha=1}^{N'}\frac{Z_\alpha}{|\underline{r}_i - \underline{R}_\alpha|} + \sum_{\substack{i,j=1\\ j>i}}^{N}\frac{1}{|\underline{r}_i - \underline{r}_j|}\right) \tag{3}$$

with positions $\underline{r}_i$ and $\underline{R}_\alpha$ of N electrons and of N' frozen nuclei with charges Z_α, respectively. The coefficients $C_i(t)$ correspond to populations $P_i(t) = |C_i(t)|^2$ of the electronic states $|\Psi_i\rangle$. They are calculated, using the semiclassical dipole approximation, as solutions of the time-dependent Schrödinger equation

$$i\hbar\left|\dot{\Psi}(t)\right\rangle = \left(H_{el} - \underline{M}\cdot\underline{\mathcal{E}}_{\pm}(t)\right)\left|\Psi(t)\right\rangle \tag{4}$$

where $\underline{M}\cdot\mathcal{E}_{\pm}(t)$ denotes the coupling of the dipole operator

$$\underline{M} = -e\sum_{i=1}^{N}\underline{r}_i + e\sum_{\alpha=1}^{N'}Z_\alpha\underline{R}_\alpha \tag{5}$$

and the right (+) or left (–) circularly polarized electric field, with field amplitude $\mathcal{E}_0$ and carrier frequency ω,

$$\underline{\mathcal{E}}_{\pm}(t) = \mathcal{E}_0 s(t)\left(\cos(\omega t)\underline{e}_x \pm \sin(\omega t)\underline{e}_y\right) \tag{6}$$

and shape function

$$s(t) = \sin^2\left(\frac{\pi t}{t_p}\right), \qquad 0 \le t \le t_p \tag{7}$$

where t_p denotes the total duration of the laser pulse. The full width at half maximum (fwhm) for the corresponding intensity $I(t) = c\varepsilon_0|\mathcal{E}_{\pm}(t)|^2 = c\varepsilon_0\mathcal{E}_0{}^2s(t)^2$ is $\tau = 2/\pi\arccos(1/2^{1/4})t_p \approx 0.364\,t_p$, related to the spectral width Γ according to $\Gamma\tau \approx 3.295\,\hbar$. Below we consider laser pulses with $\tau \approx 3.64$ fs, $\Gamma \approx 0.60$ eV so that the nuclei may be assumed to be frozen in their equilibrium positions during the laser pulse. As a convenient alternative, we solve the equivalent equations for the coefficients

$$i\hbar\dot{C}_j(t) = -\underline{\mathcal{E}}_{\pm}(t)\sum_i C_i(t)\left\langle\Psi_j\left|\underline{M}\right|\Psi_i\right\rangle e^{-i\varpi_{ij}t} \tag{8}$$

with transition frequencies

$$\varpi_{ij} = \frac{E_i - E_j}{\hbar}. \tag{9}$$

In the subsequent application, the molecule is initially in the electronic ground state $|X\rangle$, and the laser frequency ω is tuned resonant, or near-resonant to a particular transition frequency, specificly for excitation from $|X\rangle$ to a pair of degenerate excited states with E symmetry, $|j\,^1E_{u+}\rangle$ or $|j\,^1E_{u-}\rangle$, denoted for simplicity as $\varpi_{jX} = (E_j - E_X)/\hbar$. These complex wave functions $|j\,^1E_{u+}\rangle$ and $|j\,^1E_{u-}\rangle$ may be expressed in terms of two real ones,

$$\left|j\,^1E_{u\pm}\right\rangle = \frac{1}{\sqrt{2}}\left(\left|j\,^1E_{ux}\right\rangle \pm i\left|j\,^1E_{uy}\right\rangle\right), \tag{10}$$

similar to the relations of complex atomic orbitals $|2p_+\rangle$ and $|2p_-\rangle$ and real ones $|2p_x\rangle$ and $|2p_y\rangle$ [26].

For reference, let us now consider the hypothetical yet illuminating limit of right circularly polarized pulses with sufficiently long total durations t_p such that the spectral half width $\Gamma/2$ is smaller than the energy gap between the target level and neighbouring states with the same symmetry E_{u+}. For values of maximum intensities $I_{max} = c\varepsilon_0\mathcal{E}_0^2$ below $6 \cdot 10^{12}$ Wcm^{-2}, the solution of the Schrödinger equation (4) or (8) then reduces to a coherent wave packet which consists of a superposition of just two components, i.e. the ground and target electronic states

$$|\Psi(t)\rangle = C_X(t)|X\rangle e^{-iE_X t/\hbar} + C_j(t)\left|j\ {}^1E_{u+}\right\rangle e^{-iE_j t/\hbar} \tag{11}$$

with coefficients

$$C_X(t) = \cos\left(\frac{M\mathcal{E}_0}{\hbar}\int_0^t s(t')dt'\right), \tag{12}$$

$$C_j(t) = i\sin\left(\frac{M\mathcal{E}_0}{\hbar}\int_0^t s(t')dt'\right). \tag{13}$$

Apparently, the corresponding populations $P_j(t)$ yield complete population transfer at $t = t_p$ if

$$\frac{|M\mathcal{E}_0 t_p|}{\hbar} = \pi \tag{14}$$

where

$$M = \frac{1}{\sqrt{2}}\left\langle j\ {}^1E_{ux}\right|M_x\left|X\right\rangle = \frac{1}{\sqrt{2}}\left\langle j\ {}^1E_{uy}\right|M_y\left|X\right\rangle. \tag{15}$$

Likewise, half population transfer is achieved if

$$\frac{|M\mathcal{E}_0' t_p'|}{\hbar} = \frac{\pi}{2} \tag{16}$$

Expressions (6) and (7), together with the conditions (14) and (16), may be considered as definitions of right circularly polarized π and $\pi/2$ pulses designed for the $|X\rangle \to |j\ {}^1E_{u+}\rangle$ transition, in the limit of long pulse durations t_p. Equivalent expressions are obtained for the corresponding left circularly polarized π and $\pi/2$ laser pulses yielding complete or half population transfers from $|X\rangle$ to $|j\ {}^1E_{u-}\rangle$, respectively.

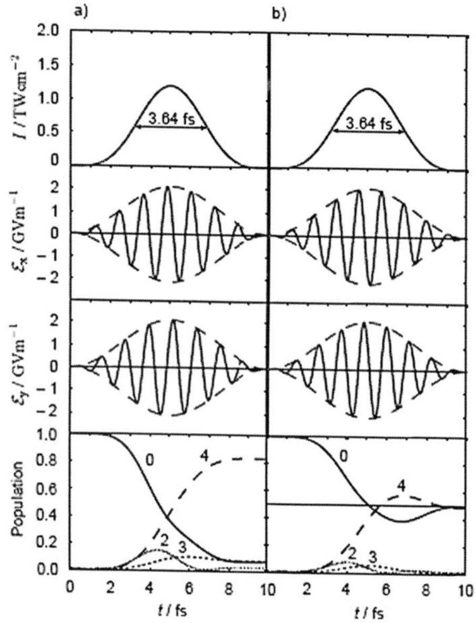

Figure 2. Approximately complete (a), and almost perfect half populations transfers (b) from the ground state $|X\rangle = |1\ {}^1A_{1g}\rangle$ to the target state $|4\ {}^1E_{u+}\rangle$ of oriented Mg-porphyrin (cf. Figure 1), by means of π laser pulses without optimization (a) and with optimization of the carrier frequency ω. The laser parameters are $\omega = 3.42$ and 3.64 eV $/\hbar$, respectively, and $\mathcal{E}_0 = 2.12$ GVm^{-1}, $t_p = 10$ fs, cf. Eqns. (6), (7), (14) and (17). The top, second, third and bottom panels show the time evolutions of the laser intensities, the x- and y- components of the electric fields, and the resulting populations of levels denoted $j = 0$ for the electronic ground state $|X\rangle$ and $j = 1, 2, \ldots$ for excited states $|j\ {}^1E_{u+}\rangle$, respectively.

3. Results and Discussion

Our design of circularly polarized laser pulses for periodic electron circulation in oriented Mg-porphyrin by means of half population transfer is demonstrated, exemplarily, for the transition from the electronic ground state $|X\rangle$ to the target hybrid state $|X, 4\ {}^1E_{u+}\rangle$. The molecular parameters are adapted from quantum chemistry CASPT2 and TD-DFT results of Rubio et al. [27] and Sundholm [28], respectively. Specificly, the energy gap of the target transition is $\hbar\varpi_{4X} = E_4 - E_X = 3.42$ eV, and the value M for the transition dipole matrix element is $M = -1.84\, ea_0$, cf. Eqn. (15). For reference, we consider first the effect of a resonant π pulse, with frequency $\omega = \varpi_{4X}$, pulse duration $\tau = 3.64$ fs corresponding to the total pulse duration $t_p = 10$ fs, and with non-optimized parameter for the field amplitude $\mathcal{E}_0 = 2.12$ GVm^{-1} in accord with the condition (14), similar to the parameters used in Ref. [22]. This non-optimized π pulse is shown in Figure 2a, together with the resulting population dynamics.

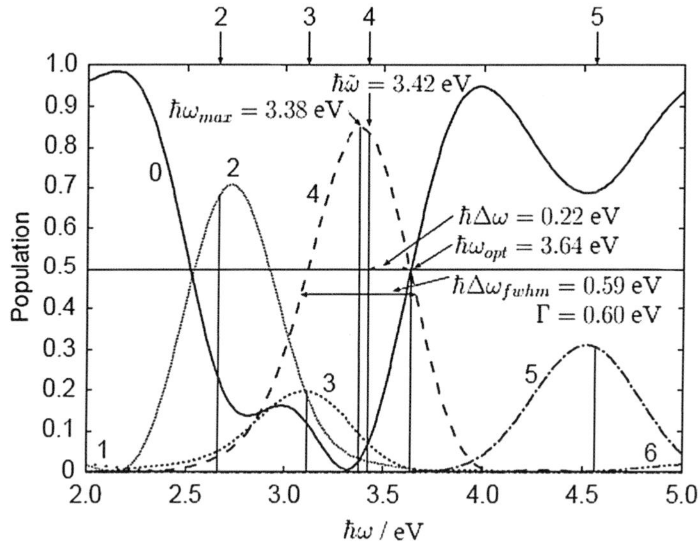

Figure 3. Population transfers from the electronic ground state $|X\rangle$ (labelled 0) to excited states $|j\ ^1E_{u+}\rangle$ (labelled $j = 1, 2, ...$) with energies E_j of oriented Mg-porphyrin (cf. Figure 1) by means of right circularly polarized laser pulses, versus carrier frequency ω. The laser parameters $t_p = 10$ fs and $\mathcal{E}_0 = 2.12\ \mathrm{GVm^{-1}}$ are the same as for the π pulse for population transfer from $|X\rangle$ to $|4\ ^1E_{u+}\rangle$, as shown in Figure 2a, with resonant frequency ϖ_{4X} and spectral width Γ. The resonance frequencies $\varpi_{jX} = (E_j - E_X)/\hbar$ for transitions to levels E_j are indicated by vertical lines labeled $j = 2, ..., 5$. The meaning of the other frequencies ω_{max}, $\Delta\omega$, ω_{opt} and $\Delta\omega_{fwhm}$ is explained in the text.

As anticipated, it transfers population preferably from $|X\rangle$ to the target state $|4\ ^1E_{u+}\rangle$, but significant populations $P_j(t = t_p)$ are also observed for other levels with the same symmetry, $|j\ ^1E_{u+}\rangle$. The reason for this deviation from perfect state-selectivity becomes obvious from a systematic comparison of the final state populations for laser pulses with the same parameters $\mathcal{E}_0$ and t_p, cf. criterion (14), but different carrier frequencies ω, cf. Figure 3. The population of the target state $P_4(t_p;\ \omega)$ versus ω resembles the Fourier transform of the intensity of the laser pulse, i.e. it has bell shape centered at $\omega_{max} = 3.38$ eV $/\hbar$ close to the resonance frequency $\varpi = \varpi_{4X}$, and widths $\Delta\omega_{fwhm} = 0.59$ eV $/\hbar$ close to the spectral width $\Gamma = 0.60$ eV. Similar bell shaped populations are also observed for the populations $P_j(t_p;\ \omega)$ of complementary states $|j\ ^1E_{u+}\rangle$ – these are centered close to corresponding resonance frequencies $(E_j - E_X)/\hbar$ and have similar widths. As a consequence, the spectral width of the short laser pulse overlaps not only with the target state but also with neighboring levels of the same symmetry. The transfer to the target state is associated, therefore, with transitions to those neighboring levels. Nevertheless, condition (14) still guarantees that the population of the corresponding target state is dominant, $P_4(t_p;\ \varpi_{4X}) = 0.84 \approx P_4(t_p;\ \omega_{max}) = 0.85$.

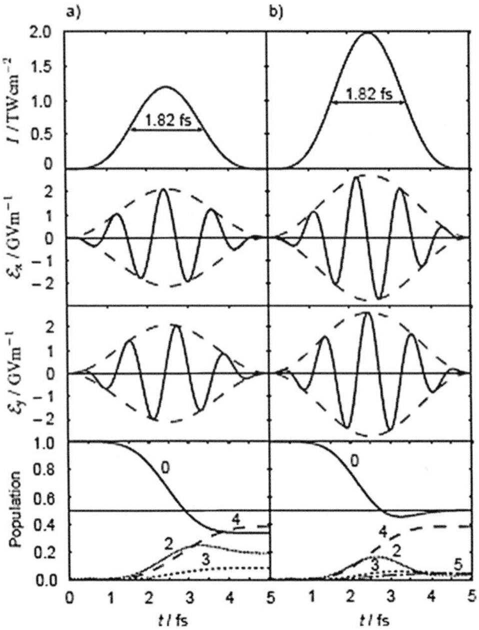

Figure 4. Non-optimized (a) and re-optimized (b) right circularly polarized $\pi/2$ laser pulses yielding approximate (a) and improved (b) half population transfers from the electronic ground state $|X\rangle$ to the target state $|4\,^1E_{u+}\rangle$ of oriented Mg-porphyrin (cf. Figure 1). The laser parameters are $\mathcal{E}_0 = 2.12$ and 2.73 GVm^{-1}, $\omega = 3.42$ and 3.78 eV $/\hbar$, respectively, and $t_p = 5$ fs, compare with Figure 2. The notations are as for Figure 2.

Comparison of the criteria for π pulses (14) and for $\pi/2$ pulses (16) suggests various strategies for the construction of $\pi/2$ pulses for half population transfers from the electronic ground state $|X\rangle$ to the target state $|4\,^1E_{u+}\rangle$. As a first example, let us use the same field strength $\mathcal{E}_0' = \mathcal{E}_0$ but half duration $t_p' = t_p/2$. Alternatively, as a second example we set $\mathcal{E}_0' = \mathcal{E}_0/2$ but $t_p' = t_p$. The resulting non-optimized $\pi/2$ pulses together with the population dynamics are shown in Figures 4a and 5a, respectively. As anticipated, both non-optimized $\pi/2$ laser pulses yield near equal populations of the electronic ground $|X\rangle$ and excited state $|4\,^1E_{u+}\rangle$, but significant populations are also observed in neighboring levels for the same reason of broad spectral widths as explained above for non-optimized π pulses. The deviations from ideal half population transfers are indeed larger for the shorter $\pi/2$ pulse shown in Figure 4a due to its broader spectral width compared to the longer $\pi/2$ pulse shown in Figure 5a.

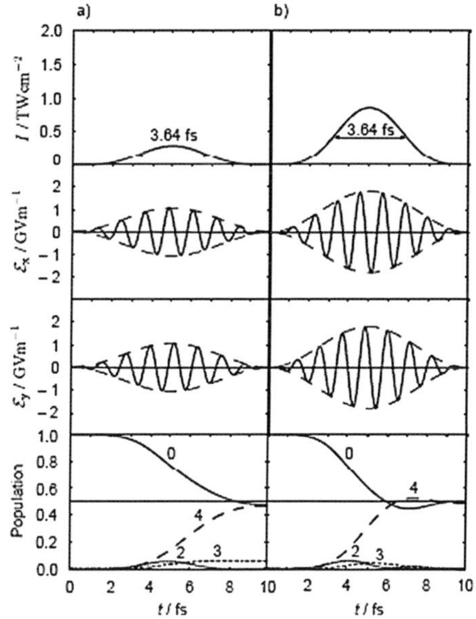

Figure 5. Non-optimized (a) and re-optimized (b) right circularly polarized $\pi/2$ laser pulses yielding approximate (a) and nearly perfect (b) half population transfers from the electronic ground state $|X\rangle$ to the target state $|4\ ^1E_{u+}\rangle$ of oriented Mg-porphyrin (cf. Figure 1). The laser parameters are $\mathcal{E}_0 = 1.06$ and 1.80 GVm^{-1}, $\omega = 3.42$ and 3.64 eV $/\hbar$, respectively, and $t_p = 10$ fs, compare with Figure 2. The notations are as for Figure 2.

The apparent "enemy" of broad spectral widths of short non-optimized π or $\pi/2$ laser pulses can also be turned into an advantage. Accordingly, our third approach to half population transfer is based on inspection of the populations $P_j(t_p;\ \omega)$ versus ω, cf. Figure 3. Apparently, the resulting decrease of $P_4(t_p;\ \omega)$ versus ω is associated with the corresponding increase of $P_0(t_p;\ \omega) = P_X(t_p;\ \omega)$ yielding nearly perfect half population transfer at $\hbar\omega_{opt} = 3.64$ eV. The corresponding laser pulse and population dynamics are illustrated in Figure 2b. This nearly perfect result is also supported by the fact that $\hbar\omega_{opt}$ is outside the domain of the widths of any other π pulses designed for population transfers to neighboring levels with the same symmetry; thus all other populations $P_j(t_p;\ \omega_{opt})$ are negligible. The "optimal" frequency ω_{opt} is rationalized easily, i.e. it is equal to the frequency where $P_4(t_p;\ \omega_{opt}) \approx 0.5$, and ω_{opt} is, therefore, close to but slightly less than the frequency where $P_4(t_p;\ \omega) = P_4(t_p;\ \omega_{max})/2$; thus

$$\omega_{opt} \approx \omega_{\max} + \Delta\omega_{fwhm} \approx \omega_{max} + \frac{\Gamma}{2\hbar}. \tag{17}$$

The corresponding detuning from the resonance frequency is $\Delta\omega = 0.22$ eV $/\hbar$. The previous three approaches to laser pulses for half population transfer employ parameters which are given by analytical expressions (16) or (17) based on simple models for $\pi/2$ or π pulses and their spectral widths. To compensate for the fact that the underlying model assumption of sufficiently long laser pulse durations is inadequate, we suggest a second step, i.e. re-optimization of the "analytical" values of the parameters ω and $\mathcal{E}_0$, if necessary. The corresponding re-optimized laser pulses and populations dynamics are shown in Figures 4b and 5b, corresponding to the non-optimized pulses shown in Figures 4a and 5a, respectively. The optimized parameters are given in the Figure legends, and the final populations $P_X(t_p)$ and $P_4(t_p)$ are indeed slightly closer to the ideal values (= 0.5) than for the non-optimized $\pi/2$ laser pulses. The laser pulse based on the "near-analytical" frequency detuning (17) already yields nearly perfect half electronic population transfer, cf. Figure 2b, because it is close to the optimized one, cf. Figure 5b.

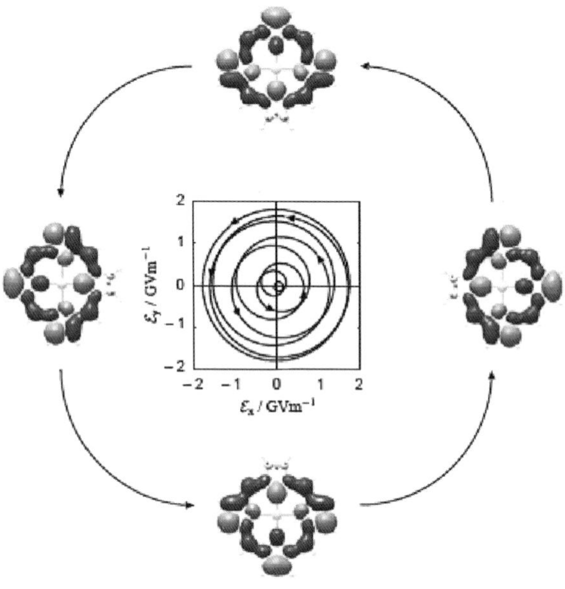

Figure 6. Re-optimized right circularly polarized $\pi/2$ laser pulse designed for half population transfer from the electronic ground state $|X\rangle$ to the target state $|4\,^1E_{u+}\rangle$ of oriented Mg-porphyrin (cf. Figures 1 and 5b), and the resulting periodic right electron circulation. The x- and y-components of the electric fields are shown as polar plot. The electron circulation is illustrated by snapshots of the difference $\Delta\rho(t)$ of the single electron density $\rho(t)$ minus the initial density $\rho(0)$, at time $t = 10.3$ fs and $t + k\tau_{4X}/4$, ($k = 1, 2, 3, \ldots$) with $\tau_{4X} = 1.2$ fs. The target state $|4\,^1E_{u+}\rangle$ corresponds to $3a_{2u}$ to $4e_{g+}$ orbital transition (weight =57 % [27]). As a consequence, $\Delta\rho(t)$ may be approximated in terms of the (partial) transition of a single active electron, essentially from $3a_{2u}$ to $4e_{g+}$.

The laser pulses designed in Figures 2b, 4b, or 5b yield nearly perfect preparations of the target hybrid states

$$\left|X, j\ {}^1E_{u+}\right\rangle = \frac{1}{\sqrt{2}}\left|X\right\rangle e^{-iE_X t/\hbar} \pm i\frac{1}{\sqrt{2}}\left|j\ {}^1E_{u+}\right\rangle e^{-iE_j t/\hbar} \tag{18}$$

with corresponding half populations of the ground and excited target states. These hybrid states may be modified by a phase shift for the two components, depending on re-optimization, but this will not affect the subsequent results, i.e. they represent periodic electron circulations, with period

$$\tau_{jX} = \frac{h}{E_j - E_X}. \tag{19}$$

The electron circulation is illustrated in Figure 6 by means of snapshots of the difference of the corresponding one-electron density at time t minus the original one at time $t = 0$. Apparently, the chirality of the right circularly polarized laser pulse is transferred into net right electron circulation in the pre-oriented Mg-porphyrin, with period $\tau_{4X} = 1.2$ fs. Likewise, left circularly polarized laser pulses with the same parameters yield equivalent net left electron circulation, with the same period.

The results (18) and (19) suggest extensions to the design of laser pulses for preparations of more general excited electronic hybrid states

$$\left|j\ {}^1E_{u+}, k\ {}^1E_{u+}\right\rangle = \frac{1}{\sqrt{2}}\left|j\ {}^1E_{u+}\right\rangle e^{-iE_j t/\hbar} \pm i\frac{1}{\sqrt{2}}\left|k\ {}^1E_{u+}\right\rangle e^{-iE_k t/\hbar} \tag{20}$$

with periods

$$\tau_{jk} = \frac{h}{\left|E_j - E_k\right|}. \tag{21}$$

By analogy with our previous approaches to preparations of various hybrid states of nuclear motions by means of series of sequential π and $\pi/2$ laser pulses, cf. Refs. [9, 11], see also Ref. [10], this goal may be achieved by means of series of selective pulses for complete and half electronic population transfers. This strategy is illustrated in Figure 7, exemplarily for the preparation of the hybrid state $|5\ {}^1E_{u+}, 4\ {}^1E_{u+}\rangle$. Specificly, we combine the re-optimized $\pi/2$ laser pulse shown in Figure 5b for half population transfer from $|X\rangle$ to $|4\ {}^1E_{u+}\rangle$ with the sequential pulse optimized for population transfer from the hybrid state $|X, 4\ {}^1E_{u+}\rangle$ to the target hybrid state $|5\ {}^1E_{u+}, 4\ {}^1E_{u+}\rangle$. This series of two laser pulses then prepares the final state which is close to the target hybrid state $|5\ {}^1E_{u+}, 4\ {}^1E_{u+}\rangle$.

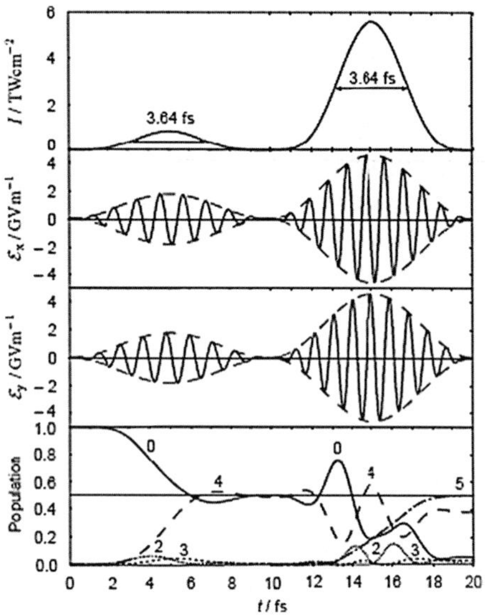

Figure 7. Series of sequential laser pulses for preparation of the target hybrid state $|5\,^1E_{u+}, 4\,^1E_{u+}\rangle$ in oriented Mg-porphyrin (cf. Figure 1). The first laser pulse prepares the hybrid state $|X, 4\,^1E_{u+}\rangle$ (adapted from Figure 5b), the second one (with laser parameters t_p = 10 fs, $\mathcal{E}_0$ = 4.59 GVm^{-1}, ω = 4.40 eV /$\hbar$) induces the transition to the target hybrid state $|5\,^1E_{u+}, 4\,^1E_{u+}\rangle$. The notations are as in Figure 2.

4. Conclusions

The present design of short right circularly polarized laser pulses allows to prepare various electronic hybrid states which consist of equal contributions of non-degenerate molecular eigenstates, e.g. $|X\rangle$ and $|j\,^1E_{u+}\rangle$ or $|j\,^1E_{u+}\rangle$ and $|k\,^1E_{u+}\rangle$, representing electron circulation. Different choices of target states correspond to different patterns of wavepackets (compare e.g. Fig. 6 and Fig. 2 of Ref. [23]), and to different periods $\tau_{jk} = h/|E_j - E_k|$ of electron circulation. The present approachs thus serve to control preparation of specific properties of electron circulation, e.g. with various periods in the femtosecond or attosecond time domains.

The results presented in Section 3 allow to extrapolate some general trends, e.g. shorter pulses call for larger intensities, and the associated broader spectral widths imply that it is more difficult to achieve perfect selectivities than for longer laser pulses. Nevertheless, re-optimization of the laser parameters of very short laser pulses allows to compensate some loss of selectivity. For the present examples, we have employed just two parameters for re-optimization, i.e. the carrier frequency and the field amplitude. In principle, one may also re-optimize additional parameters such as the phase, cf. [29–31], or chirps, cf. [19,

20], or one could employ optimal control, starting from the present restricted optimizations of π or $\pi/2$ pulses as a reference [32], see also Refs. [2, 3, 33]. These methods should also allow to shorten the overall duration of series of laser pulses such as those shown in Figure 7, e.g. by allowing the individual pulses to overlap, with simultaneous re-optimization of the parameters.

The present approach may, of course, also be applied to induce similar electron circulation in other ring-shaped molecules or toroidal type electron circulation around linear molecules. It may be considered as molecular and femto-to-attosecond scale analogon of previous preparations of Rydberg electrons circulating along Kepler orbits around atoms, with diameters of hundreds of angstroms and periods beyond nanoseconds, see the pioneering work of Yeazell and Stroud [34, 35] and more recently by Bucksbaum and coworkers [36, 37].

After selective ignition of electron circulation, the onset of coupled nuclear motions will cause competing processes such as intramolecular vibrational redistribution (IVR) or internal conversion (IC). As a consequence, periodic electron circulation will persist only with certain lifetimes, depending on the efficiencey of the competing processes. For example, if IVR or IC will take longer than say 50 fs, one should be able to observe at least ca 50 periods of electron circulation with periods in the present domain of ca 1 fs or even faster. These observations could be made by means of high harmonic generation (HHG), cf. [25, 38–40], depending on special selection rules due to the symmetry of the molecule [41], or asymmetric photoelectron spectra as suggested in Refs. [16, 30].

Acknowledgement

We would like to express sincere thanks to Drs. M. Kitzler (Technische Universität Wien), Y. Shigeta and K. Yagi (Tokyo University) for discussions of related experiments and ignition of electronic ring currents [19]. Financial support by Deutsche Forschungsgemeinschaft (project Ma 515/23-1) and Fonds der Chemischen Industrie is also gratefully acknowledged.

References

1. A. H. Zewail, Angew. Chem. Int. Ed. 39 (2000) 2586.
2. M. Shapiro and P. Brumer, Principles of the quantum control of molecular processes, VCH, Weinheim, 2003.
3. S. A. Rice and M. Zhao, Optical control of molecular dynamics, Wiley, New York, 2000.
4. G. K. Paramonov and G. K. Savva, Phys. Lett. A 97 (1983) 340.
5. T. Joseph and J. Manz, Molec. Phys. 58 (1986) 1149.
6. M. Holthaus and B. Just, Phys. Rev. A 49 (1994) 1950.
7. J. E. Combariza, B. Just, J. Manz, and G. K. Paramonov, J. Phys. Chem. 95 (1991) 10351.
8. M. V. Korolkov, J. Manz, and G. K. Paramonov, Adv. Chem. Phys. 101 (1997) 327.

9. Y. Fujimura, L. González, K. Hoki, D. Kröner, J. Manz, and Y. Ohtsuki, Angew. Chem. Int. Ed. 39 (2000) 4586.
10. L. González, D. Kröner, and I. R. Solá, J. Chem. Phys. 115 (2001) 2519.
11. K. Hoki, D. Kröner, and J. Manz, Chem. Phys. 267 (2001) 59.
12. M. Hentschel, R. Kienberger, Ch. Spielmann, G. A. Reider, N. Milosevic, T. Brabec, P. Corkum, U. Heinzmann, M. Drescher, and F. Krausz, Nature 414 (2001) 509.
13. R. Kienberger, M. Hentschel, M. Uiberacker, Ch. Spielmann, M. Kitzler, A. Scrinzi, M. Wieland, Th. Westerwalbesloh, U. Kleineberg, U. Heinzmann, M. Drescher, and F. Krausz, Science 297 (2002) 1144.
14. P. Agostini and L. F. DiMauro, Rep. Prog. Phys. 67 (2004) 813.
15. T. Brixner and G. Gerber, Opt. Lett. 16 (2001) 557.
16. A. D. Bandrauk and H. S. Nguyen, Can. J. Chem. 82 (2004) 831.
17. M. Kitzler, K. O'Keeffe, and M. Lezius, J. Mod. Opt. (2005), in press.
18. P. Johnsson, R. López-Martens, S. Kazamias, J. Mauritsson, C. Valentin, T. Remetter, K. Varjú, M. B. Gaarde, Y. Mairesse, H. Wabnitz, P. Salières, Ph. Balcou, K. J. Schafer, and A. L'Huillier, Phys. Rev. Lett. 95 (2005) 013001.
19. J. Cao, C. J. Bardeen, and K. R. Wilson, Phys. Rev. Lett. 80 (1998) 1406.
20. K. Mishima and K. Yamashita, J. Chem. Phys. 109 (1999) 1801.
21. P. Krause, T. Klamroth, and P. Saalfrank, J. Chem. Phys. 123 (2005) 074105.
22. I. Barth, J. Manz, Y. Shigeta, and K. Yagi, J. Am. Chem. Soc. (2005), in press.
23. I. Barth and J. Manz, Angew. Chem. Int. Ed. (2005), in press.
24. H. Stapelfeldt, T. Seideman, Rev. Mod. Phys., 75 (2003) 543.
25. J. Itatani, J. Levesque, D. Zeidler, H. Niikura, H. Pépin, J. C. Kieffer, P. B. Corkum, and D. M. Villeneuve, Nature 432 (2004) 867.
26. I. N. Levine, Quantum chemistry, Prentice Hall, New Jersey, 2000.
27. M. Rubio, B. O. Roos, L. Serrano-Andrés, and M. Merchán, J. Chem. Phys. 110 (1999) 7202.
28. D. Sundholm, Chem. Phys. Lett. 317 (2000) 392.
29. S. Chelkowski, H. S. Nguyen, and A. D. Bandrauk, Laser Phys. 13 (2003) 871.
30. D. B. Milošević, G. G. Paulus, and W. Becker, Phys. Rev. A 71 (2005) 061404(R).
31. S. Chelkowski and A. D. Bandrauk, Phys. Rev. A 71 (2005) 053815.
32. W. Jakubetz, J. Manz, and H. J. Schreier, Chem. Phys. Lett. 165 (1990) 100.
33. R. S. Judson and H. Rabitz, Phys. Rev. Lett. 68 (1992) 1500.
34. J. A. Yeazell and C. R. Stroud, Phys. Rev. A 35 (1987) 2806.
35. J. A. Yeazell and C. R. Stroud, Phys. Rev. Lett. 60 (1988) 1494.
36. T. C. Weinacht, J. Ahn, and P. H. Bucksbaum, Phys. Rev. Lett. 80 (1998) 5508.
37. T. C. Weinacht, J. Ahn, and P. H. Bucksbaum, Nature 397 (1999) 233.
38. P. B. Corkum, Phys. Rev. Lett. 71 (1993) 1994.
39. A. D. Bandrauk, S. Chelkowski, and H. S. Nguyen, J. Mol. Struc. 735-736 (2005) 203.
40. M. Kitzler and M. Lezius, Phys. Rev. Lett. (2005), in press.
41. O. E. Alon, V. Averbukh, and N. Moiseyev, Phys. Rev. Lett. 80 (1998) 3743.

Femtochemistry VII
Fundamental Ultrafast Processes in Chemistry, Physics, and Biology
A.W. Castleman, Jr. and M.L. Kimble

Learning Feedback Control Analysis

J.L. White[a], B. J. Pearson[b], E. C. Carroll[a], A. C. Florean[a], P. H. Bucksbaum[a], R. J. Sension[a]

[a]*FOCUS Center, University of Michigan, 500 East University Ave., Ann Arbor, MI, 48109-1120 USA*
[b]*Physics Department, Stony Brook University, Stony Brook, NY 11794-3800, USA*

1. Introduction

The central challenge of coherent control is this: Given a quantum system in a given initial state, how can we put the system into a desired final state? Advances in our ability to control light have led to much experimental and theoretical activity in this area, and learning feedback algorithms have emerged as an important tool. In recent years, learning algorithms have successfully controlled a wide variety of systems [1-3].

The success of these algorithms has not been unqualified, however. While the algorithms have proven effective at controlling a variety of experimental systems, the mechanism of control is often unclear. The optimal pulse shapes discovered by such learning algorithms are usually too complicated to be easily interpreted and often contain unnecessary features that further obscure their interpretation. Many suggestions have been made to determine which features of the pulse are necessary and sufficient for control [5-8], and there has been some success with this approach. We have developed statistical data analysis techniques to learn more about the important features of the control pulses: Principal Control Analysis (PCA) and Most Correlated Feature Analysis (MCFA).

In this paper we first briefly review learning algorithms and then discuss PCA and MCFA in the context of two control experiments: stimulated Raman scattering in liquid methanol and the multiphoton photoisomerization of 1,3-cyclohexadiene (CHD) in solution.

2. Experiment

The two experiments, SRS in liquid methanol and the CHD ring-opening, have been described in detail elsewhere [3,9]. Briefly, 800 nm pulses from a titanium sapphire oscillator are shaped with an acousto-optical modulator (AOM) and then directed into a cell containing the liquid sample. The methanol experiments were conducted on a laser system operating at 10 Hz with 8 nm of bandwidth. The CHD experiments were conducted on a kHz laser system with 20 nm of bandwidth. In both experiments the pump pulses had a maximum energy less than 1 mJ and only the phase of the pulse was modified by the AOM.

Methanol has two closely spaced vibrational levels corresponding to the symmetric and antisymmetric C-H stretches. Shaped pump pulses give rise to nonimpulsive stimulated Raman scattering, which was collected with a spectrometer. Phase shaping of the pump pulse can selectively enhance or suppress the amount of light in either Raman mode.

The ring-opening of CHD forms 1,3,5-cis-hexatriene (Z-HT) with quantum yield ca. 0.4. The photoproducts Z-HT and CHD are distinguishable by their ultraviolet absorption spectra. CHD was initially excited by multiphoton absorption of the shaped pulses. A time-delayed probe pulse consisting of ultraviolet continuum was used to measure the signature absorption of Z-HT.

3. The Problem of Control

When light is interacting with an atom or molecule, the total Hamiltonian of the time-dependent Schrodinger equation can be written

$$H = H_{system} + H_{interaction}[\varepsilon(t)] \tag{1}$$

where H_{system} is the Hamiltonian of the system in the absence of the external field $\varepsilon(t)$ and $H_{interaction}$ is the part of the Hamiltonian that (somehow) depends on $\varepsilon(t)$. Since we can control $\varepsilon(t)$, we can evidently influence the evolution of the system. We say that a system can be controlled if there exists some external field $\varepsilon(t)$ such that the system evolves from a given initial state to a given final state. The study of which final states can be reached from which initial states is the study of the controllability of the system.

Given a control problem, we can in principle determine the necessary control field $\varepsilon(t)$ by solving the Schrodinger equation. In practice, however, most systems are too complicated to allow calculation of $\varepsilon(t)$. While some theoretical progress has been made [4], we now turn to experimental methods of determining the control field $\varepsilon(t)$ and using it to learn about the Hamiltonian.

4. Learning Algorithms

We rely on an experimental search to determine the optimal pulse shape. The search space of all possible pulse shapes is usually very large. Even if we neglect amplitude shaping and specify the phase to a precision of 10 degrees at *n* discrete frequencies, we still have some 2^{5n} possible pulse shapes in our search space. For a typical experiment, $n \sim 25$, so there are 4×10^{37} pulse shapes to consider. Such experiments clearly require a good search algorithm. Learning algorithms such as the *genetic algorithm* (GA) are widely used. An important feature of the algorithms is that they require no knowledge of the Hamiltonian.

Learning algorithms are loosely based on a model of biological evolution. The algorithm starts with a random collection of pulse shapes to test. The initial collection of pulse shapes is the first *generation.* Each shape is represented as a vector (*genome*) $\mathbf{x} = (x_1, x_2, \ldots, x_n)$ of numbers (*genes*). In our experiments, each gene represents the phase of the light at a specified frequency. Each pulse shape is experimentally evaluated and assigned a fitness. The fitness is a number indicating how well the pulse shape achieved the given control goal, i.e., how close the actual final state of the system was to the desired final state. New pulse shapes are then created by combining different pulse shapes (*mating*) or modifying a pulse shape (*mutation*), with a bias towards allowing fitter pulse shapes to produce more children. These new pulse shapes are the second generation. The pulse shapes of the second generation are then evaluated and used to create a third generation, and so forth. After many generations, the fitness stops improving; the genetic algorithm is unable to improve upon the pulse shapes of the current generation, and the search is over. The pulse with the highest fitness is declared the optimal pulse.

During the search, learning algorithms may use a variety of mathematically-defined operators to create new pulse shapes from old ones. The canonical genetic algorithm operators are crossover and mutation. Crossover combines segments of different genomes to create a new genome, while mutation randomly changes some gene values. We have also used problem-specific (physically motivated) operators such as smoothing, time-domain crossover, and polynomial phase mutation. Using different operators can change both the rate at which the algorithm converges and the maximum fitness attained. Additionally, our implementation of the learning algorithm adjusts the probability that each operator is used, so that operators that contribute more to increasing fitness are used more often [3].

5. Experiments in Methanol

Learning algorithm experiments were conducted in liquid methanol. The goal was to maximize the scattered light at either the symmetric or antisymmetric Stokes shift frequency while simultaneously suppressing the other mode. The phase profile of the optimal pulse shapes found by the algorithm are shown in Figure 1. It is unclear which features are important. From the optimal pulse alone, we cannot determine which features are sufficient and which necessary.

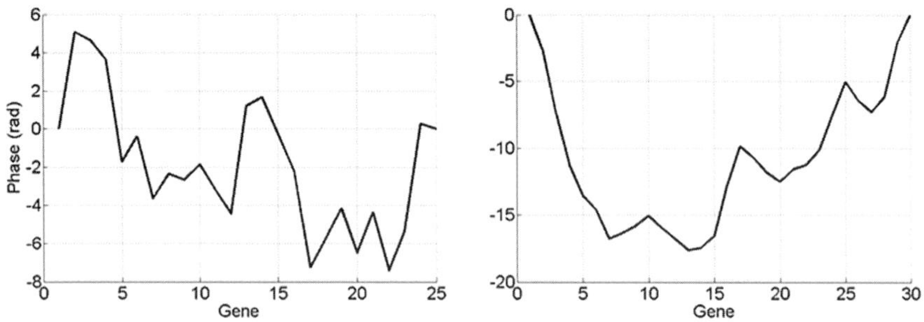

Figure 1: Genome (phase profile) of pulse shape maximizing the symmetric mode (left) and antisymmetric mode (right) Raman shift while suppressing the other mode.

6. Principal Control Analysis

During an experimental search, the learning algorithm tests hundreds or thousands of pulse shapes. Analyzing this collection of pulse shapes can reveal information about the search space that is not obtainable from the optimal pulse alone. If there are n genes, the search space may be considered an n-dimensional vector space, and statistical methods for analyzing high dimensional spaces may be used. These analyses can tell us about important features, the degrees of freedom the search touches upon, and whether there are any trends in the fitness landscape.

Principal Control Analysis is a statistical covariance analysis of the collection of pulse shapes [10]. Covariance techniques are a way to reduce the dimensionality of high dimensional data sets while retaining the essential information. In PCA, the basic idea is to look for a few linear combinations of genes that varied together during the search and which also correlate well with fitness.

We start by calculating the covariance matrix C_{ij} of the entire collection pulse shapes, defined as

$$C_{ij} = \langle \delta_i \delta_j \rangle - \langle \delta_i \rangle \langle \delta_j \rangle = \begin{bmatrix} C_{11} & C_{12} & \cdots & C_{1m} \\ C_{21} & C_{22} & & \\ \vdots & & \ddots & \\ C_{m1} & & & C_{mm} \end{bmatrix} \tag{2}$$

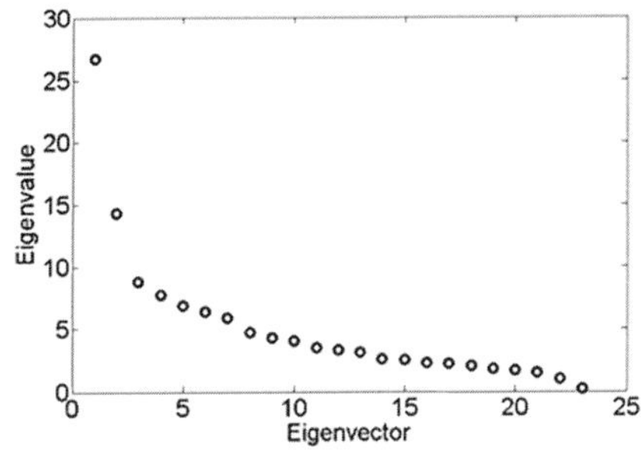

Figure 2: Eigenvalues of the covariance matrix from an experiment maximizing the symmetric mode in methanol. The shape is typical.

where $\delta_i = x_{i+1} - x_i$, $i = 1, \ldots, n-1$ are the phase differences between adjacent genes x_i, and the averages are over the collection of pulse shapes evaluated during the search. We use phase differences to eliminate the unimportant global phase. (The correlation matrix should be used instead of the covariance matrix if the genes are of different types.)

We diagonalize the covariance matrix to find the eigenvectors u_k and eigenvalues λ_k. Each eigenvector is a direction in the search space, and its eigenvalue is the variance of the data along that direction. We order the eigenvectors by eigenvalue from greatest to least. The first eigenvector is the direction in the search space along which the data varied the most, i.e., the search algorithm explored the most along this direction. The second eigenvector is the direction, orthogonal the first, along which the variance was largest, and so on. Eigenvectors with small eigenvalues correspond to directions that were not explored during the learning search. Typically, only a few eigenvectors account for most of the variance in the data (Figure 2).

The dimensionality of the data may be reduced by projecting the data onto eigenvectors with large eigenvalues. Dimensions corresponding to eigenvectors with small eigenvalues may be projected out of the data without losing important information about the search space. However, since we are interested in how the search space relates to the control goal, we also consider the fitness of each pulse shape.

The correlation R of fitness f with the projections η_i of the data onto each eigenvector u_i is

$$R_i = \frac{\langle \eta_i f \rangle - \langle \eta_i \rangle \langle f \rangle}{\sigma_{\eta_i} \sigma_f} . \tag{3}$$

The correlations R_i for methanol are shown in Figure 3, along with the correlation of fitness with the original δ_i basis for comparison. We call the k eigenvectors with the highest (absolute value) fitness correlation the principal controls. Since the data is from a fitness-directed search, the principal controls tend to also be eigenvectors with large eigenvalues, in other words, directions in the search space that the algorithm explored more. The principal controls are directions that correlate with the fitness and also along which the data varied more.

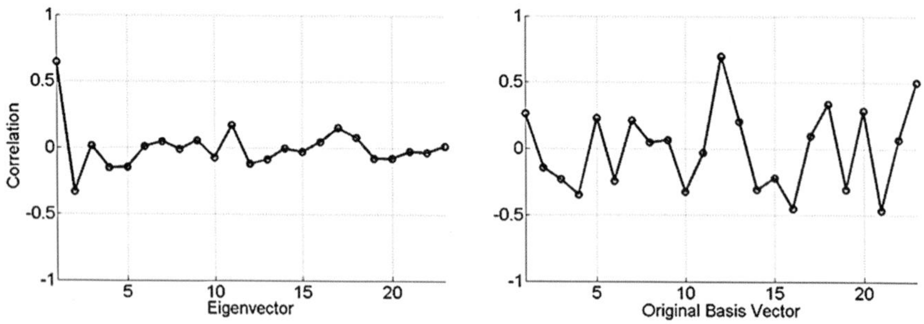

Figure 3: Correlations of fitness with projection onto eigenvector (left) and δ_i (right) for a methanol learning search maximizing the symmetric mode. The eigenvectors are in eigenvalue order from greatest to least.

The result of applying PCA to learning searches in methanol is given in detail in reference (6). Briefly, three principal controls were calculated for each learning search. Projection of the optimal pulse shapes onto the principal controls enhanced the Fourier transform of $I(t)$ around 3 THz, which is similar to the mode spacing.

7. Most Correlated Feature Analysis

In PCA, we determined which eigenvectors of the covariance matrix correlated well with fitness. In Most Correlated Feature Analysis, we instead determine which feature (direction) in the search space correlates best with fitness. We begin the MCF analysis as we began PCA, that is, by finding the eigenvectors and eigenvalues of the covariance matrix and calculating the correlation of fitness with projection onto each eigenvector. We define a *feature* to be a linear combination of the

$$feature = \sum_{k=1}^{N} \alpha_k \mathbf{u}_k \tag{4}$$

where the feature has been expressed in terms of the eigenvectors to simplify the calculation below. The correlation of fitness f with projection a of the data onto an arbitrary feature is

$$\begin{aligned} R &= \frac{\langle af \rangle - \langle a \rangle \langle f \rangle}{\sigma_a \sigma_f} \\ &= \frac{\left\langle \sum_{k=1}^{N} \alpha_k \eta_k f \right\rangle - \left\langle \sum_{k=1}^{N} \alpha_k \eta_k \right\rangle \langle f \rangle}{\sigma_a \sigma_f} \\ &= \frac{\sum_{k=1}^{N} \alpha_k Cov(\eta_k, f)}{\sigma_f \sqrt{\sum_{i,j=1}^{N} \alpha_i \alpha_j Cov(\eta_i, \eta_j)}} \end{aligned} \tag{5}$$

In the eigenvector basis, the covariance matrix is diagonal, and Eq. (5) becomes

$$R = \frac{\sum_{k=1}^{N} \alpha_k Cov(\eta_k, f)}{\sigma_f \sqrt{\sum_{i=1}^{N} \alpha_i^2 Var(\eta_i)}}. \tag{6}$$

We find the most correlated feature by setting the derivative of the fitness correlation (with respect to the linear combination coefficients of the feature) equal to zero and solving for the (unnormalized) feature coefficients. We find

$$\alpha_k = \frac{Corr(\eta_k, f)}{\sigma_{\eta_k}}. \tag{7}$$

A basis of most correlated features may be iteratively calculated by subtracting the projection of the data onto the most correlated feature from the data and repeating the MCF analysis. This basis can be used for eliminating dimensions with no fitness dependence or for choosing a few highly fitness-correlated dimensions to aid in visualizing the fitness landscape.

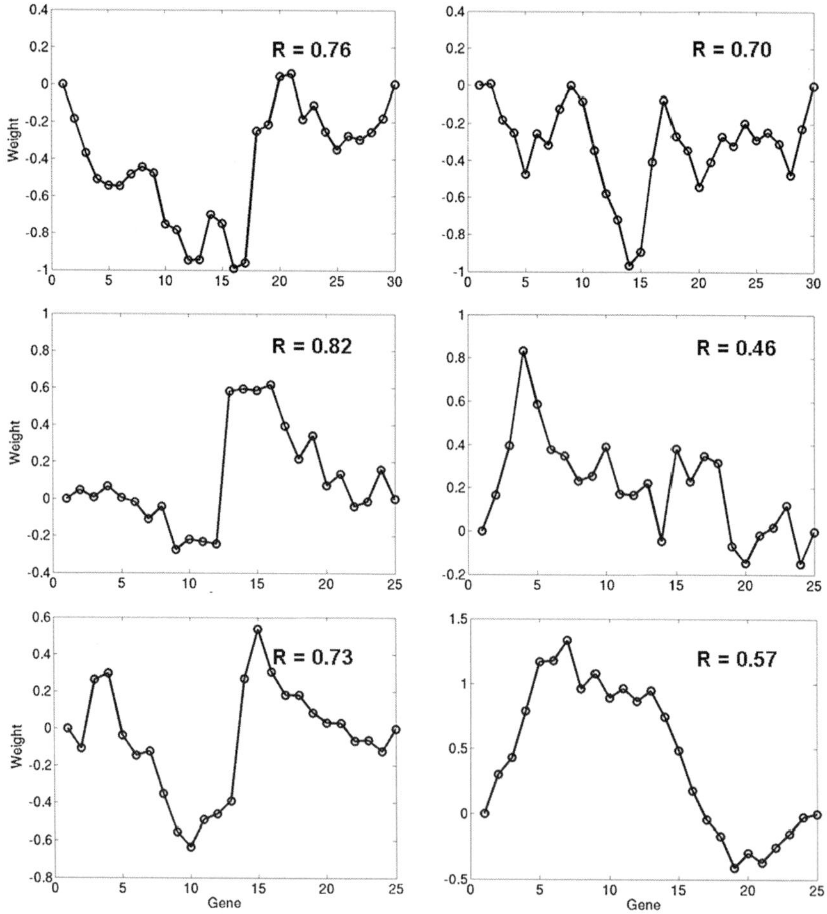

Figure 4: Left: Most correlated features for three experiments maximizing the symmetric Stokes mode, normalized and expressed in the gene basis (phase profile). The correlation R with fitness is given. Right: Same for experiments maximizing the antisymmetric Stokes mode.

The most correlated features for several representative methanol experiments are shown in Figure 4. For experiments enhancing the antisymmetric mode, the most correlated features show no obvious commonalities. For the symmetric stretch mode, the most correlated feature for seven searches includes a phase jump near the center of the laser bandwidth. This phase jump appears in the optimal pulses (see Figure 1, for example) but is obscured by other structure.

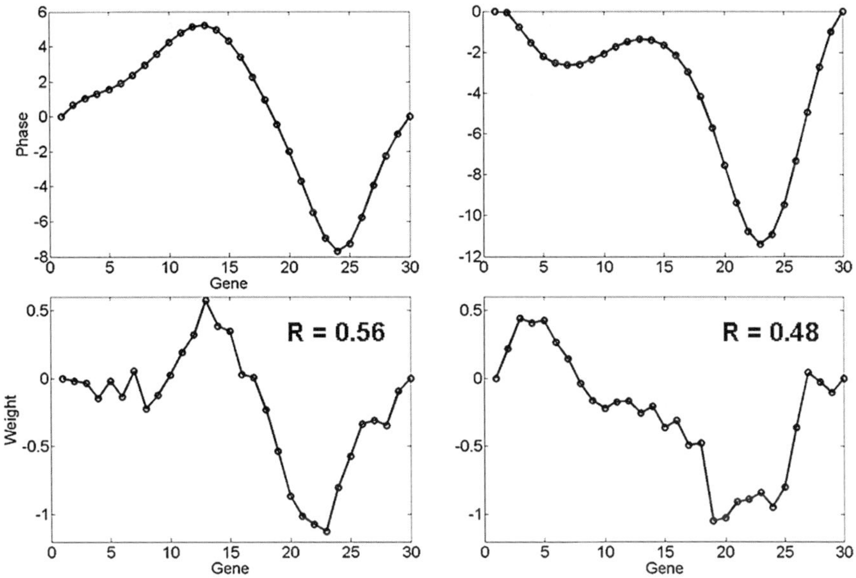

Figure 5: Top row: Genome of optimal pulse shape for two CHD experiments. Bottom row: Normalized MCF calculated for each experiment, expressed in the gene basis (phase profile). The correlation R with fitness is given.

Other experimental and theoretical work has suggested a similar control feature. Experiments have shown that the ratio of Stokes light in the two modes can be controlled by varying the phase between two frequency-shifted Gaussian pulses [11]. A theorctical model of those experiments indicates that the control mechanism is based on nonlinear optical modulation between the pump and Stokes pulses [12]. Specifically, the model agreed qualitatively with experiment when third-order nonlinear effects were included, namely saturated Raman scattering and self- and cross-phase modulation of the pump and Stokes light.

8. CHD Experiments and Analysis

This analysis of genetic algorithm searches to find the features most correlated with fitness can often provide support to physical models of the photochemical or physical process under investigation. Another example is the ring opening of 1,3-cyclohexadiene to 1,3,5-hexatriene under nonlinear excitation by intense shaped 800 nm pulses. This process was recently found to be enhanced by the imposition of negative quadratic phase (i.e., negative chirp) on the pulse [9]. A systematic scan to find the value of chirp that maximizes the hexatriene yield consistently shows that a transform-limited pulse performs less well than a negatively chirped pulse. This kind of limited scan cannot go further to investigate the effects of higher-order dispersion, for example. For this, we turned to a genetic learning algorithm to investigate the same problem. The GA goal was to maximize certain features of the transient absorption indicating the presence of 1,3,5-hexatriene. The experiments confirmed that a non-transform limited pulse with negative chirp optimizes the transient absorption signal and also showed that the best results included additional small cubic phase. Thus the optimal pulse shapes confirm that negative chirp may be an important feature.

For the CHD experiments, the MCFs are similar to the optimal pulse shapes (Figure 5). They display the same approximate magnitude and the same sign of quadratic phase and also bring out some of the cubic phase, which is also present in the optimal pulse found by scanning the chirp. While the MCF analysis does not appear to have emphasized any radically different pulse features, the results suggest that, in these experiments, the optimal pulse shapes do not contain much unnecessary structure.

9. Conclusion

Learning algorithms can be used to find optimal control fields and to investigate the dynamics of physical systems. An increasing emphasis on understanding the control mechanisms has led to recent improvements to the algorithms and how they are used. Principal Control Analysis and Most Correlated Feature Analysis are statistical techniques that use information collected during a learning search to determine important features of the control pulses and search space. The techniques are general and should be applicable to any fitness-directed search.

References

1. A. Assion *et al.*, Science 282 (1998) 919.
2. D. Meshulach and Y. Silberberg, Nature 396 (1998) 239.

3. B. J. Pearson *et al.*, PRA 63 (2001) 063412/1.
4. A. P. Peirce, M. A. Dahleh, and H. Rabitz, PRA 37 (1988) 4950.
5. T. Hornung *et al.*, Chem Physics Lett. 326 (2000) 445.
6. J. L. White *et al.*, J. Phys. B: At. Mol. Opt. Phys. 37 (2004) L399-L405.
7. A. Mitra and H. Rabitz, PRA 67 (2003) 33407.
8. T. C. Weinacht *et al.*, Chem. Phys. Lett. 344 (2001) 333.
9. E. C. Carroll *et al.*, submitted to J. Chem. Phys.
10. I. T. Jolliffe, Principal Component Analysis 2nd edn., Berlin, Springer, 2002.
11. B. J. Pearson and P. H. Bucksbaum, Phys. Rev. Lett. 92 (2004) 24300.
12. M. Spanner and P. Brumer, private communication.

Femtochemistry VII
Fundamental Ultrafast Processes in Chemistry, Physics, and Biology
A.W. Castleman, Jr. and M.L. Kimble

Analysis and Control of Ultrafast Dynamics in Clusters

Vlasta Bonačić-Koutecký[*a], Roland Mitrić[a], Jiri Pittner[b]

[a]Institut für Chemie, Humboldt Universität zu Berlin, Brook-Taylor-Straße 2, 12489 Berlin, Germany,[b]J. Heyrovský Institute of Physical Chemistry, Academy of Sciences of the Czech Republic, Dolejškova 3, 18223, Prague, Czech Republic

1. Introduction

Theoretical exploration of ultrafast processes in elemental clusters and their control by tailored laser pulses offers an excellent opportunity to determine how the interplay of size, structures, and light fields can be used to manipulate optical properties and chemical reactivity of these systems [1-5]. For this purpose the size regime of clusters in which each atom counts is particularly suitable due to the pronounced selectivity of properties by changing number of atoms [6-12]. In this manner laser-selective femtochemistry can be combined with the functionalism of nanostructures, opening new perspectives for the basic research as well as for technological applications.

The role of theory has been essential from conceptual as well as from a predictive point of view. Time resolved observation, in particular separation of time scales of different processes, are strongly dependent on the experimental conditions such as laser wavelengths, duration of pulses and their shapes, strength of electric field, competition between one- and many-photon processes, etc. [5,12-19]. Therefore, theory has to be directly involved in the conceptual planning of experiments, since it is not only needed to provide insight into time-dependent processes but also to identify conditions under which they can be experimentally observed.

Another important issue is the development of theoretical approaches for the optimal control of complex systems which allow to establish the connection between the underlying dynamical processes and the shapes of pulses. In this case, the optimal control can be used as a tool for the analysis [2,20-22].

Therefore, in this contribution we wish to address two fundamental aspects. First, what can we learn from i) fs-pump-probe signals and ii) from optimized pulse shapes? Second, under which conditions can complex systems be controlled by tailored laser fields? We also wish to show that for answering these questions the semiclassical limit of the Liouville formulation of quantum mechanics, based on the Wigner-Moyal representation of the vibronic density matrix combined directly with adiabatic and nonadiabatic molecular dynamics (MD) "on the fly" (without precalculation of the energy surfaces) developed in our laboratory [5], offers a methodological approach suited for accurate simulations of pump-probe signals and for developing new control strategies for complex systems [5,23]. This approach is characterized by the conceptual simplicity of classical mechanics and by the ability to approximately describe quantum phenomena such as optical transitions by means of the averaged ensemble over the classical trajectories. Moreover, the systematic introduction of quantum effects is possible although computationally demanding and still in development [24]. Our ab initio Wigner distribution approach is an appropriate choice to study ultrafast processes in elemental clusters with heavy atoms for which, in the first approximation, a classical description of the nuclear motion is acceptable and all degrees of freedom have to be considered because usually these clusters do not contain a "chromophore type" subunit nor do they obey regular growth patterns.

After outlining the principles of our method in Section II, the analysis of the simulated fs pump-probe signals will be presented for two examples for which experimental data are also available in Section III: a) In the framework of the negative ion-to-neutral-to-positive ion (NeNePo) pump-probe spectroscopy [25], the analysis of ultrafast processes of $Ag_2Au^-/Ag_2Au/Ag_2Au^+$ will be presented. This involves adiabatic dynamics in the ground state after vertical one-photon detachment, preparing a nonequilibrium state in the neutral ground state. This technique subsequently investigates the dynamics in the neutral ground state by two-photon ionization. Since dynamics of the nonequilibrium states pertain to transition state spectroscopy, this approach bridges the cluster dynamics with the real time exploration of chemical reaction pathways. The timescales for geometry relaxation and internal vibrational energy redistribution (IVR) will be determined. b) Fs pump-probe signals of Na_3F based on nonadiabatic dynamics for all degrees of freedom involving excited electronic states will be presented from which nonradiative lifetimes have been determined for the first time. Finally, a new strategy for optimal control in complex systems will be outlined in Section IV. Tailored pump-dump pulses will be used to drive the photoisomerization process in the Na_3F_2 cluster, populating only one selected isomer and avoiding the pathway with high excess energy involving the conical intersection. This is achieved by introducing a new strategy for the optimal control based on the intermediate target in the excited state corresponding to a localized ensemble that provides a connective pathway between the initial state and the target in the ground state. The connection between the shapes of the optimized pulses and the underlying processes will be established. A short summary and an outlook will be given in Section V pointing out that the ab initio Wigner distribution approach based on MD "on the fly" is suitable to identify different ultrafast processes in pump-probe signals to determine their timescales and to allow for developing of new strategies for the optimal control of complex systems using pulse shapes as tools for analysis of underlying processes.

2. Semiclassical Dynamics and Signals

Ab initio MD "on the fly" is now applicable to relatively large systems and classical trajectories can be used as an input in semiclassical approaches for the simulation of observables [5]. Moreover, we focus on the approaches that are able to make use of classical adiabatic ab initio MD and time dependent quantum chemistry.

The time evolution of the density operator $\hat{\rho}(t)$ is given by the quantum mechanical Liouville equation

$$i\hbar\frac{\partial\hat{\rho}}{\partial t} = \left[\hat{H},\hat{\rho}\right] \tag{1}$$

where $\hat{H}$ is the Hamiltonian of the system. Eq. (1) has the well known classical limit corresponding to the classical Liouville equation of nonequilibrium statistical mechanics:

$$\frac{\partial\rho}{\partial t} = \{H,\rho\} \tag{2}$$

where $\rho = \rho(q,p,t)$ and $H = H(q,p,t)$ are functions of classical phase space variables (q,p) and

$$\{H,\rho\} = \frac{\partial H}{\partial q}\frac{\partial\rho}{\partial p} - \frac{\partial\rho}{\partial q}\frac{\partial H}{\partial p} \tag{3}$$

is the Poisson bracket. The classical limit can be derived from the quantum mechanical Liouville equation using a Wigner-Moyal expansion in terms of $\hbar$, which emerges from

$$\left[\hat{A},\hat{B}\right] \rightarrow i\hbar\{A,B\} + O(\hbar^3) \tag{4}$$

if the expansion is terminated to the lowest order of $\hbar$. Higher terms in $\hbar$ serve for the introduction of quantum effects in dynamics.

For the formulation of the pump-probe or the pump-dump signals, we combine the ab initio MD "on the fly" with the vibronic density matrix approach in classical Wigner-Moyal representation [13]. In addition, for the lower intensities only first order optical transitions can be taken into account. Using short laser pulses which can be described with Gaussian pulse envelopes, the following analytic expression can be derived:

$$
\begin{aligned}
S[t_d] &= \lim_{t\to\infty} P_{22}^{(2)}(t) \\
&\approx \int dq_0 dp_0 \int_0^{\infty} d\tau_1 \exp\left\{ -\frac{(\tau_1 - t_d)^2}{\sigma_{pu}^2 + \sigma_{pr}^2} \right\} \\
&\times \exp\left\{ -\frac{\sigma_{pr}^2}{\hbar^2} [\hbar\omega_{pr} - V_{bc}(q_1(\tau_1; q_0, p_0))]^2 \right\} \\
&\times \exp\left\{ -\frac{\sigma_{pu}^2}{\hbar^2} [\hbar\omega_{pu} - V_{ab}(q_1(\tau_1; q_0, p_0))]^2 \right\} P_{00}(q_0, p_0)
\end{aligned}
\tag{5}
$$

In expression (5) $\sigma_{pu}(\sigma_{pr})$ and $E_{pu} = \hbar\omega_{pu}(E_{pr} = \hbar\omega_{pr})$ are the pulse durations and excitation energies for the pump and probe step with a time delay t_d. The Eq. (5) is valid for the NeNePo scheme as well as for the pump-probe signals involving ground and excited states NeExPo. In the case of NeNePo the quantity $V_{bc} = V_{IP}(q_1(\tau_1; q_0, p_0))$ labels the time dependent energy gaps between the neutral and cationic ground state calculated at coordinates $q_1(\tau_1)$ on the neutral ground state with initial coordinates and momenta q_0 and p_0 given by the anionic thermal Wigner distribution $P_{00}(q_0, p_0)$, and $V_{ab} = V_{VDE}(q_0)$ are the vertical detachment energies of the initial ensemble. Therefore, the first step for the simulation of signals involves the generation of $P_{00}(q_0, p_0)$, which is calculated either for individual vibronic states or for the thermal ensemble assuming the harmonic approximation in the case of low initial temperature for which this condition is valid. The Wigner distribution for each normal mode is given by

$$P(q,p)=\frac{\alpha}{\pi\hbar}\exp\left[-\frac{2\alpha}{\hbar\omega}\left(p^2+\omega^2q^2\right)\right] \tag{6}$$

where $\alpha = \tanh(\hbar\omega/2k_BT)$ and ω is the normal mode frequency, corresponding to the full quantum mechanical density distribution. From sampling of the phase space distribution using Eq. (6), the ensemble of initial conditions is formed which is needed for the MD on the neutral ground state. The expression for signals given by Eq. (5) is easy to interpret. In the case of NeNePo, the last exponential represents the Franck-Condon transition probability after the initial ensemble is photodetached. Then, the propagation occurs on the neutral state in terms of MD "on the fly", which gives rise to the time dependent ionization energies V_{IP}. The transition to the cationic ground state involves the probe step represented by the window function given by the second exponential of Eq. (5). The signal is obtained by the summation over the whole ensemble. The time resolution of the signal is given by the pump-probe correlation function with the probe window located around the time delay between two pulses represented by the first exponential of Eq. (5). The duration of both pulses determines the spectral distribution. The expression (5) is, in general, valid for adiabatic MD. In the case of the NeExPo scheme, the initial conditions involve the ground state, and the dynamics on the excited electronic state are probed in cationic states.

Additional modifications of expression (5) have to be introduced for the following situations: i) if the emitted electrons carry away some amount of kinetic energy. For example, for the simulations of the transient photoionization NeNePo signal, the integrations of the populations of the anionic and cationic states over the entire range of possible excess energies E_0 and E_2 have to be performed in order to include approximately the continuum (cf. Ref [13]). ii) If nonadiabatic processes have to be considered. In this case, in addition to nonadiabatic dynamics "on the fly" and to the choice of the treatment of nonadiabaticity (e.g. fewest switches surface hopping method introduced by Tully[25]), the determination of the pump-probe signal accounting for the passage through the conical intersection requires a modification of the expression for the cationic occupation in expression (5). This is due to the necessity in considering that the propagation of the ensemble starts in one excited state but can hop to the other excited states or to the ground state according to the fewest-switches surface hopping algorithm. Therefore, not only the common averaging over the whole ensemble of the initial conditions due to the Wigner approach is required, but also, for a given initial condition, an averaging over trajectories obtained from different random numbers according to the hopping algorithm must be carried out. Moreover, if more than two states are involved, the sum of weighting factors corresponding to transition moments between these electronic states, for which also energy gaps have to be calculated, are required (cf. Ref. [15]).

3. NeNePo signals for Ag_2Au

The scheme of the multistate fs-dynamics for NeNePo pump-probe spectrocopy of $Ag_2Au^-/Ag_2Au/Ag_2Au^+$ with structures and energy intervals for pump and probe steps given on the left side of Fig. 1 allows to choose appropriate energy intervals for observing different types of processes. Comparison between simulated signals according to Eq. (5) and experimental ones for three different probe energies are given on the right side of Fig. 1.

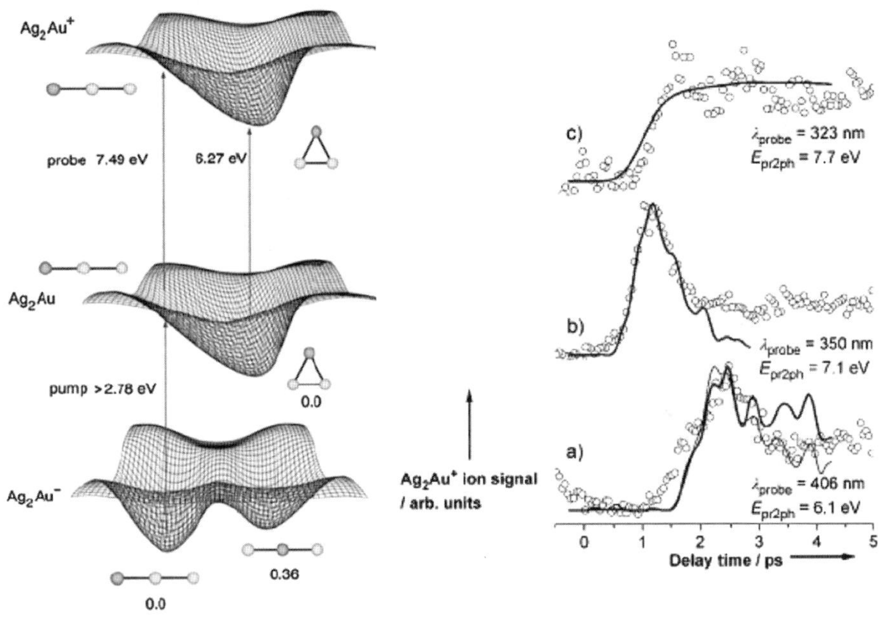

Figure 1. (Left Side) Scheme of the multistate fs-dynamics for NeNePo pump-probe spectroscopy of $Ag_2Au^-/Ag_2Au/Ag_2Au^+$ with structures and energy intervals for pump and probe steps. (Right side). Comparison between simulated (full line) and experimental (circles) pump probe signals at different probe pulse energies.

An initial temperature of 20 K, as in experiments, was selected for simulations. Under these conditions, only the lowest energy anionic isomer is populated and the harmonic approximation is valid. Therefore, the initial conditions for the MD simulations were sampled from the canonical Wigner distribution of each independent normal mode according to Eq. 6. Two different detection scenarios can be distinguished. First is NeNePo-ZEKE with defined probe energy, in which no energy will be transferred from the photon into the kinetic energy of the detached electron. The second scenario corresponds to the classical NeNePo experiment in which excess of energy is accounted for. Such NeNePo transients are obtained theoretically by integrating over the populations of anionic and cationic states over the entire excess of energy. Consequently, at high ionization energy E_{pr2ph}=7.7 eV the theory would predict no observable time dependent NeNePo signal because the photon energy is high enough to ionize the neutral cluster in any geometrical arrangement. However, weak experimental transient signal was detected which is in agreement with simulated NeNePo signal at a probe energy of 7.41 eV. This is just below the highest theoretically predicted ionization energy which corresponds to the linear transition-state structure. Thus the experiment at E_{pr2ph}=7.7 eV monitors the system when it leaves the transition state region. However, at later times all the structures with smaller bending angles can be ionized, and the signal becomes constant as does the simulated NeNePo transient. The comparison of simulated NeNePo-ZEKE and experimental transient at 7.1 eV probe photon energy are in perfect agreement. Therefore, the experimental conditions allow in this case for direct probing exclusively the geometric relaxation of Ag_2Au at the signal onset at ~500 fs, at the signal maximum ~ 1 ps, and at the offset at ~ 2 ps. Moreover, for E_{pr2ph}=6.1 eV the onset of the simulated NeNePo-ZEKE starts at ~ 2 ps, reaches the maximum at ~ 2.4 ps, and decreases slightly after ~ 2.5 ps. This corresponds to the onset of the internal vibrational energy redistribution (IVR) and the dynamics of Ag_2Au in the potential minimum of the neutral triangular geometry initiated by the collision between the terminal Au and Ag atoms. The good agreement between simulated and experimental transients indicates that in this case exclusively IVR has been observed (for details cf. Ref. [18]). This means that the experimental conditions have been close to the ZEKE conditions. In fact, for the probe energy of 6.1 eV the continuum energy is so low that the simulated NeNePo-ZEKE and NeNePo signals exhibit very similar features as shown on Fig. 1. Comparison with the experiment shows explicitly in which regime ZEKE conditions hold so that under these circumstances the time scales for different processes such as geometric relaxation versus IVR can be resolved.

4. Simulation of non-adiabatic processes in Na_3F

In this section we present a theoretical simulation of the ultrafast processes and non-radiative relaxation of the Na_3F cluster, involving both excited and ground electronic states. Furthermore, we present the simulation of the pump-probe signal and compare it with the experimental signal and with previous quantum and classical simulations of the adiabatic dynamics.

For this purpose we employ the non-adiabatic MD "on the fly" in the framework of the Full-CI method for two excess electrons which are not involved in ionic bonding together with the Tully's surface hopping algorithm. Non-adiabatic coupling matrix elements which are necessary to perform nonadiabatic dynamics are calculated also "on the fly" according to the analytic formulation presented in Ref. [28].

The initial conditions for the simulation are prepared by sampling the canonical Wigner distribution function in the harmonic approximation (cf. eq. 6) corresponding to the temperature of 50 K. At t=0 fs an ensemble of 200 trajectories has been excited to the third $^1B_1$ excited state and propagated, allowing for non-adiabatic transitions in the manifold consisting of the ground state and three lowest excited electronic states. The populations of the involved states as a function of time during the dynamics are shown in the Fig. 2b. The population of the initially excited $^1B_1$ state is decaying due to nonadiabatic transitions with a time constant of ~ 1 ps. Initially, the population transfer occurs to both second and first excited state so that the population in the second excited state reaches the maximum at ~1.2 ps. The population of the first excited state reaches the maximum at ~ 2 ps. Simultaneously, the population of the ground state starts to grow at ~ 500 fs and eventually dominates at later times. After 3 ps the populations of excited electronic states have almost completely decayed, and the system resides in the ground electronic state with a large excess of energy which partly leads to the fragmentation of the cluster. In order to get a direct comparison with the experiment [29] we have also simulated pump-probe spectra according to Eq. 5. The energy of the probe pulse used in the simulation was E_{pr}=2.05 eV, and the pulse duration was σ_{pr}=100 fs. For the pump-pulse it is assumed that the pulse duration is zero.

Comparison between the simulated and experimental pump-probe spectrum is shown in Fig. 2a. The simulated signal exhibits oscillations at early times which correspond to the periodic motion of the F-atom with the period of ~ 380 fs, as can be determined from the analysis of the dynamics. At later times, nonadiabatic transitions start to dominate leading to the decay of the signal. It is particularly interesting to compare this signal with the signals obtained from the adiabatic simulations from Ref. [29]. While signals obtained from classical and quantum dynamics simulations exhibit oscillations of the F-atom at early times, they remain constant at later times and cannot explain the decay in the experimental pump-probe signal (cf. Fig 2c and 2d). The reason for the decrease of the intensity of adiabatically simulated signals is mainly the spreading of the ensemble (or wavepacket).

In summary, the simulations of the nonadiabatic dynamics allowed us to determine the nonradiative lifetime for an electronically excited state of the Na_3F cluster to be ~1 ps. Moreover, our approach allows to simulate the population dynamics involving several excited states, which offers an opportunity to study photochemical properties of excited states in complex systems. This is particularly important for determination of emissive properties and fluorescence which are directly dependent on the rate of nonadiabatic processes.

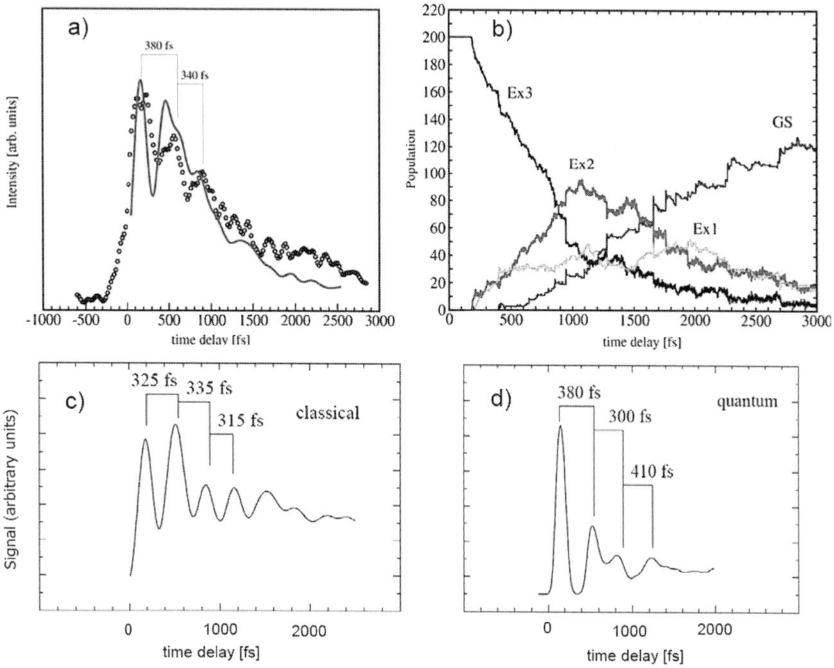

Figure 2. (a) Comparison between the theoretical pump-probe spectrum for Na_3F obtained from the nonadiabatic dynamics in the framework of the Full-CI method and experimental signal [27]. (b) Populations of the three excited and the ground state as function of time. Ex1, Ex2, Ex3 label the first, second and third excited state, respectively. GS labels the ground state. (c) Theoretical pump-probe signal obtained from semiclassical approach employing the adiabatic dynamics in the third excited state in the framework of the TDDFT method [29]. (d) Theoretical pump-probe signal obtained from a quantum dynamics simulation for reduced dimensionality [29].

Furthermore, we have shown that our approach allows for accurate simulations of fs pump-probe signals in the case that the radiationless relaxation processes are involved and permits the identification of underlying dynamical processes.

5. New strategy for optimal control of pump-dump processes

In this section we present our strategy for the control of pump-dump processes based on ab initio MD "on the fly", applicable to complex systems. The strategy will be illustrated on the example of the isomerization of the Na_3F_2 cluster (cf. Refs. [5] and [23]). The isomerization in Na_3F_2 through conical intersection between the first excited state and the ground state provides a system with high excess of internal energy which prevents selective population of the second ground state isomer through this process (cf. scheme on the left part of Fig. 3). Therefore, we developed an optimal control strategy based on the intermediate target concept as an adequate tool to find the optimal pathway which allows to selectively populate isomer II and suppress the pathway through the conical intersection.

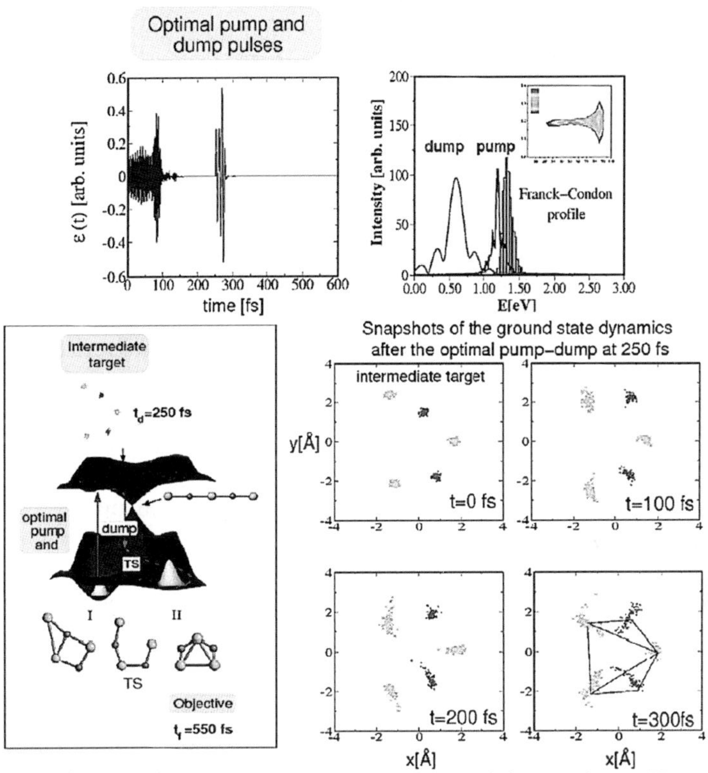

Figure 3. (Upper panel, left side) The optimal pump and dump pulses. (Upper panel , right side) Fourier transforms of the optimal pump and dump pulses and the Franck-Condon profile for the first excited state corresponding to an excitation energy T_e=1.33 eV. Wigner transform of the optimal pump pulse is shown in the insert. (Lower panel, left side). Scheme for pump-dump optimal control in the Na_3F_2 cluster with geometries of the two ground-state isomers and the transition state separating them, the conical intersection, and the intermediate target. (Lower panel, right side) Snapshots of the dynamics obtained by propagating the ensemble corresponding to the intermediate target after the optimized pump-dump at 250 fs.

For this purpose, the initial ensemble corresponding to the 50 K canonical ensemble has been generated by sampling the canonical Wigner distribution function in the harmonic approximation (cf. Eq. 6). After the pump step, the ensemble has been propagated on the excited state for 300 fs. In order to determine the intermediate target and the optimal time delay at which the dump should occur the system has been dumped to the ground state (in regular time steps of 25 fs) and propageted subsequently. The isomer II is optimally reached by the ensemble dumped at t_d=250 fs, and therefore this ensamble (cf. scheme on the left part of Fig. 3) is chosen to construct the intermediate target. The role of the intermediate target is to insure the connective pathway between the initial state (isomer I) and the objective (isomer II). The optimal pump and dump pulses have been obtained by solving the decoupled Fredholm-type integral equations (cf. Ref. [23]) and are presented on the upper part of Fig. 3. The shape of the optimal pump pulse exhibits two portions with durations of 70 and 10 fs respectively. The Fourier analysis of the optimal pump pulse and the comparison with the Franck-Condon profile reveals that the excitation of low lying vibrational modes at ~ 1.2 eV of the isomer I is responsible for driving the dynamics to the intermediate target. The Wigner-Ville transform of the optimal pump pulse shows that this energetically sharp transition corresponds to the first temporal portion of ~ 70 fs of the pump pulse. In contrast, a very short second portion after 80-90 fs is energetically much wider and corresponds to the tails of the Fourier transform which are symmetric with respect to the 1.2 eV transition. The optimal dump pulse is very short (~20 fs), leading to a short time window for the depopulation of the excited state. The dynamics after the dump pulse, illustrating the efficiency of the optimized pulses, is presented on the lower left side of Fig 3. It can be clearly seen that the phase space density is localized in isomer II after ~450 fs.

In summary, our strategy for optimal control based on the concept of the intermediate target allows us to design pump and dump pulses which drive the isomerization process in the desired objective involving the excited and ground electronic states. Furthermore, the analysis of the underlying dynamics and of the optimal pulses allows to identify the mechanism responsible for the optimal control and permits to establish conditions under which optimal control can be achieved.

6. Summary and Outlook

Analysis and control of ultrafast processes in atomic clusters in the size regime in which "each atom counts"are important from a conceptual point of view and for opening new perspectives for applications. At the same time, this research area represents a challenge for developing theoretical and computational methods including quantum chemistry, molecular dynamics, and optimal control theory. We have illustrated that theoretical methods which combine MD "on the fly" with the Wigner distribution approach represent an important theoretical tool for analysis and control of ultrafast processes in complex systems. Furthermore, due to the structure-reactivity relationships of clusters, the tailored laser fields might be used to drive selected chemical reactions towards organic and inorganic molecules. Control of their functional properties might invoke impact in different application areas.

7. Acknowledgements

This work has been supported by the Deutsche Forschungsgemeinschaft (DFG) SFB 450 "Analysis and control of ultrafast photoinduced reactions". We wish to thank Mr. Gracian Tejral for technical help.

References

1. A. H. Zewail, Femtochemistry, World Scientific, Singapore, 1994.
2. J. Manz and L. Wöste, Eds. Femtosecond Chemistry, VCH Verlagsgesellschaft mbH, Weinheim, Germany, 1995; Vols. 1 and 2.
3. A. H. Zewail, J. Phys. Chem. A 104 (2000) 5660.
4. T. Brixner and G. Gerber, ChemPhysChem. 4 (2003) 418.
5. V. Bonačić-Koutecký and R. Mitrić, Chem. Rev. 105 (2005) 11.
6. V. Bonačić-Koutecký, P. Fantucci, and J. Koutecký, Chem. Rev. 91 (1991) 1035.
7. A. W. Castleman, Jr. and K. H. Bowen, J. Phys. Chem. 100 (1996) 12911.
8. J. Jortner, Faraday Discuss. 108 (1997) 1.
9. U. Landman, Int. J. Mod. Phys. B6 (1992) 3623.
10. H. Haberland, Ed., Clusters of Atoms and Molecules; Springer Series in Chemical Physics 52, Springer-Verlag, Berlin, Heidelberg, New York, 1994.
11. W. Ekardt, Ed., Metal Clusters, John Wiley & Sons, Chichester, 1999.
12. P. Jena, S. N. Khanna, and B. K. Rao, Eds., Cluster and Nanostructure Interfaces, World Scientific, Singapore, 1999.
13. M. Hartmann, J. Pittner, V. Bonačić-Koutecký, A. Heidenreich, and J. Jortner, J. Chem. Phys. 108 (1998) 3096.
14. M. Hartmann, J. Pittner, and V. Bonačić-Koutecký, J. Chem. Phys. 114 (2001) 2106.
15. M. Hartmann, J. Pittner, and V. Bonačić-Koutecký, J. Chem. Phys. 114 (2001) 2123.
16. R. Mitrić, M. Hartmann, B. Stanca, V. Bonačić-Koutecký, and P. Fantucci, J. Phys. Chem. A 105 (2001) 8892.
17. S. Vajda, C. Lupulescu, A. Merly, F. Budzyn, L. Wöste, M. Hartmann, J. Pittner, and V. Bonačić-Koutecký, Phys. Rev. Lett. 89 (2002) 213404.
18. T. M. Bernhardt, J. Hagen, L. D. Socaciu, R. Mitrić, A. Heidenreich, J. Le Roux, D. Popolan, M. Vaida, L. Wöste, V. Bonačić-Koutecký, and J. Jortner, ChemPhysChem. 6 (2005) 243.
19. N. E. Henriksen and V. Engel, Int. Rev. Phys. Chem. 20 (2001) 93.
20. D. J. Tannor and S. A. Rice, J. Chem. Phys. 83 (1985) 5013.
21. M. Shapiro and P. Brumer, J. Chem. Phys. 84 (1986) 4103.
22. A. P. Peirce, M. A. Dahleh, and H. Rabitz, Phys. Rev. A 37 (1988) 4950.
23. R. Mitrić, M. Hartmann, J. Pittner, and V. Bonačić-Koutecký, J. Phys. Chem. A 106 (2002) 10477.
24. A. Donoso and C. C. Martens, Phys. Rev. Lett. 87 (2001) 223202.
25. S. Wolf, G. Sommerer, S. Rutz, E. Schreiber, T. Leisner, L. Wöste, and R. S. Berry, Phys. Rev. Lett. 74 (1995) 4177.

26. J. C. Tully, in: Modern Methods for Multidimensional Dynamics Computations in Chemistry, D. L. Thompson, Ed., World Scientific, Singapore, 1998.
27. J. M. L'Hermite, V. Blanchet, A. Le Padallec, B. Lamory, and P. Labastie, Eur. Phys. J. D 28 (2004) 361.
28. H. Lischka, M. Dallos, P. G. Szalay, D. R. Yarkony, and R. Shepard, J. Chem. Phys. 120 (2004) 7322.
29. M. C. Heitz, G. Durand, F. Spiegelman, C. Meier, R. Mitrić, and V. Bonačić-Koutecký, J. Chem. Phys. 121 (2004) 9906.

Femtochemistry VII
Fundamental Ultrafast Processes in Chemistry, Physics, and Biology
A.W. Castleman, Jr. and M.L. Kimble

Controlling the dynamical pathways in the ionization processes of NaK dimers

Albrecht Lindinger, Stefan M. Weber, Andrea Merli, Franziska Sauer, Mateusz Plewicki, and Ludger Wöste

Institut für Experimentalphysik, Freie Universität Berlin, Arnimallee 14, D-14195 Berlin, Germany; E-Mail: lindin@physik.fu-berlin.de

1. Introduction

The development of pulse shapers, which allow to modulate laser pulses in the amplitude and phase domain, has opened fascinating new perspectives for driving molecular reaction dynamics in real time. This scientific topic became most exciting when self-learning feedback loop algorithms were employed [1]. In this way, tailored laser pulses can be generated, which drive the photo-induced processes at a maximum yield along desired paths via the potential energy surfaces. A variety of optimization experiments were performed by several groups [2-5]. An important issue in this regard is the question concerning the information coded in the optimized laser pulse shape. The extraction and interpretation of the information from the acquired optimal laser field is a very relevant aspect, since it leads to a conceptionally new approach to the investigation of molecular dynamics. Employing evolutionary strategies – nature's concept itself – not only allows for an effective control of the dynamics of a molecular system by optimizing a problem of high dimensionality but also supplies valuable information about the involved molecular potential energy surface(s). In this regard, it is most important to begin the investigations on model systems which are simple and where the underlying molecular dynamics can still be understood on the basis of available experimental and theoretical data. Based on these findings, more complex systems with more degrees of freedom may then become experimentally controllable and understandable as well.

Small alkali clusters are suitable model systems for the investigation of optimal control scenarios since they have low ionization energies and exhibit a number of bound-bound electronic transitions, which allows to study resonant enhanced multi-photon ionization. Moreover, isotopomer selective ionization can be investigated on alkali dimers and evolution strategies can be applied in a feedback loop in order to optimize their selectivity [6]. Novel approaches to gain further information from the optimized laser field can favorably be implemented for alkali model systems. One approach is to analyze the results of the individuals during the entire optimization run regarding the evolution of intuitive and physically relevant pulse features [7]. Two other prospective methods will be reported in this contribution. The first addresses parametric optimization [8] where the search space is narrowed by introducing a few pulse parameter optimization. The suggested parameters in the time- and frequency domain are: time, distances, intensities, zero order spectral phase, chirps of different subpulses, and spectral peak patterns. In this manner, the relevance of certain structural features of optimal pulses can be investigated. The second novel method examines the implementation of genetic pressure within the algorithm for performing control pulse cleaning [9]. The aim is to remove extraneous pulse features in order to expose the most relevant structures and thus reveal mechanistic insight.

2. Experimental

Our experimental setup has been described in detail elsewhere [10]. Briefly, the molecular beam is coexpanded in an adiabatic expansion of alkali metal vapor from a heated oven together with argon carrier gas across a nozzle of 70-80 μm diameter into the vacuum. In order to form cold NaK dimers with no larger clusters present in the beam, the oven temperature is set to about 650 °C and the argon pressure to 2 bars. After passing a skimmer, the molecular beam is directed between the electrical lenses of a quadrupole mass spectrometer (Balzers QMG 420), which are oriented perpendicularly both to the propagation of the cluster beam and to the laser beam in order to extract the resulting photo-ions. A commercial femtosecond laser (Spectra Physics 3960 Tsunami), pumped by a Nd:YLF laser (Spectra Physics Millenia X), produces pulses of 120 fs duration, 6 nm spectral width, and 10 nJ energy per pulse at a central wavelength of 780 nm.

In the multi-photon optimization experiment, a single laser beam passes through a pulse shaper setup, which allows for a simultaneous phase- and amplitude modulation of the laser pulses by applying voltages to a liquid crystal spatial light modulator (SLM-256; CRI) consisting of 2x128 pixels. The shaper is placed in the center of a linear 4f arrangement. By computer control, pulses of arbitrary form can be generated. These modulated pulses are then focussed on the molecular beam and the desired mass-selected product ion current is taken as a reference feedback signal. The optimization algorithm alters the pulse shaper settings iteratively in order to maximize the desired fitness. We apply an algorithm based on evolutionary strategies to find the pulse shape which yields the highest feedback signal. The optimization procedure continues until convergence is achieved. For the isotopomer selective measurements the $^{23}Na^{39}K/^{23}Na^{41}K$ ion ratio is used as the new fitness by sequentially recording ion yields of the two isotopomers in time steps of less than half a second. The generated optimized pulse form is measured by acquiring the pulse spectra, the cross-correlation, and the frequency resolved cross-correlation (XFROG).

3. Results and discussion

3.1. Phase-only optimization of the ionization of $^{23}Na^{39}K$

When maximizing the ionization of NaK we observed an almost monotonic rise of the ion yield during the optimization run. After approximately 120 iterations the signal could not be increased any more which means that convergence has been reached. In the presented optimization only the phase has been modified in order to keep the pulse energy constant.

The comparison of the ion yields between optimized and unshaped pulses shows an improvement of the NaK^+ intensity by a factor of $I_{opt}/I_{unshaped}=1.6$. An important feature of the molecular beam, visible in the recorded mass spectra, is the lack of any larger molecules which means that NaK ions originate only from NaK dimers and not from fragmented trimers, tetramers, or larger clusters. This result demonstrates the feasibility of optimizing the ionization process; however, we intended to move a step ahead and tried to learn from the optimized pulse shape about the underlying process. Therefore we recorded the intensity cross-correlation and XFROG traces of the optimized pulse (Fig.1). It contains three dominant and some smaller subpulses with time separations of almost 1.5 times the oscillation

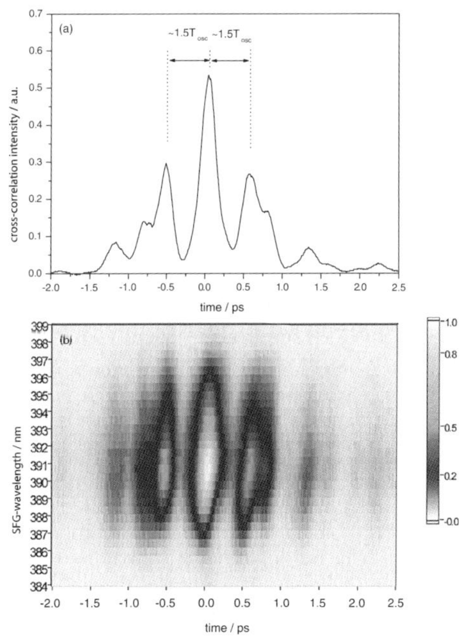

Fig. 1. Intensity cross-correlation (a) and XFROG trace (b) of the optimized pulse.

period of the first excited state (T_{OSC}= 440 fs) [11]. The central pulse is thereby the most intense.

The relative intensities of the subpulses and their position in time suggest a simple scheme of ionization (see Fig. 2). The first peak transfers some population from the ground state to the first excited $A(2)^1\Sigma^+$ state. Due to the Franck-Condon principle and resonance conditions, the wavepacket is created at the inner turning point. After 1.5 oscillation periods in the $A(2)^1\Sigma^+$ state, it is located at the outer turning point where from it can be effectively transferred to the ionic state. The second subpulse ionizes the molecule in a resonant two photon process and additionally generates a secondary wavepacket on the first excited state. Again, after the second wavepacket is located at the outer turning point, it is being transferred to the ionic state by the third subpulse. The second excited state $B(3)^1\Pi$ state will also be populated by the second and third subpulse. A fourth weak subpulse may transfer the residual population from the first and second excited state to the ionic state. Also the propagation in the $B(3)^1\Pi$ state with an oscillation period of about 700 fs is included in the ionization process. This explanation is supported by full quantum-mechanical calculations of the optimization process undertaken in the group of Prof. Bonačić-Koutecký for the slightly different center wavelength of 770 nm. They solved the optimization functional by iteration until self-consistency was achieved, and we observed a qualitative agreement between the theoretically and experimentally acquired optimized pulses [12]. Snapshots of the wavepacket propagations revealed the explained mechanism and showed that the wavepacket oscillation in the $B(3)^1\Pi$ state occurs as well within the ionization process. Moreover it was found that later smaller subpulses contribute much to the overall ion yield since they are responsible for the last excitation step to the ionic state. In our experiment we expect a similar behavior for the only slightly shifted center wavelength. Constructive interferences of the molecular wavepackets in the excited states as mentioned in [12] are presumably also utilized in the optimized ionization path for 780 nm center wavelength. Thus, it is found that basically a stepwise excitation takes place in the optimized ionization process.

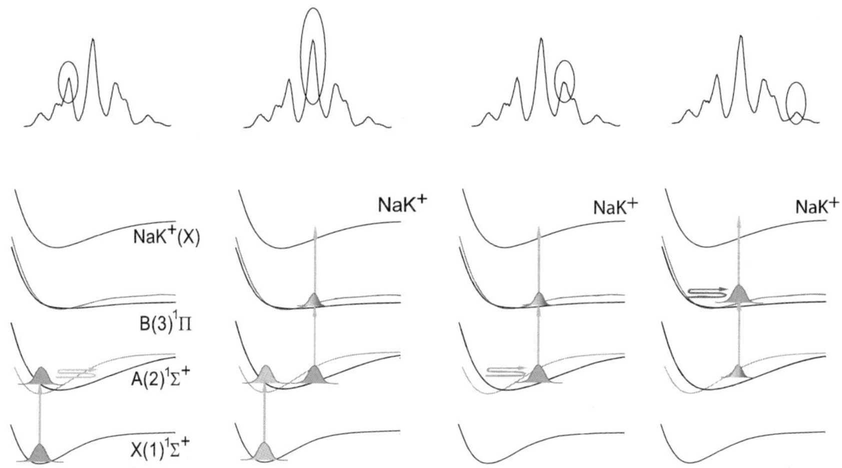

Fig. 2. Optimized photo-induced transition from the neutral ground state of NaK to the ionic ground state.

In the following, a different approach to receive further information will be applied by performing a few parameter optimizations, where the search space is reduced to permit only simple pulse trains or spectral patterns. In this manner the relevance of certain structural features of the optimized pulses can be investigated.

3.2. Parametric optimization

A further advantage of parametric optimization, besides search space reduction and application of physically relevant and intuitive parameters, is the possibility to define the 'framework' of an experiment. One can for example choose a parameter set for an intended purpose and will receive the optimized result within this parameter set. Here, we asked the question, what will be the result if we allow different numbers of subpulses, which will tell us the relevance of the additional subpulses. Thereto, we constructed the complex electrical field of a pulse train after passing the shaper by

$$\tilde{E}_{out}(t) = e^{i\omega_0 t}\sum_n \varepsilon_n(t-t_n)e^{i\varphi_n(t)} = e^{i\omega_0 t}\sum_n \tilde{\varepsilon}(t-t_n) \quad ,$$

with ε and $\tilde{\varepsilon}$ being the field envelope and complex field envelope of the subpulses, respectively, and t_n being the time distances between the subpulses. The resulting temporal field is then Fourier-transformed to receive an electrical field in the frequency description. The desired output for the electrical field is hence produced from the modulator by writing the filter function

$$\tilde{H}(\omega) = \tilde{E}_{out}(\omega) / \tilde{E}_{in}(\omega)$$

The experiments were performed by pure phase modulation in order to keep the total pulse energy constant. We reduced the full search space to an 11 or 23 parameter space for three or six subpulses, respectively, where the subpulse distances, intensities, constant phase differences, and linear chirps were subject to optimization.

Fig. 3 shows the SFG cross-correlation results for optimizing three (a) or six (b) subpulses. The results of three different runs are shown in (a) in order to test the robustness of the optimized pulse shape. Subpulse distances of 650 fs corresponding to $1.5T_{osc}$ are visible in all cases. However, the optimization factor increased from approximately 1.2 for three subpulses to about 1.5 for six subpulses, which comes closer to the value of 1.6 for free optimization. This indicates that apparently more than three subpulses are required for an efficient transfer to the ionic state, which is in concordance to the model explanation given above. The unrestricted optimization produced a comparable waveform except for one more subpulse at around 0.8 ps. The remaining difference between the parametric and the free optimization may further orginate from the search space restriction since some pulse components were still not modulated in the parametric case. Yet, this example shows the potential of the introduced method to gradually gain knowledge about the molecular system.

Another possibility is to perform spectral parametrization, where narrow Gaussian amplitude distributions are subject to modulation in order to find certain vibronic transitions. Parameters are the distances and intensities of these distributions. The phase can thereby be modulated freely to allow unrestricted temporal pulse modulation, as realized here. For the NaK model system we were able to find some involved transitions to the first and second excited state (see Fig 3 (c)). This 'transition finder' can generally be applied to search for those transitions in molecular systems which are utilized by the optimized path.

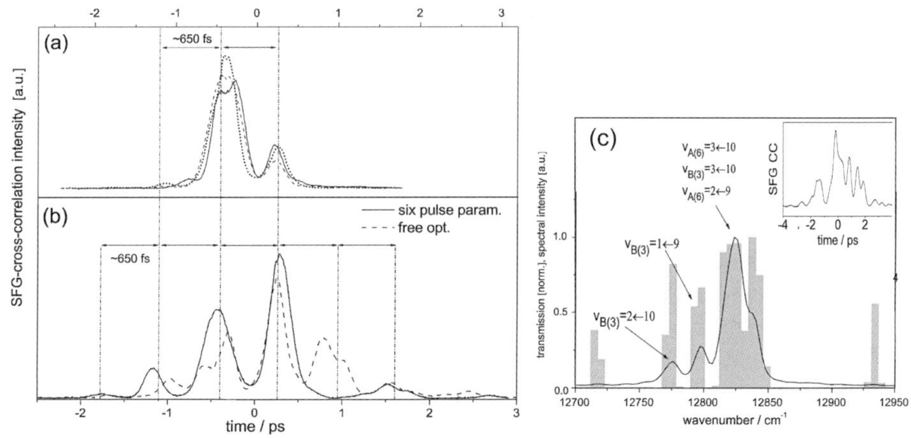

Fig. 3: The SFG cross-correlation results are shown for optimizing three (a) or six (b) subpulses (solid line). The results of three different runs are shown in (a) in order to test the robustness of the optimized pulse shape. Subpulse distances of 650 fs corresponding to $1.5T_{osc}$ are visible in all cases. The optimized pulse for free optimization is shown in (b) as a dashed line. The spectrum of the optimized pulse for parametrization in the frequency domain employing 11 narrow Gaussian distributions is shown in (c) (the inset depicts the cross-correlation trace). The spectral peaks are assigned to vibronic transitions.

3.3. *Control pulse cleaning for $^{23}Na^{39}K$*

A novel method in coherent control named control pulse cleaning describes the process of removing extraneous control field features in a closed-loop quantum dynamics optimization experiment. This is accomplished by applying genetic pressure during the optimization run in order to simplify the optimized pulses and to reveal the most relevant features. Thereto, a new fitness function has to be formed where the target goal f is optimized and simultaneously genetic pressure is implemented on certain pulse components. We solved this for the NaK model system by introducing the fitness function

$$F = \frac{f}{\left(\sum_{l=1}^{N} T_l \Big/ N\right)^{\gamma}}$$

with the transmissions T_l with l=1,2,...N for all N=128 pixels. The weighting exponent γ can be freely chosen to set the desired degree of genetic pressure.

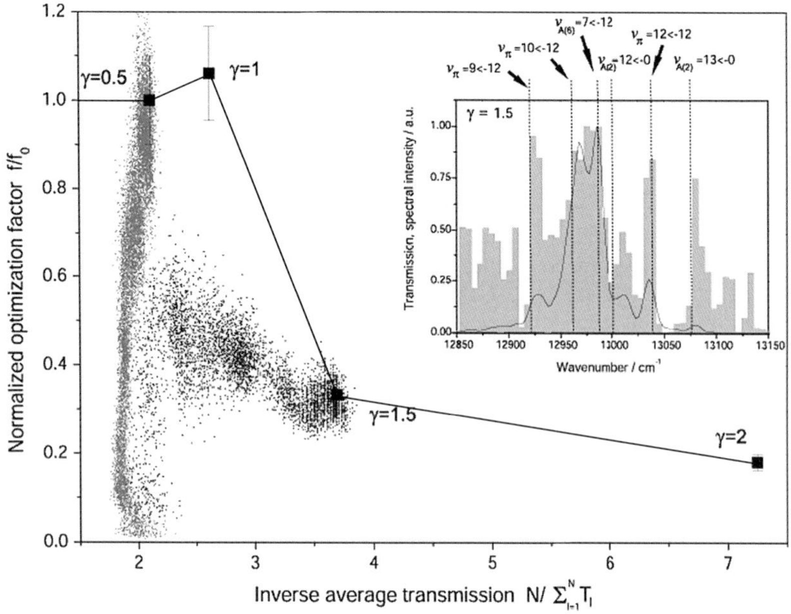

Fig. 4: Correlation plot of the normalized optimization factor f/f_0 versus the inverse average transmission. The solid line depicts the Pareto-optimal front drawn through the optimal solutions for different γ. The optimization courses for γ=0.5 (gray points) and γ=1.5 (black points) are shown by plotting the values of every individual. The inset shows the spectrum (solid line), the pixel transmissions (grey bars) of the optimized pulse for γ=1.5, and the cross-correlation trace (inset). Vibronic transitions to the first and second excited states are denoted.

The inset of Fig. 4 shows the spectrum (solid line) and the pixel transmissions (grey bars) of the optimized pulse with γ=1.5 for the slightly different center wavelength of 770 nm. It exposes otherwise not visible vibronic transitions to the first and second excited states. This demonstrates the feasibility of control pulse cleaning and shows that otherwise hidden information can be unravelled.

Fig. 4 depicts a correlation plot of the normalized optimization factor f/f_0 and the inverse average transmission. Plotted is the so called Pareto-optimal front (solid line) drawn through the optimal solutions for different γ. As examples, the optimization courses for γ=0.5 (grey points) and γ=1.5 (black points) are shown by plotting the values of every individual. Both optimizations evolve from the lower left to the upper right. While the ion yield increases at almost constant transmission for the optimization with γ=0.5, the optimization for γ=1.5 first tends to increase f/f_0 and then decreases by proceeding to lower average transition which indicates the enhanced efficiency of cleaning at the end of the optimization.

3.4. Isotope selective optimization of $^{23}Na^{39}K/^{23}Na^{41}K$

The learning curves of the phase and amplitude modulated optimization measurements for maximization and minimization of the ^{23}Na^{39}K/^{23}Na^{41}K isotopomer ratio level off at about 100 iterations, providing the optimized pulse shapes. The obtained optimization factor between minimization and maximization amounts to 10.4. For comparison we also performed pure amplitude and pure phase optimization for maximization and minimization, respectively, and achieved smaller total optimization factors of 2 and 1.6, respectively. This indicates that combined phase and amplitude modulation is necessary for distinct selectivity in this case. Compared to K_2, where the spectral components were most decisive, for NaK, both phase and amplitude have almost the same significance. This distinction is probably due to different involved excited states with the particular isotope shifts of their vibrational levels.

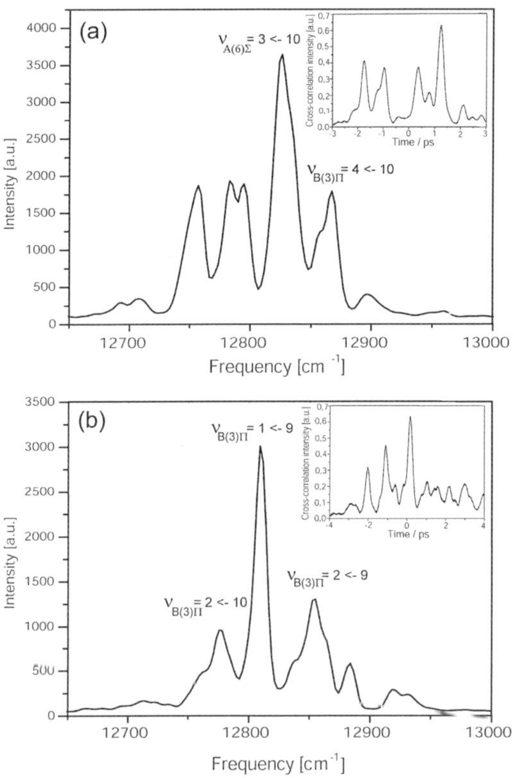

Fig. 5: Optimized pulse spectra for maximization (a) and minimization (b) of the isotope ratio ^{23}Na^{39}K/^{23}Na^{41}K. Transitions between particular vibrational energy levels of the different electronic states X(1)$^1\Sigma^+$, A(2)$^1\Sigma^+$, B(3)$^1\Pi$, and A(6)$^1\Sigma^+$ of the ^{23}Na^{39}K isotope in (a) and of the ^{23}Na^{41}K isotope in (b) are denoted. The insets exhibit the corresponding cross-correlation traces for maximization and minimization, respectively.

Fig. 5 depicts the optimized pulse spectra for maximization (a) and minimization (b) of the isotope ratio ^{23}Na^{39}K/^{23}Na^{41}K. Vibronic transitions between the different electronic states X(1)$^1\Sigma^+$, A(2)$^1\Sigma^+$, B(3)$^1\Pi$, and A(6)$^1\Sigma^+$ of the ^{23}Na^{39}K isotopomer in (a) and of the ^{23}Na^{41}K isotopomer in (b) the are denoted. The frequency shift between the two NaK isotopomers, visible in the pulse spectra, is utilized for optimized selection. Yet, this is not the only effect for enhanced selection since phase modulation contributes as well with a similar amount.

In order to receive the full description also in the time domain, cross-correlation traces for maximization and minimization of the isotope ratio by phase and amplitude modulation were recorded (insets in Fig. 5). Until now, we were not able to give a concise explanation of the complex pulse shapes, but some general aspects can be extracted. It is noticeable that the optimized pulse for maximization and minimization reveals several subpulses with almost equal distances in between. Thus, we propose a repeated stepwise excitation for ratio minimization whereby the excitation at the outer turning point of the $A(2)^1\Sigma^+$ state potential is utilized. This interpretation is backed by the fact that the Franck-Condon window via the resonant $B(3)^1\Pi$ state is favorable at this distance. However, for maximization we also find a transition to the vibrational levels in the electronic $A(6)^1\Sigma^+$ state which is occurring favorably at smaller nuclear distances due to resonance conditions. Certainly, further experimental and theoretical investigations are required to better understand the optimization process.

4. Conclusion

Active control by means of feedback optimization using tailored femtosecond laser pulses was carried out in a resonant 3-photonic ionization process for optimizing the $^{23}Na^{39}K$ isotopomer. The acquired optimized pulses were analyzed regarding their nuclear dynamics and a simple model that is based on full quantum-mechanical calculations at similar conditions was proposed. Novel coherent control methods like parametric optimization and control pulse cleaning were applied to receive information about the involved oscillations and the vibronic transitions within the optimized ionization path. When optimizing the isotopomer ratio $^{23}Na^{39}K/^{23}Na^{41}K$, considerable optimization factors were obtained and intrinsic properties of the photo-induced process emerged in the resulting pulse shapes. This latter experiment reveals how the algorithm precisely addresses the vibrational wave functions both spectrally and temporally by coherent interaction with the shaped light field.

Acknowledgement

The DFG supported this work in the frame of the SFB 450. The authors thank Boris Schäfer-Bung, Vlasta Bonačić-Koutecký and Herschel Rabitz for stimulating ideas and discussions.

References

1. R. S. Judson and H. Rabitz, Phys. Rev. Lett. 68 (1992) 1500.

2. A. Assion, T. Baumert, M. Bergt, T. Brixner, B. Kiefer, V. Seyfried, M. Strehle, and G. Gerber, Science 282 (1998) 919.
3. H. Rabitz, R. de Vivie-Riedle, M. Motzkus, and K. Kompa, Science 288 (2000) 824.
4. R. J. Levis, G. M. Menkir, and H. Rabitz, Science 292 (2001) 709.
5. R. Bartels, S. Backus, E. Zeek, L. Misoguti, G. Vdovin, I. P. Christov, M. M. Murmane, and H. C. Kapteyn, Nature 406 (2000) 164.
6. A. Lindinger, C. Lupulescu, M. Plewicki, F. Vetter, A. Merli, S. M. Weber, and L. Wöste, Phys. Rev. Lett. 93 (2004) 33001.
7. S. M. Weber, A. Lindinger, M. Plewicki, C. Lupulescu, F. Vetter, and L. Wöste, Chem. Phys. 306 (2004) 287.
8. T. Hornung, R. Meier, and M. Motzkus, Chem. Phys. Lett. 326 (2000) 445.
9. J. M. Geremia, W. Zhu, and H. Rabitz, J. Chem. Phys. 113 (2000) 10841.
10. Š. Vajda, A. Bartelt, C. Kaposta, T. Leisner, C. Lupulescu, S. Minemoto, P. Rosendo-Francisco, and L. Wöste, Chem. Phys. 267 (2001) 231.
11. L.-E. Berg, M. Beutter, and T. Hansson, Chem. Phys. Lett. 253 (1996) 327.
12. B. Schäfer-Bung, R. Mitric, V. Bonačić-Koutecký, A. Bartelt, C. Lupulescu, A. Lindinger, Š. Vajda, S. M. Weber, and L. Wöste, J. Phys. Chem. A 108 (2004) 4175.

Femtochemistry VII
Fundamental Ultrafast Processes in Chemistry, Physics, and Biology
A.W. Castleman, Jr. and M.L. Kimble

Coherent Control of Isotope Selective Ionization and Fragmentation of Potassium

Franziska Sauer[a], Andrea Merli[a], Mateusz Plewicki[a], Stefan M. Weber[a], Ludger Wöste[a], and Albrecht Lindinger[a]

[a] *Institut für Experimentalphysik, Freie Universität Berlin, Arnimallee 14, 14195 Berlin, Germany*

Abstract

By applying evolution strategies we performed isotope selctive ionization and fragmentation of potassium dimers. The optimized pulse structures revealed the ionization process. For optimizing the fragmentation of potassium dimers to atoms we performed a new two-criterium optimization.

1. Introduction

Application of evolution strategies in a feedback loop for coherent control of molecules has gained considerable interest in recent years. Various optimal control experiments have been performed so far since in 1992 Judson and Rabitz [1] proposed closed loop adaptive feedback experiments. By using an iterative process, the method enables one to find an optimal pulse form which forces the system into the target state, whereby pulse shapers [2] were combined with optimization algorithms.

We report on an approach to apply feedback control for optimizing isotopomer ion ratios and fragmentation by means of shaped fs-laser pulses. Without any prior knowledge of the system, it is possible to significantly enhance and/or decrease the ion ratio of the isotopomers with coherent control. Furthermore, we show an isotope selective fragmentation in a multi-criterion optimization, where the evolution algorithm attempts to fulfill different control aims simultaneously. This is realized through the weighted sums of two goals, the fragment isotope ion ratio and the dimer isotopomer ion ratio.

2. Experimental

Cold K_2 clusters are produced in an adiabatic co-expansion of potassium vapor and argon carrier gas through a nozzle of 70 μm diameter into the vacuum. The stagnation conditions of the molecular beam are chosen to produce K_2 with no larger clusters present. Thereto, the oven temperature is set to about 400 °C and the argon pressure to 2.5 bar. The laser beam is focused onto the cluster beam and the produced ions are extracted into a quadrupole mass spectrometer (QMS) and detected by a secondary electron multiplier. A Ti:sapphire oscillator (Tsunami; Spectra Physics) provides pulses of low energy (weak field regime). Pulses of 120 fs duration (FWHM), a spectral width of $\Delta\lambda = 8$ nm, an energy of 10 nJ per pulse, and a central wavelength between 810 and 833 nm are generated with a repetition rate of 80 MHz, which is sufficiently high to irradiate every molecule. The pulse shaper used for the experiments consists of two liquid crystal arrays (CRI; SLM-256) [2] with a discretization of 128 pixels each, placed in the Fourier plane of a zero dispersion compressor [3], and allows independent and simultaneous phase and amplitude modulation. Our feedback loop experimental design [4] combines the ion detection with the pulse shaper driven by a self-learning optimization algorithm based on evolution strategies. The ion signals for the different isotopes of potassium atoms and/or potassium dimers were acquired sequentially in time steps of about half a second [5] and the desired fitness function is calculated from these data.

3. Results and discussion

3.1. Isotopomer Selective Ionization of Potassium Dimers

To control the selective ionization of potassium dimers we performed maximization and minimization of the isotopomer ratio $^{39,39}K_2/^{39,41}K_2$ at different center wavelengths (810, 820 and 833 nm) in order to learn about the molecular system and the chosen optimal ionization paths. We found surprisingly large factors of about R_{max}/R_{min} = 14, 50, and 140 between minimization and maximization for the center wavelength 810, 820, and 833 nm, respectively. With a transform-limited pulse the regular isotopomer ratio is about $R_n = 6.9$ [5].

To decipher the underlying isotopomer selection process we recorded the pulse spectra and the XFROG traces (Fig. 1). The pulse spectra for minimization and maximization of the isotopomer ratio display several distinct sharp peaks with differing intensities and intensity attenuation to almost zero between the peaks. The peaks in the spectra for the minimization cases are shifted by -11 to -16 cm^{-1} compared to the maximization cases with rising shift at rising frequency. The peaks are situated at positions where the spectral intensity in the maximization cases is low and vice versa, which predominantly accounts for the high isotopomer selectivity. In the maximization case the most pronounced peaks in the spectra can be assigned to vibrational transitions from v''=0 in the electronic ground state to v'=12-19 (11922, 11988, 12049, 12117, 12182, 12249, 12312, 12373 cm^{-1}) of the $A^1\Sigma_u^+$ state in the $^{39,39}K_2$ isotopomer (see scales in Fig. 1). For the minimizing optimizations, the same transitions of the $^{39,41}K_2$ isotopomer are visible in the spectra. Because of the isotopomer shift are the transitions v'=12-19 ← v''=0 of the $^{39,41}K_2$ isotopomer at the following frequencies: 11899, 11973, 12038, 12103, 12168, 12235, 12300, 12359 cm^{-1}.

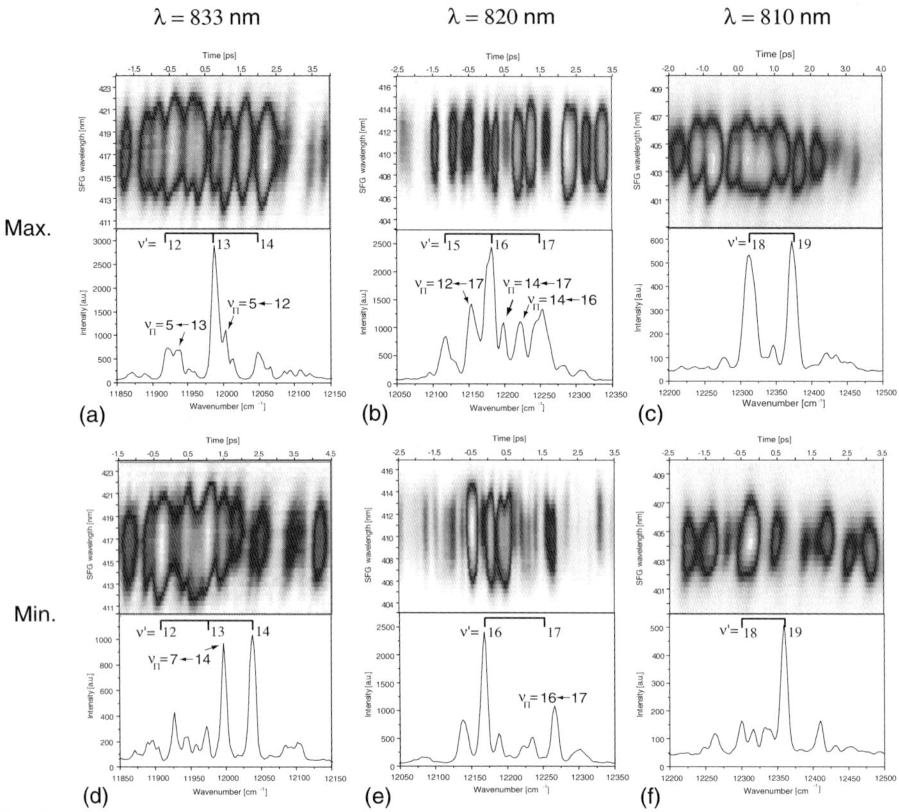

Fig. 1. Optimized Pulse Spectra and XFROG traces for maximization and minimization at three different central wavelengths.

We could assign other peaks in the pulse spectra with transitions from the $A^1\Sigma_u^+$ to the $2^1\Pi_g$ state (marked in Fig. 1). The $2^1\Pi_g$ state thereby serves as a resonant transition state for the ionization step. This transition occurs at the favorable Franck-Condon window at the outer turning point of the wave packet in the $A^1\Sigma_u^+$ state.

The recorded cross-correlation frequency resolved optical gating (XFROG) traces shown in Fig. 1 reveal for maximization and minimization sequences of subpulses. It can be noted that the subpulse distances often amount to multiples of 250 fs corresponding to $n/2 \cdot T_{osc}$ in the $A^1\Sigma_u^+$ state. Thus, we can propose a repeated stepwise excitation to the inner turning point of the $A^1\Sigma_u^+$ potential followed by an excitation at the outer turning. The reason for the reduced optimization factors for shorter wavelengths is not yet clear. As the selection in the first transition from the $X\ ^1\Sigma_g^+$ to the $A^1\Sigma_u^+$ state is expected to be even enhanced at shorter wavelengths due to the larger isotope shifts, the second excitation from the $A^1\Sigma_u^+$ to the $2^1\Pi_g$ state is assumed to be responsible for the decreased total selectivity.

3.2. Fragmentation Selective Control

As continuative experiment we performed a two-criterium optimization in order to control the fragmentation and to learn more about the specific process. The fragment $I(^{39}K^+)/I(^{41}K^+)$ and the dimer ion ratio $I(^{39,39}K_2^+)/I(^{39,41}K_2^+)$ are chosen as the two goals of the optimization. We realize this through the weighted sums by employing the new fitness function

$$F = (1-\beta)\,[I(^{39,39}K_2^+)/I(^{39,41}K_2^+)] + \beta\,[I(^{39}K^+)/I(^{41}K^+)]$$

where β is the weighting factor. The Fitness function has three different regimes. For $\beta < 0$, the isotopomer ion ratio is optimized at the cost of the fragmentation ion ratio, for $0 < \beta < 1$ both goals are optimized simultaneously, and for $\beta > 1$ the fragmentation is favored at the cost of dimer isotopomer ionization.

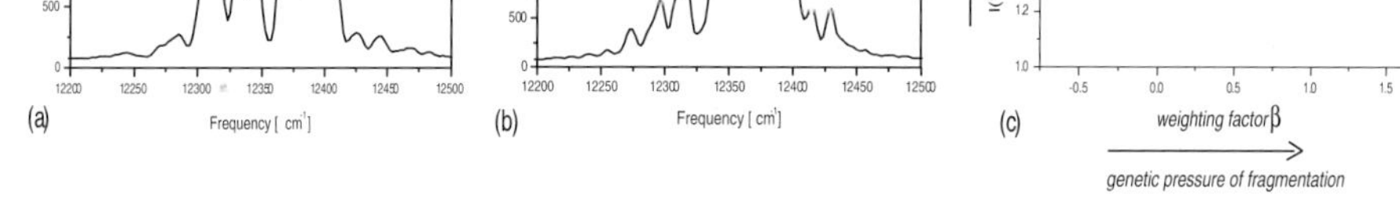

Fig. 2. Spectra and XFROG traces of the optimized pulses for β = -0.5 (a) and β = 1.5 (b). (c) shows the correlation diagram between the ion ratios $I(^{39}K^+)/I(^{41}K^+)$and $I(^{39,39}K_2^+)/I(^{39,41}K_2^+)$ and the weighting factor β.

We performed several optimizations with phase and amplitude modulation for different values of β at 810 nm center wavelength (chosen because of its highest fragmentation factor [6]), whereby the fitness function F was maximized. Fig. 2 (a) and (b) show the optimized spectra and XFROG traces for the most extreme cases of β = -0.5 and β = 1.5, respectively. Most peak positions are similar for both spectra; however the peaks which do not belong to the first $A^1\Sigma_u^+ \leftarrow X^1\Sigma_g^+$ transition are tentatively stronger for β = 1.5 because of the aim of the algorithm to produce as many $^{39}K^+$ fragments as possible. A tendency to increase the peaks at higher frequencies can also be made out, which leads to an increase of the fragmentation in the ionic state.

Fig. 2 (c) shows the ratio between the ion isotope ratio and the ion isotopomer ratio as a function of β. An increase of the ratio with increasing β is visible. This demonstrates that the processes are influenced as intended by the choice of β. Apparently, three-photonic ionization is sufficient for efficient fragmentation.

4. Conclusion

We show coherently controlled isotope selective ionization and fragmentation of potassium dimers. At different central wavelengths a considerably large enhancement of one isotopomer compared to the other was obtained. From the optimized spectra, we were able to extract information about the ionization paths. The demonstration of a multi-optimization of physically relevant observables can also be applied to other goals in differing systems, with the aim of yielding insight into the relation between these goals.

Acknowledgements

The presented work is supported by the Deutsche Forschungsgemeinschaft (SfB 450). Franziska Sauer thanks the Studienstiftung des deutschen Volkes for financial support and Andrea Merli acknowlegdes the Cusanuswerk.

References

1. R. S. Judson and H. Rabitz, Phys. Rev. Lett. 68 (1992) 1500.
2. A. M. Weiner, D. E. Leaird, J. S. Patel, and J. R. Wullert, IEEE J. Quantum Electron. 28 (1992) 908.
3. O. E. Martinez, IEEE J. Quantum. Eletron. 23 (1987) 59.
4. A. Bartelt, S. Minemoto, C. Lupulescu, S. Vajda, and L. Wöste, Eur. Phys. J. D 16 (2001) 127.
5. A. Lindinger, C. Lupulescu, F. Vetter, M. Plewicki, S. M. Weber, A. Merli, and L. Wöste, J. Chem. Phys. 122 (2005) 024312.
6. A. Lindinger, A. Merli, M. Plewicki, F. Vetter, S. M. Weber, and L. Wöste, Chem. Phys. Lett. 413 (2005) 315.

Femtochemistry VII
Fundamental Ultrafast Processes in Chemistry, Physics, and Biology
A.W. Castleman, Jr. and M.L. Kimble

Understanding Learning Control in a Molecular Family

David Cardoza[a], Mark Baertschy[b], Thomas Weinacht[a]

[a] *Stony Brook University, Department of Physics, Stony Brook, NY 11794-3800, USA*
[b] *University of Colorado, Denver, Department of Physics, Denver, CO 80217, USA*

1. Introduction

Recent experiments have demonstrated coherent control over a wide variety of atomic and molecular systems using shaped ultrafast laser pulses. In many cases, where the system Hamiltonian is not known well enough to predict optimal control fields *a priori*, learning algorithms have been used [1]. While learning algorithms have been successful at discovering optimal laser pulses for control, it is generally difficult to understand the physical mechanisms underlying control. Although a few pioneering experiments have been able to uncover the control mechanism exploited by an optimal pulse, the lack of systematic techniques to gain insight into control mechanisms and to develop predictive models for control remains a significant barrier to achieving the ultimate goal of laser selective chemistry .

Using a combination of learning algorithm modifications, pump–probe spectroscopy, *ab initio* molecular structure calculations, and quasi-static molecular ionization (QSMI) calculations, we are able to decipher the mechanism underlying control in 1,1,1-trifluoroacetone (CH_3COCF_3, TFA). The insight gained from this analysis allows us to predict control in similar molecules such as tri-deuterated acetone (CH_3COCD_3, TDA) and 1,1,1-trichloroacetone (CH_3COCCl_3, TCA). Extending the control experiments to include dibromo-trifluoroacetone ($CHBr_3COCF_3$, Br2TFA) yields an interesting variation on the control mechanism which relies on both electronic and vibrational coherences.

2. Experimental

We use an amplified Titanium:Sapphire laser system that produces 30 fs pulses with 1 mJ of energy at a repetition rate of 1 kHz. The laser pulses are shaped in an acousto-optic modulator pulse shaper controlled by a Genetic algorithm (GA).

Shaped laser pulses were directed into a vacuum chamber, where they were focused into an effusive molecular beam. Ionic molecular fragments formed in the laser focus were directed into a time-of-flight mass (TOFM) spectrometer.

A recent publication discusses the details of feedback experiments where control was demonstrated over the CF_3^+/CH_3^+ ratio [2]. These publications also present a model for the control mechanism. Briefly, the model consists of three parts: 1) Launching of a vibrational wave packet on a dissociative potential energy surface (PES) through ionization, 2) Wave packet evolution, and 3) Enhanced ionization at a critical distance of 3.6 Å, leading to increased CF_3^+ production. We arrived at this model through a combination of the optimal pulse shapes for control, pump-probe measurements of the fragment yield, molecular structure calculations, and QSMI calculations.

3. Results and Discussion

We made several tests of the control mechanism, but the most substantial one was to make and test predictions for control in similar molecules, TCA and TDA [3]. Structure calculations performed for these molecules indicated that TDA parent ion has a substantial barrier to dissociation, and the wave packet launched through ionization never reaches the C-CD_3 bond length required for enhanced ionization. Similar calculations for TCA indicated that the molecular ion is not stable to dissociation along the C-CCl_3 bond. These calculations led to the prediction that control should be possible in TCA, with the same mechanism as in TFA, while there should be no control in TDA. Figure 1(a) shows control data for TFA, TCA, and TDA. The white bars show the CX_3^+/CH_3^+ for an optimally shaped pulse found from our closed loop learning control experiment. The black bars show this ratio for a transform limited (unshaped) laser pulse.

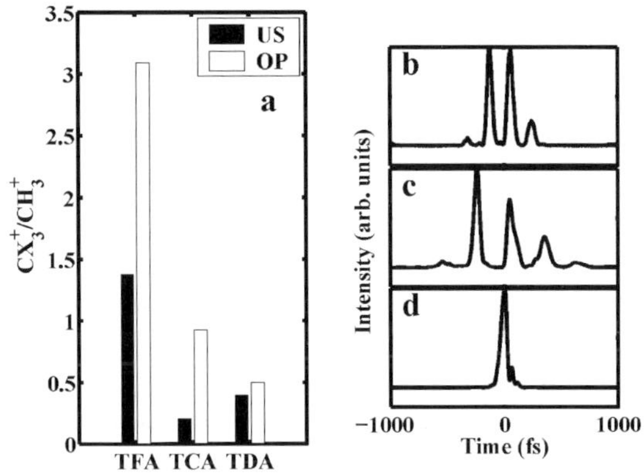

Fig.1: a) Control ratios for TFA, TCA, and TDA, with black being the CX_3^+/CH_3^+ ratio for an unshaped pulse and white the same ratio for an optimally shaped pulse. On the right are optimally shaped pulses for b)TFA, c)TCA, and d)TDA.

The figure demonstrates that control was possible in TCA but not in TDA. Figure 1(b), (c), and (d) show intensity profiles for optimally shaped laser pulses for control in TFA, TCA, and TDA, respectively. While the optimal pulses for TFA and TCA are similar, the optimal pulse for TDA is clearly qualitatively different. The optimal pulses for TFA and TCA are consistent with the control model. The spacing between the main subpulses in the optimal pulses is equal to the time that we calculate the dissociative wave packet takes to reach the C-CX_3 bond length for enhanced ionization (170 fs for TFA and ~350 fs for TCA). In the case of the TFA, the subpulse separation for the two largest pulses was consistently around 170 fs for many different optimizations.

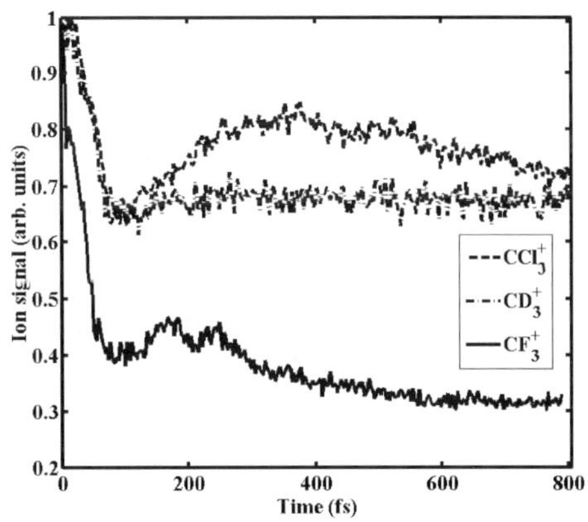

Fig. 2: Pump-Probe ion signals for CCl_3^+ (dash), CD_3^+ (dash-dot), and CF_3^+ (solid).

However, for TCA, the subpulse separation over many optimizations had some spread. These times and variations in optimal subpulse spacings are consistent with the calculated shape of the dissociative PES's.

Pump-probe measurements of the fragment yields for all three molecules further confirmed the control model. Figure 2 shows the pump-probe ion signal for CCl_3^+ (dash), CD_3^+ (dash-dot), and CF_3^+ (solid). The breadth of the ionization enhancement for the CCl_3^+ is consistent with a broad wave packet reaching the enhanced ionization distance and corroborates our control data. There are no modulations in the CD_3^+ signal, which is also consistent with our interpretation. We attribute the double peak in the CF_3^+ as due to bifurcation of the wave packet on its dissociative potential along the C-CF_3 coordinate.

The electronegativity imbalance in the halogenated acetones discussed above is responsible for much of the systematic behavior we observe. In light of this, we extend this molecular family by considering a molecule with an intermediate asymmetry in electronegativity, $CHBr_2COCF_3$ (Br_2TFA). Control experiments in this case yielded results similar to TFA and TCA, with a factor of roughly six in the $CF_3^+/CHBr_2^+$ ratio comparing optimal pulses for maximizing either methyl group ion. An important difference was measured in the corresponding pump-probe measurements. Figure 3 shows the pump-probe ion signals for several Br2TFA fragments. What is striking about this data is that the $CHBr2^+$ and the $CHBr_2CO^+$ display modulations which are anti-correlated with the CF_3^+ signal. No fragments other than CX_3^+ showed any modulation in TFA and TCA. The pump-probe results point to a charge transfer mechanism rather than enhanced ionization. This mechanism is discussed in detail in a forthcoming publication.

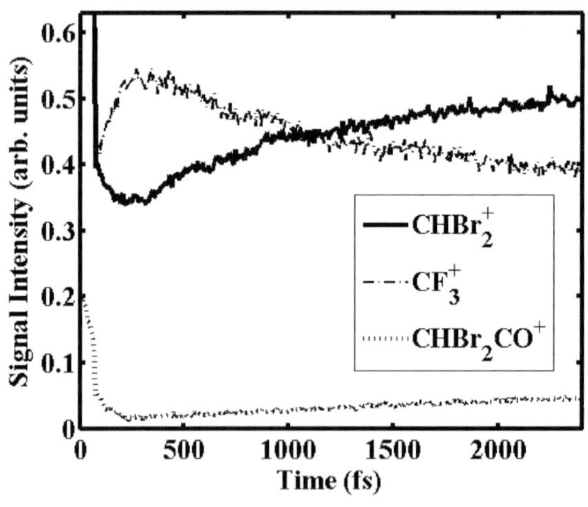

Fig. 3 Pump-probe ion signals for fragments of Br_2TFA.

We have deciphered the physical mechanism underlying control in a molecular family. This result motivates extending the idea of homologous series to chemistry driven by photonic reagents.

References

1. R. Judson and H. Rabitz, Phys. Rev. Lett. 68 (1992) 1500.
2. D. Cardoza, M. Baertschy, and T. Weinacht, J. Chem. Phys. 123 (2005) 074315.
3. D. Cardoza, M. Baertschy, and T. Weinacht, Chem. Phys. Lett. 411 (2005) 311.

Femtochemistry VII
Fundamental Ultrafast Processes in Chemistry, Physics, and Biology
A.W. Castleman, Jr. and M.L. Kimble

Controlling the dissociative ionization of DCl by femtosecond laser pulses

Hans Georg Breunig, Alexandra Lauer, Karl-Michael Weitzel

Philipps Universität Marburg, Fachbereich Chemie, Hans-Meerweinstr., D – 35032 Marburg, Germany

1. Introduction

The enormous technical development of recent years has enabled not only valuable progress in understanding chemical dynamics but has also triggered efforts for controlling chemical reactions [1,2]. Understanding such control schemes ultimately requires knowledge of all molecular quantum states involved. In the current work, we investigate deuterium chloride ions (DCl^+) for which a wealth of theoretical electronic structure data [3] and experimental spectroscopic data [4] is available.

We have recently investigated the prospect of controlling the photofragmentation of deuterium chloride ions (DCl^+) via strong ultra short IR laser pulses by numerical solution of coupled Schrödinger equations [5]. The calculations provide evidence that the ratio of product ion yields, i.e. Cl^+ versus D^+, can be manipulated by appropriate choice of laser pulse parameters, in particular central laser wavelength, pulse duration, intensity, and chirp. The analysis of time dependent populations reveals competition between intra- and inter-electronic state excitation, enabling the understanding of quantum control at the molecular level. In the current work we present first experimental results for the dissociative ionization of DCl in the near IR.

2. Experimental Details

DCl gas is introduced into a high vacuum chamber as a thermal sample at typical pressures of 10^{-6} mbar. Ions are formed by the interaction of the gas with femtosecond laser pulses in the source of an ion spectrometer. All ions are mass analyzed in a two stage linear time of flight mass spectrometer (TOF-MS) and detected by micro channel plates. Typical acceleration fields are on the order of a few hundred V/cm (e.g. 156 V/cm for Figs. 1 and 2, and 313 V/cm for Figs. 3 and 4). The length of the drift tube is 30 cm.

Femtosecond laser pulses are generated by a 1 kHz multi-pass amplifier which is seeded by a Ti:sapphire oscillator operating at a central frequency of about 806 nm. Available pulse energies are in the range up to 1.5 mJ. All data shown in this work have been recorded with 500 μJ/pulse. The pulse duration ranges from 45 fs (chirp free) to 150 fs FWHM. Frequency chirp is introduced into the laser pulses by changing the length of the compressor stage in the amplifier. To first approximation, this gives rise to a linear chirp. The value of the chirp is determined from measured autocorrelation times using the relation [6]

$$\tau^2 = \tau_o{}^2 + (8\alpha \cdot \ln 2 / \tau_o)^2 \ ,$$

where α is the chirp parameter. The chirp has been calibrated and its sign derived by introducing a 100 mm glass rod (SF14) into the optical path. All chirp information given in the results section refers to the actual value inside the experimental chamber assuming linear dispersion of quartz windows. The laser light was linearly polarized with the polarization plane either parallel to the TOF spectrometer axis or perpendicular. The laser light is focussed into the ion spectrometer by a f = 300 mm lens. Time of flight spectra are obtained by averaging typically 1000 sweeps.

3. Results

The dissociative ionization of DCl has been investigated employing fs pulses at 806 nm as a function of laser intensity, polarization, and chirp.

A typical TOF spectrum recorded at pulse energy of 500 μJ is shown in Fig. 1. Under these conditions, the formation of parent ions is by far dominating. We also observe the formation of fragment ions D^+ and Cl^+ with a total fraction of about 5%. The fraction of doubly charged ions (Cl^{++} and to a lesser extend also DCl^{++}) is below 2%. With increasing intensity, the fraction of D^+ and Cl^+, but also that of doubly charged ions, increases steeply. We will come back to this point below. Fig. 2 shows an enlarged view of the TOF distribution of D^+ ions again at 500 μJ/pulse. In principle the width of TOF distributions reflects the kinetic energy of the particles. However in the current experiment the actual width is subject to space charge effects. This effect scales with the number of ions formed in a given volume. For parallel polarization, a modest splitting is observed. This is most likely due to the anisotropy of the process, which favors formation of fragment ions with velocity vectors aligned parallel to the polarization plane of the laser field. For perpendicular polarization, the TOF distribution is slightly narrower, again due to the anisotropy. In this case the small dip at the peak center is due to angular discrimination of the spectrometer.

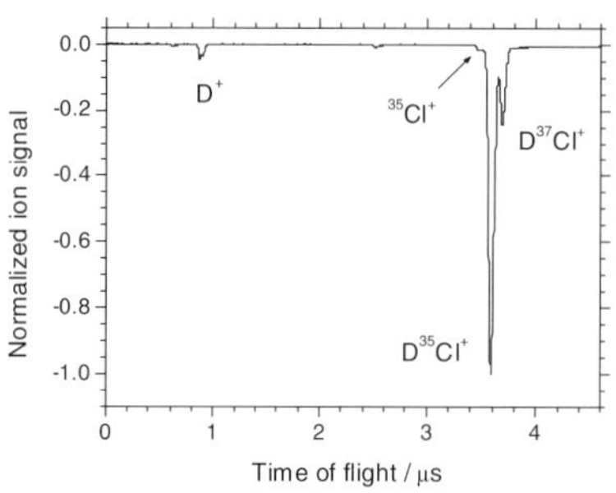

Fig. 1 TOF-MS spectrum of DCl (parallel polarization)

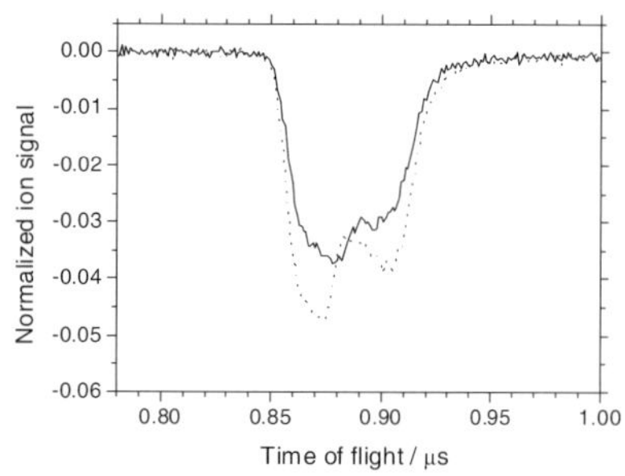

Fig. 2 D^+ ion TOF distribution for perpendicular (solid) and parallel polarization (dots).

At the highest pulse energies employed (1.2 mJ/pulse), the D^+ TOF distribution exhibits distinctly separated peaks for forward and backward scattered signal (parallel polarization). There, the kinetic energy of D^+ ions is significantly higher than at 500 μJ/pulse. In part, this could be caused by space charge effects due to the concomitantly increased ion signal. From the difference between the TOF distribution of parent and fragment ions, however, we conclude that a major fraction is due to the release of kinetic energy. This, in turn, may well be caused by 'Coulomb explosion' of doubly charged DCl^{++}. The 500 μJ/pulse data, shown in this work, are believed to be basically free from the influence of doubly charged ions.

Alternatively, singly charged atomic ions ($D^+$and Cl^+) could also originate from the dissociation of neutral DCl into D and Cl with subsequent ionization of the atoms. Given the rather short laser pulses (typically less than 100 fs) and the time inherently required for any dissociation process, we believe that the latter process cannot be the origin of the D^+ and Cl^+ ion signal. For D_2^+ e.g. Corkum *et al.* have shown that the internuclear distance changes at a rate of 0.1 Å / fs under similar conditions [7]. Reaching a distance of 10 Å, where the bond can be considered broken, thus requires about 100 fs.

This leaves us with the third possibility of single ionization of DCl with subsequent (photo)dissociation of the DCl^+, which we believe is the dominant source of our D^+ and Cl^+ signal at 500 μJ/pulse, in line with previous reports [8]. In the following, we discuss the influence of the chirp on the fragmentation yields. Fig. 3 and 4 show the TOF spectra of DCl for differently chirped laser pulses with parallel polarization. All spectra are normalized to the $D^{35}Cl^+$ peak. Evidently the ratio of ion yields D^+ / Cl^+ is similar for a modest down chirp (Fig. 3) and an up chirp of +459 fs^2 (Fig. 4). However, the fragmentation ratio is a factor of 2 smaller for a down chirp of -468 fs^2. Note that the two spectra shown in Fig. 4 refer to identical intensity. More specifically, the yield of D^+ is strongly disfavored by the down chirp.

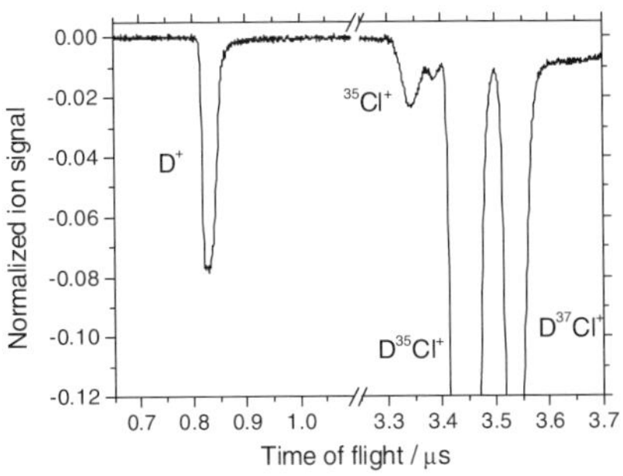

Fig. 3 TOF-MS spectra of DCl recorded for small negative chirp (-100fs^2).

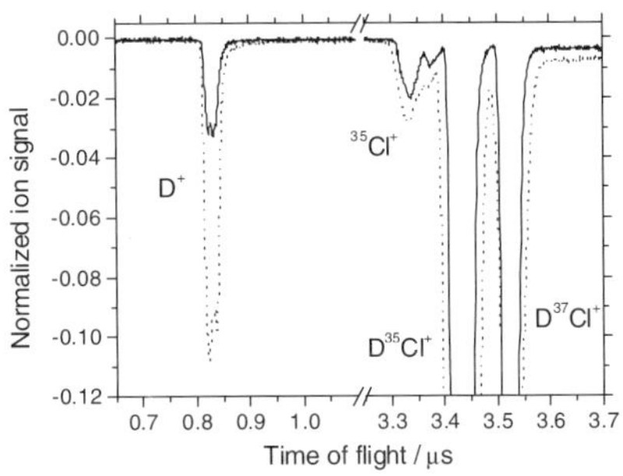

Fig. 4 TOF-MS spectra of DCl recorded for chirp +459fs^2 (dots) and -468fs^2 (solid).

Assuming that the formation of D^+ is dominated by fragmentation of DCl^+ ions, this would be compatible with the situation where a down chirp tends to favor vibrational excitation of molecules within the anharmonic manifold of vibrational states. In the ion ground state, this would lead to the formation of Cl^+ [4]. The effective dissociation of neutral molecules by down chirped mid IR laser pulses has been reported for a number of cases [9]. An up chirped laser pulse on the other hand favors electronic excitation of the molecular ion ultimately leading to the formation of D^+ fragments [5]. In conclusion, we have shown that the branching ratio of competing reaction channels can be controlled by the frequency chirp of the laser pulse.

We thank Prof. M. Motzkus for the loan of a SF14 rod.

References

1. M. Shapiro, P. Brumer, Principles of Quantum Control of Molecular Processes, Wiley, Hoboken, New Jersey 2003.
2. S. A. Rice, M. Zhao, Optimal Control of Molecular Dynamics, Wiley-VCH, Weinheim 2001.
3. A. D. Pradhan, K. P. Kirby, and A. Dalgarno, J. Chem. Phys. 95 (1991) 9009.
4. M. Michel, M. V. Korolkov, and K. M. Weitzel, J. Phys. Chem. A 108 (2004) 9924.
5. M. V. Korolkov and K. M. Weitzel, J. Chem. Phys. 123 (2005) in press.
6. P. Balling, D. J. Maas, and L. D. Noordam, Phys. Rev. A: Atomic, Molecular, and Optical Physics 90 (1994) 4276.
7. H. Niikura, F. Legare, R. Hasbani, M. Y. Ivanov, D. M. Villeneuve, and P. B. Corkum, Nature 421 (2003) 826.
8. A. Conjusteau, A. D. Bandrauk, and P. B. Corkum, J. Chem. Phys. 106 (1997) 9095.

9. see e.g. T. Witte, T. Hornung, L. Windhorn, D. Proch, R. Vivie-Riedle, M. Motzkus, and K. L. Kompa, J. Chem. Phys. 118 (2003) 2021.

Femtochemistry VII
Fundamental Ultrafast Processes in Chemistry, Physics, and Biology
A.W. Castleman, Jr. and M.L. Kimble

Spectrum Control in Femtosecond Time-Resolved Spectroscopy

J. Konradi, A.V. Scaria, A. K. Singh, and A. Materny*

International University Bremen, Campus Ring 1, 28759 Bremen, Germany
e-mail: a.materny@iu-bremen.de

1. Introduction

Femtosecond pulses are spectrally broad, and therefore, a selective excitation of specific vibrational modes of molecules is usually not achieved in ultrafast spectroscopy. Coherent control techniques have been applied over the last few years to a variety of systems in order to influence the multi-mode dynamics. The shaping of the femtosecond laser pulses by variation of the phases and/or amplitudes of their different frequency components proved to have considerable influence on the excitation of the molecular modes. The fact that a prediction of the pulse shape required for a desired control of molecular or reaction dynamics becomes nearly impossible for complex systems (including conditions imposed by the experimental setup) leads to the introduction of self-learning optimization techniques [1]. Here, the experimental result serves as a feedback signal for an optimization algorithm, *e.g.* based on evolutionary strategies.

Femtosecond time-resolved coherent anti-Stokes Raman scattering (fs-CARS) is a powerful technique to obtain information about the coherence of vibrational excitations. The anti-Stokes signal is poorly resolved due to the spectrally broad excitation and probing inherent to the fs-CARS. Recently, we could demonstrate that the introduction of a small chirp for one of the exciting laser pulses (pump or Stokes, both resonant with the absorption of the molecules) can result in drastic changes of the CARS spectra and transients [2]. While transform-limited laser pulses excited a variety of vibrational modes in polydiacetylene molecules, a small linear chirp of the Stokes laser pulse resulted

---

* Author to whom correspondence should be directed.

in a dominant contribution of only one mode in the transient signal. In order to perform a coherent control of the mode excitation, we decided to apply the self-learning loop technique. The results were most encouraging [3]. Each of the modes within the spectral excitation range could be selectively enhanced or suppressed (relative intensities) by only varying the phases of either pump or Stokes laser pulses. The mode selection persisted in the CARS spectrum for the full coherence lifetime.

The observed spectrum control is of considerable interest. On the one hand, the selective excitation of specific molecular modes is an important stepping stone towards the realization of the control of chemical reactions. On the other hand, the targeted suppression of Raman lines in the CARS spectrum would have considerable advantages for frequency-resolved spectroscopy. Here, one would still be able to obtain mode-specific information with femtosecond time resolution.

In the following, we shortly introduce two experiments, where we have tried to answer mainly three questions: (i) Is this control relying on the involvement of dynamics in excited electronic states? (ii) Is the control also involving intermolecular dynamics? (iii) Is it perhaps an interference effect within the coherent signal which causes the spectral changes, *i.e.*, do we really control the mode excitation?

2. Experimental

For the experiments, the output of a Clark-MXR CPA-2010 femtosecond laser in combination with two OPAs was used to create the three femtosecond pulses required for the four-wave mixing process (pulse duration ≈ 80 fs; for Stokes laser wavelengths, see Fig. 1). The Stokes laser was phase modulated in a feedback-controlled loop by means of a folded 4*f*-arrangement of two gratings and a liquid crystal spatial light modulator (SLM-640/d, Jenoptik). The pixels were addressed using a polynomial representation to reduce the number of free parameters. The pulses interacted with the sample in a folded BoxCARS geometry. The anti-Stokes signal was measured as a function of probe delay time by means of a single monochromator equipped with a CCD camera for broadband detection. During the mode optimization, the nonresonant background in the CARS signal was permanently removed by introducing a fixed delay time (500-1200 fs) between the probe pulse and the stimulated Raman excitation process.

3. Results and Discussion

In the first experiment, fs-CARS was applied to liquids consisting of one molecular species. In the following, results from optimizations of the nonlinear spectrum obtained from toluene are discussed. Toluene does not absorb in the visible spectral region. Therefore, the CARS process does not involve any excited electronic state. Centering the stimulated Raman excitation at approximately 1090 cm^{-1} results in an excitation of a variety of vibrational modes (for details, see *e.g.* reference 4). In the CARS spectrum three bands can be detected, which, for simplicity, are labeled by ① ($\approx$1000 cm^{-1}), ② ($\approx$800 cm^{-1}), and ③ ($\approx$1200 cm^{-1}) (left part of Fig. 1). The fs-CARS spectrum obtained for transform-limited laser pulses is dominated by band ① and only shows weak shoulders at the positions of lines ② and ③ (see panel A of Fig. 1). Performing an optimization of the Stokes phase shape resulted in dramatic changes of the anti-Stokes spectra. Here, the Raman band ratio served as the fitness function for the evolutionary algorithm used to control the spectral phase of the Stokes pulse.

Panels B and C of Fig. 1 depict the spectra obtained after optimization of bands ② and ③, respectively. In both cases, the formerly very weak bands clearly dominate the spectra. The corresponding Stokes laser pulses are shown next to the spectra. They were measured using a transient grating frequency resolved optical gating (TG-FROG) setup. The spectral selection was observed to persist for the full coherence lifetime (not shown here).

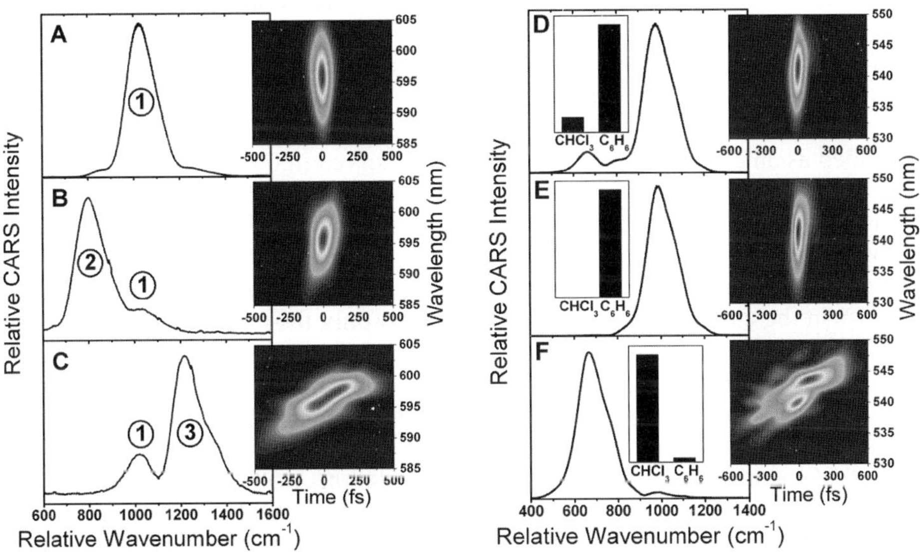

Figure 1: fs-CARS spectra taken with transform-limited (A,D) and phase-shaped (B,C,E, and F) Stokes pulses from toluene (left side) and a mixture of chloroform and benzene (right side). For each spectrum the TG-FROG trace of the associated Stokes laser pulse is shown.

The second experiment introduced here demonstrated the control of CARS spectra obtained from mixtures of different molecules. As an example, in the right part of Fig. 1 the result for a binary mixture of benzene and chloroform is shown. Again, the CARS excitation is not resonant with the molecular electronic transitions. The difference between pump and Stokes laser wavenumbers was set to 800 cm^{-1}. The spectrum displayed in panel D of Fig. 1 was obtained with transform-limited laser pulses. It shows a strong line at 992 cm^{-1}, which can be assigned to the ring breathing mode of benzene. The weaker line at 670 cm^{-1} belongs to chloroform. Again, an optimization is possible, which is shown in panels E and F. Here, a selection of the benzene and the chloroform bands was successfully performed, respectively (see also bar diagrams comparing the CARS intensities of both molecules). The FROG traces of the corresponding Stokes laser pulses are given for each spectrum. Like for toluene, also for the mixture, the spectral selection persists during the observable lifetime of the signal. Additionally, the optimized laser pulses were also applied to the pure substances. In each case, the pulses suppressing the respective substance contribution in the mixture also resulted in a drastic reduction of the signal intensity in the liquid containing only these molecules.

A careful analysis of the optimization results was performed. As a short summary of our findings, the following points should be emphasized. (i) It is obvious that no excited electronic states have to be involved for the control of the mode excitation. (ii) The experiments performed on mixtures show that no intermolecular process contributes to the control mechanism. (iii) The application of the pulses optimized for the spectrum control in mixtures to pure substances provides additional information. These experiments demonstrate that interference effects of different components of the coherent signal can be excluded as an explanation for the changes observed in the spectra. (iv) Some of the known mechanisms of mode control (impulsive stimulated Raman, π phase steps, variation of the excitation frequency by temporal shift of chirped pulses; for a discussion, see reference 4) do not apply as exclusive interpretation of our experimental results, but they clearly contribute to varying degrees.

In the electronically non-resonant case, only those components of pump and Stokes lasers play a role, which are within the temporal overlap range of both pulses. A simple explanation for a spectrum control therefore would be a temporal shift of "unfavorable" pulse components out of the interaction window. By calculating the excitation spectrum using the measured pulse fields as well as by directly obtaining the cross-correlations from measurements of the non-resonant CARS spectra from quartz plates, we could rule out this trivial solution as the sole mechanism. Clearly, the mechanism involves a control of the ground state dynamics, which also explains the molecule specificity.

Since the experimental results are of considerable interest for applications like the mode-selective investigation of molecular dynamics or background suppression in time and frequency domain spectroscopy, further experiments are underway in our laboratories. Applying amplitude modulation of the pulse spectrum will give us further insight into the mechanism of this spectrum control.

References

1. C. J. Bardeen, V. V. Yakovlev, K. R. Wilson, S. D. Carpenter, P. M. Weber, and W. S. Warren, Chem. Phys. Lett. 280 (1997) 151.
2. T. Chen, A. Vierheilig, P. Waltner, M. Heid, W. Kiefer, and A. Materny, Chem. Phys. Lett. 326 (2000) 375.
3. D. Zeidler, S. Frey, W. Wohlleben, M. Motzkus, F. Busch, T. Chen, W. Kiefer, and A. Materny, J. Chem. Phys. 116 (2002) 5231.
4. J. Konradi, A. K. Singh, and A. Materny, Phys. Chem. Chem. Phys. 7 (2005) 3574.

Femtochemistry VII
Fundamental Ultrafast Processes in Chemistry, Physics, and Biology
A.W. Castleman, Jr. and M.L. Kimble

Pulse shaping control of spatially aligned rotational wavepackets of N_2 and O_2

Rebeca de Nalda[a], Christian Horn[b], Marc Krug[b], Florian Ausfelder[a], Matthias Wollenhaupt[b], Luis Banares[a] and Thomas Baumert[b]

[a]*Dpto. Química Física I, Facultad de Ciencias Químicas, Universidad Complutense de Madrid, Avda. Complutense s/n, 28040 Madrid, Spain.*
[b]*Experimentalphysik III, Universität Kassel, Heinrich-Plett-Strasse 40, 34132 Kassel, Germany.*

1. Abstract

Control on the structure of the laser-induced field-free alignment revivals in diatomic molecules at room temperature is demonstrated. The control is achieved by temporally shaping the laser field, which modifies the excitation of the rotational Raman wave packets and their subsequent evolution.

2. Introduction

Laser-induced molecular alignment is a powerful tool to create spatially aligned distributions in an otherwise randomly oriented, gaseous sample of molecules [1]. In the regime called "impulsive alignment", it is possible to obtain field-free molecular alignment at the times when the "revivals" occur. In those, diatomic molecules oscillate between the situation of alignment and *anti*alignment in tens to hundreds of femtoseconds - faster than their rotational period. Alignment refers to a situation where the internuclear axis is preferentially aligned along the direction of the polarization vector of the electric field of the laser. In *anti*alignment the internuclear axis is preferentially aligned on the plane perpendicular to the polarization vector.

The phenomenon of laser-induced molecular alignment is currently receiving a great deal of attention, with most recent progress concentrating on the effort to control the dynamics of the rotational wave packets by use of tailored laser fields or sequences of pulses. Despite a considerable body of theoretical work on the subject [2-4], experimental data are scarce. Two recent papers [5,6] have reported the enhancement of the degree of alignment by applying pairs of pulses. Renard *et al.* [7] applied a pulse shaping technique that allowed them to control the symmetry of the wave packet and thus the rephasing period.

In this contribution, we will show that it is possible to control the structure of the revivals using pulses with a modified temporal structure. We show control on the structure of the revivals in N_2 and O_2 at room temperature, for instance, enhancement of alignment at the expense of antialignment or vice versa. The results are discussed with reference to theoretical simulations that imply solving the time-dependent Schrödinger equation for a quantum rotor.

3. Simulations

In the non-resonant case for linearly polarized light interacting with a linear molecule, the effective Hamiltonian is given by [1,8,9]

$$H(t)=BJ^2-1/2\,\varepsilon^2(t)\,[(\alpha_{\parallel}-\alpha_{\perp})\cos^2\theta+\alpha_{\perp}] \tag{1}$$

where ε is the electric field, θ is the angle between the molecular axis and the field polarization vector and $\alpha_{\parallel}$, $\alpha_{\perp}$ are the components of the polarizability tensor parallel and perpendicular to the molecular axis.

The signal stems from the birefringence of the aligned molecules, which induces some ellipticity on the initially linearly polarized probe pulse and is proportional to

$$I_{signal} \propto [\ \langle\cos^2\theta(t)\rangle - 1/3 + C\ ]^2 \tag{2}$$

where C is due to a residual elliptical polarization of the probe pulse. If C=0, the sign of $\langle\cos^2\theta(t)\rangle$ - 1/3 is lost (homodyne detection). For $C \gg \langle\cos^2\theta(t)\rangle - 1/3$, the sign can be retrieved from the data (heterodyne detection) [10].

The calculation is performed in the basis of field-free stationary rotational states $|\,J, M\,\rangle$ of the molecule. Only the matrix elements $\langle J, M\,|\cos^2\theta|\,J, M\rangle$ and $\langle J, M\,|\cos^2\theta|\,J\pm2, M\rangle$ are non-zero. For a given start wave function $\psi_0 = |\,J, M\rangle$, M remains fixed, and the final wave function after the interaction consists of a superposition of either only odd or even J states. With increasing laser intensity, more J states become significantly populated. Finite temperature is considered by a thermal ensemble of initial wave functions with a weight given by the Boltzmann factor and the multiplicity due to nuclear spin statistics.

4. Experimental details

The experimental approach follows the scheme of [10]. The output beam from an amplified 1 kHz Ti:sapphire laser system (785 nm center wavelength, 30 fs FWHM duration, 0.8 mJ per pulse) is split in two branches that act as a strong alignment beam and a weak probe beam. The temporal shape of the alignment beam is controlled with a home built pulse shaper [11]. Its polarization is then rotated by 45°, and the beam is focused onto the air sample with a 100 mm focal length lens. The alignment is probed by a weak-field polarization technique that is sensitive to the instantaneous birefringence that appears in the gas sample at a revival time. To that end, the frequency-doubled probe beam is delayed with respect to the alignment beam using a motorized linear delay stage.

5. Results and discussion

By the application of the pulse shaping technique, control on the revival structures of N_2 and O_2 has been achieved. In this report, we will concentrate on the influence of third order dispersion (TOD) on the (half-) revival structure, where molecules switch from a situation of alignment to *anti*alignment. We can show, both in simulations and experimentally, that it is possible to enhance alignment at the expense of *anti*alignment or vice versa at the half revival by choosing the appropriate pulse shapes.

The results can be seen in Figure 1. A phase retardation in the shape of a cubic polynomial (TOD) was applied in the Fourier plane of the pulse shaper (left panels). The variation of the sign and magnitude of this cubic phase affected the temporal shape of the pulses (middle panels). Fig. 1(c) shows the structure of the first half revival in N_2 when a positive chirp of +20000 fs^3 was applied. This, in the time domain, corresponds to a pulse train with the highest intensity at the leading edge. Fig. 1(f) shows the result obtained for a pulse where a cubic phase of the same magnitude but opposite sign (reversed temporal structure) was applied. In both (c) and (f), the first peak corresponds to alignment and the second, to *anti*alignment.

A theoretical simulation was run for a room temperature sample of N_2 molecules for an electric field shaped as in the experiment. In the simulation, $(\langle\cos^2\theta(t)\rangle - 1/3)^2$ is evaluated and scaled to the signal. As can be seen in Figure 1, agreement with the experimental results was obtained.

In conclusion, the control of one parameter (TOD) allowed us to promote either alignment or *anti*alignment at will in a half revival of a diatomic molecule. Even though the inverse problem is harder to tackle (obtaining a pulse shape that produces a given desired structure for a revival), the versatility of pulse shapers, allowing to finely control the wavepacket rephasing dynamics, was demonstrated and shows our increasing ability to manipulate the external degrees of freedom of molecules.

This work was supported by the Spanish-German project *Acciones Integradas* HA2003-0047. RdN thanks MEC (Spain) for a *Ramon y Cajal* research contract.

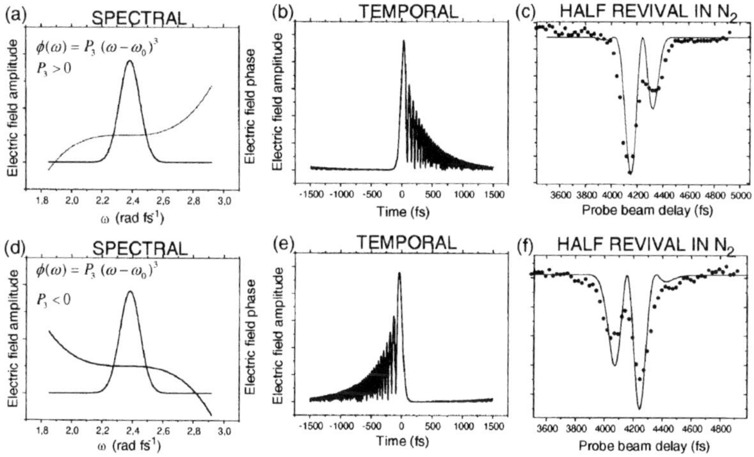

Figure 1. Control of the magnitude of alignment (*anti*alignment) in the first half revival of N_2 by applying pulses with a positive (negative) TOD. (a), (d) Amplitude and phase of the electric field in the spectral domain. (b), (e) Envelope of electric field in the time domain. (c), (f) The dots correspond to the experimental results: total homodyned signal detected around the time of a half revival in N_2. The lines correspond to the theoretical simulation.

References

1. H. Stapelfeldt and T. Seideman, Rev. Mod. Phys. 75 (2003) 543.
2. M. Leibscher, I.Sh.Averbukh, and H. Rabitz, Phys. Rev. Lett. 90 (2003) 213001.
3. M. Spanner, E. A. Shapiro, and M. Ivanov, Phys. Rev. Lett. 92 (2004) 093001.
4. C. M. Dion, A. Ben Haj-Yedder, E. Cancès, C. Le Bris, A. Keller, and O. Atabek, Phys. Rev. A 65 (2002) 063408.
5. K. F. Lee, I. V. Litvinyuk, P. W. Dooley, M. Spanner, D. M. Villeneuve, and P. B. Corkum, J. Phys. B: At. Mol. Opt. Phys. 37 (2004) L43.
6. C. Z. Bisgaard, M. D. Poulsen, E. Peronne, S. S. Viftrup, and H. Stapelfeldt, Phys. Rev. Lett. 92 (2004) 173004.
7. M. Renard, E. Hertz, B. Lavorel, and O. Faucher, Phys. Rev. A 69 (2004) 043401.
8. J. Ortigoso, M. Rodriguez, M. Gupta, and B. Friedrich, J. Chem. Phys. 110 (1999) 3870.
9. J. G. Underwood, M. Spanner, M. Yu. Ivanov, J. Mottershead, B. J. Sussman, and A. Stolow, Phys. Rev. Lett. 90 (2003) 223001.
10. V. Renard, M. Renard, S. Guerin, Y. T. Pashayan, B. Lavorel, O. Faucher, and H. R. Jauslin, Phys. Rev. Lett. 90 (2003) 153601.
11. A. Präkelt, M. Wollenhaupt, A. Assion, Ch. Horn, C. Sarpe-Tudoran, M. Winter, and T. Baumert, Rev. Sci. Instr. 74 (2003) 4950.

Femtochemistry VII
Fundamental Ultrafast Processes in Chemistry, Physics, and Biology
A.W. Castleman, Jr. and M.L. Kimble

"Radiationless" photon echo

Boris D. Fainberg

Department of Physics, Holon Academic Institute of Technology, 52 Golomb St., Holon 58152, Israel
School of Chemistry, Tel-Aviv University, Tel-Aviv 69978, Israel

1. Introduction

In this work, I propose a new type of a photon echo experiment based on *non-radiative* transitions controlled with strong electromagnetic field. The point is that the analysis of controlling the long range electron transfer (ET) with laser pulses [1] shows that radiationless processes controlled with strong electromagnetic field behave in some extent similar to optical processes. In particular, it is possible to realize a photon echo based on non-radiative transitions.

This is an usual photon echo experiment from an experimental standpoint. The point is that an object must posses a large difference in permanent dipole moments between donor and acceptor states. For example, for mixed-valence transition metal ET complexes, the permanent dipole moment difference between donor and acceptor electronic states can be very large (~70*D*) [2]. At the same time the magnitude of an electronically off-diagonal matrix element in such systems is much smaller and can be omitted [2]. In spite of existing the radiationless coupling between donor and acceptor states, the radiationless transition in such a system does not occur without electromagnetic perturbation due to the absence of resonance.

Interaction of strong electromagnetic field with the systems under consideration leads to modulation of their energetic spectrum by the field frequency similar to the spectrum of a frequency-modulated wave [1-3]. A new arrangement of the configurational surfaces can correspond to a resonance between donor and acceptor states that enables us to realize a radiationless transition during the optical pulse action. It is worthy to note that the last transition is induced by a radiationless matrix element, since direct optical transition between donor and acceptor state is forbidden. The electromagnetic field does not induce the transition. It changes only the energy gap between the donor and acceptor states. However, such a changing occurs during the short optical pulse action only. By this means one can realize a "non-radiative" analogy to photon echo spectroscopy.

2. Basic equations and results

Let us consider a donor/acceptor system in a solvent described by the zero order Hamiltonian

$$H_0 = \sum_n |n\rangle[E_n + W_{BOn}(\mathbf{Q})]\langle n| \qquad (1)$$

where E_n is the energy of state n, $W_{BOn}(Q)$ is the Born-Oppenheimer Hamiltonian of reservoir R (the vibrational subsystems of a solute and a solvent interacting with the electron system under consideration in state n). The electronic state before the charge transfer is denoted as donor $|1>$. We address our theory to the long range electron transfer systems like metal-metal charge transfer complexes. We study the direct transition $|1> \to |2>$ where $|2>$ denotes the state describing the electron on the acceptor site.

The molecule is affected by electromagnetic radiation of three non-overlapping beams

$$\mathbf{E}(\mathbf{r},t) = \mathbf{E}^+(\mathbf{r},t) + \mathbf{E}^-(\mathbf{r},t) = \frac{1}{2}\sum_{m=1,2,3} \vec{\varepsilon}_m(t)\exp[i(\mathbf{k}_m\mathbf{r} - \omega_m t) + c.c. \qquad (2)$$

Electromagnetic field (2) induces an optical polarization in the medium $\mathbf{P}(\mathrm{t})$:

$$\mathbf{P}(\mathrm{t}) = NTr_R(\mathbf{D}\rho(t)) \qquad (3)$$

where N is the density of particles in the system; $\mathbf{D}$ is the dipole moment operator of a solute molecule; ρ is the density matrix of the system, which satisfies the Liouville equation:

$$i\hbar\frac{\partial\rho}{\partial t} = [\mathbf{H}_0 + \mathbf{H}'(t), \rho] \qquad (4)$$

Here $H'(t) = W(t) + V'$ and $W = -\mathbf{DE}(\mathbf{r},t)$ is the operator of electric dipole interaction; V' describes non-radiative ET between localized diabatic states 1 and 2.

As has been mentioned before, for mixed-valence transition metal ET complexes the permanent dipole moment difference between donor and acceptor electronic states can be very large ($\Delta D \sim 70\ D$). At the same time the magnitude of an electronically off-diagonal matrix element in such systems is much smaller and can be omitted. Therefore, for the "non-radiative" (NR) systems under consideration the polarization is

$$\mathbf{P}_{NR}(t) = NTr(\mathbf{D}\rho(t)) = NTr_R(\mathbf{D}_{11}\rho_{11}(t) + \mathbf{D}_{22}\,\rho_{22}(t)) \qquad (5)$$

The positive frequency component of the polarization, which changes with the signal frequency ω_s, is given by

$$\mathbf{P}^{+}_{NR}(\mathbf{r},t) = \overset{\rightarrow +}{P}_{NR}(t)\exp[i(\mathbf{k}_s\mathbf{r} - \omega_s)]$$

$$= N(\mathbf{D}_{11} - \mathbf{D}_{22})\frac{V'_{12}}{\hbar\omega_{21}}J_0(z_1(t))J_0(z_2(t))J_0(z_3(t))Tr_R[\rho^{\text{int}}_{21}(t)] \qquad (6)$$

where J_k is the kth-order Bessel function, $z_m(t) = \Delta D\varepsilon_m(t)/(\hbar\omega_m)$, $\Delta D \equiv D_{22} - D_{11}$ and $\rho^{\text{int}}_{12}(t)$ is the density matrix in the interaction representation [1,3]. In the following, we shall impose the one-photon resonant condition, i.e. only ω_m be close to the frequency of the transition $1 \rightarrow 2$.

In a stimulated photon echo measurement, three short laser pulses with wavevectors $\mathbf{k}_1$, $\mathbf{k}_2$, and $\mathbf{k}_3$ are sequentially applied to the system, and the signal is generated in the direction $\mathbf{k}_s = \mathbf{k}_3 + \mathbf{k}_2 - \mathbf{k}_1$. In these experiments, the signal power I_s at time t, is proportional to:

$$I_s \sim \left|P^{+}_{NR}(t)\right|^2 \qquad (7)$$

When the stimulated photon echo is time gated, for instance, by mixing the echo signal with an ultrashort gating pulse in a nonlinear crystal, the signal is proportional to the echo profile at time t_g of the arrival of the gating pulse: $J_{gated}(t_g,T,\tau) \sim \left|P^{+}_{NR}(t_g,T,\tau)\right|^2$, where τ and T are delay times of the second and third pulses, respectively. Calculating $\rho^{\text{int}}_{21}(t)$

with diagram technique and using Eq.(6), we obtain in the limit of short pulses

$$P_{NR}^{+}(t)=\left(\frac{i}{\hbar}\right)^{3}N\frac{(D_{22}-D_{11})}{\hbar\omega_{21}}\int\int_{0}^{\infty}\int d\tau_1 d\tau_2 d\tau_3 J_1(z_3(t-\tau_3))$$
$$\times J_1(z_2(t+T-\tau_3-\tau_2))J_1(z_1(t+T+\tau-\tau_1-\tau_2-\tau_3))$$
$$\times\left[\tilde{R}_3(t,T,\tau)+\tilde{R}_2(t,T,\tau)\right] \qquad (8)$$

Here the four-time correlation functions $\tilde{R}_i(\tau_3,\tau_2,\tau_1)$ are similar to those $R_i(\tau_3,\tau_2,\tau_1)$, used for the allowed optical transition ($D_{12}\neq 0$) [4,5] with the only difference that D_{12} must be substituted by V'_{12}.

Using Eqs. (7) and (8), one can evaluate the ratio of the photon echo signals for "non-radiative" and "radiative" (R) [4,5] systems:

$$\frac{\left|P_{NR}^{+}(t)\right|^2}{\left|P_{R}^{(3)+}(t)\right|^2}\sim 64\left(\frac{\Delta D}{D_{12}}\right)^2\left(\frac{V'_{12}}{\hbar\omega_{21}}\right)^2\left(\frac{(V'_{12})^3}{(D_{12})^3\varepsilon_1\varepsilon_2\varepsilon_3}\right)^2 J_1^2(z_3)J_1^2(z_2)J_1^2(z_1)$$

Putting $\Delta D\approx 70D$, $\varepsilon_{1,2,3}=8.5\bullet 10^6$ V/cm, $\omega_{21}\sim 10^4 cm^{-1}$, V'_{12} ` $10^2 cm^{-1}$, and $D_{12}\approx 1D$, and bearing in mind that $J_{1max}\approx 0.6$, we obtain $\left|P_{NR}^{+}(t)\right|^2/\left|P_{R}^{(3)+}(t)\right|^2\sim 0.2$, i.e. the echo signal for a "non-radiative" system may be as much as 10% of that for a "radiative" system.

References

1. B. D. Fainberg, V. A. Gorbunov, and S. H. Lin, Chem. Phys.307 (2004) 77.
2. Y. Dakhnovskii, V. Lubchenko, and R. D. Coalson, Phys. Rev. Lett. 77 (1996) 2917.
3. V. L. Bogdanov and B. D. Fainberg, Opt. Spectrosc. 41 (1976) 472.
4. S. Mukamel, Principles of Nonlinear Optical Spectroscopy, Oxford University Press, New York, 1995.
5. B. D. Fainberg, in: Advances in Multiphoton Processes and Spectroscopy, S. H. Lin, A. A. Villaeys, and Y. Fujimura (Eds.) Vol. 15, World Scientific, Singapore, New Jersey, London, 2003, p. 215-374.

Femtochemistry VII
Fundamental Ultrafast Processes in Chemistry, Physics, and Biology
A.W. Castleman, Jr. and M.L. Kimble

Exciton Control in the FMO Complex: Influence of Laser Pulse Polarization

Ben Brüggemann,[a] Tõnu Pullerits[a] and Volkhard May[b]

[a]*Department of Chemical Physics, Lund University, P.O. Box 124, SE-22100 Lund, Sweden*
[b]*Department of Physics, Humboldt University at Berlin, Newtonstr. 15, D-12489 Berlin, Germany*

1. Introduction

We suggest the creation of a coherent excitonic wavepacket by a shaped laser pulse in close analogy to the well-established approach to control vibrational wavepackets, which has been used to break chemical bonds or to trigger conformational changes. Using a density matrix version of the optimal control theory, the first task has been to localize an excitonic wavepacket at a certain chromophore in the FMO complex [1]. (See Figure 1). Today's pulse shaping technology enables one to simultaneously and independently manipulate two different polarizations of the incoming laser [2]. We will investigate the potential of the polarization control in comparison to the

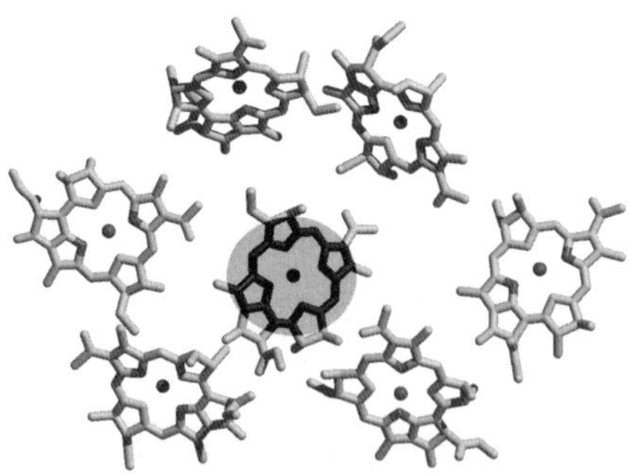

Fig. 1. Arrangement of the seven Bacteriochlorophyll in the FMO complex. At the end of the optimized pulse, the excitation energy should be localized at the central chromophore.

approach with only linear polarized light. Earlier simulations of exciton localization in the PS1 core antenna have shown that the control efficiency is strongly temperature dependent, and that for each temperature a different optimal pulse length exists [3]. Furthermore for the FMO complex, the influence of higher excited states and exciton-exciton annihilation has been discussed, as well as that of orientational and energetic inhomogeneity [1].

2. Polarization Shaping in Optimal Control Theory

The optimal control theory generates the optimized laser pulse shape by selfconsistently solving a set of differential equations, coupled by the electric field. The general theory we use here can be found in [1]. However the inclusion of two different polarization directions leads to a slight modification. Additionally, the equations are we use here optimize a laser field for N chromphore complexes at the same time. The forward propagation for the reduced density matrix $\hat{\rho}^{(k)}(t)$ of the k-th complex from the ground state at t_0 ($\hat{\rho}^{(k)}(t_0)=|0\rangle^{(k)}\langle 0|^{(k)}$) to t_f is given by

$$\frac{\partial}{\partial t}\hat{\rho}^{(k)}(t)=(-iL^{(k)}-D^{(k)}-iL^{(k)}_{field}(t))\hat{\rho}^{(k)}(t).$$

It includes the Liouville operators of the free time evolution $L^{(k)}$, the dissipation $D^{(k)}$, and the coupling to the electric field $L^{(k)}{}_{field}$ [1,3]. The backward propagation for the operator $\hat{\sigma}^{(k)}(t)$ from the target chromophore m at t_f ($\hat{\sigma}^{(k)}(t_0)=|m\rangle^{(k)}\langle m|^{(k)}$) to t_0 reads

$$\frac{\partial}{\partial t}\hat{\sigma}^{(k)}(t)=(-iL^{(k)}+D^{(k)+}-iL^{(k)}_{field}(t))\hat{\sigma}^{(k)}(t).$$

The coupling is given by the electric field, now for two polarizations (x and y)

$$E_{x,y}(t)=\frac{i}{\hbar\lambda(t)}\sum_{k=1}^{N}tr\{\hat{\sigma}^{(k)}(t;\mathbf{E})[\hat{\mu}^{(k)}_{x,y},\hat{\rho}^{(k)}(t,\mathbf{E})]\}_{-},$$

it enters the field part of both equations of motion

$$L^{(k)}_{field}(t)\hat{\rho}^{(k)}(t)=\frac{1}{\hbar}[\hat{\rho}^{(k)},(\hat{\mu}^{(k)}_{x}E_{x}+\hat{\mu}^{(k)}_{y}E_{y})]_{-},$$

where $\hat{\mu}^{(k)}_{x}$ and $\hat{\mu}^{(k)}_{y}$ are the dipole operators of the k-th FMO complex along the x- and y-polarization direction of the laser pulse, and $\lambda(t)$ is the Lagrange parameter used to couple the field equation to the equation of motion. Its time dependency assures that the pulse vanishes at t_0 and t_f. Now the set of equations is solved iteratively, starting with a back propagation using a test pulse, until convergence is reached.

3. Results and Discussion

The following simulations are carried out at 4K and the optimization interval of 600 fs. All definitions and parameters are the same as in Ref. [1]; two- and higher exciton states are neglected. For the polarization sensitive optimization, the outcome strongly depends on the orientation of the complex. Therefore we optimize for 10 complexes at the same time. They are randomly oriented, and the chromophores exhibit an energetic disorder of 100 cm^{-1} (FWHM).

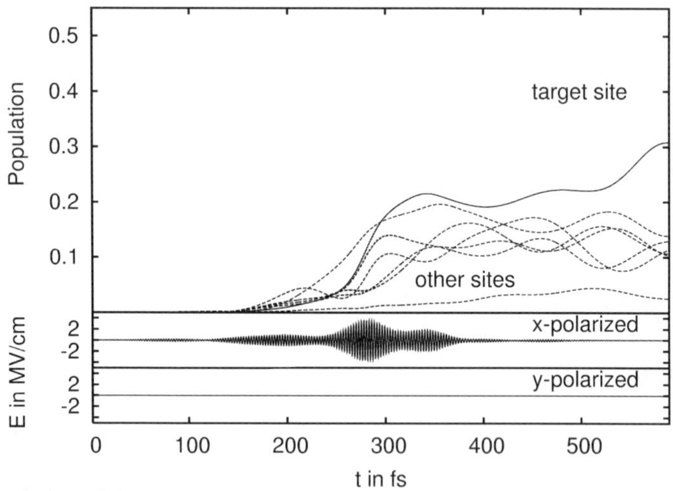

Fig. 2. Time evolution of the populations of the single chromophores in the FMO complex. The populations have been averaged over the respective chromophores in the 10 randomly oriented and energetically disordered complexes, and divided by the total excited state population at 600 fs. The target site population is given by a solid line, all other by dashed lines (Fig. 1). The optimization was restricted to a linear polarized pulse (in x-direction, lower part of the figure).

As it can be seen in Fig. 2, the average population of the central chromophore in the FMO complex is coherently driven by the optimized laser pulse up to a level of 0.32, similar to our earlier results in [1]. Compared to the population of about 0.7 that could be achieved for a properly oriented single FMO [1], the level of control is heavily reduced, but still reasonable.

Now we enable polarization control, but leave all other parameters untouched. For a single complex, the target site population went up to 0.8 (not shown), and the optimization gain does not depend as much on the orientation as in the previous case. The result of optimizing for 10 randomly oriented and energetically disordered complexes is shown in Fig. 3.

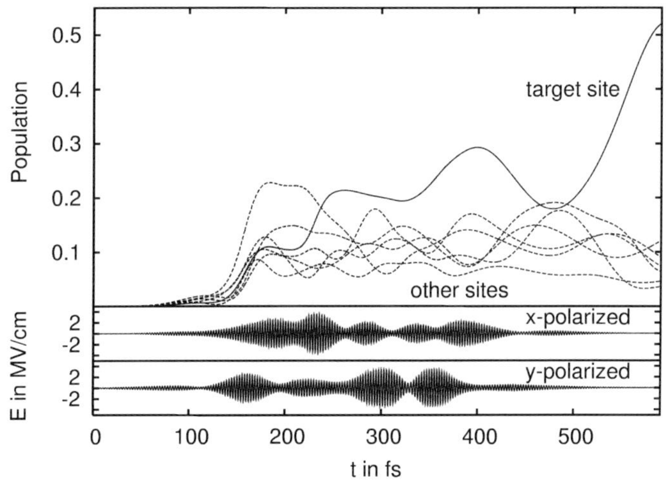

Fig. 3. Same as Fig. 2, but polarization shaping has been used. The x- and y-polarized parts of the optimal laser pulse are shown in the lower part of the figure.

The level of control could be drastically enhanced by an optimal pulse consisting of an *x*- and *y*-polarized part. The average population of the target state went up to 0.52 at the target time.

4. Conclusion

We have shown the potential of polarization shaping in exciting localized excitonic wavepackets. Further simulations will be necessary to clarify how many complexes have to be included in order to repeat the experimental conditions [4].

References

1. B. Brüggemann and V. May, J. Phys. Chem. B 108 (2004) 10529.
2. T. Brixner, G. Krampert, T. Pfeifer, R. Selle, G. Gerber, M. Wollenhaupt, O. Graefe, C. Horn, D. Liese, and T. Baumert, Phys. Rev. Lett. 92 (2004) 208301.
3. B. Brüggemann and V. May: Chem. Phys. Lett. 400 (2004) 573.
4. B. Brüggemann, T. Pullerits, and V. May, to be submitted.

Femtochemistry VII
Fundamental Ultrafast Processes in Chemistry, Physics, and Biology
A.W. Castleman, Jr. and M.L. Kimble

Spin-orbit coupling under control

Jesús González-Vázquez, Ignacio R. Sola, Jesus Santamaria

Departamento de Química Física I, Universidad Complutense, 28040 Madrid, Spain.

1. Introduction

Non-adiabatic transitions are ubiquitous on the excited state dynamics of polyatomic molecules [1]. The fate of many important reactions depends on the characteristics of these intramolecular couplings. Particularly, in heavy-atom molecules where relativistic effects are not negligible, the spin-orbit coupling must be taken into account, inducing population exchange between potentials of different multiplicity. In this work we will use ultrafast laser pulses in order to control the singlet-triplet transition between two excited states of the Rb_2 dimer.

The population exchange between two discrete quantum states |1> and |2> coupled by a spin-orbit term V_{SO} follows the famous Rabi formula,

$$P_2(t) = \frac{4V_{12}^2}{4V_{12}^2 + \Delta E^2} \sin^2\left(\frac{t}{2}\sqrt{\Delta E^2 + 4V_{12}^2}\right) \quad (1)$$

where $E_2 - E_1 = \Delta E$ is the energy difference and $V_{12} = <1|V_{SO}|2>$ the state-to-state coupling. Full population transfer will only occur whenever $\Delta E \approx 0$ after a minimum time $t = \pi/2V_{12}$. The two parameters that dominate the transition are molecule-specific, and therefore they are typically considered uncontrollable. However, under the influence of an external control field, ΔE can be manipulated by Stark-shift in second order dependence to the field, while by changing the wave function, V_{12} can be controlled in fourth order in the field. Additionally, the time duration of the field sets up a third control parameter.

The same ideas can roughly be extended to molecular potentials instead of discrete quantum levels. Thus, the mechanism to control the spin-orbit switch

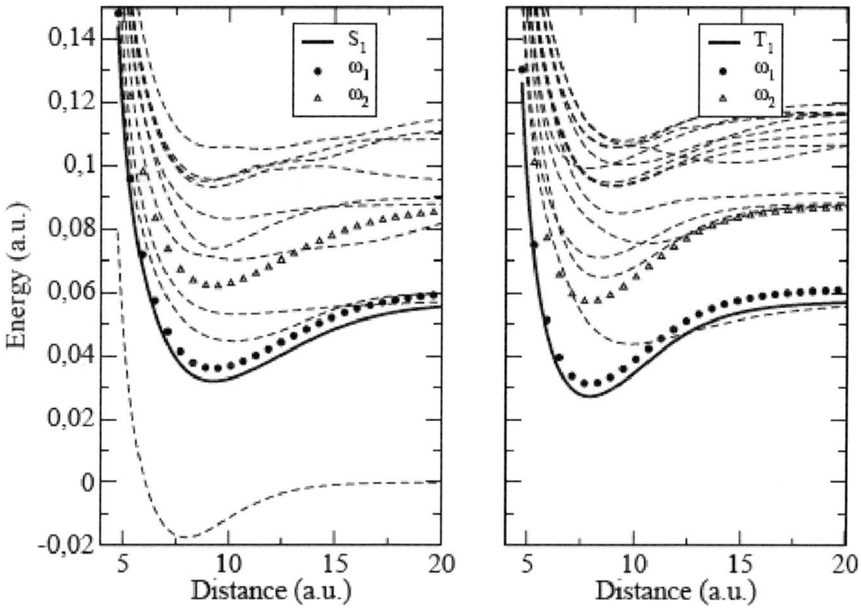

Figure 1: Singlet and triplet electronic curves included in the simulation. The S_1 and T_1 potentials are shown dressed by the energy of the two field choices ω_1 and ω_2.

basically consists of controlling the direction, magnitude, and duration of the (strong) Stark-shift that moves the energy of the potentials. Two limiting situations occur. If initially the spin-orbit coupling between a singlet and triplet potential is very strong, one may want to block the transfer of population. To that end a pulse must work, assuring that $\Delta E^2 >> 4V_{12}^2$ for as long as the transition needs or wants to be blocked. At the same time, the pulse must avoid population transfer to other potentials. On the other limit, one may want to induce full population transfer between weakly coupled potentials. To that end, a pulse must be chosen so that $\Delta E \approx 0$ during the minimum time that assures the population transfer according to the Rabi formula. In this work, we will show results of the scheme applied to the first case.

2. Model

We will use the Rb_2 molecule as a test system with large spin-orbit coupling between the electronic potentials $1^1\Sigma_u$ and $1^3\Pi_u$ (hereafter S_1 and T_1, respectively) [2-4]. Zhang and co.[5] estimated the coupling to be V_{SO}=75cm^{-1}. We will assume that V_{SO} is coordinate independent. Since S_1 and T_1 are roughly

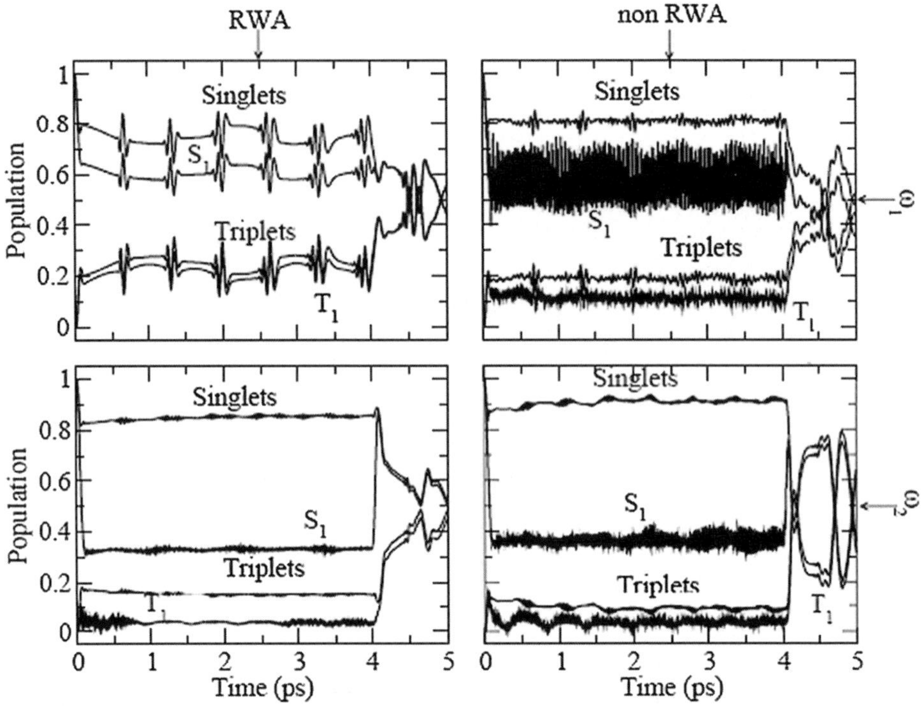

Figure 2: Population dynamics in the singlets and triplets and specifically in S_1 and T_1 for the two different laser choices (ω_1 and ω_2) assuming or not the RWA.

in resonance, V_{SO} assures a fast and strong population transfer between the potentials.

To perform our simulations, we solve the time dependent Schrödinger equation of the system using the Split-operator method with Fast-Fourier transform techniques. In the molecular model, we include all the 24 *ab initio* potential energy curves (PES) calculated by Park *et al.* [5] that can be coupled via dipole moment with the S_1 and T_1 potentials, that is, all the singlets and triplets shown in Fig.1. A single spin-orbit coupling (between S_1 and T_1) is considered, and we assume the Condon limit (with unit dipole moment) for all the dipole allowed transitions. The results are calculated with and without the rotating wave approximation (RWA) [6]. Finally, the initial condition of the dynamics is always the $v = 0$ eigenstate of S_1. That is, we assume that the population has been previously excited from the ground potential.

3. Results and Discussion

The method relies on shifting the potentials so that S_1 and T_1 are clearly out of resonance so that the coupling is weak. To induce the Stark effect, we need a non-resonant laser. The amplitude and frequency of the laser must be chosen so that the Stark shift effectively decouples S_1 and T_1 but the population is not excited and spread over the many electronic states accessible. We have found that a field amplitude of 0.01 a.u. (intensity $\approx 3.5\,TW/cm^2$) provides excellent results, while two off-resonant frequencies, $\omega_1 \approx 900\,cm^{-1}$ and $\omega_2 \approx 6600\,cm^{-1}$, avoid, in principle, large population transfer by resonant excitation to the excited potentials. In Fig.1 we show the S_1 and T_1 potentials dressed by the energy of one (ω_1 and ω_2) photons. All other parameters of the pulse are less important. We fix the shape of the control laser, with 0.1 ps switch on, followed by a plateau region of 3.9 ps, followed by a 0.1 ps switch off. The relatively slow switch on/off of the laser allows ensuring the adiabaticity of the dynamics.

The dynamics of the population of singlets and triplets are shown in Fig.2. In the left column are represented the results assuming the RWA. Our simple mechanism stops the spin-orbit coupling while the laser is on with both frequencies, although the results using ω_1 are less perfect than those using ω_2. During the first 0.1 ps (the duration of the switch on), there is some fast population exchange between the S_1 and T_1. Then the population remains basically locked among the set of singlets and triplets. Finally, after 4.1 ps, the laser is switched off, and one can observe first the almost adiabatic recovery of the population to the original S_1 and T_1 (the superposition present after the 0.1 ps) and then the subsequent laser-free spin-orbit coupling dynamics.

The RWA is a powerful tool to understand the mechanism of the scheme in the adiabatic frame, but it must be applied carefully [6]. Firstly, it only proves useful when one can disregard several competing multi-photon pathways, as in our Rb_2 model. Secondly, it starts to fail for low frequencies (as ω_1) especially when the energy difference between the rotating and counter-rotating terms is smaller than the field amplitude, as in our case (see Fig.2).

In summary, we have developed a method to manipulate the spin-orbit coupling, and we have shown in a numerically complex model of Rb_2 how a strong singlet-triplet transition can be stopped by a non-resonant field. The method is quite simple and robust and can be easily extended to control other types of non-adiabatic crossings.

4. Bibliography

1. A. Douhal and J. Santameria (Eds.) Femtochemistry & Femtobiology, World Scientific, 2002.
2. B. Zhang, L.-E. Berg, and T. Hansson, Chem. Phys. Lett. 325 (2000) 577.
3. N. Gador *et al.*, Chem. Phys. Lett. 368 (2003) 202.
4. B. Zhang, N. Gador, and T. Hansson, Phys. Rev. Lett. 91 (2003) 173006.
5. S. J. Park, S. W. Suh, Y. S. Lee, and G.-H. Jeung,, J. Mol. Spec. 207 (2001) 129.
6. B. W. Shore, Theory of Coherent Atomic Excitation, Wiley, New York, 1990.

Femtochemistry VII
Fundamental Ultrafast Processes in Chemistry, Physics, and Biology
A.W. Castleman, Jr. and M.L. Kimble

Quantitative phase characterization and compensation by multiphoton intrapulse interference phase scan (MIIPS)

Jess M. Gunn, Bingwei Xu, Johanna M. Dela Cruz, Vadim V. Lozovoy, Marcos Dantus

Department of Chemistry, Michigan State University
East Lansing, MI 48824, USA

1. Introduction

Accurate and reproducible research in the field of femtochemistry requires both the measurement and then subsequent correction of phase distortions inherent in femtosecond laser pulses. Spectral phase distortions, which are introduced by the transmission of an ultrashort laser pulse through a lens, optical fiber, or microscope objective, or by interaction with an optical surface such as a dielectric mirror, have a negative effect both on the ability to carry out and the ability to reproduce experiments in which pulse duration is of importance, or non-linear optical processes are utilized. Here we present a quantitative study on the ability of multiphoton intrapulse interference phase scan (MIIPS) [1,2] to both characterize phase distortions with better precision and reproducibility than other established methods, such as FROG [3,4] and SPIDER [5], and to correct them to obtain transform limited pulses. The theory and setup for this method have been laid out elsewhere [1,2,6].

2. Accuracy and Reproducibility of MIIPS

The reproducibility of MIIPS was tested by retrieving the phase ten successive times and analyzing for accuracy and reproducibility as described elsewhere [6]. The results are shown in Fig. 1. The reproducibility of MIIPS is seen to be excellent over a range well exceeding that of the FWHM, as shown in the right panel.

To allow for a quantitative comparison of the reproducibility of MIIPS with other methods, the statistical phase error was calculated [6,7]. Over the entire range of the spectrum, analyzing the full set of data, the statistical phase error was 0.013 rad. Using a reduced set of five scans, the statistical phase error was 0.011 rad. These values are superior to the results reported by Gallman, et. al. for FROG and SPIDER [7]. In the range over which the retrieved phase was flat and near zero (i.e. MIIPS has successfully compensated), shown in the right panel of Fig. 1, the statistical phase error is reduced to 0.004 rad. Over the FWHM of the pulse, the statistical phase error is reduced further to 0.0028. This data leads us to conclude that MIIPS is a method with high reproducibility, exceeding that of FROG and SPIDER.

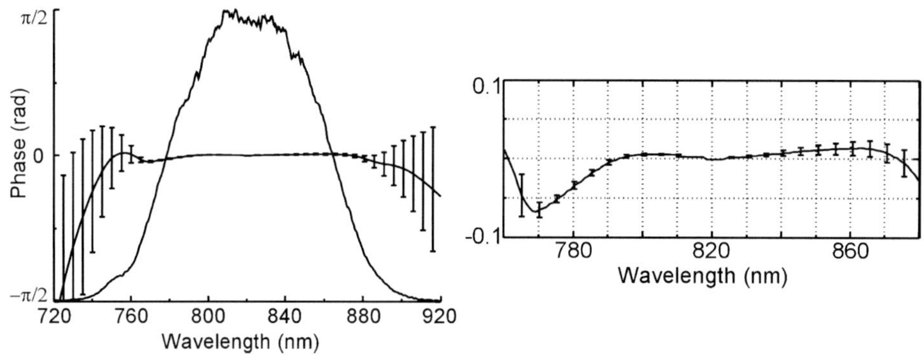

Fig. 1. Reproducibility of MIIPS. The solid lines show the average of ten independently retrieved phases, while the error bars show ± 1 standard deviation for every fifth point. The left panel shows the full range of collected data. The right panel shows a closer view of the region over which MIIPS can compensate (760-880 nm).

3. Absolute Accuracy (GVD)

To quantitatively test the accuracy of the phases retrieved by MIIPS, the group velocity dispersion (GVD) of quartz was measured. The beam was passed through quartz windows of varying thicknesses before its incidence on an SHG crystal. MIIPS was used to quantify the spectral phase of those pulses, and the GVD introduced by the quartz windows was measured to be 36.2 ± 0.5 fs^2/mm at 800 nm, which is in good agreement with the 35.92 ± 0.05 fs^2/mm measured by Diddams and Diels [8], using white-light interferometry. The GVD of quartz at 800 nm was calculated using the Sellmeir formula and the constants from Malitson [9] and found to be 36.162 fs^2/mm, which is in excellent agreement with our measured value.

4. Retrieval of Arbitary Phase

The ability to retrieve arbitrary phase distortions was evaluated by applying known phases to a TL pulse, running MIIPS, and comparing the retrieved phase with the applied phase. Two functions were analyzed: a sine function and a double Gaussian (Fig. 2). For these measurements, the pulses were first compensated to get TL. The applied synthetic phase function was introduced, and the retrieved function was obtained. Note the good agreement between both the retrieved and applied phases in each case.

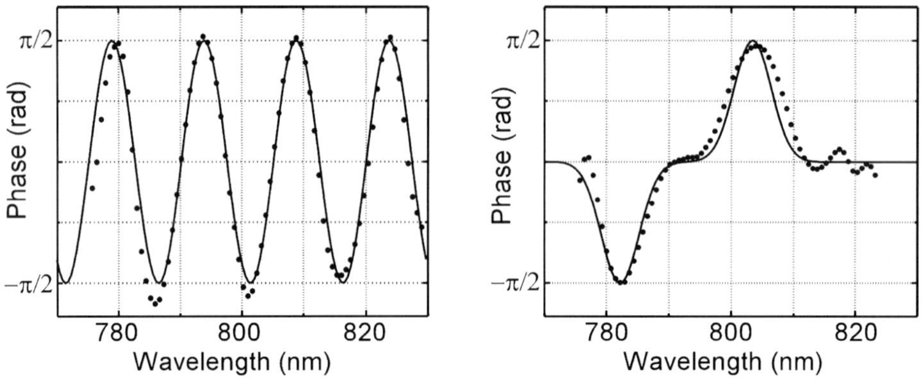

Fig. 2. Ability of MIIPS to retrieve arbitrary phase functions. The solid lines indicated the applied phase. The dots show experimental retrieved phases.

5. High Numerical Aperture (NA) Objective

One obstacle of multiphoton imaging is the significant phase distortion introduced by high numerical aperture (NA) microscope objectives. Here we demonstrate that MIIPS can successfully quantify and compensate all orders of phase distortions introduced into a sub-10 fs pulse by a 60x, 1.45 NA microscope objective.

The large quadratic chirp was pre-compensated by a pair of SF10 prisms used for pre-compression. Fig. 3 shows the compensation result. The left panel of Fig. 3 shows the full set of acquired data for the objective. The right panel shows the corresponding data over the FWHM with a reduced scale. Note that for this 1.45 NA objective, the phase is compensated to within 0.1 rad over the entire bandwidth of the pulse (770-860 nm), indicating that the distortions introduced by high NA objectives are well compensated over that range. This compensation is also characterized by a high precision; the statistical phase error was calculated to be 0.026 rad for an initial distortion of ~100 rad.

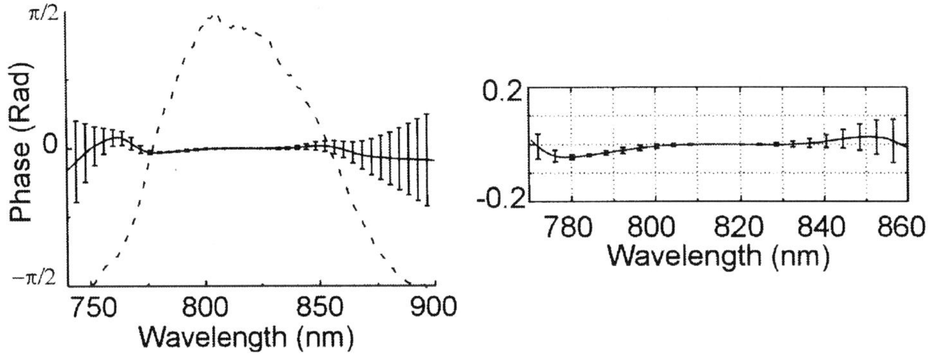

Fig. 3. Compensation of phase distortions introduced by a 60x/1.45 NA microscope objective.

6. Conclusion

The MIIPS method provides accurate phase retrieval and phase correction at the sample. These are important functions for reproducible femtosecond laser pulse experiments, especially those involving nonlinear optics. The MIIPS setup can also be used to deliver accurate synthetic wave functions at the sample as would be required for laser control experiments.

References

1. V. V. Lozovoy, I. Pastirk, and M. Dantus, Opt. Lett. 29 (2004) 775.
2. J. M. Dela Cruz, I. Pastirk, V. V. Lozovoy, K. A. Walowicz, and M. Dantus, J. Phys. Chem. A 108 (2004) 53.
3. R. Trebino and D. J. Kane, J. Opt. Soc. Am. A 10 (1993) 1101.
4. R. Trebino, K. W. DeLong, D. N. Fittinghoff, J. N. Sweetser, M. A. Krumbugel, B. A. Richman, and D. J. Kane, Rev. Sci. Instrum. 68 (1997) 3277.
5. C. Iaconis and I. A. Walmsley, Opt. Lett. 23 (1998) 792.
6. B. Xu, J. M. Gunn, J. M. Dela Cruz, V.V. Lozovoy, and M..Dantus, J. Opt. Soc. Am. B. (2005) accepted.
7. L. Gallmann, D. H. Sutter, N. Matuschek, G. Steinmeyer, and U. Keller, Appl. Phys. B-Lasers O. 70 (2000) S67.
8. S. Diddams and J. C. Diels, J. Opt. Soc. Am. B 13 (1996) 1120.
9. I. H. Malitson, J. Opt. Soc. Am. 55 (1965) 1205.

8. Intense Laser – Matter Interactions

Femtochemistry VII
Fundamental Ultrafast Processes in Chemistry, Physics, and Biology
A.W. Castleman, Jr. and M.L. Kimble

Ultrafast cluster electron and Coulomb explosion dynamics driven by ultraintense laser fields

Isidore Last, Andreas Heidenreich, and Joshua Jortner

School of Chemistry, Tel Aviv University, 69978 Tel Aviv, Israel

1. Prologue

Ultraintense table top lasers are characterized by a maximal intensity of $\sim 10^{20}$ Wcm^{-2}, which constitutes the highest light intensity on earth. Novel features of light-matter interactions emerge from the interaction of clusters with ultrashort (pulse temporal width τ = 10-100 fs) and ultraintense (peak intensity I_M = 10^{15}-10^{20} Wcm^{-2}) laser fields [1-14]. The modern research area of femtosecond cluster dynamics on the time scale of nuclear motion is transcended by moving towards attosecond-femtosecond electron dynamics in ultraintense laser fields. The response of clusters to ultraintense lasers induces well characterized ultrafast dynamics of electrons (on time scales of ~ 1-100fs) and of ions (on time scales of 5-100 fs) in these large finite systems [7-14].

We shall address some of the novel physical features of the response and dynamics of homonuclear and heteronuclear clusters in ultraintense laser fields, which pertain to:

(1) New mechanisms of cluster extreme multielectron ionization.

(2) The consequences of the interaction of an ultraintense laser with a spatially confined cluster. A cluster of n constituents (each with m atomic constituents) initially results in confined mn ions of extremely high charges, as well as bare nuclei, e.g., H^+, D^+, C^{4+}, C^{6+}, O^{6+}, O^{8+}, Xe^{q+} (q = 8-36), I^{q+} (q = 7-35) [1-14].

(3) The implications of Coulomb instability of highly multicharged clusters. These result in the interplay between cluster fission [15,16], i.e., instability towards cluster dissociation into two (or a small number of) large ionic fragments and cluster Coulomb explosion (CE) [16,17] into a large number of small ionic species (Fig. 1).

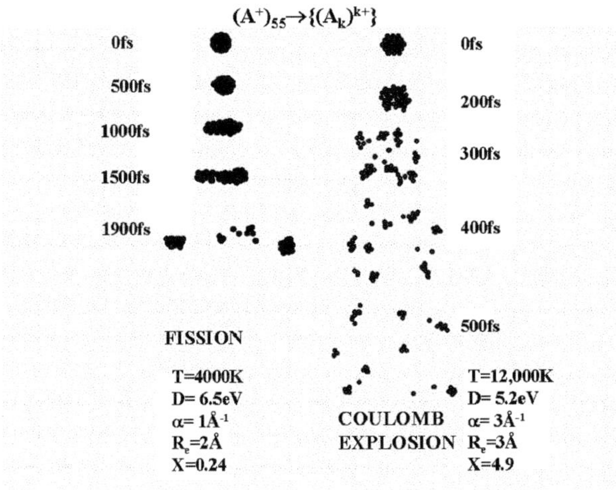

Fig. 1. Molecular dynamics snapshots of fragmentation of highly charged $(A^+)_{55}$ Morse clusters (adopted from data of ref. 16). The potential parmeters were varied, changing the fissibility parameter X = E(Coulomb)/ 2E(surface). Note fission for X < 1 and CE for X > 1.

(4) The formation of high-energy, highly charged ions and bare nuclei in the energy domain of nuclear physics (1keV – 1meV) via cluster CE [2-4,6-14].

(5) New effects of boosting of the energy by energetic driving and kinematic runover effects of the light ions (e.g., D^+) by highly charged heavy ions, e.g., C^{q+} (q = 4,6 or I^{q+} (q = 7-35), in the CE of heteroclusters [8-12].

(6) The dramatic effects of nuclear fusion driven by cluster CE [1-4,9-14].

2. Extreme Cluster Multielectron Ionization

The ionization of a single q-fold charged atomic or molecular ion with an ionization potential P_q in a laser electric field F_ℓ, whose peak is $|F_\ell^M| = 2.745\text{x}10^{-7}(I_M/\text{Wcm}^{-2})^{1/2}$ (eVÅ^{-1}), can be described in terms of the barrier suppression ionization (BSI) mechanism [6], which (in the absence of electron tunneling effects) results in the threshold field for inducing classical ionization is $|eF| = P_q^2/4B(k+1)$, where $F = F_\ell$ and B = 14.4eV. The BSI model predictions [6] for multielectron ionization of Xe atoms estimated with $F = F_\ell^M$ are in accord with experiment (Fig. 2) in the range $I_M = 10^{16}$–10^{18} Wcm^{-2}, where highly charged Xe^{q+} (q = 8–16) single ions are produced.

Extreme cluster multielectron ionization in ultraintense laser fields (involving the stripping of all electrons from light, first row atoms, and the formation of heavily charged ions, e.g., Xe^{26+} or I^{25+} at $I_M = 10^{19}$ Wcm^{-2}) is

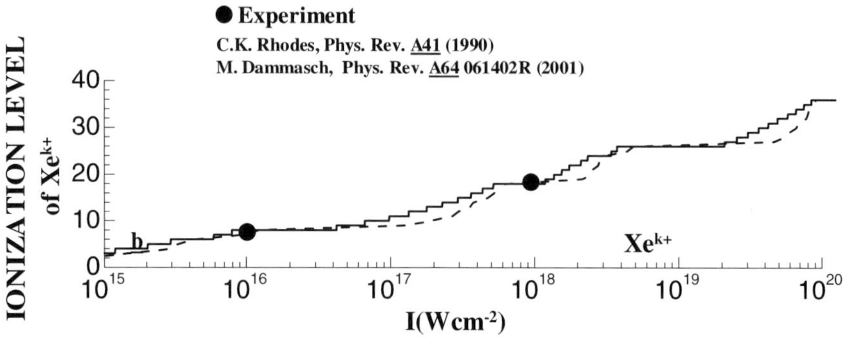

Fig. 2. The laser peak intensity dependence of the ionization level of a single Xe atom [6]. BSI (solid line) and ADK model (dashed line). The dots represent experimental data [6].

distinct, both from the electron dynamic response in ordinary fields ($I_M \leq 10^{13}$ Wcm^{-2}), where perturbative quantum electrodynamics is applicable, and from the response of a single atomic or molecular species in terms of mechanisms, the ionization level, and time scales for electron and nuclear dynamics. On the basis of our recent analyses and simulations [6-14,18], the electron dynamics of elementary and molecular clusters, e.g., $(Xe)_n$, $(D_2)_n$, $(CD_4)_n$, $(AI)_n$, $(CA_3I)_n$ (with A = H, D or T), in ultraintense laser fields involves three sequential–parallel processes of inner ionization, the nanoplasma formation, and outer ionization [6-9].

The laser electric field acting on the elemental or molecular cluster is taken as $F_\ell(t) = F_{\ell 0}(t)\cos(2\pi\nu t)$, where ν is the laser frequency and $F_{\ell 0}(t)$ is the pulse envelope function. Molecular dynamics simulations (including magnetic field and relativistic effects) [6-9,14] and analyses of high-energy electron dynamics and nuclear dynamics in a cluster interacting with a Gaussian laser field $F_{\ell 0}(t) = F_\ell^M \exp[-2.773(t/\tau)^2]$, with $\nu = 0.35\ fs^{-1}$ (photon energy 1.44eV and pulse temporal width (FWHM) $\tau = 25$fs), elucidated the time dependence of inner ionization, the formation, persistence, and decay of the nanoplasma and of outer ionization.

The cluster inner ionization [5,6,18] is driven by two processes: (A) The BSI mechanism, which is induced by a composite field $\mathbf{F} = \mathbf{F}_\ell + \mathbf{F}_i$, where $\mathbf{F}_i$ is the inner field generated by electrostatic interactions with the ions (ignition effects) and with the nanoplasma electrons (screening effects). The BSI level and time-resolved dynamics were evaluated from the expression for the threshold field F given above [6]. (B) Electron impact ionization (EII), which involved inelastic, reactive impact ionization of ions by the nanoplasma electrons [5,6,18]. EII in Xe_n clusters was explored using experimental data for the energy dependence of ionization cross sections of Xe^{j+} ions (j = 1–10), which were fit by a three-parameter Lotz-type equation [18]. The relative yields of EII are lower than those for BSI, increasing with increasing the cluster radius R_0 at fixed I_M and decreasing with increasing I_M at fixed R_0 and n. For

Xe_{1061} at $I_M = 10^{15}$ Wcm^{-2}, the ratio of the yields for EII and for BSI (at $\tau = 25fs$) is ~50% [18]. The relative EII yield increases markedly with increasing τ, opening avenues for the control of the ionization level [18].

The ionization level of Xe_n clusters (characterized by the average charge q_{av} = <q> on each ion), which manifests the production of $\{Xe^{q+}\}_n$ multicharged ions is presented in Fig. 3, where we also portray the ionization levels of a single Xe atom at each intensity (marked by horizontal arrows). In the highest intensity domain of $I_M = 10^{19}–10^{20}$ Wcm^{-2}, q_{av} = 26–36 per ion is very large, being independent of the cluster size. For $I_M = 10^{18}$ Wcm^{-2}, q_{av} increases with increasing R_0, manifesting ignition effects. For $I_M = 10^{17}$ Wcm^{-2}, q_{av} first increases and then decreases with increasing R_0, exhibiting the interplay between ignition effects and nanoplasma screening effects. For $I_M = 10^{17}–10^{20}$ Wcm^{-2}, the q_{av} for the clusters converges to the single atom ionization level with decreasing R_0. For $I_M = 10^{15}$ Wcm^{-2}, q_{av} for small clusters is considerably larger than the single atom ionization level, due to the contribution of inner field effects and EII. These results unveil new features of cluster multielectron ionization, which gives rise to a large number (n) of highly charged atoms or nuclei (e.g., up to Xe^{36+}, I^{35+}, C^{6+}, O^{8+}), for novel chemical and physical applications.

Simulation results [7] for electron dynamics (for τ = 25fs), which are expressed in terms of the number of electrons depleted per constituent atom, i.e., n_{ii} (n_{oi}) for inner (outer) ionization (Figs. 4 and 5), exhibit a gradual rise and long-time saturation. The characteristic times for inner/outer ionization of Xe_n (n = 479,1061) clusters are (7–30fs)/(5–13fs) and decrease with increasing I_M. Quasi-resonance effects in the nanoplasma are manifested in the temporal oscillations of the n_{oi} and in the center of gravity of the nanoplasma electrons (Fig. 4 and 5), driving outer ionization.

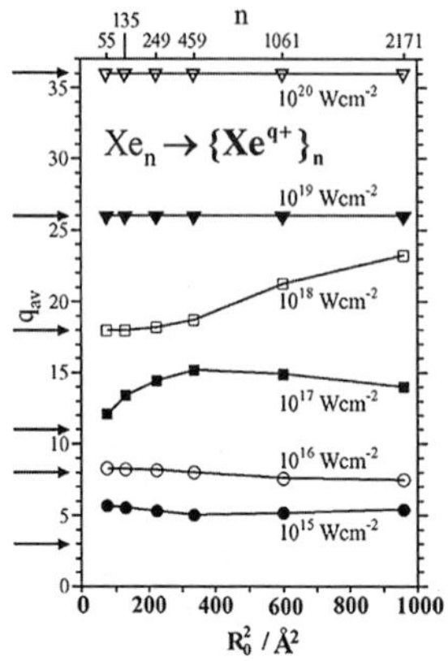

Fig. 3. Cluster size and laser intensity (marked on the curves) dependence of the average inner ionization level of each atom for Xe_n clusters ($\tau = 25$fs). The ionization levels of a single atom are marked by horizontal arrows.

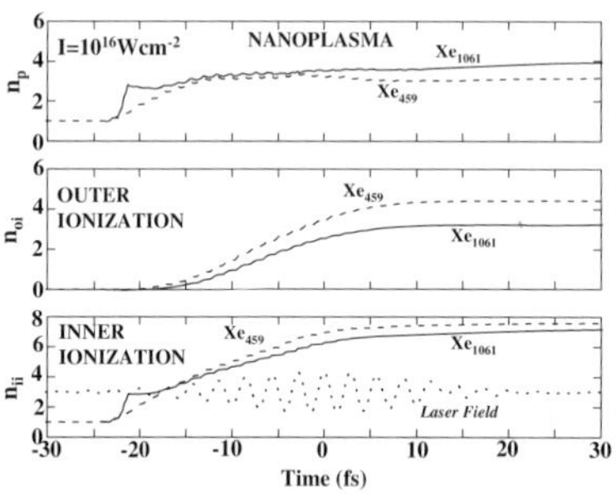

Fig. 4. Time-resolved electron dynamics in Xe_n ($n = 1061,459$) clusters driven by a Gaussian laser field with $I_M = 10^{16}$ Wcm^{-2} (represented on arbitrary scale in the lower panel). The numbers of electrons n_{ii}, n_{oi} and n_P per atom (see text) are presented.

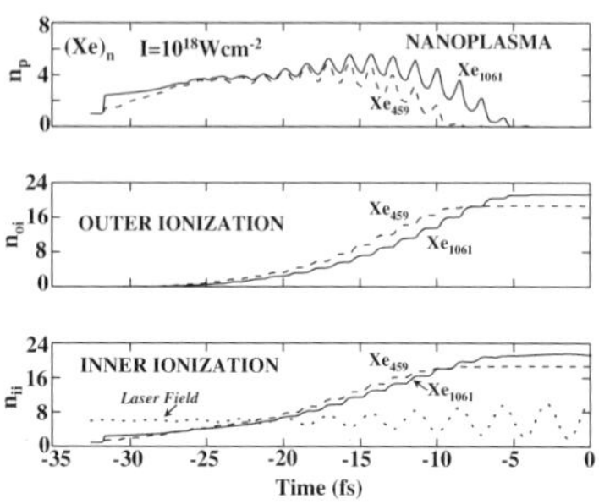

Fig. 5. Same as for Figure 2 with $I_M = 10^{18}$ Wcm^{-2}.

3. The Nanoplasma

The nanoplasma is formed from the unbound electrons generated by inner ionization, which are confined to the cluster and to its vicinity, and by the cluster ions. The number of electrons in the nanoplasma (per atomic constituent) is $n_p = n_{ii} - n_{oi}$. n_p first increases and subsequently either reaches saturation for lower intensities $I_M = 10^{16}$ Wcm^{-2} (Fig. 4) or is completely depleted at $I_M = 10^{18}$ Wcm^{-2} (Fig. 5). The spatial structure and response of the nanoplasma (Figs. 6 and 7) reveals that for $I_M = 10^{16}$ Wcm^{-2}, the electron clouds are nearly spatially homogeneous with the majority of the electrons being located within the cluster (Fig. 6), while for $I_M = 10^{18}$ Wcm^{-2}, the electron spatial distribution is inhomogeneous, assuming a "sausage type" structure along the laser electric field (Fig. 7). General characteristics of the nanoplasma are [7-9]:

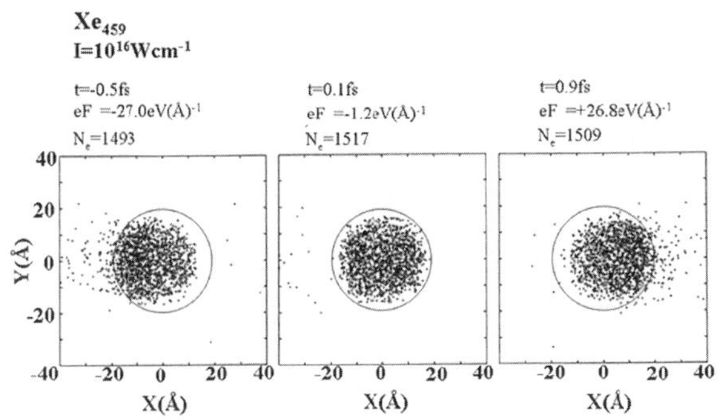

Fig. 6. The response of the electron cloud, which contains Ne electrons (black points) in the nanoplasma in a Xe_{459} cluster at $I_M = 10^{16}$ Wcm^{-2}. Circles represent the cluster size. Laser propagation direction along the X axis, with the electric field F_ℓ at time t.

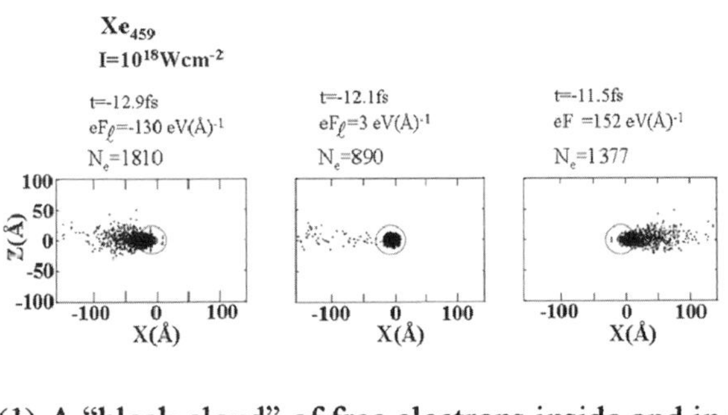

Fig. 7. Same as for Fig. 4 with $I_M = 10^{18}$ Wcm^{-2}.

(1) A "black cloud" of free electrons inside and in the vicinity of the cluster. (2) The nanoplasma is positively charged. (3) High average electron densities of ~(2-3)10^{22} cm^{-3}, being comparable to electron densities in metals. (4) High average electron energies of ~50eV-1keV, which increase with increasing I_M. (5) Spatial inhomogeneity and angular anisotropy induced by the laser field (Figs. 6 and 7). (6) The existence of a persistent nanoplasma with a lifetime τ_P > 100fs at $I_M = 10^{15}$-10^{16} Wcm^{-2} and a transient nanoplasma with τ_P < 10fs at I_M

= 10^{18} Wcm^{-2} (Figs. 4 and 5). (7) Partial or complete depletion by outer ionization.

4. Energetics and Dynamics of Uniform and Nonuniform Coulomb Explosion

Cluster electron dynamics triggers nuclear dynamics, with the outer ionization being accompanied by CE [8-14,16,17]. The traditional view of CE involves a uniform explosion of a homonuclear cluster in the limit of cluster vertical ionization (CVI). The CVI time scales for multielectron ionization are short on the time scale of nuclear motion and outer ionization is complete [7-11]. We advanced [8,9] a border cluster radius $R_0^{(I)} \propto I_M^{1/2}$, with the CVI being applicable for $R_0 < R_0^{(I)}$. Under CVI conditions, the maximal ion energy for CE of $(A^{q_A+})_n$ is [8,9] $E_M = (4\pi/3)\bar{B}\rho_0 q_A^2 R_0^2$, the average ion energy is E_{av} = $(3/5)E_M$, while the distribution of the ion kinetic energies is $P(E) = (3/2E_M)(E/E_M)^{1/2}$ for $E \leq E_M$, where ρ_0 is the constituent density and $\bar{B}$ = 14.40eV. Uniform CE is well obeyed for elemental and homonuclear molecular clusters under sufficiently high I_M where the CVI conditions prevail. The average energies for CE of $(D_2)_{n/2}$ (n = 459-3.3·10^4) clusters obey the CVI relation $E_{av} \propto R_0^2$ for $I_M > 10^{17}$ Wcm^{-2} (Fig. 8), while the energy distribution obeys the $P(E) \propto E^{1/2}$ relation (Fig. 9 and its left inset).

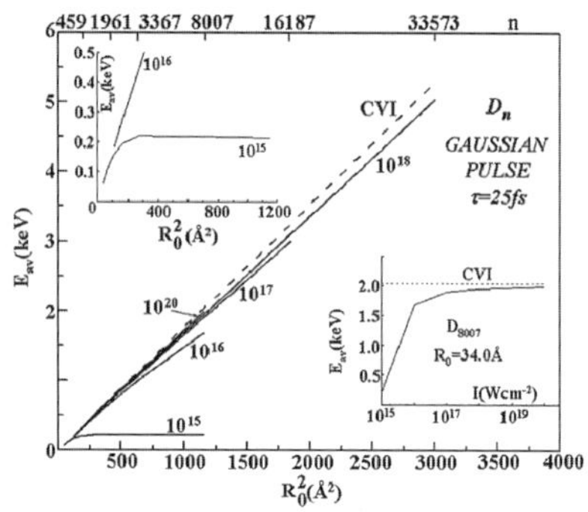

Fig. 8. Cluster size dependence of D^+ energies from CE of $(D_2)_{n/2}$ clusters (n = 354-33573) induced by Gaussian laser pulses with $I_M = 10^{15}$-10^{20} Wcm^{-2}, which are marked on the curves. The result of the CVI is also presented. The insets show severe deviations from the CVI at $I_M \leq 10^{17}$ Wcm^{-2}.

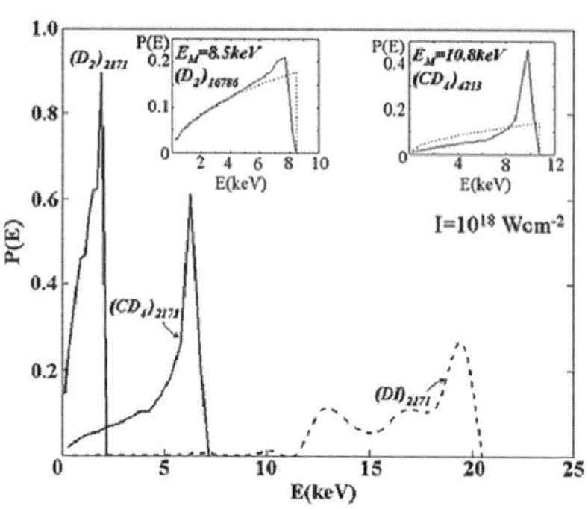

Fig. 9. Kinetic energy distribution P(E) of D^+ ions in the CE of $(D_2)_n$ and deuterium containing hetero-clusters at I_M = 10^{18} Wcm^{-2}. The insets show the $P(E) \propto E^{1/2}$ CVI relation, which acocunts well only for the CE of the $(D_2)_n$ homocluster (left and inset).

The uniform CE picture breaks down for the explosion of $(A_k^{q_A+}B_\ell^{q_B+})_n$ heteroclusters consisting of A^{q_A+} ions of masss m_A and heavy $B_\ell^{q_B+}$ ions of mass m_B, e.g., $(H^+T^+)_n$, $(D_2^{++}O^{q+})$ (q = 6,8), $(C^{q+}D_4^{4+})_n$, $(D^+I^{q+})_n$ (q = 5-35), or $(C^{6+}D_3^{+3}I^{q+})_n$ (q = 5-35) heteroclusters [9,13,14]. Two major new effects in the CE of multicharged heteroclusters were predicted [9-14]:

(A) Energetic boosting of the energies of low-charged, light ions driven by the heavy ions. The energetics E_M and E_{av} of D^+ ions from the CE of $(D_2O)_n$, $(CD_4)_n$, or $(DI)_n$ clusters are expected to be considerably larger than for D^+ energies from $(D_2)_{n/2}$ clusters of the same size (Fig. 10).

(B) Kinematic driving effects, which are determined by the kinematic parameter $\eta_{AB} = q_A m_B / q_B m_A$. For $\eta_{AB} > 1$, a runover of the light ions over the heavy ions prevails, resulting in two distinct spatial distributions of the light/heavy ions [9,13,14]. At higher ion energies, the marked deviation of P(E) for CE of $(CD_4)_n$ heteroclusters (Fig. 9) is due to the runover effects. This process results in an isotope effect in the energetics of light ions from heterocluster CE (e.g., for $(CD_4)_n$ and for $(CH_4)_n$) [9], which was experimentally confirmed [19]. Extreme runover effects are manifested for CE of extremely charged light-heavy heteroclusters with large η_{AB} (e.g., $(H^+I^{25+})_n$ or $(D^+I^{25+})_n$), which reveal expanding, 2-dimentional spherical thin shells of light ions [13,14] (Fig. 11). The light ion energy distribution is narrow, falling in the vicinity of the high values of E_M (Fig. 9 for $(DI)_n$). These manifest the attainment of transient self-organization in complex systems driven by repulsive Coulomb interactions.

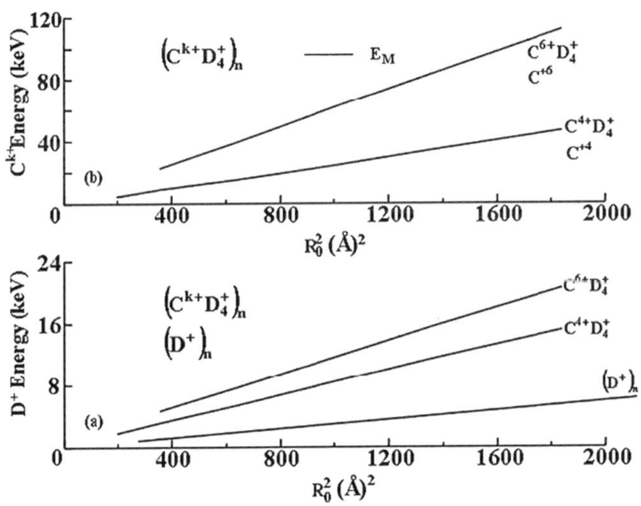

Fig. 10. Cluster size dependence of the maximal energy of ions in the CE of $(D_2)_n$ and $(CD_4)_n$, obeying the CVI relation. The marked D^+ energy boosting effects on the energy of D^+ ions in the CE of the heterocluster is manifested.

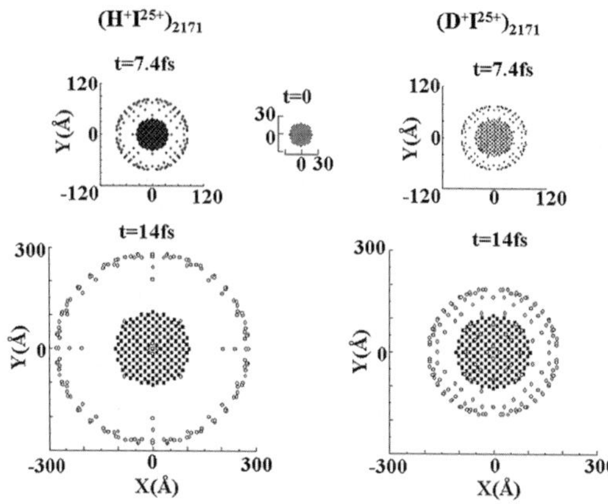

Fig. 11. 2-dimensional pictures of the time-dependent spatial configuration of ions in CE of $(H^+I^{25+})_{2171}$ and $(D^+I^{25+})_{2171}$ heteroclusters, which manifest extreme kinematic runover effects. (■) I^{25+} and (O) H^+ or D^+.

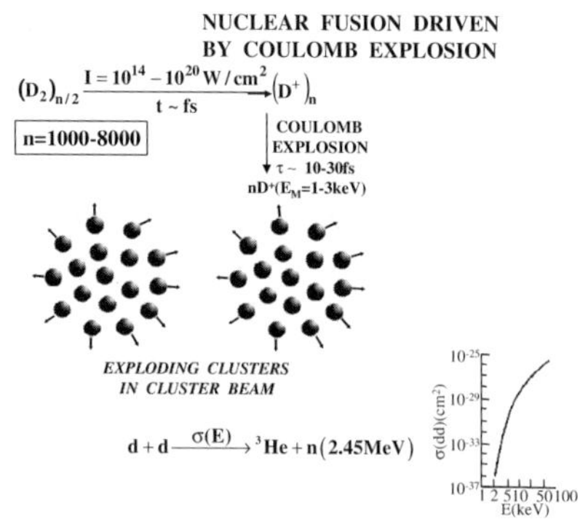

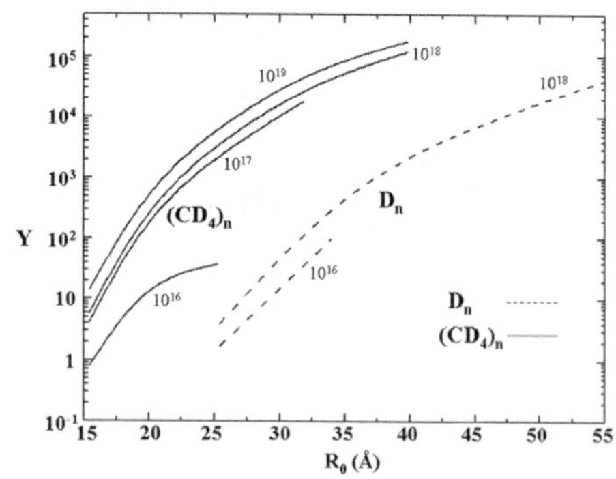

Fig. 12. Artist's view of NFDCE in an assembly of $(D_2)_n$ clusters, with energetic deuterons from different clusters undergoing dd fusion within the macroscopis plasma filament with the cross sections for dd fusion being represented in the lower part of the figure.

Fig. 13. Neutron yields pere laser pulse for dd NFDDCE of $(D_2)_n$ and $(CD_4)_n$ clusters (see text) at intensities $I_M = 10^{16}\text{-}10^{18}$ Wcm^{-2} (marked in the curves).

5. Nuclear Fusion Driven by Cluster Coulomb Explosion

The energies of D^+ ions from CE of $(D_2)_{n/2}$ clusters in the range $E_M \leq 10$ keV (Fig. 8) fall in the domain of nuclear physics. Compelling experimental and theoretical evidence [2,4,9,20] was advanced for dd nuclear fusion driven by Coulomb explosion (NFDCE) in an assembly of deuterium $(D_2)_n$ clusters (Fig. 12). We predicted [9-14,20] that the CE of highly charged heteronuclear deuterium containing molecular clusters will result in considerably higher D^+ energies due to energetics boosting effects (section 4 and Fig. 9 and 10) and a narrower energy distribution in the vicinity of E_M due to kinematic runover effects (section 4 and Fig. 9). The neutron yields Y (per laser pulse) calculated under the conditions of the Lawrence - Livermore experiment [2,4] for $I_M > 10^{17}$ Wcm^{-2} are higher by 2-3 orders of magnitude for CE of $(CD_4)_n$ clusters than for $(D_2)_n$ clusters of the same size (Fig. 13). These theoretical predictions [9-11] were experimentally confirmed in Sacley [3], the Lawrence-Livermore laboratory [4], and in the Max-Born Institute [21]. Moving to NFDCE of extremely charged light-heavy hetoercluster, e.g., $(D^+I^{25+})_n$ (Fig. 9), Y can be enhanced by another 2-3 orders of magnitude over $(CD_4)_n$ clusters of the same size. We are currently exploring further perspectives of nuclear reactions of astrophysical interest, e.g., in the NCO cycle in hot stars [22], which can be driven by table-top nucleosynthesis induced by heteroclusters CE.

During the last decade, cluster science explored new and fascinating territories, bridging between cluster electron-nuclear dynamics and table-top nuclear fusion.

Acknowledgement. This research was supported by the Deutsche Forschungs-gemeinschaft (DFG) SFB 450 on "Analysis and Control of Ultrafast Photo-induced Reactions".

1. T. Ditmire *et al.*, Phys. Rev. Lett. 78 (1997) 2832.
2. J. Zweiback *et al.*, Phys. Rev. Lett. 84 (2000) 2634.
3. G. Gullon *et al.*, Phys. Rev. Lett. 89 (2002) 065005.
4. R.W. Madison *et al.*, Phys. Plasmas 11 (2004) 270.
5. I. Last and J. Jortner, Phys. Rev. 64 (2001) 063201.
6. I. Last and J. Jortner, J. Chem. Phys. 120 (2004) 1336.
7. I. Last and J. Jortner, J. Chem. Phys. 120 (2004) 1348.
8. I. Last and J. Jortner, J. Chem. Phys. 121 (2004) 3030.
9. I. Last and J. Jortner, J. Chem. Phys. 121 (2004) 8329.
10. I. Last and J. Jortner, Phys. Rev. Lett. 87 (2001) 033401.
11. I. Last and J. Jortner, J. Phys. Chem. A 106 (2002) 10877.
12. J. Jortner and I. Last, ChemPhysChem 3 (2002) 845.
13. I. Last and J. Jortner, Proceed. Natl. Acad. Sci. USA 102 (2005) 1291.
14. I Last and J. Jortner, Phys. Rev. A 71 (2005) 063204.
15. C. Bréchignac *et al.*, Phys. Rev. Lett. 64 (1990) 2893.
16. I. Last, Y. Levy, and J. Jortner, J. Chem. Phys. 123 (2005) 154301.

17. O. Zhong and A.W. Castleman, Jr., Chem. Rev. 100 (2000) 4039.
18. A. Heidenreich, I. Last, and J. Jortner, Eur. Phys. J. D 35 (2005) 567.
19. M. Hohenberger *et al.*, Phys. Rev. Lett. 95 (2005) 195003.
20. I. Last and J. Jortner, Phys. Rev. A 64 (2001) 063201.
21. S. Ter-Avetisyan *et al.*, Phys. Plas. 12 (2005) 12702.
22. G. Wallerstein *et al.*, Rev. Mod. Phys. 69 (1997) 995

Femtochemistry VII
Fundamental Ultrafast Processes in Chemistry, Physics, and Biology
A.W. Castleman, Jr. and M.L. Kimble

Nonadiabatic multielectron dynamics in (moderaterately) strong laser fields: C_{60} a model case for large finite system

Tim Laarmann, Ihar Shchatsinin, Mark L. Boyle, Gero Stibenz, Günter Steinmeyer, C.P. Schulz, and Ingolf V. Hertel*

Max Born Institute, Max-Born-Str. 2a, D-12489 Berlin, Germany
**also: Free University Berlin, Dept. of Physics, Arnimallee 14, D-14195 Berlin, Germany*

1. Introduction

Large molecular systems in strong and ultrashort laser fields are a subject of great current interest [1-3]. They offer considerable potential for photodynamical control of their reactive and dynamical behavior [4-6]. The aim of such studies is the manipulation of potential energy surfaces by the strong laser fields to force the molecular system into specific reaction pathways. For example, the possibility for selective bond dissociation and rearrangement in polyatomic molecules has recently been demonstrated [7]. This opens the door to new, exciting experiments in photochemistry with interesting potential applications in biology or medicine – keeping in mind that large molecules mediate fundamental processes in living organisms.

The success of such experiments depends on a detailed knowledge of the fundamental processes initiated when focusing intense, short laser light pulses onto large finite molecules. In this context, C_{60} is a very special example for which numerous studies on photoinduced reactions have been performed, focusing on energy redistribution [8], ionization, and fragmentation dynamics [9-12]. With its 60 essentially equivalent, delocalized π electrons it may be seen as a prototype of a large, finite system for studies of the photodynamics in strong laser fields [13-15]. Additionally, there is hope that C_{60} – albeit very large – can nevertheless be handled by theory [16,17] due to its high symmetry, which in turn triggers detailed experimental studies and vice versa. The exploration of the electronic and nuclear dynamics in C_{60} has a long history. One astonishing early observation was the delayed ionization on a microsecond time scale upon irradiation with nanosecond laser pulses [18]. This has been explained by thermionic electron emission from vibrationally excited C_{60} fullerenes. Strong electron-phonon coupling leads to efficient heating of the nuclear motion during nanosecond laser excitation, and the subsequent ionization in turn is one important energy relaxation (cooling) channel [19]. It has been found that the ionization behavior depends sensitively on the excitation time scale, i.e. on the laser pulse duration τ [13]. The subject has recently been reviewed [20]. We summarize here only some key aspects. For pulse durations above 1 ps one observes a characteristic bimodal fragmentation pattern of heavy fullerenes C_{60-2m} separated by C_2 units ranging down to $60-2m \geq 32$ and a series of small carbon clusters C_n below $n \leq 28$. In contrast, for pulse durations below 500 fs the excitation energy tends to remain in the electronic system (the electron-phonon coupling time being ≈ 250 fs), and multiply charged clusters are observed. Thermalization within the electronic system due to electron-electron scattering occurs on a time-scale of about 100 fs, and ionization due to statistical electron emission after equilibration among the electronic degrees of freedom may be observed [21]. Direct multiphoton ionization dominates for very short laser pulses $\tau < 70$ fs [11]. As a fingerprint of the multiphoton process, the photoelectron spectra of C_{60} exhibit a characteristic above threshold ionization (ATI) structure [13] in which the active electron absorbs more laser photons than necessary to overcome the ionization potential. This results in a kinetic energy distribution of photoelectrons, which exhibits a series of equally spaced maxima separated by the photon energy $h\nu$, as well known from atomic systems [22]. In addition, sharp peaks were discovered in the photoelectron spectra of C_{60} on top of the ATI series and the thermal electron contribution [14,23,24].

In this contribution, we present results from pump probe studies using ultrashort ($\tau \approx 9$ fs) and longer ($\tau \approx 50$ fs) laser pulses at 800 nm. We will show that in both cases multi electron dynamics govern the excitation process, but the additional "heating" with longer pulses or the absence of it at ultrashort pulses leads to substantially different ionization and fragmentation dynamics.

2. Experimental

The experimental setup has been described in detail in a previous publication [24]. Here a short outline will be adequate. The molecular beam is generated from gold grade C_{60} powder heated to 770 K in a small oven. It is collimated to about 6°, and its diameter in the interaction region is about 4 mm. The photo electrons and ions generated at the intersection volume are detected using time-of-flight (TOF) spectrometers equipped with multi-channel plates (MCP) and counted using FAST multi-scalar cards. The mass resolution $m/\Delta m$ of the TOF spectrometer is about 2900, hence enabling isotope identification in C_{60}. Photo ion spectra are averaged typically over 5000 to 20000 laser pulses using a standard PC programmed with LabVIEW.

In both experiments presented here the laser pulses are generated by a 1 kHz Ti:sapphire laser system. The Gaussian laser beam is split in two beams of different intensity. They are mildly focused onto the molecular beam. The time delay between the pulses is set by a translation stage. A collinear pump-probe geometry is used, perpendicular to the C_{60} molecular beam and to the TOF spectrometer. Time-zero was determined by an auto-correlation measurement detecting Xe^+. The intensity of the laser beam (maximum value in the focus) was calibrated by a standard knife edge method.

For the experiments with ultra short pulses, laser pulses from a commercial multipass Ti:sapphire amplifier system were compressed from 35 fs down to 9 fs pulse duration. To achieve this, the 35 fs pulses are focused into a hollow capillary of 48 cm length and 250 μm inner diameter, which is filled with 500 mbar of argon gas. Due to the strong self-phase modulation, the spectrum is strongly broadened at the output of the fiber. The spectrally broadened pulses are recompressed by 4 bounces off specially designed double-chirped mirrors with a group delay dispersion of -60 fs^2 per bounce [25]. The number of bounces has been carefully chosen to precompensate for the 350 μm thick entrance window to the experimental chamber. Due to the high reflectivity of the chirped mirrors, pulse energies up to 300 μJ are available after compression.

3. Results and Discussion

3.1. Pump-Probe with 50fs pulses

We will first discuss the data taken with 50 fs pulses at 800 nm. A full account of this work is given in [26]. Figure 1 gives an overview of the mass spectra and compares excitation by the pump (a, intensity 5×10^{13} W/cm^{-2}) or probe pulse (d, 1.8×10^{13} W/cm^{-2}) only. Excitation by two pulses is shown in Figures 1b and c; in (b) C_{60} is first excited by the pump pulse and probed 50 ps later while in (c) the pump pulse follows the probe. Figure 2a-c show the corresponding spectra expanded in the region of small masses for 12-36 u and 84-180 u. For each of the single pulses, very little if any small fragments, C^+, C_2^+, C_3^+ etc., are produced. However, when adding the second, delayed pulse, a significant signal is observed for these small fragments. When the pump pulse arrives first, these signals are substantially larger than for the opposite sequence. We also note a significant reduction of the parent ions, in particular C_{60}^+, without a corresponding strong enough increase of the heavy fragments C_{60-2m}^{r+}.

Several scenarios can in principle explain these observations, and no final conclusion will be reached presently. We propose, however, as one plausible chain of processes fast fragmentation of neutral C_{60} and/or C_{60}^{r+} by the first laser pulse. The neutral fragments are then ionized (and possibly further fragmented)

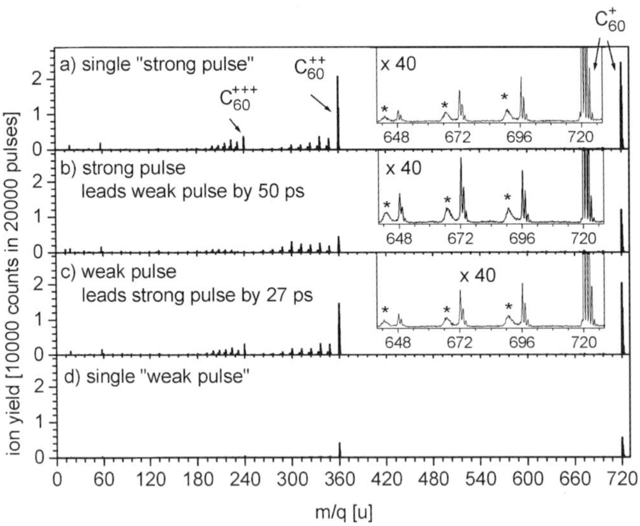

Figure 1: Overview mass spectra of C_{60} irradiated with 50 fs pulses at 800 nm: a) excitation by pump pulse only (5×10^{13} W/cm^2); b) probe pulse (1.8×10^{13} W/cm^2) delayed by 50 ps with respect to pump pulse (5×10^{13} W/cm^2); c) pump delayed by 25 ps with respect to probe; d) weak pulse only (2.5×10^{13} W/cm^2). Metastable fragments marked *.

by the second laser pulse. This is consistent with the work of Lykke and Wurz [27,28], who used nanosecond-laser pulses to preexcite and/or ionize C_{60} and probed the interaction products with a second postionizing laser. We recall here that using single pulse excitation in combination with "standard" time of flight-techniques allows one to detect only charged products, while information on neutral fragmentation can be gleaned only indirectly. Little is known about neutral fragments which are nevertheless expected either following the excitation of neutrals or as byproducts of the formation and destruction of large cationic fragments $C_{60-2m}{}^{r+}$. The exciting laser pulse initiates a (multielectron) dynamics followed by a variety of nonadiabatic transitions (electron-phonon coupling). The second pulse in principle probes the changes of the system - albeit often only indirectly, i.e. by fragmentation processes occurring long after the probe pulse interaction with the system. In contrast, for the proposed neutral fragmentation and post-ionization we expect immediate response.

For all charge states, the $C_{60}{}^{r+}$ ion abundance decreases substantially upon irradiation with the second pulse. This can be seen by comparing e.g. the $C_{60}{}^{+}$ signal in Figures 1b and 1c. In contrast, the signal from all singly and doubly charged heavy fragments increases while the signal of triply charged fragments decreases again. We have to distinguish this final response from an initially fast internal dynamics which obviously enables the system to absorb energy more readily from the second pulse. Unfortunately, a detailed study of the transients on a sub ps time scale is hampered by interference fringes characteristic to 1-C

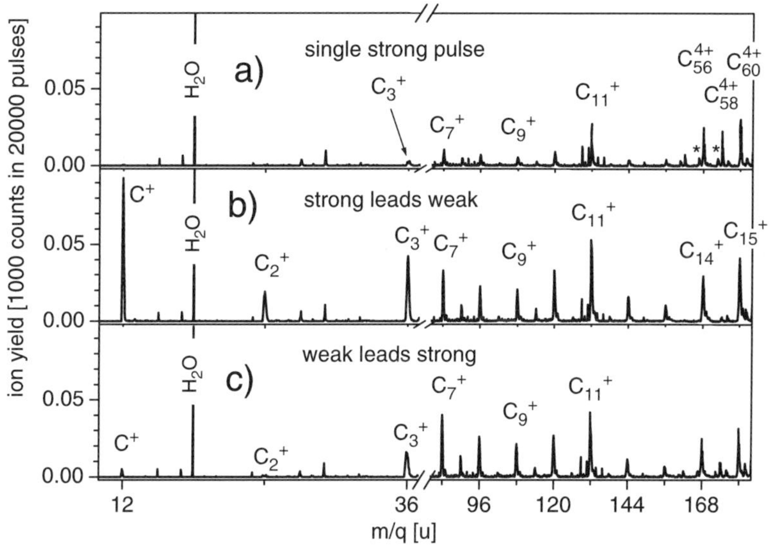

Figure 2: Mass spectra of C_{60} as shown in Fig. 1a-c, but blown up in the small mass range. Note the break at mass 36, showing the fragments C_3^+, C_2^+ and C^+ when the pump pulse arrives first. The broadening of the mass peaks indicates initial kinetic energies of several eV.

pump-probe experiments. A closer inspection shows that all changes in the heavy mass signals occur with a time constant of about 250 fs or shorter – characteristic for electron phonon coupling. Hence, the first pulse excites the system which then relaxes in such a way that additional energy deposition by the second pulse is facilitated. This is followed by fragmentation on a time scale typically ≥ ns, i.e. much longer than our pump-probe window. Apparently this sequence of processes is the more efficient, the more energy is already deposited by the pump process, i.e. the larger the structural changes of the system are.

Figure 3 shows the measured time-dependence for the small fragments. In contrast to the heavy fragments, the smallest fragments C^+, C_2^+, C_3^+ exhibit a dynamics on the 10 to 50 ps time scale. When the pump pulse initiates the multielectron dynamics in the parent (probably C_{60}), a dramatic increase in the C^+ - C_3^+ ion yield is observed as the delay between the pulses is increased. This dynamics can be modelled by single exponential rise functions with time constants of 11 ps, 12 ps, and 18 ps for C^+, C_2^+ and C_3^+, respectively, while C_4^+ shows essentially no dynamics at these conditions. Intermediate fragments such as C_7^+ and C_{11}^+, shown in Figure 3a, give some indication of an opposite behavior; a fraction of the signal decreases as the delay increases, indicating that these clusters may in part be transient products. In contrast, no indication of

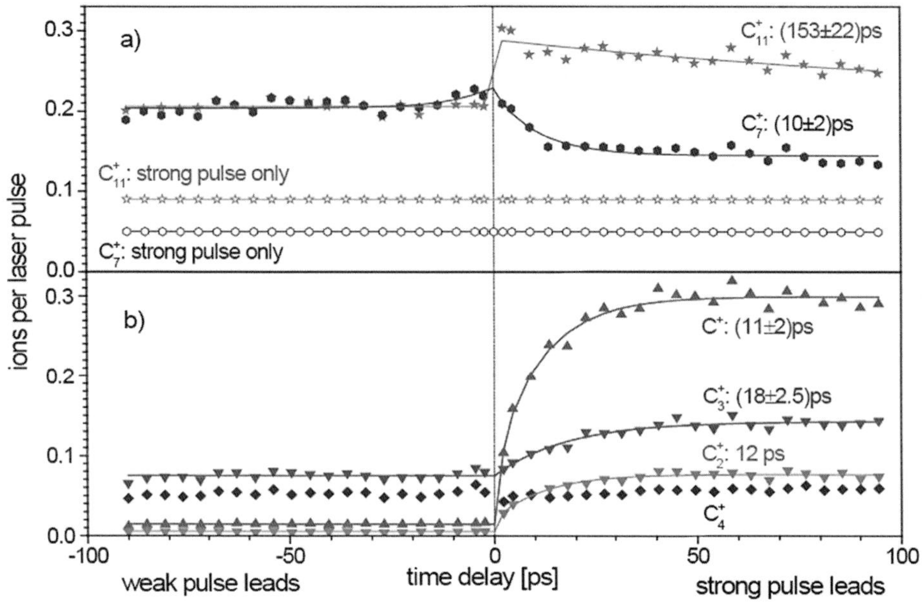

Figure 3: Delay time dependent ion signals for medium and small masses from C_{60} after irradiation with 800 nm pump (5×10^{13} W/cm^2) and probe (1.8×10^{13} W/cm^2) pulses. Empty symbols show the ion yield for the pump pulse only. Numbers at the fits give exponential rise and decay times.

a correspondingly delayed decrease for any of the heavy fragments C_{60-2m}^{r+} or parent ions C_{60}^{r+} is observed. Thus, we propose that the small cations arise from neutral parents – most likely from highly excited neutral C_{60} – which fragments more or less directly (as opposed to the statistical behavior held responsible for the heavy fragments). These fragments are then post-ionized by the probe laser pulse or – if larger intermediates are involved – may be generated by dissociative ionization processes. Alternative explanations, such as a specific change of absorption bands on a ps time scale can be ruled out in view of the time constant (250 fs) for such relaxation and rearrangement established for the heavy fragments.

Clearly, the observed fragmentation times on the order of some 10 ps indicate a non statistical decay; for comparison we estimate from simple RRK considerations [29] that e.g. a unimolecular C_2 evaporation from C_{60} would require internal energies as high as about 200 eV, while the very low abundance of C_{60-2m}^+ fragments detected shows that only a small part of all parent molecules contains energies above 100 eV.

3.2. Pump probe spectroscopy with ultrashort pulses

In the following, time-resolved one-color pump-probe mass spectroscopy with ultrashort pulses will be applied to learn more about energy relaxation in the initially excited C_{60}. The goal is to observe directly fingerprints of multi-electron effects (MAE/NMED) in the initial excitation steps of C_{60} irradiated with ultrashort 9 fs pulses. More specifically, we want to see in more detail how long energy is stored in the excited electron system after absorption of the first pulse - or alternatively see to what extent ionization occurs through a genuine multiphoton process once the pulse duration is short enough. In general, the coupling of excited electrons to atomic motion leads to nuclear rearrangement in the ionic or in the neutral molecular system. According to recent theoretical work [17,30], this results in characteristic oscillations which may be observed with time-resolved mass spectroscopy. Both the density of excited electrons and the nuclear geometry is expected to effect the photoionization yield of C_{60} in a time-dependent study. For pump-probe

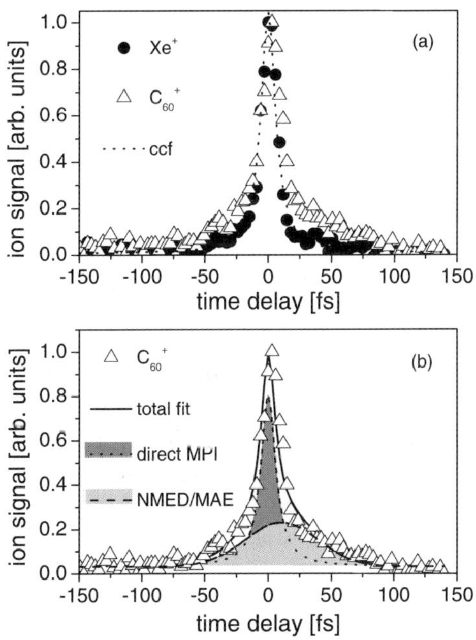

Figure 4: (a) C_{60}^+ ion yield (open triangles) as a function of the time-delay between pump (7.9×10^{13} W/cm^2) and probe pulse (6.8×10^{13} W/cm^2, normalized to the maximum signal. t = 0 has been determined by the autocorrelation of Xe^+ (closed circles) measured simultaneously. (b) Contributions from direct MPI (dark grey shaded) and NMED/MAE (light grey shaded) refer to our tentative deconvolution of the C_{60}^+ photoion yield, for details see the text.

spectroscopy, we split the radiation behind the hollow-fibre compressor and the double-chirped mirrors into two beam paths without lengthening the light pulses. The ultrashort pump pulse of 7.9×10^{13} W/cm^2 intensity deposits energy solely in the electronic systems during the interaction. The dynamics of the energy redistribution within the electronic and nuclear degrees of freedom is then probed by a delayed, slightly less intense probe pulse (6.8×10^{13} W/cm^2). To avoid the typical interference fringes at short delay times, both laser pulses have perpendicular polarization. Figure 4 shows the measured time dependence of the normalized C_{60}^+ ion signal. Since both pulses, pump and probe individually acting on the molecule, already cause some ionization at these intensities, the sum of single pulse signals has been tentatively subtracted from the pump-probe ion yield. We are aware of the limitations of this procedure, as many photons are involved even in the first excitation step. However, the graph gives first insight into the time evolution of the photoinduced processes. In particular, the comparison with the simultaneously measured autocorrelation of Xe^+ included in Figure 4a is instructive, a genuine direct MPI process with one active electron determining the system's response,.

The C_{60}^+ ion signal is clearly broadened at the bottom of the spectrum. As shown in Figure 4b, the deconvolution of the total ion yield results into two main contributions: i) direct MPI of C_{60} from the neutral ground state to the continuum state (dark grey shaded), which essentially follows the Xe autocorrelation and ii) a significant contribution exhibiting dynamics on a sub-100 fs time scale (light grey shaded) slightly shifted towards positive time delays (where the stronger pump pulse leads the weaker probe pulse). These findings give some evidence to multi-electron excitation in a sub-ensemble of C_{60} in the laser fullerene interaction. Supported by recent theoretical work [16,17,30], the present results indicate that, in addition to the direct MPI process, there is a probability to initially excite several electrons into the t_{1u} resonant state, which in turn might act as a bottleneck (doorway) to ionization. The observed dynamics are comparable to the characteristic time for thermalization within the electronic system due to electron-electron scattering, previously concluded from single pulse experiments [13,21]. As intuitively expected, the density of the hot electron cloud depends on the laser intensity and its time evolution on the electron-electron scattering time constant [17]. The excited electron density in the doorway state determines the transition probability into the ionic continuum. This explains why the ion distribution due to doorway state excitation is slightly shifted to positive time-delays, as shown in Figure 4b; the stronger pulse helps to generate sufficient population in the doorway state. From the comparison of the maximum signals due to direct MPI (SAE) and MAE driven ionization processes, it can be concluded that the latter covers less than 25% due to the ultrashort laser pulses.

Another motivation for this one color pump-probe experiment was to find evidence for exciting a specific mode in the t_{1u} state – i.e. see oscillations in the pump-probe response. No evidence for such a process is seen above noise level. The reason might be two-fold. First, 9 fs laser pulses are so short that the coupling to either electron gas or nuclear body is weak - hence no "memory" of the pump pulse remains in the molecular system; only the electronic system is excited. Alternatively one might argue that for any nuclear motion induced, the strong coupling of various vibrational modes in intense fs laser fields mixes many rovibrational states, and no specific oscillation can be observed. We note that recent experiments in our laboratory using 2-color blue pump/red-probe scheme with slightly longer laser pulses of 27 fs reveal some distinct ultrafast dynamics initiated by the blue pump pulse [31].

4. Conclusion

Evidence for the influence of multi electron dynamics in ultrafast fragmentation and ionization dynamics in C_{60} has been presented. Fragmentation initiated by 50 fs pulses is substantially enhanced if the ionizing pulse is preceded by a strong pump pulse; after excitation of several electrons, the energy deposited into the electronic system is redistributed among many degrees of freedom. For large fragments, the fragmentation process occurs on a time scale of nanoseconds to microseconds in the ionic system. In contrast, small fragments are created on a femto- and picosecond time scale, most probably by dissociative processes in the neutral system and subsequent ionization.

Using still shorter pulses of 9 fs, energy absorbed by the electronic system is redistributed within sub 100 fs among the electronic system, and no evidence for electron-phonon coupling is observed at intensities of about 7×10^{13} W/cm^2. Further experimental and theoretical efforts are needed to fully understand the complex charge and energy deposition and redistribution process in large finite systems in strong laser fields.

Acknowledgement
This work has been financially supported by the Deutsche Forschungsgemeinschaft through Sfb 450, Subproject A2.

References

1. A. N. Markevitch, D. A. Romanov, S. M. Smith, H. B. Schlegel, M. Y. Ivanov, and R. J. Levis, Phys. Rev. A 69 (2004) 013401.
2. M. Lezius, V. Blanchet, M. Y. Ivanov, and A. Stolow, J. Chem. Phys. 117 (2002) 1575.
3. T. E. Dermota, Q. Zhong, and A. W. Castleman, Chem. Rev. 104 (2004) 1861.
4. A. Assion, T. Baumert, M. Bergt, T. Brixner, B. Kiefer, V. Seyfried, M. Strehle, and G. Gerber, Science 282 (1998) 919.
5. S. A. Rice, Nature 409 (2001) 422.
6. M. Shapiro and P. Brumer: Principles of the Quantum Control of Molecular Processes, Wiley, New York, 2003.

7. R. J. Levis, G. M. Menkir, and H. Rabitz, Science 292 (2001) 709.
8. A. G. Stepanov, M. T. Portella-Oberli, A. Sassara, and M. Chergui, Chem. Phys. Lett. 358 (2002) 516.
9. S. C. O'Brien, J. R. Heath, R. F. Curl, and R. E. Smalley, J. Chem. Phys. 88 (1988) 220.
10. H. Hohmann, C. Callegari, S. Furrer, D. Grosenick, E. E. B. Campbell, and I. V. Hertel, Phys. Rev. Lett. 73 (1994) 1919.
11. M. Tchaplyguine, K. Hoffmann, O. Dühr, H. Hohmann, G. Korn, H. Rottke, M. Wittmann, I. V. Hertel, and E. E. B. Campbell, J. Chem. Phys. 112 (2000) 2781.
12. V. R. Bhardwaj, P. B. Corkum, and D. M. Rayner, Phys. Rev. Lett. 93 (2004) 043001.
13. E. E. B. Campbell, K. Hansen, K. Hoffmann, G. Korn, M. Tchaplyguine, M. Wittmann, and I. V. Hertel, Phys. Rev. Lett. 84 (2000) 2128.
14. M. Boyle, K. Hoffmann, C. P. Schulz, I. V. Hertel, R. D. Levine, and E. E. B. Campbell, Phys. Rev. Lett. 87 (2001) 273401.
15. V. R. Bhardwaj, P. B. Corkum, and D. M. Rayner, Phys. Rev. Lett. 91 (2003) 203004.
16. B. Torralva, T. A. Niehaus, M. Elstner, S. Suhai, T. Frauenheim, and R. E. Allen, Phys. Rev. B 64 (2001) 153105.
17. G. P. Zhang, X. Sun, and T. F. George, Phys. Rev. B 68 (2003) 165410.
18. E. E. B. Campbell, G. Ulmer, and I. V. Hertel, Phys. Rev. Lett. 67 (1991) 1986.
19. F. Rohmund, M. Hedén, A. V. Bulgakov, and E. E. B. Campbell, J. Chem. Phys. 115 (2001) 3068.
20. I. V. Hertel, T. Laarmann, and C. P. Schulz, in Adv. At. Mol. Opt. Phys., B. Bederson, H. Walther (Eds.), Elsevier, Amsterdam, 2005, p. 219.
21. K. Hansen, K. Hoffmann, and E. E. B. Campbell, J. Chem. Phys. 119 (2003) 2513.
22. P. Agostini, F. Fabre, G. Mainfray, G. Petite, and N. K. Rahman, Phys. Rev. Lett. 42 (1979) 1127.
23. M. Boyle, M. Hedén, C. P. Schulz, E. E. B. Campbell, and I. V. Hertel, Phys. Rev. A 70 (2004) 051201.
24. M. Boyle, T. Laarmann, K. Hoffmann, M. Hedén, E. E. B. Campbell, C. P. Schulz, and I. V. Hertel, Eur. Phys. J. D 36 (2005) 339.
25. G. Sansone, G. Steinmeyer, C. Vozzi, S. Stagira, M. Nisoli, S. De Silvestri, K. Starke, D. Ristau, B. Schenkel, J. Biegert, A. Gosteva, and U. Keller, Appl. Phys. B 78 (2004) 551.
26. M. Boyle, T. Laarmann, I. Shchatsinin, C. P. Schulz, and I. V. Hertel, J. Chem. Phys. 122 (2005) 181103.
27. K. R. Lykke and P. Wurz, J. Phys. Chem. 96 (1992) 3191.
28. K. R. Lykke, Phys. Rev. A 52 (1995) 1354.
29. H. Hohmann, R. Ehlich, S. Furrer, O. Kittelmann, J. Ringling, and E. E. B. Campbell, Z. Phys. D 33 (1995) 143.
30. G. P. Zhang and T. F. George, Phys. Rev. Lett. 93 (2004) 147401.
31. I. Shchatsinin, A. Stalmashonak, C. P. Schulz, T. Laarmann, N. Zhavoronkov, and I. V. Hertel, to be published.

Femtochemistry VII
Fundamental Ultrafast Processes in Chemistry, Physics, and Biology
A.W. Castleman, Jr. and M.L. Kimble

Chirped photoelectron attochemistry

Andre D. Bandrauk[1], Szczepan Chelkowski[1], and Gennady L. Yudin[1,2]

[1]*Laboratoire de Chimie Th´eorique, Universit´e de Sherbrooke, Sherbrooke, Qu´ebec J1K 2R1, Canada*
[2]*National Research Council of Canada, Ottawa, Ontario K1A 0R6, Canada*

Abstract

A theoretical study is presented suggesting photoelectron spectroscopy measurement of electron dynamics in molecules on "attosecond" time scales. We study analytically the photoionization of a coherent superposition of electronic states and show that chirped pulses can measure attosecond time scale electron dynamics just as effectively as transform-limited attosecond pulses of the same bandwidth. Molecular interference modulates the time-dependent photoelectron spectrum and asymmetry. We show that the important parameter for attosecond science is not the pulse duration, but the bandwidth of phased radiation.

1. Introduction

Advances in current laser technology allow experimentalists to have access to new laser sources for examining molecular structure and dynamics on the natural time scales for atomic (vibrational) motion in molecules, the femtosecond (fs, 10^{-15} s) [1]. These new sources are ultrashort (t_p <10 fs) and intense ($I \geq 10^{14}$ W/cm^2) [2]. Such pulses lead to a new regime of laser-matter interaction, the nonlinear, non-perturbative regime. Thus since the atomic unit (a. u.) of electric field strength $E_0 = e/a_o^2 = 5 \times 10^9 V{=}cm$ (a_0=1 a. u. = 0.0529 nm) corresponds to an a. u. of intensity, $I_0 = cE_o^2/8\pi = 3.5 \times 10^{18}$ W/cm^2, laser intensities approaching I_0 have led to the discovery of new nonlinear phenomena in laseratom experimental and theoretical studies [2, 3], such as above-threshold ionization (ATI), the multiphoton equivalent of single-photon ionization, and high-order harmonic generation (HHG).

One of the seminal ideas emanating from theories of intense laser atom interactions is a simple quasistatic model of electron tunnel ionization [2] followed by *recollision* of the ionized electron with its parent ion [4]. The returning energy of the electron at the nuclei can be shown to satisfy a simple recombination law, which explains the maximum energy of harmonics emitted by such a recolliding electron with its parent ion,

$$N_m\hbar\omega = I_p + 3.17U_p. \quad (1)$$

N_m is the maximum harmonic order, I_p is the ionization potential, and U_p is the ponderomotive or oscillatory energy of an electron of mass m in a pulse of frequency ω and intensity I,

$$U_m[a.u.] = \frac{I}{4m\omega^2} = 3.4 \times 10^{21} I[W/cm^2]\lambda^2[nm]. \quad (2)$$

The corresponding field induces displacement of such an ionized electron and is given by the expression

$$\alpha[a.u.] = \frac{eE}{m\omega^2} = 2.4 \times 10^{-12}(I[W/cm^2])^{1/2}\lambda^2[nm]. \quad (3)$$

Thus at an intensity $I = 10^{14}$ W/cm^2 and wavelength λ= 1064 nm, U_p=10.5 eV and α=1.2 nm. In molecules, electron collisions can occur with neighboring ions, and this generally leads to even larger harmonic orders with photon energies of $6U_p$ and even up to $126U_p$, thus extending the atomic recollision maximum energy cut-off law 1 [5].

Using ultrashort intense laser pulses ($t \leq 5$ fs) results in continuum HHG in the regime of the cut-off law 1 from which, by selecting a slice of the spectrum, one can synthesize "*attosecond*" (asec) pulses. First experimental results on the production and measurements of sub-fs, asec, pulses from high-intensity ionization of Ar atoms was reported in the last few years [6, 7]. This opens up new possibilities for studying and controlling [8] electron dynamics in atoms and molecules on the electron, i. e., asec time scales. We have recently shown that asec pulses can be used to measure the absolute electric field of intense few-cycle laser pulses [9]. Furthermore this leads to "*asec control*" of ionization and thus to increased HHG intensities, allowing to create new single X-ray or VUV pulses as short as 250 asec [10]. Even shorter asec pulses have been predicted from exact numerical solution of coupled Maxwell - time-dependent Schrödinger equations (TDSE) for the molecular ion H_2^+ which represents a first complete treatment of photon-electron-nuclear dynamics [11]. We were able to demonstrate from solutions of such TDSE simulations how pump-probe experiments of the photoioionization of H_2^+ using asec pulses can allow one to study the evolution of the movement of localized electron wave packets, thus leading to new asec molecular spectroscopy [12]. Such experiments should now be feasible with the recent discovery of very large HHG from ions themselves at very high intensities [13, 14], thus extending asec pulse into the soft X-ray region.

Recent advances in ultrashort laser-pulse technology enables experimentalists to synthesize and characterize ultrashort pulses in the attosecond time regime [15–20]. The development of femtosecond technology has been an interplay between the needs of science, demanding ever faster measurements, and technology, producing ever shorter pulses [21]. Chirped pulses play an important role in technology; transform-limited pulses are used for measurement. Attosecond technology is a radical departure from the ultrafast science that proceeded it, see review [22]. Chirp is inherent in the attosecond generation process [4] except near the cut-off frequency [16, 17]. Pulses with the broadest phased bandwidth will have substantial chirp [18].

Measuring electron dynamics in atoms and molecules [23–25] is a natural initial target for attosecond science. When measurement is via photoelectrons, chirped pulses allow the same resolution limit as transform-limited pulses of the same bandwidth. What is essential is that the measurements are differential. That is, the photoelectron spectrum is resolved in momentum and in space.

Our method is based on broad bandwidth photoionization. Single-photon ionization from a stationary state produces photoelectrons that contain information on the optical pulse and the phase and structure (through the transition moment) of the initial state. If the state is known, measurement of the photoelectron spectra produced in the presence of a phased infrared field fully characterizes the attosecond pulse (see references in review [19] and more recent papers [26–28]); if, on the other hand, the attosecond pulse is known, measurement of photoelectron spectra produced from two or more coherently excited states determines all information on the associated bound-state wave packet.

Photoionization is related to a proposal to measure bound-state electron dynamics during the attosecond generation process itself [23]. Attosecond single-photon ionization has three advantages over strong-field techniques [24, 25]. Specifically, we can: (i) measure a bound electron state that is not perturbed by a strong laser field, (ii) correlate photoelectrons to ion fragments so that we can specify the molecular internal states and alignment direction, and (iii) precisely take into account the Coulomb potential in the molecular continuum.

2. Attosecond Molecular Photoelectron Spectroscopy

A. General results

We assume that a pump pulse has prepared a coherent superposition of two bound states,

$$\Psi(\mathbf{r},t) = \alpha_1\Psi_1(\mathbf{r},t) + \alpha_2\exp(-i\beta)\Psi_2(\mathbf{r},t), \tag{4}$$

where $\alpha_{1,2}^2$ are their populations ($\alpha_1^2+\alpha_2^2=1$), and β is an adjustable initial phase that depends on the excitation scheme. The coherently prepared state is the simplest wave packet that can be formed. The pump pulse does not need to be short.

Now we consider single photon photoionization induced by a probe attosecond x-ray pulse. The probe pulse is precisely timed with respect to the phase of the pump pulse. Phasing occurs naturally in attosecond technology if the initial wave packet is excited by the fundamental pulse or its harmonics. We define the vector potential of such a pulse as a linearly chirped Gaussian

$$\mathbf{A}(t) = \mathbf{e}(A_0/2)\exp[-i\Omega t - (t-t_0)^2/2(1-i\xi)\tau^2], \tag{5}$$

where Ω is the central frequency of the attosecond pulse, e, A_0, and t_0 are the linear polarization vector, amplitude, and peak of the pulse. Positive dimensionless chirp ξ corresponds to the instantaneous frequency increasing with time. The pulse duration is $\tau_{FWHM} = 2\tau\sqrt{\ln 2}\sqrt{1+\xi^2}$. Eq. (5) describes a pulse whose bandwidth is invariant to the pulse chirp.

The photoionization amplitude for the coherently coupled state (4) and three-fold differential angle resolved photoelectron spectrum are given by

$$M_{coh} = M_0\left[\gamma_1 M^{(1)} + \gamma_2 e^{-i[\delta(t_0)+\phi(\xi,\varphi)]} M^{(2)}\right], \tag{6}$$

$$\frac{d^3P_{coh}(p,\theta,\varphi)}{dp\,d\Omega_e} \equiv S(p,\theta,\varphi) = p^2 \mid M_{coh} \mid^2, \tag{7}$$

where $M_0 = A_0\varepsilon\tau[\pi(1-i\xi)/2]^{1/2}$, $M^{(1)}$ and $M^{(2)}$ are the photoionization amplitudes for two bound states in the coherent superposition (4), ε is a complex phase factor, $|\varepsilon|=1$, $\gamma_{1,2} = \alpha_{1,2}\exp(-\omega_{1,2}^2\tau^2/2)$, $\omega_{1,2} = p^2/2 + I_p^{(1,2)} - \Omega$, $I_p^{(1,2)}$ are the ionization potentials of states $\Psi_1(\mathbf{r},t)$ and $\Psi_2(\mathbf{r},t)$, and $d\Omega_e = \sin\theta d\theta d\varphi$ is the solid angle (unless stated otherwise, we use atomic units $e = m_e = \hbar=1$).

Eq. (6) and Fig. 1 show that there are two contributions to the photoelectron spectrum. The bandwidths $\Delta_{1,2}$ of electron wave packets in Fig. 1 are defined by functions $\gamma_{1,2}$ in the amplitude (6). It is of importance that both $\gamma_{1,2}$ and $\Delta_{1,2}$ do not depend on linear chirp of the x-ray pulse (for example, the transform-limited τ_{FWHM} = 24.19 asec (i. e., 1 a. u.) pulse and chipred τ_{FWHM} = 250 asec pulse at $\xi \approx 10$ generate electron wave packets with the same bandwidths). For broad-bandwidth pulses, a photoelectron wave packets overlap and this spectral overlapping *is the same* for a given transform-limited pulse ($\xi \approx 0$) and its chirped modifications.

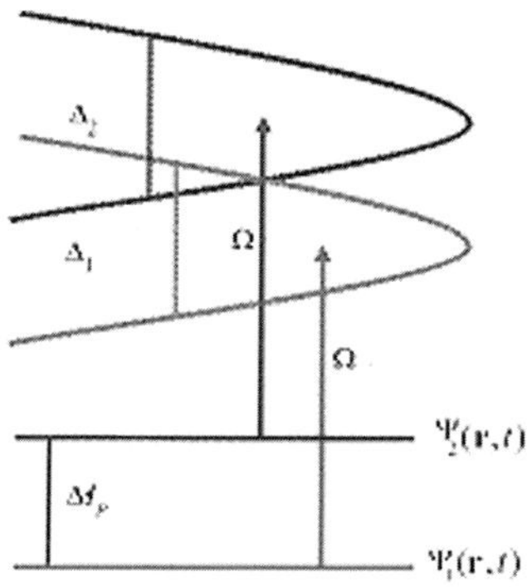

FIG. 1. A sketch of an electron wave packet generated by photoionization of a coherent superposition of electronic states. The bandwidths $\Delta_{1,2}$ and spectral overlapping of wave packets do not depend on linear chirp of x-ray pulse.

The resolution limit of a wave packet measurement is independent of the chirp. The interference, however, is different in details, and this difference offers a new possibility for using chirped pulses to make single-time delay attosecond dynamics measurements. The interference structure is determined by the full phase shift between amplitudes in (6). This phase is universal and does not depend on structure of the photoionization amplitudes $M^{(1,2)}$. However it depends on the chirp. The time-dependent phase shift $\delta(t) = \beta + \Delta I_p t$ parameterizes the bound electron motion. The chirp-dependent part of the full phase shift [25]

$$\phi(\xi,p) = \xi\tau^2\Delta I_p[p^2/2 + I_p^{(av)} - \Omega] \tag{8}$$

describes the chirped interference inside the photoelectron spectrum. $\Delta I_p = I_p^{(1)} - I_p^{(2)}$ is the level separation which depends on R in molecular case and $I_p^{(av)} = (I_p^{(1)} + I_p^{(2)})/2$ is the average ionization potential of coherently coupled electronic states.

Both phase factors are periodic. The factor exp $[-i\delta(t_0)]$ which corresponds to electron "*hopping*" [12, 24] is the periodic function of time with the period $T_0 = 2\pi/\Delta I_p$, and the factor exp $[-i\,\phi(\xi,p)]$ is a periodic function of the photoelectron energy $E_p = p^2/2$ with the period [25]

$$E_0 = T_0/\mid\xi\mid\tau^2 = 2\pi/\mid\xi\mid\tau^2\Delta I_p. \tag{9}$$

Experimentally it may be more useful to measure the momentum asymmetry than the momentum spectrum because the momentum asymmetry is self calibrating. In analogy with the integral asymmetry [32] (see also [24]), the absolute (Δ) and normalized (Δ_N) differential asymmetries are defined as

$$\Delta(p,\theta,\varphi) = S(p,\theta,\varphi) - S(p,\pi-\theta,\varphi), \tag{10}$$

$$\Delta_N(p,\theta,\varphi) = \frac{S(p,\theta,\varphi) - S(p,\pi-\theta,\varphi)}{S(p,\theta,\varphi) + S(p,\pi-\theta,\varphi)}. \tag{11}$$

The asymmetry (10) is independent of the azimuthal angle φ since the molecule is aligned along the z-axis.

B. Attosecond photoelectron spectra

So far, our description is completely general. However, for illustrative purposes it is useful to select a specific case. As a theoretically tractable example that captures the essence of our approach, we consider the H_2^+ molecular ion with or without a coherent superposition of two electronic states, $\sigma_g 1s$(state|1>) and $\sigma_g 2p_O$(state|2>). The electron and nuclear time scales in the molecule remain well separated so the nuclei are considered fixed. We assume that the molecular ion is aligned. Alignment can be effectively achieved experimentally for unstable molecules since, following ionization, the molecular fragments determine the alignment with high accuracy [29–31]. In our three-dimensional model, the molecular ion and electric field of the attosecond pulse are aligned along the z axis in Cartesian space.

We describe the initial H_2^+ wave functions as a linear combinations of atomic orbitals centered on each nucleus:

$$\Psi_b(\mathbf{r}) = \sum_i c_i \phi_i^{(1)}(\mathbf{r}_1) + \sum_j c_j \phi_j^{(2)}(\mathbf{r}_2), \tag{12}$$

where vectors $\mathbf{r}_{1,2} = \mathbf{r} \pm \mathbf{R}/2$, $\mathbf{r}$ is the electron coordinate, and $\mathbf{R}$ is directed from the nucleus 1 to the nucleus 2. The number of the atomic basis functions in the linear combination and their variational Slater parameters $a = 1/Z_{eff}$ depend on internuclear distance R. In homonuclear molecules at large R, Eq. (12) can be reduced to a linear combination of two identical atomic orbitals:

$$\Psi_b^{\pm}(\mathbf{r}) \approx c_{\pm}[\phi(\mathbf{r}_1) \pm \phi(\mathbf{r}_2)]. \tag{13}$$

We calculate the molecular photoionization amplitude using the final two-center Coulomb wave function. For the simplest choice (13), the molecular ionization transition amplitudes are determined by atomic amplitudes:

$$M_{mol}^{\pm} \approx \chi_{\pm} c_{\pm} M_{at}, \tag{14}$$

where $\chi_{\pm}$ are the molecular interference factors [24, 25]:

$$\chi_{\pm} = N_p[\exp(i\mathbf{pR}/2)G(-\mathbf{R}) \pm \exp(-i\mathbf{pR}/2)G(\mathbf{R})]. \tag{15}$$

$N_p = \exp(\pi/2p)\Gamma(1 + i/p)$ is the normalization factor, and $G(\mathbf{R})$ is the confluent hypergeometric function, $G(\mathbf{R}) = F(i/p;\ 1;\ i(pR + \mathbf{pR}))$.

For isolated initial molecular states, the attosecond photoionization spectra are displayed in Figures 2 and 3. The appropriate dipole atomic photoionization amplitudes are calculated in [24, 25]. The spectra are generated by the chirped $\tau_{FWHM} = 100\sqrt{1+\xi^2}$ asec pulse. The angle between the vectors **e** and **p** is θ_e=30°, Ω= 100eV . With changing of R, the molecular ionization potentials slightly change, but interference patterns are defined mostly by the values of pR and **pR** in the interferences factors (15). The photoelectron signal for ionization of the $\sigma_g 1s$ initial state at R=6.8 (Figure 2) and for the $\sigma_g 2p_O$ initial state at R=8.0 (Figure 3) are suppressed due to destructive molecular interference. The insets correspond to the different interference factors, χ_+ in the case of $\sigma_g 1s$ initial state and χ_- in the case of $\sigma_g 2p_O$ initial state. Note that the spectra are very sensitive to all parameters of the attosecond pulse and molecular geometry.

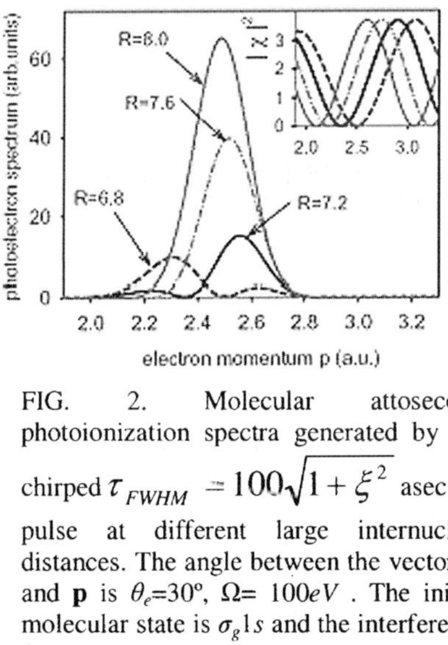

FIG. 2. Molecular attosecond photoionization spectra generated by the chirped $\tau_{FWHM} = 100\sqrt{1+\xi^2}$ asec pulse at different large internuclear distances. The angle between the vectors **e** and **p** is θ_e=30°, Ω= 100eV . The initial molecular state is $\sigma_g 1s$ and the interference factor $\chi = \chi_+$.

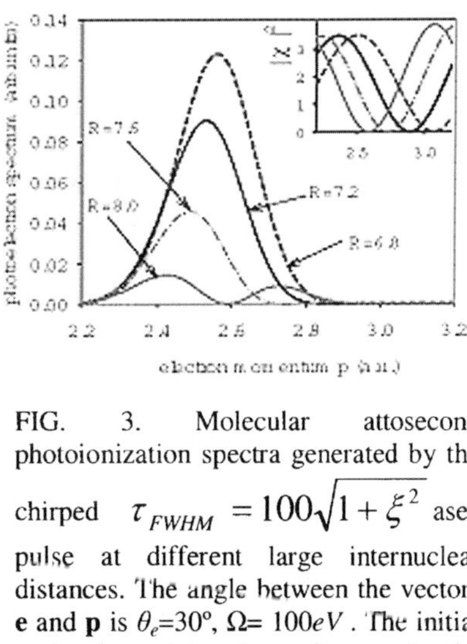

FIG. 3. Molecular attosecond photoionization spectra generated by the chirped $\tau_{FWHM} = 100\sqrt{1+\xi^2}$ asec pulse at different large internuclear distances. The angle between the vectors **e** and **p** is θ_e=30°, Ω= 100eV . The initial molecular state is $\sigma_g 2p_0$ and the interference factor $\chi = \chi_-$.

The absolute molecular asymmetries for the case of $\sigma_g 1s + \sigma_u 2p_O$ coherently coupled states are displayed in Fig. 4. The populations of coherently coupled states are equal, $(\alpha_1^2 = \alpha_2^2)$, central photon energy Ω = 100 eV, and angle between polarization and photoelectron momentum $\theta = \pi/3$. The photoelectron signal and, therefore, the absolute asymmetry at R=12 is suppressed due to destructive molecular interference. The minimum value of $|\chi_+|^2$ in Fig. 4 is in the range of the photoelectron momenta.

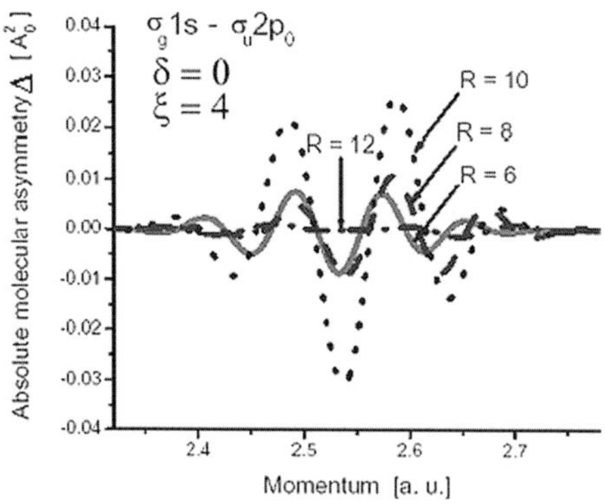

FIG. 4. Absolute molecular asymmetries for the case of $\sigma_g 1s + \sigma_u 2\ p_0$ coherently coupled states. The populations of coupled states are equal, ($\alpha_1^2 = \alpha_2^2$), central photon energy Ω= 100 eV, chirp ξ=4, and angle between polarization and photoelectron momentum $\theta = \pi/3$.

3. Conclusions

In conclusion, attosecond pulses, because of their large energy bandwidth and high frequency, project bound-state electron wave packets into the continuum. There, the wave packet can be fully characterized through its angle-resolved photoelectron spectrum. In practice, current attosecond laser sources are not intense enough and operate at a low repetition rate, making a full reconstruction impractical. However, they are perfectly adequate to resolve attosecond dynamics by measuring the changing momentum asymmetry of the spectrum [24, 25]. This would allow attosecond electron dynamics experiments that are analogous to vibrational dynamics experiments, which have been so influential in photochemistry [1]. As much higher repetition rate systems are developed [33], full characterization of both the spatial and temporal structures of electron wave packets will become possible.

Long chirped attosecond pulses can measure attosecond time scale electron dynamics. The dynamics can be measured just as effectively as if the pulses were transform-limited. Pump-probe time delay spectroscopy is possible, but not necessary if chirped pulses are used [25]. The oscillations in the photoelectron spectra are described by the universal phase shift between the interfering photoionization amplitudes. The phase shift allows us to read the dynamics at a single time delay. The differential normalized asymmetry of the photoelectrons reaches 100% [25]. All electrons in some momentum windows are sent in a given direction. The direction changes with time delay.

Until now figure-of-merit of attosecond pulses has been the pulse duration. Our results imply that the emphasis of attosecond technology should shift from producing transformlimited pulses to producing broad bandwidth pulses. The pulses can be chirped or unchirped. All that is needed is that the chirp is well characterized.

Acknowledgements

We are grateful to P. B. Corkum and M. Yu. Ivanov, and F. Krausz for valuable discussions.

References

1. A. H. Zewail, J. Phys. Chem. A 104 (2000) 5660.
2. T. Brabec and F. Krausz, Rev. Mod. Phys. 72 (2000) 545.
3. M. Gavrila, Atoms in Intense Laser Field, Academic Press, N.Y., 1992.
4. P. B. Corkum, Phys. Rev. Lett. 71 (1993) 1994.
5. A. D. Bandrauk and H.Yu, J. Phys. B 31 (1998) 4243; Phys. Rev. A 59 (1999) 539.
6. M. Nentschel, *et al.*, Nature (London) 414 (2001) 509.
7. R. Kienberger, *et al.*, Science 297 (2002) 1144.
8. A. D. Bandrauk, Y. Fujimura, and R. Gordon (Eds.), Laser Control and Manipulation of Molecules, ACS symposium series, Washington, DC, v. 821, 2002.
9. A. D. Bandrauk, S. Chelkowski, and H. S. Nguyen, Phys. Rev. Lett. 89 (2002) 283903.
10. A. D. Bandrauk and H. S. Nguyen, Phys. Rev. A 66 (2002) 031401.
11. A. D. Bandrauk, S. Chelkowski, and H. S. Nguyen, J. Mol. Struct. 735C (2004) 203.
12. A. D. Bandrauk, S. Chelkowski, and H. S. Nguyen, Int. J. Quant. Chem. 100 (2004) 834.
13. E. A. Gibson, *et al.*, Phys. Rev. Lett. 92 (2004) 033001.
14. S. Seres, et al., Phys. Rev. Lett. 92 (2004) 163002.
15. M. Drescher, *et al.*, Nature (London) 419 (2002) 803.
16. A. Baltuška, *et al.*, Nature (London) 421 (2003) 611.
17. R. Kienberger, *et al.*, Nature (London) 427 (2004) 817.
18. Y. Mairesse *et al.*, Science 302 (2003) 1540.
19. P. Agostini and L. F. DiMauro, Rep. Progr. Phys. 67 (2004) 813.
20. R. López-Martens *et al.*, Phys. Rev. Lett. 94 (2005) 033001.
21. J.-C. Diels and W. Rudolph, Ultrashort Laser Pulse Phenomenon: Fundamentals, techniques and applications on a femtosecond time scale, Academic Press, Boston, 1996.
22. Levesque and P. B. Corkum, Can. J. Phys. (to be published).
23. H. Niikura, D. M. Villeneuve, and P. B. Corkum, Phys. Rev. Lett. 94 (2005) 083003; H. Niikura *et al.*, J. Mod. Optics 52 (2005) 453.
24. G. L. Yudin, *et al.*, Phys. Rev. A 72 (2005) 051401(R).
25. G. L. Yudin, A. D. Bandrauk, and P. B. Corkum, (submitted).
26. J. Itatani, *et al.*, Laser Phys. 14 (2004) 344 (2004).
27. Y. Mairesse and F. Quéré, Phys. Rev. A 71 (2005) 011401(R); F. Quéré, Y. Mairesse, and J. Itatani, J. Mod. Optics 52 (2005) 339.
28. E. Cormier, *et al.*, Phys. Rev. Lett. 94 (2005) 033905.
29. H. Niikura, *et al.*, Nature (London) 417 (2002) 917; 421 (2003) 826.
30. J. Ullrich, *et al.*, Rep. Prog. Phys. 66 (2003) 1463; C. Dimopoulou, *et al.*, J. Phys. B 38 (2005) 593.
31. T. Weber, *et al.*, Phys. Rev. Lett. 92 (2004) 163001; Nature (London) 431 (2004) 437; A. L. Landers, *et al.*, Phys. Rev. A 70 (2004) 042702.

32. A. D. Bandrauk, S. Chelkowski, and N. H. Shon, Phys. Rev. Lett. 89 (2002) 283903; Phys. Rev. A 68 (2003) 041802(R).
33. R. J. Jones, *et al.*, Phys. Rev. Lett. 94 (2005) 193201.

Femtochemistry VII
Fundamental Ultrafast Processes in Chemistry, Physics, and Biology
A.W. Castleman, Jr. and M.L. Kimble

Intense Femtosecond Laser Driven Explosions of Heteronuclear Clusters

T. Ditmire, M. Hohenberger, D. R. Symes, K. W. Madison, F. Buersgens, R. Hartke, J. Osterhoff, A. Henig, A. Edens

Texas Center for High intensity Laser Science, Dept. of Physics, The University of Texas, Austin, TX 7871

1. Introduction

Recently, there has been intense interest in the explosions of atomic and molecular clusters ionized by intense ultrashort pulses at intensity 10^{16} – 10^{18} W/cm^2 [1-3]. In large high Z clusters, space charge forces confine the electrons to the charged ion sphere, and a micro-plasma is created by field ionization [4]. In smaller clusters, or lower Z clusters, such as few nm diameter hydrogen or deuterium clusters, the laser field at these intensities is strong enough to remove field ionized electrons directly from the cluster [5]. In this case, the cluster explodes by the Coulomb repulsion of the closely spaced cluster ions. This effect has been exploited in gases of deuterium clusters irradiated by intense fs laser pulses to produce plasmas of multi-keV average ion energy which emit a burst of neutrons from DD fusion subsequent to the clusters' explosions [6-7]. This effect may be of interest for the development of compact, laser driven fast neutron sources.

One interesting and unexplored aspect of strong field driven Coulomb explosions in large clusters are explosions of clusters formed from heteronuclear molecules. Some initial experimental studies of explosions in large heteronuclear molecular clusters (like HI, NH_3, and pyridine) driven with modest intensity (~10^{14} W/cm^2) laser pulses have been presented recently [8-10]. The motivation for studying Coulomb explosions of mixed species clusters in the context of exploding cluster fusion was advanced recently by Last and Jortner [11]. They showed, through a series of molecular dynamic simulations, that the energies of deuterons from exploding heteronuclear clusters, like D_2O, could exhibit enhancements in ion energy through a dynamical effect in the

Coulomb explosion. The Coulomb explosion of a single species cluster, like D_2 clusters, will eject deuterons with an energy given directly by the potential energy of the ion as it is initially in the fully stripped cluster. In this simple picture, if higher Z ions are present and ionized to higher charge states by the laser, the presence of the more highly charged ions should increase the stored potential energy of the deuterons and increase the subsequent energy of the ejected deuterons. Last *et al.* furthermore showed that, provided that the kinematic parameter $\eta = q_D m_D / q_B m_B$ is greater than 1 (where q_B and m_B are the charge and mass of the heavier, high Z ion and m_D, q_D are the mass and charge of the light ion), then the light deuterons will outrun the heavier ion, B, and explode in an out shell with a higher average energy than would be expected from the naïve estimate of ion energy based on initial potential energy. Ref. [12] showed that fusion yield could be substantially increased in plasmas formed from explosions of heteronclear clusters over that of neat D_2 clusters of the same size because of this kinematic enhancement of ion energies in the mixed ion case. However, such a kinematic enhancement is an *isotope* effect and cannot be identified by a comparison of CD_4 and D_2 clusters as in previous studies.

In this paper, we present an experimental confirmation of the prediction of this theory by measuring energy spectra from laser induced explosions of the methane isotopes CH_4 and CD_4. We observed a large difference in the energy spectra of the light ions (H or D). Since these clusters are produced under otherwise identical conditions and differ only in the η_A parameter, these data provide unambiguous evidence of a dynamic acceleration effect during the ion expansion since the initial potential driving the cluster explosion is the same in both cases.

In addition we have studied the fusion resulting from the explosions of a gas of CD_4 clusters. We have conducted an experimental comparison of ion energies and fusion yields from plasmas produced from the explosions of both CD_4 clusters and D_2 clusters ionized by 100 fs, 10 J laser pulses. We have found that CD_4 clusters exhibit higher ion energies than comparably sized D_2 clusters. However, while fusion yields from CD_4 cluster plasmas are higher than D_2 cluster plasmas at low laser energy (0.1 J), the opposite is true with laser energy of a few joules. Finally, we have conducted careful measurements of the angular distribution of fusion neutrons ejected from exploding deuterated methane clusters. We find that there is anisotropy in the neutron spatial distribution, though the angular distribution we observe differs markedly from that reported by Grillon *et al.* [13]. We see a peaking of the neutron yield along the axis of laser propagation and no enhancement in the perpendicular direction.

2. Measurements of light ions from exploding CH_4 and CD_4 clusters

In our first experiment we produced clusters with a pulsed supersonic gas jet directed through a skimmer and produced a low density cluster beam.

The experimental apparatus is illustrated in Figure 1. These clusters were irradiated by the THOR laser at the University of Texas, which delivers 40fs, 800nm pulses with energy up to 0.75J at 10Hz. The laser was focused using a *f*/4.9 refractive graded index aspheric lens into the cluster beam. The energy was varied between 21mJ and 70mJ, providing intensities between $\approx 9.6\text{x}10^{16}\text{W/cm}^2$ and $\approx 3.2\text{x}10^{17}\text{W/cm}^2$. The energies of the ions created by the laser cluster interaction were determined by measuring their time of flight (TOF) in a 1.14m field free drift tube. A microchannel plate (MCP) detected ions in an angular cone of $7.8\text{x}10^{-4}$sr from the focus. We estimated the cluster size to be approximately 3nm with both CH_4 and CD_4 clusters.

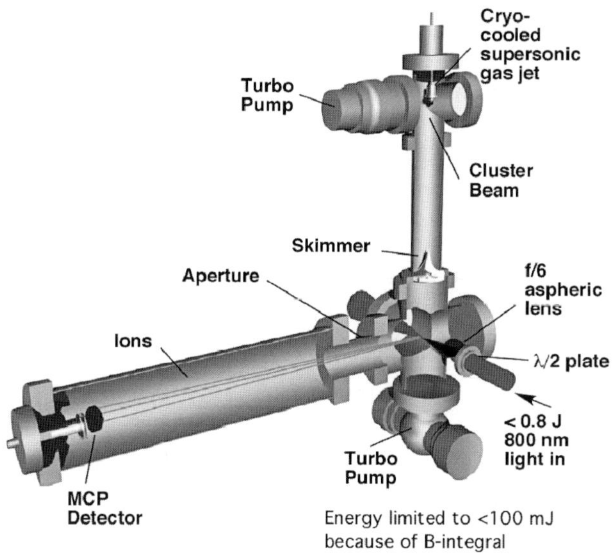

FIGURE 1. Experimental apparatus for the exploding methane cluster experiment.

A measured ion distribution from both methane and deuterated methane clusters is shown in Figure 2. The original TOF data for each were averaged over 200 shots, normalized, and converted to an energy spectrum. The shape of the spectrum is consistent with that expected from Coulomb explosion when a distribution of cluster sizes is taken into account [5]. The transformation to an energy distribution required us to assume a certain mass, which we set to the mass of H^+ or D^+, according to the gas we were shooting. The signal is a superposition from both carbon and deuterium/hydrogen arriving at the MCP, and these signals cannot be separated. We assume that the maximum energy E_C, for carbon ions of mass m_C, is approximately 2.5-3 times the maximum energy of the lighter species. We can then estimate a minimum TOF for the heavier carbon ions which corresponds to a maximum energy in the energy spectrum obtained using the mass of the lighter ion species m_A. Any signal above this energy $E_{min} = E_C \cdot m_A/m_C$, denoted in Figure 2 by the dotted lines, can only be caused by the detection of H^+ or D^+ ions.

Figure 2 shows kinetic energy spectra from CH_4 and CD_4 data. The difference in kinematic parameters between H^+ ($\eta = 3$) and D^+ ($\eta = 1.5$) causes a clearly apparent isotope effect. In the energy ranges attributed solely to the lighter ion species, the H^+ energies exceed the D^+ energies, indicating a higher average energy in the proton distribution and validating the theory predicting kinetic energy enhancement. Figure 3 illustrates similar data for a range of laser intensities. At the lowest intensities, we see very little difference between the proton and deuteron data. This would be consistent with the fact that the laser ponderomotive energy is below that required to extract all the electrons and to drive a Coulomb explosion. In this case we expect a plasma-like explosion and, as the data illustrate, expect no difference between the two isotopes. At the highest intensities, however, the Coulomb explosion mechanism dominates and the dynamic enhancement seen in Figure 2 becomes more prominent.

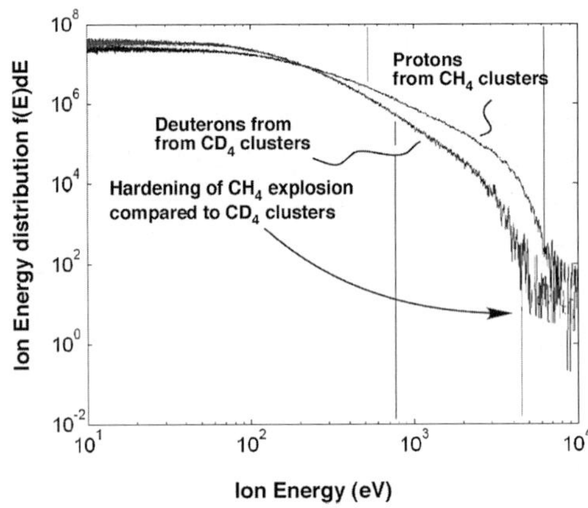

FIGURE 2. Measured ion distribution from both methane and deuterated methane clusters

3. Measurements of fusion neutron yield from CD_4 cluster targets

Our next set of experiments were devoted to study of fusion neutron yield when a gas plume of dueterated methane clusters was irradiated. Our intention is to

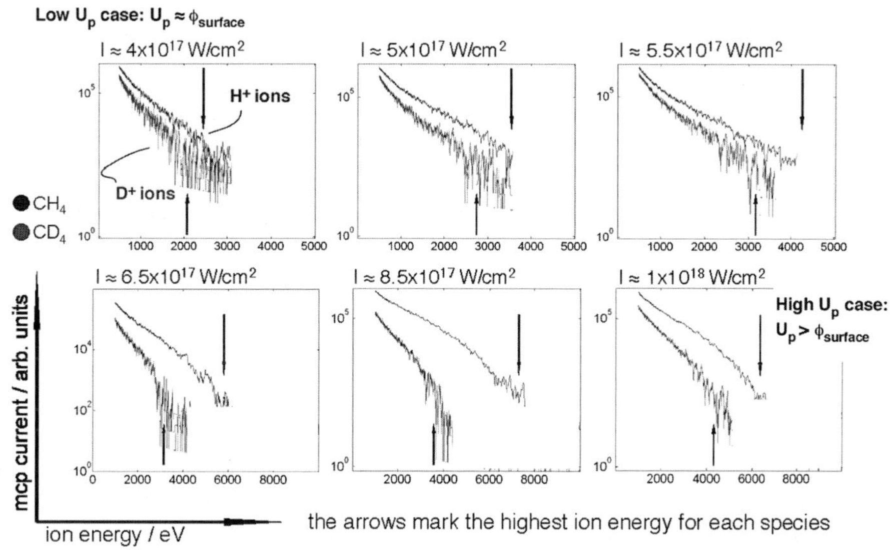

FIGURE 3. Measured ion spectra from CH_4 and CD_4 clusters at a range of laser intensities.

compare the fusion neutron yield between these CD_4 targets and neat D_2 cluster targets. In these experiments, a dense plume of either CD_4 or D_2 clusters was produced from the expansion of deuterium gas from a high pressure (70 atm), supersonic gas jet nozzle with a 10° opening angle and a 0.75 mm throat. This jet could be cryogenically cooled (-160 C) with liquid nitrogen and reached a peak atomic density of 10^{19} cm^{-3} in the jet center, as determined from time resolved interferometry measurement of the electron density immediately (20 ps) after ionization by the laser pulse. Rayleigh scatter measurements of the clusters indicate that the D_2 clusters exhibited an average size of 10 nm, while CD_4 clusters had an average diameter of 5 nm. The gas plume was then irradiated by a single pulse of 800 nm light from the Lawrence Livermore National Laboratory JanUSP laser, which delivered 100 fs pulses with energy up to 10 J per pulse. This pulse was focused into the gas plume by a 300 mm (f/#=2) parabola producing a vacuum focal spot diameter of 5 mm (FWHM), which contained 30% of the total energy. This configuration yielded a peak intensity of 2 x 10^{20} W/cm^2 in vacuum. The experimental set-up is illustrated in Figure 4.

In order to characterize the energy distribution of the deuterium ions ejected from the clusters in the laser irradiated gas plumes, we employed an array of Faraday cups placed at various distances from the interaction region. The Faraday cups yielded a time dependent current generated from the expulsion of ions from the plasma. To reject the large number of low energy electrons that originate from the focus, the Faraday cups were negatively biased

(- 50 V) . To determine the fusion yield from the laser irradiated plasmas, we counted the number of 2.45 MeV neutrons emitted into three independent solid angles with three plastic scintillating neutron detectors. The detectors consisted of a 10 cm diameter, 12 cm thick barrel of plastic scintillator coupled to a PMT. We obtained the neutron yield per shot by assuming that the neutrons are emitted nearly isotropically (see below). The three detectors were placed at different distances from the source to provide a larger dynamic range for neutron detection on each shot.

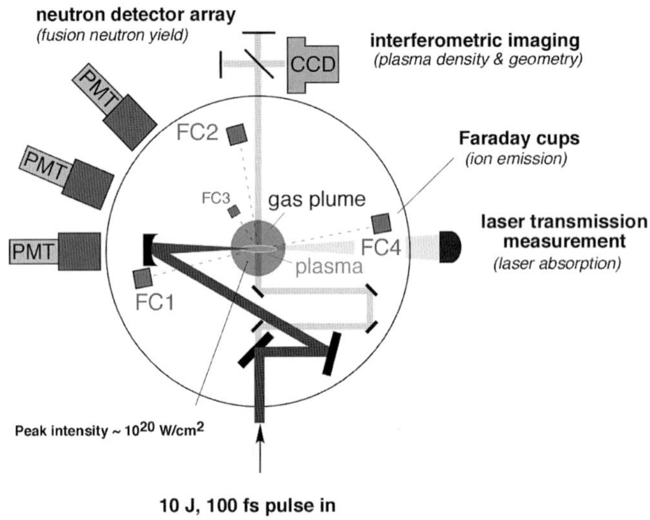

FIGURE 4. Experimental set-up for the exploding cluster fusion experiment at the JanUSP laser.

Figure 5 illustrates the measured average ion energies and total ion yields from both CD_4 clusters and D_2 clusters as a function of laser energy. We see a dramatically higher ion energy for the methane clusters. The average ion energies from CD_4 clusters exceed those of the D_2 clusters by over a factor of 2 through the laser energy range studied. On the other hand, the number of fast ions observed was somewhat less for the CD_4 target. This is a consequence of the fact that the D_2 clusters were produced by cryogenic cooling of the jet, yielding a higher average density of ions at the laser focus. The net result of these parameters on the total fusion yield of the plasma is illustrated in Figure 6. We see that the yields are very similar between the two cluster targets. The advantage gained by the higher ion energies with CD_4 is offset by the lower ion density of the plasma. These data suggest that the higher ion energies possible with CD_4 clusters will ultimately be advantageous for this fusion experiment if the average molecular density can be increased.

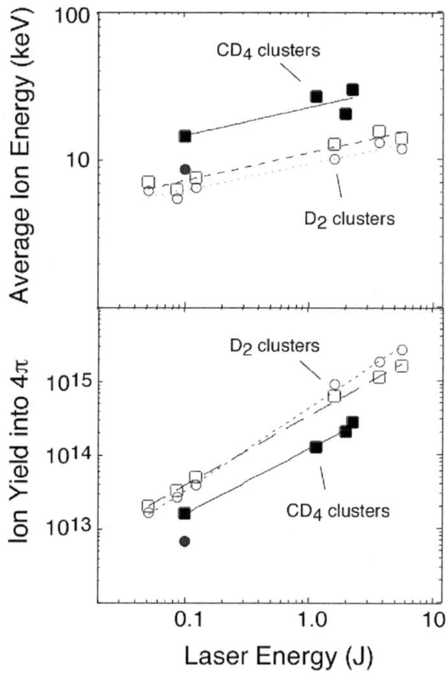

FIGURE 5. T Measured average ion energy (top) and ion yield (bottom) from deuterium and deuterated methane clusters.

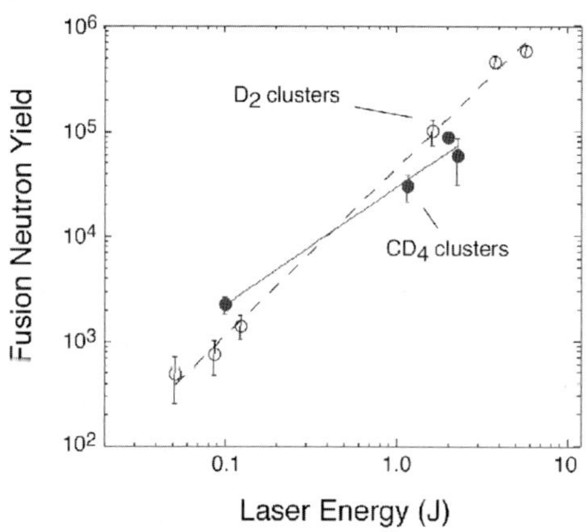

FIGURE 6. Total measured fusion neutron yield per shot from deuterium and CD_4 clusters as a function of the incident laser energy.

4. Measurements of fusion neutron angular distributions from CD_4 cluster targets

In our final set of experiments, we examined the angular distribution of neutrons from the cluster plasmas. The isotropic ion distribution resulting from a Coulomb explosion should lead to nearly isotropic neutron emission, particularly in the regime in which the deuterons have only a few tens of keV of energy. Interestingly, an angular scan of the fusion neutron yield from exploding deuterated methane clusters (CD_4) irradiated by a femtosecond laser carried out by Grillon *et al.* [13] showed, instead, a marked anisotropy. The experiment reported there observed an enhancement by nearly 30% of neutrons emitted in the direction perpendicular to the laser propagation. This large, anomalous anisotropy in the neutron distribution was explained with a model in which high energy (many hundred keV) ions are ejected radially from the plasma filament produced as the laser propagates through the cluster gas. Such a model is at odds with the supposition that the ion energies are derived from the isotropic distribution of ions from exploding clusters and, instead, requires the dominant coupling mechanism to the ions to be from a bulk radial explosion of the plasma filament, driving ions into the cold background gas.

Our experiment was performed on the THOR laser at UT. The experimental set-up for this measurement is illustrated in Figure 7. In these experiments, a dense plume of either CD_4 clusters was produced by the expansion of the gas from a high pressure (P_0 = 70 bar) reservoir into a vacuum. We used a 24 mm long conical, supersonic nozzle with a 750 μm orifice and an opening angle of 10°, the same jet used in the JanUSP fusion experiments described in the last section. In this experiment the CD_4 was pre-cooled to approximately 270K with a heat exchanger connected to the gas jet valve. The clusters were irradiated approximately 2-3mm underneath the nozzle by a 40fs, 800 nm pulse provided by the THOR laser. The pulse energies were in the range of 40-300mJ. The laser was focused by a *f*/12 spherical mirror yielding a peak intensity of $\approx 10^{17}$ W/cm^2 when 10 TW of power was employed. The molecular density of CD_4 was estimated to be $\approx 2 \times 10^{18}$ cm^{-3}.

The nuclear fusion yield was monitored with a set of four neutron detectors. All detectors consisted of a scintillating plastic rod of 12.7 cm diameter and 15 cm length attached to a fast 5" photomultiplier tube with a conical light guide made of Plexiglas. Time of flight diagnosis permitted us to distinguish 2.45 MeV fusion neutrons from the small number of hard x-rays through the stainless steel chamber wall. For this CD_4 run, the detectors were situated at an elevated position with respect to the gas jet (about 0.28 m, looking down at an angle of 30°-45°).

The azimuthal distribution of the neutron emission was measured for CD_4 clusters. In these measurements two detectors were placed at the same, fixed position as normalization. A third detector was moved around the chamber at a slightly elevated position with respect to the gas jet. Hence the data obtained at the position of the detector were not purely azimuthal but also had a very slight polar component. This was necessary to avoid the entrance flanges on the equator of the chamber and ensure that neutrons traversed the same amount of chamber materials at every point. Neutron yield as a function of azimuthal angle for this experiment is illustrated in Figure 8. The number of shots varies from 200 to 1800 shots per data point. Figure 8 indicates that a slight anisotropy clearly exists in the neutron yield. This anisotropy is not, however, consistent with the pattern observed by Grillon *et al.* [13] who saw an almost 40 % increase of the neutron yield at 54° from the forward direction. Instead, we observe a slightly higher yield in the forward and backward directions. We have modeled this distribution by Monte Carle methods and have determined that the slightly anisotropic distribution is due merely to the geometry of the gas jet and cigar shaped plasma filament (results from this model are plotted on the data points as a solid line). Both factors conspire to create a higher probability for fusion events occurring from ions emitted in the forward and backward direction, leading to the observed anisotropy. So the observed distribution appears to be consistent with the Coulomb explosion model described in the introduction.

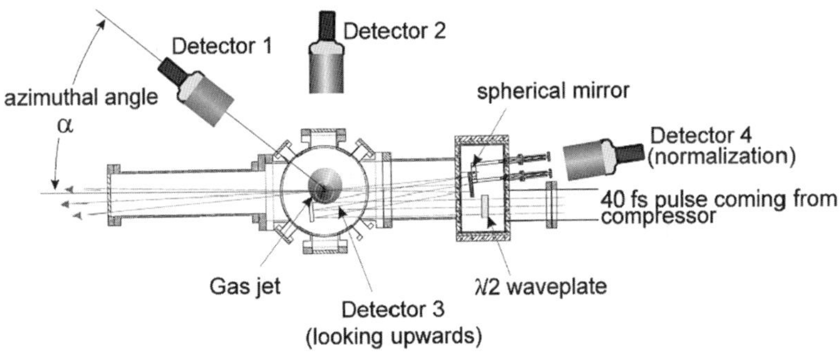

FIGURE 5. Experimental set-up for the fusion neutron anistropy measurement on the THOR laser.

5. Conclusion

In conclusion, we have presented experimental measurements of exploding heteronuclear clusters driven by intense femtosecond laser pulses. We have confirmed the prediction of the theory of Last and Jortner predicting a dynamic enhancement of ion energies in some heteronuclear clusters by measuring energy spectra from laser induced explosions of the methane isotopes CH_4 and CD_4. In addition we have studied the fusion resulting from the explosions of a gas of CD_4 clusters. We have conducted an experimental comparison of ion energies and fusion yields from plasmas produced from the explosions of both CD_4 clusters and D_2 clusters ionized by 100 fs, 10 J laser pulses. We have found that CD_4 clusters exhibit higher ion energies than comparably sized D_2 clusters. Finally, we have conducted careful measurements of the angular distribution of fusion neutrons ejected from exploding deuterated methane clusters. We find that there is anisotropy in the neutron distribution, though the angular distribution observed differs from that reported by Grillon *et al.*

This work was supported by the Dept. of Energy Office at Basic Energy Sciences, the National Science Foundation through the FOCUS Physics Frontier Center, and an NNSA Stockpile Stewardship Academic Alliances Center of Excellence, contract DE-FC52-03NA00156.

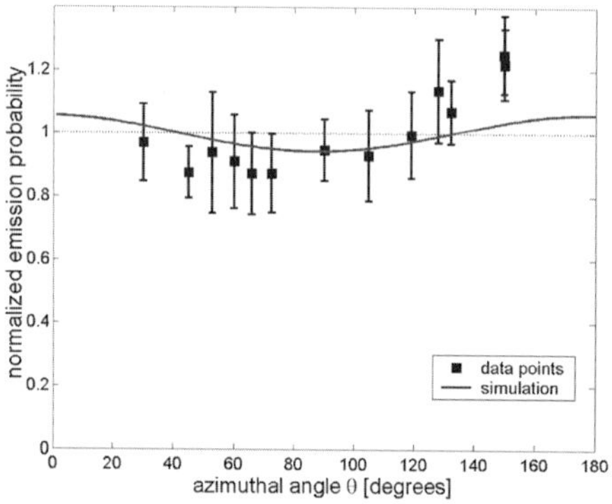

FIGURE 6. Measured neutron yield as a function of azimuthal angle from CD_4 clusters irradiated by 0.6 J pulses.

6. References

1. T. Ditmire *et al.*, Nature (London) 386 (1997) 54.
2. M. Lezius *et al.*, Phys. Rev. Lett. 80 (1998) 261.
3. V. Kumarappan, M. Krishnamurthy, and D. Mathur, Phys. Rev. Lett. 87 (2001) 085005.
4. T. Ditmire *et al.*, Phys. Rev. A 53 (1996) 3379.
5. K. W. Madison *et al.*, Phys. Plasmas 11 (2004) 270.
6. T. Ditmire *et al.*, Nature (London) 398 (1999) 489.
7. J. Zweiback *et al.*, Phys. Plasmas 9 (2002) 3108.
8. J. Purnell *et al.*, Chem. Phys. Lett. 229 (1994) 333.
9. E. M. Snyder *et al.*, Chem. Phys. Lett. 248 (1996) 1.
10. J. W. G. Tisch *et al.*, Phys. Rev. A 60 (1999) 3076.
11. I. Last and J. Jortner, Phys. Rev. Lett. 87 (2001) 033401.
12. I. Last and J. Jortner, J. Phys. Chem. A. 106 (2002) 10877.
13. G. Grillon *et al.*, Phys. Rev. Lett. 89 (2002) 065005.

Femtochemistry VII
Fundamental Ultrafast Processes in Chemistry, Physics, and Biology
A.W. Castleman, Jr. and M.L. Kimble

Spectral phase and detuning effects in high-power chirped pulse excitation of a dye solution

E. C. Carroll[a], A. C. Florean[a], K. G. Spears[a,b], P. H. Bucksbaum[a], R. J. Sension[a]

[a)]*FOCUS Center, University of Michigan, 500 East University Ave., Ann Arbor, MI, 48109-1120 USA*
[b)]*Department of Chemistry, Northwestern University, 2145 Sheridan Road Evanston, IL, 60208-3113 USA*

1. Introduction

Chirped laser pulses can be used to manipulate vibrational coherences and population dynamics in dye solutions. Bardeen *et al.* [1-2] have shown that high-power pulses with positive linear chirp enhance fluorescence in styryl and oxazine laser dyes, compared to a transform-limited pulse. Negative chirp, on the other hand, increases resonant stimulated Raman processes as the frequency sweep of the laser pulse follows the dynamic Stokes shift of the dye. This effect is widely viewed as an intrapulse version of the Tannor-Rice "pump-dump" scheme [1].

In a generalized picture of high-power chirped-pulse interactions in solutions, developed by Fainberg *et al.* [3], excited state population also depends on detuning from the vertical (Franck-Condon) transition of the absorber. Detuning is defined in terms of the peak of the absorption spectrum ω_{21}, and the peak of the fluorescence spectrum, ω_{st}:

$$\delta = (\omega - \omega_{21}) / \omega_{st} \quad (1)$$

The power spectrum of the ultrafast pulse determines which vibronic states are accessed on the ground and excited state manifold, as well as the relative intensity of transitions across the spectrum of the dye. We have experimentally investigated detuning and chirp effects in the laser dye LD690, specifically, comparing the optical response of LD690 at $\delta = 0$ and $\delta = +0.1$.

2. Experiment

In these experiments, ultrafast visible pulses were generated with a Ti:Sapphire pumped non-collinear optical parametric amplifier (NOPA). The NOPA spectrum is tunable between 470-740 nm. The visible pulse passes through a prism pair set in a 4-f configuration. Spectral phase was controlled with a 128-pixel liquid crystal modulator (LCM). A second prism pair compensated for second-order dispersion such that the pulse was nearly transform limited when no voltages were applied to the LCM pixels. Pulses were characterized by SHG-FROG to have ~25 fs duration at transform limit. The LCM was programmed to apply positive and negative linear chirp.

A solution of LD690 (Exciton) in ethanol was flowed through a quartz cell with a 100 micron path length and ca. 0.8 optical density. The transmitted pump light was collected with a multichannel spectrometer. Fluorescence collected 30 degrees from the forward scattering direction was focused onto a photodiode. A long-pass filter reduced pump scatter at the photodiode. Time-resolved transient absorption measurements were obtained by adding a time-delayed probe pulse. The probe consisted of supercontinuum generated in sapphire [4], and changes in probe absorption were monitored with a multichannel spectrometer.

3. Results

3.1. Fluorescence and pump depletion

Fluorescence is proportional to the excited state population. For high-energy pulses, absorption approaches a saturation threshold where fluorescence is not linear in pulse energy. Under these conditions, the excited state population becomes chirp dependent. For zero detuning ($\delta = 0$) we recover the results reported by Bardeen *et al.* [1]: fluorescence is strongly enhanced for positive chirp (30% in our data). For a pulse detuned to higher frequencies ($\delta = +0.1$) total fluorescence is less dependent on chirp (5% increase for positive chirp), though the saturation threshold is also expected to be higher.

Alternately, information about population dynamics on the timescale of the pulse duration can be gained from the spectrum of the transmitted pump pulse. Both absorption and stimulated emission contribute to this "pump depletion" spectrum. At low power, the pump depletion spectrum matches the steady state absorption, independent of chirp. However, near the saturation threshold, stimulated emission becomes significant on the red edge in the case of zero detuning (Figure 1). For negative chirp, the absorption peak shifts to higher frequency. Positive chirp results in relatively minor modulation of the depletion spectrum.

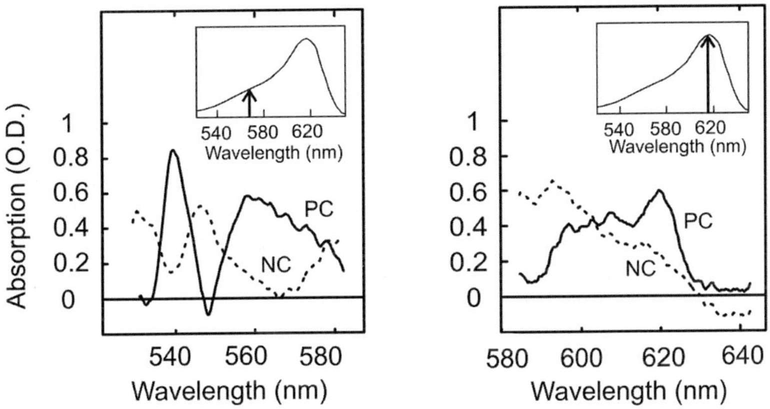

Figure 1. Depletion spectrum for chirped pulses with positive detuning (a) and zero detuning (b). The extent of the spectra shown represents the approximate FWHM of the pulse spectra. Positive chirp (PC, solid) $\phi(\nu)'' = 2 \times 10^4$ fs^2 Negative chirp (NC, dashed) $\phi(\nu)'' = -2 \times 10^4$ fs^2. The steady state absorption spectrum of LD690 in ethanol is shown for reference.

At $\delta = +0.1$, the pump depletion spectrum depends strongly on linear chirp despite the weak dependence in fluorescence. This observation suggests phase-dependent population dynamics that do not affect the final population in the first electronic excited state. It is likely that excited state absorption contributes to the pump deplection at the higher frequencies of the detuned pulse. The absorption extrema alternate as shown in Figure 1 for positive and negative chirps with $|\phi(\nu)| \approx 4\times10^4$ fs^2.

3.2. Transient absorption

Time-resolved transient absorption measurements were performed to characterize vibrational coherences generated by the chirped pulses. Impulsive stimulated resonant Raman scattering can produce a nonstationary coherent wavepacket on the ground state. Phase evolution of this wavepacket causes absorption probabilities to be modulated at the frequency of the vibration, resulting in an oscillatory transient absorption signal.

Oscillations corresponding to a 586 cm^{-1} vibrational mode are clear at both $\delta = 0$ and $\delta = +0.1$. This corresponds to the strongest mode in the LD690 Raman spectrum [2], believed to be a ring-breathing mode. Fourier power spectra are shown in Figure 2, corresponding to $\phi(\nu)'' \approx -2\times10^4$ fs^2 (left panel) and $\phi(\nu)'' \approx 2\times10^4$ fs^2 (right panel). To obtain the power spectrum, the oscillatory component of the transient absorption signal was isolated by a high pass Fourier filter and normalized to the strength of the underlying bleach.

In the vertical transition, negative chirp results in oscillations ten times stronger than the transform-limited pulse. Similar results were reported in [2]. This is consistent with the pump-dump model, where vibrational levels in the ground state are coupled through impulsive stimulated emission. Positive chirp also increases the oscillations, but to much lesser extent.

At $\delta = +0.1$, the 586 cm^{-1} oscillations are weak at all three chirps. Negative chirp produced weaker oscillations than either transform-limit or positive chirp. Greater competition between excited state absorption and stimulated emission to the ground state may result in the smaller vibratonal coherences observed for the detuned pulse compared to the $\delta = 0$ pulse. In the transient absorption spectrum, excited state absorption is stronger than the ground state bleach for wavelengths below 560 nm. More complete analysis is underway.

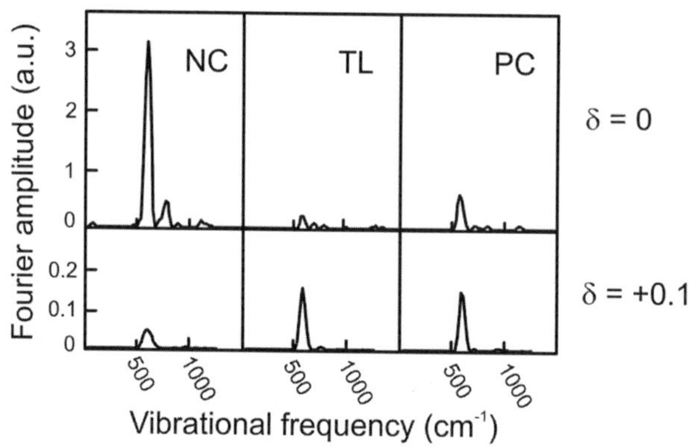

Figure 2. Normalized Fourier amplitude at probe wavelength 615 nm as a function of chirp and detuning. The scale of the Fourier amplitude in the $\delta = +0.1$ series (top row) is 1/10 that of the series at $\delta = 0$ (bottom row).

4. Conclusions

The optical response of the laser dye LD690 clearly depends on both the power spectrum and spectral phase of a high-power ultrafast pulse. In a vertical transition, chirp can control excited state population and enhance vibrational coherence in at least the ground state. Chirped pulses detuned to the blue edge of the dye's spectrum are less effective at controlling fluorescence and ground state vibrational coherences, probably due to the role of excited state absorption.

References

1. G. Cerullo, C.J. Bardeen, Q. Wang, and C.V. Shank, Chem. Phys. Lett. 262 (1996) 362.
2. C. J. Bardeen, Q. Wang, and C. V. Shank, J. Phys. Chem. A 102 (1998) 2759.
3. B. D. Fainberg and V. A. Gorbunov, J. Chem. Phys. 121 (2004) 8748.

4. S. A. Kovalenko, A. L. Dobryakov, J. Ruthmann, and N. P. Ernsting,, Phys. Rev. A 59 (1998) 2369.

Femtochemistry VII
Fundamental Ultrafast Processes in Chemistry, Physics, and Biology
A.W. Castleman, Jr. and M.L. Kimble

Squeezing Rb_2 wave packets with mixed adiabatic and dynamic strategies

Ignacio R. Sola[†], Bo Y. Chang[‡], Jesús González-Vázquez[†] and Jesus Santamaria[†]

[†]*Departamento de Química Física I, Universidad Complutense, 28040 Madrid, Spain*
[‡]*College of Environmental Science (BK21), Kyung-Hee University, Republic of Korea*

1. Introduction

Many molecular properties stem directly from the vibrational wave function; thus the control of the precise position and width of the wave packet has played such a fundamental role in Quantum control theory of molecules with laser pulses [1]. The goal of this work is squeezing a wave packet. The extreme spatial localization of a quantum motion poses an interesting limit to observe the quantum uncertainty relations and has important potential applications in time-dependent spectroscopy [2] (retrieving potential energy data) and quantum dynamics [1] (photodissociation and scattering).

The majority of the physical mechanisms suggested for laser-assisted squeezing are based on the properties of the quantum dynamics on excited states assuming first-order perturbation regime [3], or on the direct application of optimal control theory [4]. Using the properties of molecular light-induced potentials [5] (LIP), we have recently proposed a scheme for adiabatic squeezing [5]. However, it is well known that the free-dynamics of a stretched wave packet is a natural and powerful source of squeezing [6]. In this work we propose a combination of both afore-mentioned principles. Our adiabatic-diabatic squeezing (ADS) scheme involves: (1) nonresonant adiabatic excitation to a LIP where the wave packet is stretched, (2) ultrafast diabatic de-excitation of the wave packet to the ground potential, and (3) free dynamic evolution leading to "natural" squeezing. The mechanism and efficiency of the scheme will be exemplified on the Rb_2 dimer.

2. Model

We consider the three electronic states of Rb_2 shown in Fig.1(a): the ground X $^1\Sigma_g$, and two excited potentials $1^1\Pi_u$ and $2^1\Pi_g$ (hereafter V_1, V_2 and V_3 respectively) coupled by two laser fields: Ω_1 (which couples V_1 and V_2) and Ω_2 (which couples V_2 with V_3). Assuming the rotating-wave approximation (RWA), the Hamiltonian of the model system is

$$H = -\frac{1}{2m}\frac{\partial^2}{\partial x^2} I + \begin{pmatrix} V_1 - \Delta_1 & -\Omega_1/2 & 0 \\ -\Omega_1/2 & V_2 & -\Omega_2/2 \\ 0 & -\Omega_2/2 & V_3 - \Delta_2 \end{pmatrix} \tag{1}$$

where m is the reduced mass of Rb_2 and I is the unit matrix. Δ_1 and Δ_2 are the energy detunings between the potential energy minima dressed by the energy of the photons. For instance, if both transitions are resonant, then $\Delta_1 = \Delta_2 = 0$. The ADS must operate with nonresonant lasers to avoid population transfer (and Rabi cycling) to the intermediate potential V_2. The pulse shapes are parameterized as a three step function with square sine (cosine) switch on (off) of duration τ_1 (τ_3) and a plateau region of duration τ_2 and field amplitude Ω_0. We assume constant unit dipole moments.

The quantum dynamic simulations are obtained by solving the time dependent Schrödinger equation of the system using standard propagation techniques. The potential energy curves are obtained from Park *et al.* [7].

3. Results and discussion

The mechanism of the ADS scheme can be briefly summarized in the process:

$$V_1 \xrightarrow{adiabatic} U_w(V_1, V_2, V_3) \xrightarrow{diabatic} V_1$$

where the first process adiabatically moves the initial wave function to a wide LIP, U_w, so that the wave packet adjusts to the width of the potential and therefore stretches. The second process diabatically (*i.e.* almost instantaneously) transfers this wave packet back to the initial potential. Then the free-evolution of the stretched wave packet leads naturally to periodic (for harmonic potentials) oscillation and breathing of the packet. The wide U_w is a mixture of the three electronic potentials, coupled by the two nonresonant fields Ω_1 and Ω_2. In order to adiabatically transfer the initial wave function in V_1 to the bottom of the LIP, a certain pulse sequence is required. In the results of Fig.1 we use a counter-intuitive sequence (first switching on Ω_2 and then Ω_1) with negative detuning, $\Delta_1 = \Delta_2 = \Delta \approx -1100$ cm^{-1}, so that $V_1 - \Delta_1$ and $V_3 - \Delta_2$ are both above V_2.

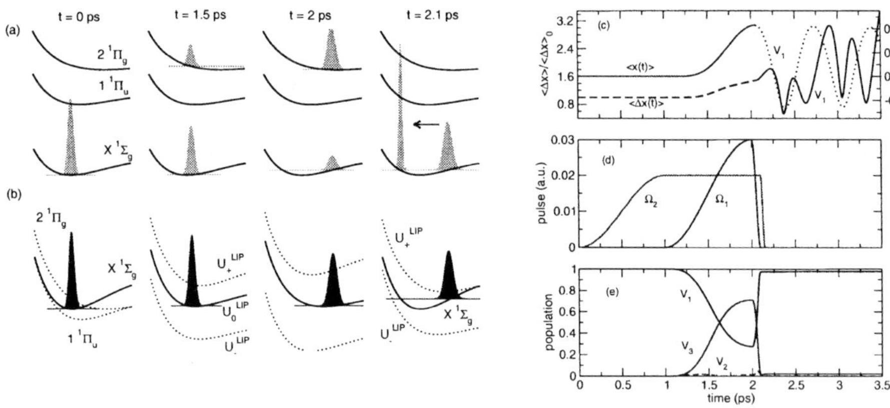

Figure 1: ADS dynamics. Snapshots of the wave pac
representation. In (c) we show the wave packet width and average position in time for the pulse sequence given in (d). In (e) we show the electronic populations.

Then U_w is usually labelled U_0. Additionally, the switch on must be slow enough to allow the wave packet to adapt to the laser-induced changing of the potential energy structure. This is achieved with τ_1 = 1ps for both pulses, which are time-delayed for 1 ps as well. The efficiency of the scheme is not very sensitive to these parameters.

When both pulses reach the plateau region, the wave packet sits in the equilibrium configuration of the prepared LIP. The width of the adiabatic packet depends only on the width of the LIP, which can be controlled by Ω_0 and Δ. For the first pulse we have chosen $\Omega_{0,1}$ = 0.03 au and for the second pulse $\Omega_{0,2}$ = 0.02 au. As Fig.1(e) shows, for these parameters, U_w has a mixture of near 70% of V_3 and 30% of V_1 with negligible contribution of V_2. Since V_3 is wider than V_1, the formed wave packet is stretched relative to the first one. In Fig.1(c) we show the evolution of the wave packet observables: the relative width, $\langle\Delta x(t)\rangle/\langle\Delta x\rangle_0$ (where $\langle\Delta x\rangle_0=\langle\psi_1|x|\psi_1\rangle$ is the width or spatial dispersion of the initial state, and $\langle\Delta x(t)\rangle$ is the width of the evolving wave packet) and the relative average position $(\langle x(t)\rangle - x_0)/d$, where x_0 is the average position of the initial state (the initial bond distance). $\langle x(t)\rangle$ is the time evolving bond distance, and d is the distance between the equilibrium configurations of the V_1 and V_3 potentials. In Fig.1(c), the scale for the width is on the left hand side of the plot, whereas the scale for the position is on the right hand side.

Finally, the quasi-instantaneous de-excitation of the wave packet back to V_1 can be achieved by a fast switch off of the pulses in reverse order of the sequence, that is, first switching off Ω_1 and then Ω_2. This diabatic or vertical transition can be efficiently performed using, for instance, τ_3 = 100 fs (or less) for both pulses with an equivalent time-delay. The stretched wave packet initiates a laser-free dynamics in V_1 from a non-equilibrium position, therefore oscillating and breathing. For this anharmonic dynamics, the period of breathing is more than two times the period of oscillation. After the first period of breathing, the wave packet can be maximally squeezed about 45% in the example shown in Fig.1 near the inner turning point of V_1. In subsequent periods, the wave packet stretches and squeezes again but achieves lower compression. Fig.1(a) and (b) show snapshots of the wave packet at the different stages of the ADS mechanism in the diabatic and adiabatic representations, respectively.

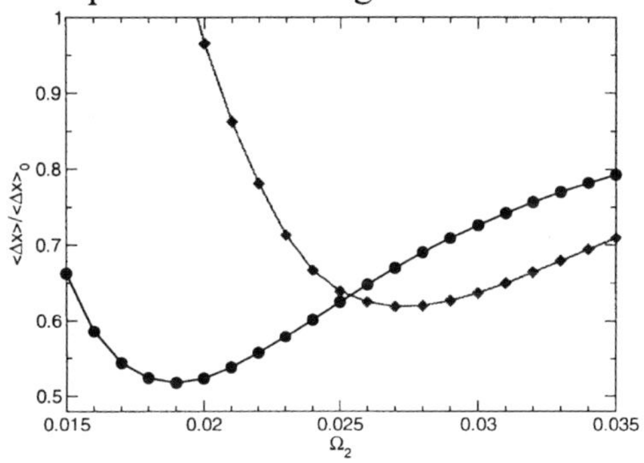

Figure 2. Maximal dynamic squeezing achieved as a function of $\Omega_{0,2}$

In order to increase the achieved squeezing, one can manipulate the laser knobs, specially harmonic, there must be a balance contribution of V_3 in U_0 increases, and therefore the wave packet is wider before its free evolution. This would guarantee a higher degree of squeezing if the dynamics were harmonic. However, at the same time, both the diabatic and the adiabatic transitions are more difficult to complete, and the stretched wave packet lands in a region of high anharmonicity when moved back to V_1, since the equilibrium configuration in U_w is close to the one in V_3. The optimal balance implies a wide LIP with equilibrium configuration not far from x_0. In Fig.2 we show the maximum squeezing achieved by changing $\Omega_{2,0}$ ($\Omega_{1,0}$ is fixed at 0.03 au) after the first (line with circles) and third (line with diamonds) period of breathing.

In summary, we have designed a physically appealing mechanism to achieve dynamic squeezing of a vibrational wave packet in the ground potential, following a sequence of adiabatic and diabatic transformations, which experimentally just requires control of the timing for the switch on and off of the pulses. The scheme seems to make maximum benefit of both adiabatic [5] and dynamic [6] squeezing. Moreover, several alternative pulse sequences can be devised along the same principles.

4. Bibliography

1. S. A. Rice and M. Zhao, Optical control of molecular dynamics, Wiley, New York 2000.
2. A. H. Zewail, Science 242 (1988) 1645.
3. J. Janszky et al., Chem. Phys. Lett. 213 (1993) 368.
4. I. Averbukh and M. Shapiro, Phys. Rev. A 47 (1993) 5086.
5. B. Y.Chang, I. R.Sola, S. Lee, and J.Santamaria, J. Chem. Phys. 122 (2005) 204316.

6. R. W. Robinett, Phys. Rep. 392 (2004) 1.
7. S. J. Park, S. W. Suh, Y. S. Lee, and G.-H. Jeung, J. Mol. Spec. 207 (2001) 129.

Femtochemistry VII
Fundamental Ultrafast Processes in Chemistry, Physics, and Biology
A.W. Castleman, Jr. and M.L. Kimble

Ultrafast conformational changes in carbonmonoxy-myoglobin after photolysis observed by time-resolved circular dichroism

Thibault Dartigalongue and François Hache

Laboratoire d'Optique et Biosciences – CNRS/INSERM/Ecole Polytechnique – 91128 Palaiseau, France

1. Introduction

Understanding the first steps of conformational changes in proteins is very challenging. In that prospect, the processes occurring in carboxy-myoglobin (MbCO) after ligand dissociation have been thoroughly studied. However, time-resolved absorption or Raman experiments [1] which probe localized electronic or vibrational modes of the protein don't provide much information on the global conformation changes. Such is not the case for circular dichoism (CD) which is very sensitive to the geometry of the whole molecule [2]. Here, we present a sub-picosecond time-resolved CD experiment on photolyzed MbCO and couple it with a CD calculation based on the polarizability theory [3] in order to access information on the geometrical relaxation of the protein. We observe conformational changes on a sub-100 ps timescale that we assign to the strain brought onto the proximal histidine by the heme doming.

2. CD calculation

Circular dichroism is the difference in absorption for a left or right circularly polarized light. When dealing with liquids, it is encountered only for chiral molecules. Most of biomolecules are chiral because these molecules are composed of several sub-units (not necessarily chiral) which are arranged in a non symmetrical manner in space. This feature has been widely used for conformational analysis and measurement of the CD can yield information on the geometry of biomolecules [2]. Myoglobin (Mb) is a chiral molecule and displays a nice CD feature in the Soret band which can be associated with its conformation. The principle of our experiment is therefore to photodissociate the CO ligand from Mb with a pump pulse and to carry out CD measurement with an optically-delayed probe. Interpretation of data requires a better knowledge of the origin of the CD in Mb and some calculations that we detail now.

Origin of the CD in MbCO has been thoroughly studied by Woody [4]. The heme moiety is a nonchiral molecule, but it displays a CD mainly due to its coupling with its environment. Interaction between the heme transitions and the $\pi-\pi^*$ transitions of surrounding aromatic residues originates from the dipole-dipole coupling and is therefore very sensitive to the relative disposition of the molecules and to their distance. However, it is necessary to develop a detailed calculation to describe the origin of the CD and to be able to understand the role of each residue.

The CD is calculated within the polarizability theory framework [3] which is well fitted for chiral molecules whose optical activity arises from the coupled-oscillator mechanism. The principle of the calculation is to divide the protein in polarizable groups (heme + surrounding aromatic residues (Figure 1)) characterized by their optical transitions (energy, direction, and magnitude of the electric dipole transition moments) and to allow the various groups to interact through dipole-dipole coupling. Calculating the normal (electronic) modes of the ensemble allows one to readily obtain the oscillator and rotational strengths of each mode and to deduce the absorption and CD spectra. The calculation is detailed in Ref [5]. The steady-state structures of MbCO and Mb were obtained from the Protein Data Bank (MbCO: 1A6G; Mb: 1A6N) [6]. This calculation allows us to determine the contribution of each amino acid to the Soret rotational strength and to analyze various protein conformations. The main information of this calculation is the very particular role played by the proximal histidine. Actually, the proximal histidine does not contribute much to the rotational strength which is surprising at first sight as it is the closest amino acid from the heme. The reason for this is that the imidazole plane is almost perpendicular to the heme plane, resulting in two opposite contributions when coupling with the heme x and y transitions is considered. However, the proximal histidine plays an essential role by lifting the degeneracy of these heme transitions and determining the normal modes. In particular, we have observed that only a rotation of this histidine about the iron-imidazole axis can result in a large drop of the CD [5], as observed experimentally (see below).

3. Experiments

MbCO samples are prepared by dissolving commercial horse heart met-myoglobin (Sigma) in a buffer (100 mM Tris-HCl, pH 8.0). After conversion to the ferrous form by addition of sodium dithionite under anaerobic conditions,

CO gas is flowed through the sample to obtain MbCO. A 230 μM solution is placed in a 1 mm thick silica cell in the optical set-up. The cell is continuously moved so as to excite a fresh sample at each pump shot.

The experimental set-up is derived from a frequency-doubled amplified Titanium-Sapphire laser. The main part of the 400 nm pulse is sent directly on the sample to initiate the dissociation of CO from the heme. The remaining part is focussed into a sapphire plate to generate a light continuum tunable between 400 and 480 nm. This pulse serves as a probe to monitor the evolution of the photolyzed sample. An optical delay is inserted on the pump path to allow time-resolved measurements. In order to access the CD, the probe passes through a longitudinal Pockels cell on which we apply a square ± 1300V voltage. The probe polarization is therefore alternately left and right circular. The CD then translates into a transmission modulation that we detect with a lock-in amplifier. A very stringent alignment procedure is applied to insure that the two circular polarizations are symmetrical and that polarization artifacts are negligible [7].

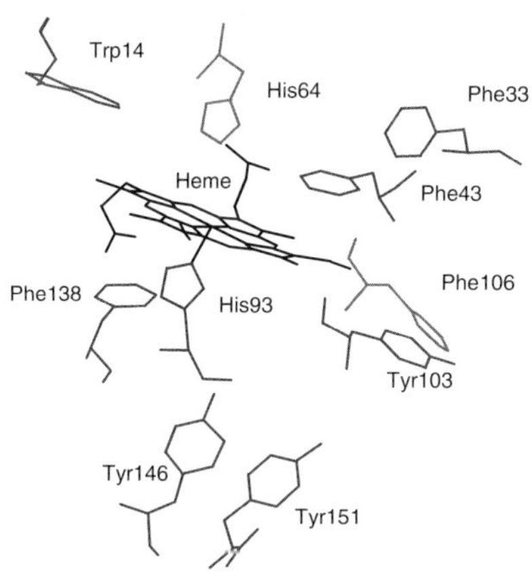

Fig. 1. Sketch of the main residues involved in the CD signal. The proximal histidine (His93) is singled out due to its particular role; it induces the lifting of the degeneracy of the heme transitions but contributes very little to the rotational strength. All the other residues yield a significant rotational strength, negative for His 64 and Phe 106, positive for all the others.

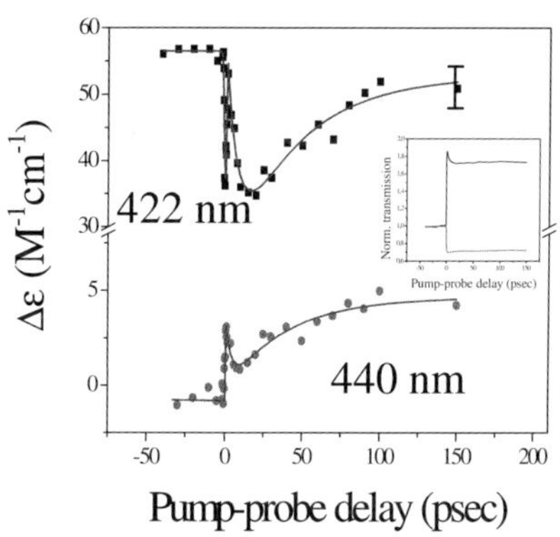

Fig. 2. Time-resolved circular dichroism for two probe wavelengths (422 and 440 nm). The peak at zero delay is an artefact which shows the time resolution of the experiment. The CD curve is fitted by two exponential functions (7 and 43 psec). The inset shows the corresponding time-resolved transmissions.

Results for a probe wavelength of 422 nm (close to the MbCO absorption peak) and 440 nm (close to the Mb absorption peak) are displayed in Figure 2. Whereas there is no visible evolution on the transmission curve (except for a 3 ps decay corresponding to excited heme electronic states [8]), a nice feature appears on the CD signals: a first ~10 ps decrease is followed by a ~50 ps relaxation towards the steady-state value. The peak at zero delay is the pump-probe cross-correlation and yields a 650 fs time resolution. It is noteworthy that the two curves display the same dynamics, revealing a global change of the CD spectrum with time.

4. Discussion

Assignment of this signal to a conformational change in the protein first requires a careful examination of the possible artifacts. Several sources of such artifacts exist. Three major ones are relevant: the change of the heme electronic state, the pump-induced anisotropy of the sample, or the temperature rise. However, all these phenomena can be ruled out, in particular, due to their response time which does not correspond to our 100 ps measurements.

We can now discuss the origin of the CD change that we observe in view of our calculations. Experiments and calculation agree to show that going from the MbCO conformation to the steady-state Mb one does not yield a noticeable CD change and cannot explain the observed CD drop. From our calculations, only rotation of the imidazole plane of the proximal histidine can bring such a large change. These results allow us to bring an interpretation of our signal in terms of conformational changes. During the first picoseconds (timescale = 7 ± 3 ps), the signal arises from the stress of the proximal histidine which gets squeezed between the domed heme and the F helix. On a longer timescale (43 ± 17 ps), global motion of the protein occurs, which translates into an increase of the CD towards its steady-state value. These timescales are in agreement with thermal phase grating [9] and time-resolved Raman [10] measurements.

References

1. J. Everse, K. D. Vandegriff, and R. M. Winslow (eds.), Methods in Enzymology, vol. 232, Academic Press, 1994.
2. G. D. Fasman (ed.), Circular dichroism and the conformational analysis of biomolecules, Plenum Press, 1996.
3. J. Applequist, K. R. Sundberg, M. L. Olson, and L. C. Weiss, J. Chem. Phys. 70 (1979) 1240.
4. M. C. Hsu and R. W. Woody, J. Am. Chem. Soc. 93 (1971) 3515.
5. T. Dartigalongue and F. Hache, to be published in J. Chem. Phys. (2005).
6. J. Vojtechovsky, K. Chu, J. Berendzen, R. M. Sweet, and I. Schlichting, Biophys. J. 77 (1999) 2153.

7. T. Dartigalongue and F. Hache, J. Opt. Soc. Am. B 20 (2003) 1780.
8. S. Franzen, L. Kiger, C. Poyart, and J. L. Martin, Biophys. J. 80 (2001) 2372.
9. R. J. D. Miller, Acc. Chem. Res. 27 (1994) 145.
10. Y. Mizutani and T. Kitawaga, J. Phys. Chem. B 105 (2001) 10992.

AUTHOR INDEX

Akahane, J. 377
Åkesson, E. 113
Ariola, F. S. 396
Arzhantsev, S. 225
Atas, E. 118
Aumanen, J. 39
Ausfelder, F. 510
Baertschy, M. 495
Baker, G. 225
Bakker, H. J. 123
Bandrauk, A. D. 553
Barbara, P. F. 284
Bañares, L. 49, 510
Bányász, Á. 254
Barth, I. 441
Baudin, J.-B. 204
Baumert, T. 510
Bdzoch, J. 415
Benesch, F. 23
Berera, R. 387
Bianco, N. J. 109
Blankenship, R. E. 331
Bonačić-Koutecký, V. 466
Boyle, M. L. 543
Bragg, A. E. 163
Bredenbeck, J. 402
Breunig, H. G. 500
Brixner, T. 331, 372
Brüggemann, B. 113, 518
Buchvarov, I. 249
Bucksbaum, P. H. 455, 573
Buersgens, F. 563
Bugár, I. 132, 420
Burghardt, I. 143
Cardoza, D. 495
Carroll, E. C. 455, 573
Castleman, A.W., Jr. 33, 80, 105, 109
Castro, J. C. 128
Cervetto, V. 402
Chakraborty, A. 66, 70
Chang, B. Y. 578
Chelkowski, S. 553
Chen, X.-H. 304
Cho, M. 331
Chorvat, D. 420
Chosrowjan, H. 424
Čík, G. 132
Comstock, M. 215
Couch, V. A. 53
Crossley, M. J. 113
Dantus, M. 434, 526
Dao, L. V. 377
Dartigalongue, T. 583
Darugar, Q. 260
Dela Cruz, J. M. 434, 526
Deligeorgiev, T. G. 391
de Nalda, R. 49, 510
Dijkhuizen, N. 136
Ditmire, T. 563
Do, M. T. T. 377
Dokter, A. M. 123
Domcke, W. 154
Douhal, A. 183
Du, L.-C. 304
Edens, A. 563
Eichberger, R. 295, 299
Eisenthal, K. 215
Eisfeld, W. 154
El-Sayed, M. 260
Engelke, M. 274
Ernstorfer, R. 295, 299
Espagne, A. 204
Fainberg, B. D. 514
Fanciulli, R. 136
Fayer, M. 195
Felber, S. 295, 299
Fiebig, T. 249
Fink, W. H. 53
Fleming, G. R. 331, 372
Florean, A. C. 455, 573
Fürstenberg, A. 391
Fuß, W. 96
Gadjev, N. I. 391
Ganteför, G. 274
Gardiner, A. T. 372
Gauden, M. 321
Gaveau, M.-A. 174
Gilb, S. 14
Gilch, P. 357
Gloaguen, E. 174

Gong, Q. 75
González-Vázquez, J. 522, 578
Gosselin, J. L. 44
Griffin, G. B. 163
Groot, M. L. 428
Gundlach, L. 295, 299
Gunn, J. M. 526
Guo, T. 53
Gust, D. 387
Gustavsson, T. 254
Hache, F. 583
Hamm, P. 402
Han, H. 62
Hannaford, P. 377
Harris, D. A. 189, 382
Hartke, R. 563
Hegemann, P. 321
Heidenreich, A. 532
Heikal, A. 396
Helbing, J. 402
Hellingwerf, K. J. 321
Henig, A. 563
Herek, J. L. 136
Herrero, C. 387
Hertel, I. V. 367, 543
Heyes, D. J. 428
Hochstrasser, R. M. 128
Hohenberger, M. 563
Horn, C. 510
Hou, Z. 337
Hsin, T.-M. 66
Huang, W. 260
Hunter, C. N. 428
Hydutsky, D. P. 109
Hynes, J. T. 143
Iglev, H. 249
Ikuta, M. 87
Improta, R. 254
Irimia, D. 49
Ito, N. 225
Izquierdo, J. G. 49
Janek, M. 420
Janssen, M. H. M. 49
Jiao, G.-S. 128
Jin, H. 225
Jones, C. E. 33
Jortner, J. 532
Julliard, M. D. 391
Kammrath, A. 163
Kao, Y.-T. 407
Karpiuk, J. 415
Kennis, J. T. M. 321, 387
Key, J. M. 321
Kim, H.-R. 396
Kim, T. G. 128
Kim, Y. D. 274
Kiviniemi, T. 39
Kjellberg, P. 57
Kleiman, V. D. 118
Knappenberger, K. L., Jr. 14, 33, 105
Kobayashi, T. 87
Kodali, G. 337
Kodis, G. 387
Koller, F. O. 311
Konradi, J. 505
Kosmidis, C. 96
Koyama, Y. 377
Krug, M. 510
Laan, W. 321
Laarmann, T. 543
Laimgruber, S. 357
Larsen, J. 113
Last, I. 532
Lau, L. 14
Laubereau, A. 249
Lauer, A. 500
Lazzarotto, E. 254
Lee, T. 23
Lee, Y. J. 284
Leone, S. 14
Lerch, E. W. 14
Li, F. 75
Lindner, J. 235
Lindinger, A. 479, 490
Liu, Y. 396
Lochbrunner, S. 415
Loudet, A. 128
Lozovoy, V. V. 434, 526
Madison, K. W. 563
Mančal, T. 372
Manz, J. 441
Markovitsi, D. 254
Maroncelli, M. 225
Martin, M. M. 204

Mataga, N. 424
Materny, A. 404
May, V. 518
McArthur, E. A. 215
Merli, A. 479, 490
Mestdagh, J.-M. 174
Minitti, M. P. 44
Mitrić, R. 466
Moilanen, D. E. 195
Moore, A. L. 387
Moore, T. A. 387
Moroder, L. 311
Myllyperkiö, P. 39
Narayanan, M. 337
Neumann, S. 154
Neumark, D. M. 163
Niemietz, M. 274
Organero, J. A. 183
Osterhoff, J. 563
Palacios, R. E. 284, 387
Palszegi, T. 420
Pan, J. 62
Panja, S. 96
Parker, D. H. 174
Pastirk, I. 434
Pearson, B. J. 455
Peng, Z. 118
Pettersson, M. 39
Pfister, R. 402
Piletic, I. R. 195
Pittner, J. 466
Plaza, P. 204
Plewicki, M. 479, 490
Poisson, L. 174
Polívka, T. 113
Prince, B. D. 66, 70
Prince, B. M. 66, 70
Pullerits, T. 57, 518
Qian, W. 260
Qiu, W. 411
Radloff, W. 367
Rao, Y. 215
Read, E. L. 372
Reich, C. 23
Renner, C. 311
Riedle, E. 415
Ritze, H.-H. 367
Rose-Petruck, C 23
Samoylova, E. 367
Sancar, A. 407
Santamaria, J. 522, 578
Santos, L. 183
Sanz, M. 183
Sauer, F. 479, 490
Savolainen, J. 136
Saxena, C. 407
Scalmani, G. 254
Scaria, A. V. 505
Schmid, W. E. 96
Schrader, T. E. 311
Schreier, W. J. 311, 357
Schultz, T. 367
Schulz, C. P. 543
Sekiguchi, K. 245
Sension, R. J. 189, 382, 455, 573
Shan, F. 53
Shchatsinin, I. 543
Sieg, A. 311
Singh, A. K. 505
Sly, J. 113
Smith, V. R. 367
Sobhy, M. A. 33, 80
Sobolewski, A. L. 154
Soep, B. 174
Sofo, J. O. 80
Sola, I. R. 522, 578
Sølling, T. I. 44
Spears, K. G. 573
Spezia, R. 143
Spry, B. 195
Stanley, R. J. 337
Stauffer, H. 66, 70
Stavros, V. G. 14
Steinmeyer, G. 543
Stenger, J. 331
Stibenz, G. 543
Stickrath, A. B. 382
Strasser, D. 14
Sundström, V. 113
Symes, D. R. 563
Sytina, O. 428
Szocs, V. 420
Tahara, T. 245
Tan, H. S. 195

Taniguchi, S. 424
Tian, J. 62
Topp, M. R. 128
Tormo, L. 183
Torres, E. A. 14
Trifonov, A. 249
Trushin, S. A. 96
van Amerongen, H. 387
van Grondelle, R. 321, 387, 428
van Stokkem, I. H. M. 321, 387
Vassilev, A. A. 391
Vaswani, H. 331
Vauthey, E. 391
Velič, D. 132, 420
Vengris, M. 387
Verlet, J. R. R. 163
Vöhringer, P. 235
Wachtveitl, J. 311
Walvick, R. 396
Wang, L. 407, 411
Wang, P. 62
Wang, S. 75
Wang, Z. 62
Weber, P. M. 44
Weber, S. M. 479, 490
Wei, Z. 62
Weinacht, T. 495
Weitzel, K.-M. 500
Weng, Y. X. 304
White, J. L. 455
Willig, F. 295, 299
Wollenhaupt, M. 510
Wöste, L. 479, 490
Woutersen, S. 123
Xu, B. 526
Yamaguchi, S. 245
Yang, H. 75
Yang, K. 337
Yang, T. 62
Yu, Q. 396
Yuasa, Y. 87
Yudin, G. L. 553
Zewail, A. H. 3
Zhang, J. 62
Zhang, L. 411
Zhong, D. 346, 407, 411
Zigmantas, D. 372
Zinth, W. 311
Žitnan, M. 132, 420